Science Teaching

'This is a transformative book. It provides an enlightening cartography of the uses of history and philosophy in the science classroom. No one interested in science teaching or science culture should be without a copy of this updated classic.'

Alberto Cordero, Philosophy Program, The CUNY Graduate Center and Queens College CUNY, USA

'This book's importance transcends science education. Its coverage of topics such as the impact of constructivism on education provides the book with a universal importance. I strongly recommend it to everyone interested in teaching and learning.'

John Sweller, School of Education, University of New South Wales, Australia

'The Pendulum chapter is a masterpiece! It should be considered obligatory reading for everyone who aims at becoming a science (especially physics) teacher.'

Ricardo Karam, Physikdidaktik, Universität Hamburg, Germany

'Science Education is a rigorous and necessary resource for science education researchers, policy makers and practitioners.'

Sibel Erduran, School of Education, University of Limerick, Ireland

Michael R. Matthews is an Honorary Associate Professor in the School of Education at the University of New South Wales, Australia. He is Founding Editor of the international journal *Science & Education*; Founding President of the International History, Philosophy and Science Teaching Group; and President of the Inter-Divisional Teaching Commission of the International Union of History and Philosophy of Science. He has trained, taught and published in science education and in history and philosophy of science.

Science Teaching

The Contribution of History and Philosophy of Science

20th Anniversary Revised and Expanded Edition

Michael R. Matthews

NEW YORK AND LONDON

Second Edition published 2015
by Routledge
711 Third Avenue, New York, NY 10017

and by Routledge
2 Park Square, Milton Park, Abingdon, Oxon OX14 4RN

Routledge is an imprint of the Taylor & Francis Group, an informa business

First edition published by Routledge 1994

Library of Congress Cataloging in Publication Data
Matthews, Michael R.
Science teaching: the contribution of history and philosophy of science, 20th anniversary revised and expanded edition/Michael R. Matthews. – Second edition.
pages cm
Includes bibliographical references and index.
1. Science – Study and teaching – History. 2. Science – Study and teaching – Philosophy. 3. Science teachers – Training of. I. Title.
Q181.M183 2014
507.1—dc23
2014009781

ISBN: 978–0–415–51933–5 (hbk)
ISBN: 978–0–415–51934–2 (pbk)
ISBN: 978–0–203–12305–8 (ebk)

Typeset in Sabon by
Florence Production Ltd, Stoodleigh, Devon, UK

Printed and bound in the United States of America by
Edwards Brothers Malloy on sustainably sourced paper

For my daughters: Clare, Alice and Amelia

Brief contents

Contents

Preface (2014)

It is a pleasure to see the twentieth anniversary of my 1994 *Science Teaching* book being celebrated by publication of an updated and enlarged edition. The book has stayed in print for 20 years, which suggests that it has some merit. The intellectual background to the book is described in the following 1994 Preface. Pleasingly, if philosophical arguments are any good, then they retain their merit for a long time. Having 'philosophical merit' is, of course, not the same as 'being correct', but it does mean being clear enough to enable readers to see where the mistakes are (this issue of clarity in communication and argument will be something returned to in Chapter 12). The central conviction of the first edition was stated in its Preface:

> For all its faults, the scientific tradition has promoted rationality, critical thinking and objectivity. It instils a concern for evidence, and for having ideas judged not by personal or social interest, but by how the world is; a sense of 'Cosmic Piety', as Bertrand Russell called it. These values are under attack both inside and outside the academy. Some educationally influential versions of postmodernism and constructivism turn their back on rationality and objectivity, saying that their pursuit is Quixotic. This is indeed a serious challenge to the profession of science teaching.
>
> The vitality of the scientific tradition, and its positive impact on society, depends upon children being successfully introduced to its achievements, methods and thought processes, by teachers who understand and value science. The history and philosophy of science contribute to this understanding and valuation.

World events and educational developments in the subsequent 20 years have only strengthened these convictions. The 'flight from science' has continued unabated and has been extensively documented in US and European government reports. There have been continuing debates over many socio-scientific issues, such as the utilisation of stem cells from manufactured human embryo cells, the control or utilisation of genetically modified crops, the reality and mitigation of androgenic global warming, harnessing or otherwise of nuclear energy, and compulsory child vaccination. With economic and cultural globalisation, serious questions have been asked about the supposed universality of science and of the justification and utility of teaching orthodox science in cultures that have their own rich lore of understandings of nature and non-scientific worldviews. After the 1960s' Kuhnian trumpet blast, various postmodernist waves have swept through the academy, including schools of education, each disputing the traditional foundations for science teaching. And there are many other such pressing issues, all of which have philosophical dimensions.

There have been constant wars in the Middle East, Africa and the Indian subcontinent, fuelled by ideology, but fought with high-tech, science-enabled weaponry. Each drone attack, each report of the use of oxygen-deprivation bombs, to say nothing of ordinary bombs and napalm, each poison-gas attack brings into focus the values of science, the responsibility of scientists and the purpose of science teaching. Understanding these events and issues, and then appropriately responding to them, requires a degree of rational, critical and objective analysis; the way forward is not advanced by embracing irrational, uncritical and subjective thinking. These intellectual and personal capacities – scientific habits of mind or scientific temper – can be developed in science classrooms, if the curriculum and pedagogy are informed by the history and philosophy of science (HPS).

Since the book's first edition, there have been considerable developments in science-education curricula that explicitly recognise the importance of teaching the philosophical, cultural and historical dimensions of science. In the United States, the first-ever National Science Education Standards were published by the National Research Council in 1996 (National Research Council 1996). These standards recognise the centrality of philosophical and historical knowledge in the teaching of science. In the UK, a group of prominent science educators, reflecting on Britain's National Curriculum and the most appropriate form of science education for the new millennium, wrote a report with ten recommendations, the sixth of which said that: 'The science curriculum should provide young people with an understanding of some key ideas about science, that is, ideas about the ways in which reliable knowledge of the natural world has been, and is being, obtained' (Millar & Osborne 1998, p.20). Different European and Asian countries have comparable statements about desired broader and deeper outcomes of school science.

Clearly, the goals of the US National Standards, the UK group and other national groups can only be realised if science teachers have some familiarity and enthusiasm for the history and philosophy of their subject. A position paper of the US Association for the Education of Teachers in Science, the professional association of those who prepare science teachers, has recognised this in its own recommendation that: 'Standard 1d: The beginning science teacher educator should possess levels of understanding of the philosophy, sociology, and history of science exceeding that specified in the [US] reform documents' (Lederman *et al.* 1997, p.236).

The arguments advanced by the above curriculum writers are basically the same as those advanced in the first edition of this book.

Along with curriculum developments, there has been, in the past 20 years, a significant amount of interdisciplinary research in the field of HPS and science teaching (HPS&ST). This research makes contributions to three categories of question faced by science teachers:

1 *theoretical* questions that impinge on science education, such as: constructivist claims about the knowledge claims of science, feminist critiques of science, the status of indigenous or local sciences and how they should or should not be taught in science programmes, science and religion, the status of models in science, scientific values and their relation to cultural values, and so on;
2 *curriculum* questions about the structure, content and scheduling of school science programmes;
3 *pedagogical* questions about how the utilisation of historical and philosophical material affects student motivation, interest and learning *of* science and *about* science.

The major development in HPS&ST research since the 1994 publication has been the establishment and continued growth of the journal *Science & Education: Contributions from History, Philosophy and Sociology of Science and Education*. The journal is now in its twenty-third year of publication, with ten issues being published per year (www.springerlink.com). About 800 research papers have been published; in 2011, there were 108,650 article downloads from the journal's website, and it is noteworthy that the most downloads are from Asia.

A core part of the HPS&ST infrastructure has been the International History, Philosophy and Science Teaching Group (IHPST) (www.ihpst.net). The group has been associated with the journal; it held its inaugural meeting in Tallahassee in 1989 and has continued to hold successful biennial conferences,[1] with select proceedings published in the journal;[2] and it has commenced a programme of biennial regional meetings in Latin America and Asia.[3] These are attended by teachers, educators, historians, philosophers and cognitive scientists.

The vitality and international reach of current HPS&ST scholarship and engagement is manifest in the three-volume, seventy-six-chapter *International Handbook of Research in History, Philosophy and Science Teaching* (Matthews 2014). It has sections on Pedagogical Studies, Theoretical Studies, National Studies and Biographical Studies and is contributed to by 125 authors from thirty countries and contains 11,000 references. Many of the issues and debates 'touched on' in this book are developed at length in chapters of the *Handbook*.

This book has three core purposes: one, to show educators that HPS is an interesting and engaging subject, and that it can usefully illuminate many of the theoretical, curricular and pedagogical issues that they encounter; two, to show historians and philosophers that their own expertise and scholarship can be utilised in science-education debates, curriculum development and classroom teaching; and three, to cultivate a sense among science teachers of belonging and contributing to the scientific and philosophical tradition that has had such enormous international social and cultural influence. Everyone should be mindful that, without science teachers, there would be no science. I have tried as much as possible to provide extended quotations from the main scholars discussed – Aristotle, Galileo, Huygens, Newton, Priestley, Mach and others – so that something of their own voice can be heard; too often, the names are known, but their voices are not heard; quotations are a meagre way of giving them some expression.

Michael R. Matthews
School of Education, University of New South Wales,
Sydney 2052, Australia
February 2014

Notes

1 These were: Minneapolis 1995, Calgary 1997, Pavia 1999, Denver 2001, Winnipeg 2003, Leeds 2005, Calgary 2007, Notre Dame 2009, Thessaloniki 2011 and Pittsburgh 2013.
2 For select proceedings, see: Pavia (Vol.10, Nos. 1–2, 2001), Winnipeg (Vol.14, Nos. 3–5, 2005), Leeds (vol.16, nos. 2–4, 2007), Calgary (Vol.18, Nos. 3–4, 2009), Notre Dame (vol.20, nos. 7–8, 2011) and Thessaloniki (Vol.22, No. 6, 2013).
3 Brazil (2010), Argentina (2012), Korea (2012) and Taiwan (2014).

References

Lederman, N.G., Kuerbis, P.J., Loving, C.C., Ramey-Gassert, L., Roychoudhury, A. and Spector, B.S.: 1997, 'Professional Knowledge Standards for Science Teacher Educators', *Journal of Science Teacher Education* 8(4), 233–240.
Matthews, M.R. (ed.): 2014, *International Handbook of Research in History, Philosophy and Science Teaching*, 3 volumes, Springer, Dordrecht, The Netherlands.
Millar, R. and Osborne, J.: 1998, *Beyond 2000: Science Education for the Future*, School of Education, King's College, London.
NRC (National Research Council): 1996, *National Science Education Standards*, National Academies Press, Washington, DC.

Preface (1994)

This book seeks to contribute to science teaching and science-teacher education by bringing the history and philosophy of science and science teaching into closer contact. My belief is that science teaching can be improved if it is infused with the historical and philosophical dimensions of science. Such contextual, or liberal, teaching of science in schools benefits both those students going on to further study of science, and those, the majority, for whom school science is their last contact with formal science instruction.

The conviction that the learning of science needs to be accompanied by learning about science is basic to liberal approaches to the teaching of science. This position has been eloquently argued by, among others, Ernst Mach, James Conant, Gerald Holton, Joseph Schwab and Martin Wagenschein. This book is a housekeeping effort in the liberal tradition: it attempts to survey the history of debate on the matter; to list the chief publications; to itemise contemporary relevant research, particularly in children's learning of science; to point to present-day practical and theoretical problems in science education to which the history and philosophy of science can contribute; to give an account of curriculum developments embodying the liberal spirit of science instruction; and to indicate ways in which the history and philosophy of science can be usefully included in teacher preparation programmes.

This book is the work of an under-labourer in the garden, to use John Locke's expression. Some furrows have been made, and some seeds planted. Hopefully, other people will water the garden, straighten the furrows, plant other seeds and remove some of the weeds. If the book stimulates science teachers at both schools and universities to be more interested in the history and philosophy of science, and encourages historians, philosophers and sociologists of science to become interested and involved with science education, then it will have achieved one purpose. If it contributes to the inclusion of HPS studies in science-teacher education programmes, it will have achieved another purpose. If it promotes an interest in educational theory among science educators, it will have achieved still another.

The theme of this book is that science teachers need three competencies: first, knowledge and appreciation of science; second, some understanding of HPS in order to do justice to the subject they are teaching and to teach it well, and in order to make intelligent appraisals of the many theoretical and educational debates that rage around the science curriculum; third, some educational theory or vision that can inform their classroom activities and relations with students, and provide a rationale and purpose for their pedagogical efforts. Science teachers contribute to the overall education of students, and thus they need some moderately well-formed view of what education is, and the goals it should be pursuing. Teachers need to keep their eyes on the educational prize, the more so when social pressures increasingly devalue the intellectual and critical traditions of education.

It is widely recognised that there is a crisis in Western science education. Levels of science literacy are disturbingly low. This is anomalous, because science is one of the greatest achievements of human culture. It has a wonderfully interesting and complex past, it has revealed an enormous amount about ourselves and the world in which we live, it has directly and indirectly transformed the social and natural worlds, and the human and environmental problems requiring scientific understanding are pressing – yet students and teachers are deserting science.

This flight from the science classroom, by both teachers and students, has been depressingly well documented. In the US in the mid 1980s, it was estimated that, each year, 600 science graduates entered the teaching profession, while 8,000 left it (Mayer 1987). In 1986, 7,100 US high schools had no course in physics, and 4,200 had no course in chemistry (Mayer 1987). In 1990, only four states required the three years of basic science recommended by the sobering 1983 report *A Nation at Risk*; the rest allowed high-school graduation with only two years of science (Beardsley 1992, p.80). Irrespective of years required, 70 per cent of all school students drop science at the first available opportunity – which is one reason why, in 1986, fewer than one in five high-school graduates had studied any physics. In 1991, the Carnegie Commission on Science, Technology and Government warned that the failings of science education were so great that they posed a 'chronic and serious threat to our nation's future' (Beardsley 1992, p.79). In the UK, recent reports of the National Commission on Education and the Royal Society have both documented similar trends. One commentator has said that, 'wherever you look, students are turning away from science . . . Those that do go to university are often of a frighteningly low calibre' (Bown 1993, p.12). In Australia, in 1989, science-education programmes had the lowest entrance requirement of all university degrees.

There are complex economic, social, cultural and systemic reasons for this rejection of science. These are beyond the scope of teachers to rectify. But there are also educational reasons for the rejection of science that are within the power of teachers and administrators to change. In 1989, for example, a disturbing number of the very top Australian school science achievers gave 'too boring' as the reason for not pursuing university science. It is these curriculum and pedagogical failings that the history and philosophy of science (HPS) can help rectify.

One part of this contribution by HPS is to connect topics in particular scientific disciplines, to connect the disciplines of science with each other, to connect the sciences generally with mathematics, philosophy, literature, psychology, history, technology, commerce and theology; and finally, to display the interconnections of science and culture – the arts, ethics, religion, politics – more broadly. Science has developed in conjunction with other disciplines; there has been mutual interdependence. It has also developed, and is practised, within a broader cultural and social milieu. These interconnections and interdependencies can be appropriately explored in science programmes, from elementary school through to graduate study. The result is far more satisfying for students than the unconnected topics that constitute most programmes of school and university science. Courses in the sciences are too often, as one student remarked, 'forced marches through unknown country without time to look sideways'.

The defence of science in schools is important, if not necessary, to the intellectual health of society. Pseudoscientific and irrational worldviews already have a strong hold on Western culture; anti-science is on the rise. It is not just the ramparts of society that have been invaded – witness the checkout-counter tabloids with their 'Elvis lives' stories, Gallup polls showing that 40 per cent of the adult US population believe that human life began on Earth just a couple of thousand years ago, and astrology columns in every newspaper. But the educational citadel has been compromised – a small, and hopefully not representative, 1988 survey of US biology teachers revealed that 30 per cent rejected the theory of evolution, and 22 per cent believed in ghosts (Martin 1994). For all its faults, the scientific tradition has promoted rationality, critical thinking and objectivity. It instills a concern for evidence, and for having ideas judged, not by personal or social interest, but by how the world is; a sense of 'Cosmic Piety', as Bertrand Russell called it. These values are under attack both inside and outside the academy. Some educationally influential versions of postmodernism and constructivism turn their back on rationality and objectivity, saying that their pursuit is Quixotic. This is, indeed, a serious challenge to the profession of science teaching.

The vitality of the scientific tradition, and its positive impact on society, depends upon children being successfully introduced to its achievements, methods and thought processes, by teachers who understand and value science. The HPS contribute to this understanding and valuation.

This book grows out of, and is a contribution to, the International History, Philosophy, and Science Teaching Group. This is a heterogeneous group of teachers, scientists, educators, historians, mathematicians, philosophers of education and philosophers of science who, over the past 5 years, have staged two conferences[1] and have arranged the publication of many special issues of academic journals devoted to HPS and science teaching.[2] Some basic papers in the field have been gathered together and published in Matthews (1991), *History, Philosophy, and Science Teaching: Select Readings* (OISE Press, Toronto, and Teachers College Press, New York, 1991). These might be useful for further reading. The International History, Philosophy, and Science Teaching Group is also associated with a new journal devoted to the subject of this book – *Science & Education: Contributions from the History, Philosophy, and Sociology of Science and Mathematics.*[3]

Notes

1 The proceedings of the 1989 Tallahassee conference are available in Herget (1989, 1990); those of the 1992 Kingston conference are in Hills (1992).
2 The journal special issues include the following: *Educational Philosophy and Theory* 20(2), (1988); *Synthese* 80(1), (1989); *Interchange* 20(2), (1989); *Studies in Philosophy and Education* 10(1), (1990); *Science Education* 75(1), (1991); *Journal of Research in Science Teaching* 29(4), (1992); *International Journal of Science Education* 12(3), (1990); and *Interchange* 23(2,3), (1993).
3 The journal is published by Kluwer Academic Publishers, PO Box 17, 3300 AA Dordrecht, The Netherlands. It is available at reduced rates through the international HPS&ST group (inquiries to the author).

References

Beardsley, T.: 1992, 'Teaching Real Science', *Scientific American* October, 78–86.
Bown, W.: 1993, 'Classroom Science goes into Freefall', *New Scientist* December, 12–13.
Herget, D.E. (ed.): 1989, *The History and Philosophy of Science in Science Teaching*, Florida State University, Tallahassee, FL.
Herget, D.E. (ed.): 1990, *The History and Philosophy of Science in Science Teaching*, Florida State University, Tallahassee, FL.
Hills, S. (ed.): 1992, *The History and Philosophy of Science in Science Education*, 2 volumes, Queen's University, Kingston.
Martin, M.: 1994, 'Pseudoscience, the Paranormal, and Science Education', *Science & Education* 3(4), 357–372.
Matthews, M.R. (ed.): 1991, *History, Philosophy and Science Teaching: Selected Readings*, OISE Press, Toronto.
Mayer, J.: 1987, 'Consequences of a Weak Science Education', *Boston Globe* September.

Acknowledgements

Acknowledgements (2014)

The bulk of personal debts for this twentieth anniversary edition of my 1994 book are the same as those for the original. First, as with all book writing, families pay a price. Since 1994, my wife, Julie, and Clare and Alice have been joined by a third daughter, Amelia, and two grandchildren, Joshua and Elenore. All have seen my time taken up with this project and have, pleasingly, taken it on faith that I have been doing something worthwhile. It is for readers to judge whether my time would have been better spent with my family.

Writing this second edition has been a wonderful opportunity to revisit and re-evaluate thoughts and arguments that were originally written in response to a 1989 invitation from Israel Scheffler to write a book on science teaching for his Routledge Philosophy of Education Research Library. Neither of us could have thought that the book would stay in print for so long, or that, 25 years later, a second edition would be warranted.

In 1994, I mentioned my debt to teachers who first introduced me to the subject matter of the book: at the University of Sydney, Wallis Suchting (Philosophy) and Bill Andersen (Education); at Boston University, Robert S. Cohen, Abner Shimony and Marx Wartofsky (Philosophy). Clearly, the debt to learned and capable early teachers always remains. In the 20 years since the first edition, I have learned things from a number of scholars whom I have had the good fortune to meet and engage with. Among these, Mario Bunge warrants particular mention. Now enjoying his ninety-fourth year, he continues to write books and articles that move easily, but with great erudition, across history of philosophy, science and philosophy of science, always with an admirable clarity of expression and a willingness to engage with serious educational issues.

In 1994, I mentioned my good fortune to edit the journal *Science & Education*, which then was in its second year of publication. Twenty years later, I am still editing the journal, and it has put me in contact with hundreds of scholars, from scores of countries around the world. These have been a great source of ideas and a privileged way of being kept abreast of current research, even if this knowledge has not always been internalised in ways that it deserved to be.

In 1994, I also mentioned my debt to the IHPST. This debt has simply grown by a further 20 years of valuable and intellectually productive friendships. Of special note have been the meetings held in Greece, Finland, Argentina, Brazil, Mexico, Denmark, Spain, India and Korea – these have all been wonderful occasions for discussing and hearing about history, philosophy and science teaching in contexts outside the dominant Anglo-American sphere. Within the latter sphere, for the past 20 years, the biennial IHPST meetings have been themselves happy and enormously productive gatherings, characterised by the admirable mix of serious scholarship and good fellowship.

I have benefited significantly from my editorship of the seventy-six-chapter, three-volume *Handbook of Research in History, Philosophy and Science Teaching* (Springer 2014), which has been contributed to by 125 authors from thirty countries. My debt is plainly visible in the Reference list for each of this book's chapters. The present book could be regarded as a 'primer' for the larger handbook; all of the arguments here, and more, are developed and documented at considerable length in the latter work.

Many friends have read and commented on different chapters of this book: Ricardo Karam, Yann Benétreau-Dupin, Colin Gauld, Robert Nola, Roland Schulz, Edgar Jenkins and Gürol Irzik. I am indebted to them, as are readers, for their suggestions and corrections. Julie House and Hans Schneider pleasingly corrected and copy-edited different chapters. A particular debt is owed to Paul McColl, who closely read, made valuable suggestions for, and carefully copy-edited the entire manuscript: a heroic task. Special thanks are due to the diligence and professional competence of Louise Smith, a Routledge-contracted UK copy-editor, who even after all the foregoing reading and corrections, nevertheless raised 110 'author queries' for me to rectify. Readers have been saved a good deal of frustration by these 110 lapses not finding their way into print. I commend her services to any author.

Finally, this book would not have happened except for the kind invitation of Naomi Silverman, the Routledge Taylor & Francis Education Editor, to write a second and enlarged edition of the 1994 book. Working with her has been a very happy and easy experience; I commend her to all authors.

Sources

In a number of places, this enlarged edition has drawn on material I have published over the past 20 years, specifically:

- Chapter 6 is partly dependent on: Matthews, M.R.: 2001, 'Methodology and Politics in Science: The Case of Huygens' 1673 Proposal of the Seconds Pendulum as an International Standard of Length and Some Educational Suggestions', *Science & Education* 10(1–2).
- Chapter 7 is partly dependent on: Matthews, M.R.: 2009, 'Science and Worldviews in the Classroom: Joseph Priestley and Photosynthesis', *Science & Education* 18(6–7).

- Chapter 10 is partly dependent on: Matthews, M.R.: 2009, 'Teaching the Philosophical and Worldview Components of Science', *Science & Education* 18(6–7).
- Chapter 11 is partly dependent on: Matthews, M.R.: 2012, 'Changing the Focus: From Nature of Science (NOS) to Features of Science (FOS)'. In M.S. Khine (ed.) *Advances in Nature of Science Research*, Springer, Dordrecht, The Netherlands.
- Chapter 12 is partly dependent on: Matthews, M.R.: 2014, 'Discipline-based Philosophy of Education and Classroom Teaching', *Theory and Research in Education* 12(1), 19–108.

I am grateful to Springer and SAGE for permission to use this material.

Acknowledgements (1994)

For the past 5 years, the writing of this book has severely encroached upon my family time. My wife, Julie House, and daughters, Clare and Alice, deserve thanks for their forbearance. Julie House is owed a significant additional debt for copy-editing and proofing numerous drafts of the book. She did her best to correct the worst of the expression, the most serious of the grammatical mistakes and frequent misspellings. In addition, she argued over most of the central points and was persistent in trying to keep the text focused. All readers are indebted to her for making their job so much easier than it otherwise would have been.

I am grateful to Professor Israel Scheffler for the invitation to write this book for his, and Vernon Howard's, Philosophy of Education Research Library, and to Jayne Fargnoli, the Routledge Education Editor, for her patience.

I am indebted to the many members of the International History, Philosophy, and Science Teaching Group, who have been generous over the past 5 years with ideas, hospitality and enthusiasm. Two major conferences, organised by Ken Tobin and David Gruender (Tallahassee 1989) and Skip Hills and Brian McAndrews (Kingston 1992), stimulated much that has gone into this book. Editorship of the journal *Science & Education*, which is devoted to the theme of the book, has enabled me to read, and benefit from, the work of a wide range of authors from all over the globe. The advice and encouragement of Martin Eger and Fabio Bevilacqua have been of special importance. Others, too numerous to mention, are, I hope, aware of my gratitude.

I have a debt to my teachers who introduced me to the history and philosophy of science. I am particularly grateful to Professor Wallis A. Suchting, formerly of Sydney University, whose standards of scholarship and breadth of knowledge are a model for those who have had the good fortune to be his students. Professor Abner Shimony, at Boston University, introduced me to the writings of Galileo, and Professors Robert S. Cohen and Marx W. Wartofsky, also of Boston University, helpfully placed science and the philosophy of science in the broader social and historical context. I am also indebted to Dr Bill Andersen, formerly of Sydney University, my first teacher in philosophy of education, who encouraged, among others, a young, naive science student to identify and grapple with philosophical questions in education.

My employers, the University of New South Wales and the University of Auckland, have made this book possible. The former's library is a cornucopia of materials in science education and the history and philosophy of science. The University of Auckland was a generous and supportive employer during my 2-year period as the Foundation Professor of Science Education. This enabled me to complete the book.

Friends have been good enough to read the penultimate version of the manuscript and suggest corrections and offer valuable advice. To Drs Michael Howard, Peter Slezak, Colin Gauld, Wallis Suchting, Richard Thorley, Fabio Bevilacqua, Harvey Siegel and James Wandersee, I am very grateful. Their scholarship and attention to detail have saved readers from the worst of my errors. Jan Duncan has been of great assistance in proofreading, checking of references, and the preparation of figures.

Finally, I am grateful to Gill Kent, Routledge's copy-editor, for her meticulous attention to detail. The book is considerably more polished for her labours.

Chapter 1

The Rapprochement Between History, Philosophy and Science Education

Science has been the foremost contributor to our understanding of the natural and social world, and, through its engagement with religion, worldviews, economies and technologies, it has been a major influence on culture. Food production, medicine, entertainment, war, industry, reproduction, transportation, accommodation, religion, space exploration, and people's self-understanding and their worldviews – their sense of place in the universe and in nature – have all been profoundly affected by science – mostly for good; sometimes for bad. Understanding the 'balance sheet' is of utmost importance, and this understanding is only possible with knowledge of the history and philosophy of science (HPS). This chapter will mention some of the elements that constitute the current rapprochement between history, philosophy and science teaching, or components of the 'HPS&ST programme', as it might be called. These include:

- the significant engagement of historians and philosophers with theoretical, curricular and pedagogical issues in science teaching;
- the growth of liberal education and the recognition of the historical and philosophical components necessary for this education;
- the acknowledgement of the requirement of basic philosophy for good technical science education;
- the recognition that HPS can contribute to ameliorating some of the widespread and well-known problems of science education;
- the realisation that HPS is a necessary condition for achieving any 'flow-on' effects from science learning for solving major issues in personal and social life;
- the realisation that HPS knowledge is required for meeting the explicit requirements of many new national and provincial science curricula.

Philosophers and Historians Engage with Science Education

Thirty-five years ago, Robert Ennis wrote a comprehensive review of the extant literature on philosophy of science and science teaching. His review listed six questions that science teachers constantly encounter in their classrooms and staffrooms, questions that the deliberations and researches of philosophers and historians of science could illuminate. These questions were:

- What characterises the scientific method?
- What constitutes critical thinking about empirical statements?
- What is the structure of scientific disciplines?
- What is a scientific explanation?
- What role do value judgements play in the work of scientists?
- What constitute good tests of scientific understanding?

These questions are of perennial concern to science teachers and science-teacher education programmes. However, Ennis made the melancholy observation that: 'With some exceptions philosophers of science have not shown much explicit interest in the problems of science education' (Ennis 1979, p. 138). Pleasingly, in recent decades, there has been a degree of rapprochement between these fields. Both the theory of science education and, importantly, science curricula and classroom pedagogy have become more informed by HPS. (These themes will collectively be referred to as history, philosophy and science teaching (HPS&ST).) This book contributes to HPS&ST by:

- outlining the arguments for the role of HPS in science education;
- reviewing the history of school science curricula in order to situate the claims of HPS-informed teaching against other approaches to science pedagogy;
- examining the successes and failures of previous efforts to bring HPS into closer connection with the science programme;
- elaborating some case studies where the contrast between HPS and 'professional' or 'technical' approaches to science teaching and curricula development can be evaluated;
- examining some instances of prominent educational debates in science education – constructivism, feminism, multiculturalism, worldviews and nature of science – that can be clarified and informed by HPS;
- outlining the contribution that HPS can make to science-teacher education.

It is hoped that the book will stimulate interest in educational matters among historians and philosophers of science, and encourage interest in historical and philosophical matters among science teachers and, particularly, the educators of science teachers.

When Ennis wrote, in the late 1970s, the exceptions among post-war historians and philosophers who had written on science education included Michael Martin, who published a series of articles (1971, 1974, 1986/1991) and wrote a popular book, *Concepts of Science Education* (1972), on philosophy and science education. Other philosophers and historians of science, who 40 years ago, had written on the subject include Stephen Brush

(1969), Robert Cohen (1964), Yehuda Elkana (1970), Herbert Feigl (1955), Philipp Frank (1947/1949), Gerald Holton (1975, 1978), Noretta Koertge (1969), Ernst Nagel (1969, 1975) and Israel Scheffler (1973). Happily, this situation of relative philosophical and historical neglect has changed, and, in the past few decades, many philosophers of science[1] and historians of science[2] have addressed different of the myriad theoretical, curricular and pedagogical problems of science teaching.

The engagement of philosophers and historians with science education can be seen in contributions to thematic issues of the journal *Science & Education*[3] and in contributions to anthologies such as *History, Philosophy and Science Teaching* (Matthews 1991), *Science, Worldviews and Education* (Matthews 2009), *Epistemology and Science Education* (Taylor & Ferrari 2011) and *Philosophy of Biology: A Companion for Educators* (Kampourakis 2013) and to the three-volume, 76-chapter *International Handbook of Research in History, Philosophy and Science Teaching* (Matthews 2014).

Ennis's six questions are perennial, but they do not exhaust the field of HPS&ST concerns, as can be quickly seen by looking at the titles of the above-cited articles. Philosophers have usefully contributed to pedagogical problems, to curricular discussions and to debate about the following theoretical issues: feminist critiques of science, multiculturalism and science, evaluation of constructivist theory, environmental ethics, the nature of science, science and religion, and so on. One of the theses of this book is that these are not extracurricular or add-on questions for science teachers: philosophy of science is part of the fabric of science teaching, and students acquire or 'pick up' a philosophy of science from their teachers. The issue is just how clearly this is recognised, and how explicitly the philosophical questions are dealt with. It is clear that all of these discussions are improved by philosophical and historical input; indeed, it is impossible to have informed and intelligent discussion of any of the listed theoretical issues without HPS.

History and Philosophy of Science: A Partnership

The conviction of this book is that the philosophy of science needs to be cognisant of the history of science, and the reverse: 'Philosophy of science without history of science is empty; history of science without philosophy of science is blind', as Imre Lakatos memorably expressed the matter (Lakatos 1978, p. 102.) This view was urged against those who saw philosophy occupying an autonomous position, such as Hans Reichenbach, who expressed this latter view in his classic distinction between the contexts of discovery and the contexts of justification in science. For Reichenbach, philosophy was concerned only with the context of justification, whereas history, sociology and psychology are concerned with the context of discovery (Reichenbach 1938).

The proper relation between the history and philosophy of science is much debated, with experts disagreeing on just how necessary the former is for the latter. Hilary Putnam at one point exclaimed that the history of science is 'irrelevant' to the philosophy of science (Suppe 1977, p. 437). The very influential positivist philosopher of science Rudolf Carnap has said of himself that he 'was as unhistorically minded a person as one could imagine' (Suppe 1977, p. 310). Carnap's student, Willard van Orman Quine, has said the same thing; his influential epistemological corpus is devoid of any historical reference (Quine 1960).

On the other side, for those wishing to keep history of science separate from philosophy, questions arise such as: How do we identify the history of science, without some philosophical presuppositions? How do we separate useful history of science from useless history of science, without some prior conception of proper method? It seems that we need to know in advance of writing a history of science what will count as science; if we do not have such a view, then we could presumably set off researching astrology, numerology and stamp collecting, rather than chemistry or geology.

As with many either/or questions, the answer lies somewhere between. The relationship between history of science and philosophy of science has to be interactive. There is ample evidence of history of science being written in the service of philosophical, political and religious commitments. It is notorious that Galileo has become a 'Man for all philosophical seasons' (Crombie 1981), with every methodologist seeing their own favoured methodology being followed by Galileo. Here, history is at best cherry-picked, and the opportunity for history of science to refine or change philosophical commitments is lost. Thomas Kuhn's story of his philosophical transformation, occasioned by having to teach a Harvard general education course on the history of science, is a well-known recent example where history transformed philosophy. Philosophy is required to begin writing history, but it should be capable of being transformed by historical study.[4]

This debate about the place of history is characteristic of many issues in philosophy of science – it would be a rash person who said that the contentious matters of realism, empiricism, causation, explanation, idealisation, truth, falsification and rationality have been settled. But some things regarding the interplay of philosophy and history are agreed upon. Clearly, the history of science should be used to illustrate positions arrived at in philosophy of science. An exposition of the nature of science, of theory evaluation or the ontological commitments of science that did not make mention of Galileo, Newton, Kepler, Lavoisier, Darwin, Mendel, Mach or Einstein, and the scientific controversies they engendered, would be very odd. Unfortunately, philosophy of science courses too often neglect the history of science. Commonly, students read of the debates over scientific methodology engaged in by Carnap, Nagel, Popper, Kuhn, Lakatos, Feyerabend, Laudan, van Fraassen and others, but have to take the contenders' historical interpretations of Aristotle, Galileo, Huygens and Newton on faith; students become spectators to an academic game. What should be a course that enhances appreciation of the scientific tradition and deeper thinking about it can, in the absence of history, become more like a catechism class. This is particularly odd in educational settings where science teachers and science students have heard of the famous names and might expect to see their work figure in any discussion of the nature of science or other philosophical issues occasioned by science.[5] This is *Bildung* in the European tradition.

Science and Liberal Education

The present rapprochement between HPS and science education represents, in part, a renaissance of the long-marginalised liberal, or contextual, tradition of science education, a tradition contributed to in the last 100 years by scientists and educators such as Ernst Mach, Pierre Duhem, Alfred North Whitehead, Frederick W. Westaway, E.J. Holmyard, Percy Nunn, James Conant, Joseph Schwab, Martin Wagenschein, Walter Jung and Gerald Holton. At its most general level, the liberal tradition in education embraces Aristotle's delineation of truth, goodness and beauty as the ideals that people ought to cultivate in their appropriate spheres of endeavour. That is, in intellectual matters, truth should be sought, in moral matters goodness, and in artistic and creative matters beauty. Education is to contribute to these ends: it is to assist the development of a person's knowledge, moral outlook and behaviour, and aesthetic sensibilities and capacities. For liberal educationalists, education is more than the preparation for work; education is valued because it contributes to the cognitive and moral development of both the individual and their culture.

The liberal tradition has a number of educational commitments.[6] One is that education entails the introduction of children to the best traditions of their culture, including the academic disciplines, in such a way that they understand the claims and theories of a specific discipline and know something about the discipline itself – its methodology, assumptions, limitations, history and so forth. A second commitment is that, as far as is possible and grade-level appropriate, the relations of particular subjects to each other, and their relation to the broader canvas of ethics, religion, culture, economics and politics, should be acknowledged and investigated. The liberal tradition seeks to overcome intellectual fragmentation. A third commitment is that education needs to be conducted in an ethical manner, and this is applicable to both classrooms and the wider institutional conduct of schooling. Ethics has both proximal and distal reach.

The liberal tradition maintains that science education should not just be an education or training in science, although of course it must be this, but also an education about science. Students educated in science should have an appreciation of scientific methods, their diversity and their limitations. They should have a feeling for methodological issues, such as how scientific theories are evaluated, how competing theories are appraised, how common controversy is in science, and how scientific argument and debate are engaged in the resolution of these controversies; they should also have an appreciation of the interrelated role of experiment, mathematics, and religious, philosophical and ideological commitment in the development of science. All students, whether science majors or others, should have some knowledge of the great episodes in the development of science and, consequently, of culture: the ancient demythologising of the world picture; the Copernican relocation of the earth from the centre of the solar system; the development of experimental and mathematical science associated with Galileo and Newton; Newton's demonstration that the terrestrial laws of attraction operated in the celestial realms; Darwin's epochal theory of evolution and his claims for a naturalistic understanding of life; Pasteur's discovery of the microbial basis of infection; Einstein's theories of gravitation and relativity; and the discovery of the DNA code and research on the genetic basis of life.[7] They should, depending upon their age, have an appreciation of the intellectual, technical, social and personal factors that contributed to these monumental achievements.

Clearly, all of these goals for general education, and for science education, require the integration of history and philosophy into the science curriculum of schools and teacher education programmes. As will be elaborated in Chapter 12, good teachers of science, and indeed of all subjects, need to know something of the history and philosophy of the discipline they are teaching and be able to enthuse students with these dimensions of science.

History, Philosophy and Technical Education

The rapprochement between HPS and science education is not only dependent on having a liberal view of science education: a good technical science education also requires some integration of history and philosophy into the programme. Knowledge of science entails knowledge of scientific facts, laws, theories – the products of science; it also entails knowledge of the processes of science – the social, technical and intellectual ways in which science develops and tests its knowledge claims. HPS is important for the understanding of these process skills. Technical – or 'professional' or 'disciplinary', as it is sometimes called – science education is enhanced if students know the meaning of terms that they are using; if they can think critically about texts, reports and their own scientific activity; if they know how certain evidence relates or does not relate to hypotheses being tested; if they can intelligently and carefully represent data and argue from data to phenomena; and if they can discuss, argue and advance thinking among their colleagues. These scientific abilities are enhanced if students have read examples of sustained enquiry, clever experimentation, insightful hypotheses and exemplary debates about hypothesis evaluation and testing. Alfred North Whitehead expressed this view of good technical education when, just after World War Two, he said:

> The antithesis between a technical and a liberal education is fallacious. There can be no adequate technical education which is not liberal, and no liberal education which is not technical: that is, no education which does not impart both technique and intellectual vision.
>
> (Whitehead 1947, p. 73)

To teach Boyle's Law without reflection on what 'law' means in science, without considering what constitutes evidence for a law in science, and without attention to who Boyle was, when he lived and what he did, is to teach in a disappointingly truncated way. More can be made of the educational moment than merely teaching, or assisting students to discover, that, for a given gas at a constant temperature, pressure multiplied by volume is a constant. This is something, but it is minimal. Similarly, to teach Darwinian evolutionary theory without considerations concerning theory and evidence, the roles of inductive, deductive and abductive reasoning, Darwin's life and times and the religious, literary and philosophical controversies his theory occasioned is also limited. Students doing and interpreting experiments need to know something of how description of data relies upon theory, how evidence relates to the inductive support or deductive falsification of hypotheses, how real cases relate to ideal cases in science, how messy 'lived experience' connects with abstracted and idealised scientific theories, and a host of other matters that all involve philosophical or methodological concerns. Science has a rich and influential history and it is replete with philosophical and cultural ramifications. An education in science should present students with something of this richness and engage them in some of the big questions that have consumed scientists. Whether these questions are regarded as extra-scientific or intra-scientific is, pedagogically, not very important.

Problems with Science Education

It is internationally recognised that there are problems with science education. Orthodox, technical, non-contextual teaching is largely failing to engage students or to promote knowledge and appreciation of science in the population. There is a well-documented crisis in contemporary science education, evidenced in the flight from the science classroom of both teachers and students, and in the appallingly high figures for science illiteracy in the Western world. This has prompted massive rethinking and reforms in national curricula and science-education policy across the world.

The Flight from Science

In the US, these reform efforts have been rolling on for the past 30 years.[8] Two decades ago, in the US, 70 per cent of all school students dropped science from their programme at the first available opportunity. The American National Science Foundation (NSF) charged that, 'the nation's undergraduate programmes in science, mathematics and technology have declined in quality and scope to such an extent that they are no longer meeting national needs. A unique American resource has been eroded' (Heilbron 1987, p. 556) Recent US reports on college science enrolments are similarly bleak (Ashby 2006). The National Research Council (NRC) says, in its *Next Generation Science Standards*, that:

> The U.S. has a leaky K–12 science, technology, engineering and mathematics (STEM) talent pipeline, with too few students entering STEM majors and careers

> at every level. . . . We need new science standards that stimulate and build interest in STEM.
>
> (NRC 2013)

In Europe, political and educational effort has gone into similar wide-ranging reform initiatives. A 1995 European Commission report said that:

> Traditional science teaching, aiming at the mastery of a strictly logic order, of the deductive system, of abstract notions among which mathematics dominate, seems to paralyse and to make a passive subject of the learner, suffocating his imagination.
>
> (EC 1995, in Dibattista & Morgese 2014)

Acknowledging the failure of science teaching and the flight from science, a 2004 European Commission report was bluntly titled 'Europe needs more scientists' (EC 2004)! The following year, the Commission commissioned a Europe-wide survey that revealed that 50 per cent of adults saw their school science courses as 'not sufficiently appealing', and curriculum and pedagogical changes were called for to redress the science literacy and engagement problems.[9]

Science Literacy

Given the amount of state and private money and resources provided for science education, the levels of adult scientific illiteracy are depressing (Roberts 2007, Shamos 1995). For over four decades, Jon D. Miller and colleagues have conducted a series of NSF-sponsored, large-scale studies on scientific literacy in the US (Miller 1983, 1987, 1992, 2007). For Miller, literacy is measured on two dimensions: knowledge of scientific content and knowledge of scientific processes. The former includes basic knowledge of the meaning of concepts such as 'atom', 'gravity', 'gene' and so forth, and basic factual knowledge. For the latter, literacy requires some knowledge of how science works, what it is to study something scientifically and some basics about experiment and hypothesis testing. In 1985, he judged only 3 per cent of high-school graduates, 12 per cent of college graduates and 18 per cent of college doctoral graduates to be scientifically literate. Among statements to which he asked a representative sample of 2,000 adults to answer true or false were, 'The earliest human beings lived at the same time as the dinosaurs' and 'Antibiotics kill viruses as well as bacteria'. Only 37 per cent of the sample answered the first question correctly, and 26 per cent the second. He concluded that 5–9 per cent of US citizens were scientifically literate (Miller 1992, p. 14). In 2005, his testing was extended to thirty-four nations; pleasingly, the US science literacy rate rose to 28 per cent, but only one country, Sweden, registered an adult science literacy rate above 30 per cent (Miller 2007).[10]

There are, of course, separate arguments about what constitutes scientific literacy[11] and why citizens and educational administrators should be concerned about low and falling levels of scientific literacy. The standard reasons for concern have been:

- *cultural* – science, like music, religion and art, is an important part of our cultural heritage and so needs to be known;
- *vocational* – science, like mathematics and computer competence, is indispensable for a wide range of contemporary occupations and so needs to be mastered;
- *disciplinary* – without a spread of basic scientific knowledge, there will not be a big enough pool of school students who might decide to pursue higher studies and careers in science, or a public supportive of their taxes funding research in scientific disciplines;
- *environmental* – people ought know something about the inhabitants, constitution and processes of natural physical, plant and animal worlds in which they live, and that need to be sustained;
- *utilitarian* – scientific knowledge is useful for myriad everyday life and decision-making.

The final reason reverts back to the 'science of everyday things' that once dominated curricular decision-making, is now making a comeback and is perhaps the most common justification for promoting science literacy and enforcing compulsory school science. As two sociologists of science ventured, science education is helpful because it helps us, among other things, 'know where in the oven to put a soufflé' (Collins & Pinch 1992, p. 150). Yet research suggests that knowledge of disciplinary science has precious little, if anything, to do with everyday decision-making in kitchens, in supermarkets, on the road, in hospitals or most other places, even when explicitly socio-scientific issues are being resolved.[12]

HPS-informed curricula and classroom teaching are surely not the sole solution to these 'problems' of science education, but assuredly they can make the subject more 'appealing', engaging and better connected with other subjects being learned – mathematics, history, philosophy, religion and so on. That it is not immediately useful in the kitchen is not a great drawback; much 'standard' science is not immediately useful either. Apart from better learning of science, a HPS-informed science curriculum can have significant impacts on people's worldviews and their religious and cultural understandings. These impacts are not useless.

Occult and Pseudoscientific Belief

The figures on scientific illiteracy are doubly depressing, as they not only indicate that large percentages of the population do not know the meaning of basic scientific concepts, and thus have little if any idea of how nature works, but because such illiteracy is linked to widespread antiscientific and illogical thought. Gallup polls consistently show that about one-third of

Americans believe in ghosts, telepathy, demonic possession, psychic powers and a range of such completely discredited and dangerous ideas (Gallup & Newport 1991). Newspaper astrology columns are read by far more people than are science columns; the tabloid press, with their Elvis sightings and Martian visits, adorn checkout counters and are consumed by millions worldwide each day. Countless thousands of Internet sites and telephone yellow-page directories offer services such as: astrological therapy, palm reading, aura readings, past-life interpretations, feng shui alignments, future-life happenings, dealing with aliens, clairvoyance, tarot-card readings and the whole gamut of such misplaced and misdirected engagements.[13]

It is unfortunate that these 'alternative' beliefs are frequently associated with artistic endeavour. Communities with the greatest concentration of artists also have the greatest concentration of 'New Age' practitioners. The only town in the Australian state of New South Wales to reject fluoridation of its water supply was the artistic hub of Byron Bay. In Arizona, the town of Sedona is deservedly famous for its scores of art galleries and hundreds of artists, but the town is also awash with purveyors of every kind of occult and psychic therapy and treatment. Everything is for sale: Chakra healing, crystal healing, spiritual acupuncture, past-life therapy, Tao-card analysis, guru sessions and so on. And there are special cosmic energy lines where, for a fee, people can sit at their precise node or vortex and absorb the energy by osmosis.[14] One of the hundreds of alternative business operations claims to:

> have discovered some of the most potent concentrated energy fields (Vortex Phenomena) in the Sedona area to reconnect you with the energetic nurturance of Mother Earth's NEMFs (natural electro-magnetic fields).

Most of the thousands of people in Sedona who, every year, pay money to charlatans and purveyors of nonsense have studied high-school science. One of the tasks of this book will be to understand how 'orthodox' school science makes possible this level of credulity, and how HPS-informed school science might make folk more informed and sceptical, more resistant to nonsense. There is ample 'mystery', wonderment and metaphysics available within science, if it is properly taught.

When thought becomes so free from rational constraints, then outpourings of racism, prejudice, hysteria and fanaticism of all kinds can be expected. For all its faults, science has been an important factor in combating superstition, prejudice and ignorance. It has provided, albeit falteringly, a counter-influence to the natural inclinations of people to judge circumstances in terms of their own experience and self-interest. When people, en masse, abandon science, or science education abandons them, then the world is at a critical juncture. At such a time, the role of the science teacher is especially vital and in need of all the intellectual and material support possible.

No one thinks that just technical science education can 'roll back' the tide of questionable, if not completely nonsensical, personal and cultural beliefs.

There is much evidence that achievement of even high-level technical competence in science is consistent with deeply held, silly beliefs. For example, Sir Oliver Joseph Lodge (1851–1940) was an eminent British experimental physicist, a contributor to the nascent science of radio transmission and creator of the first spark plug for automobiles; nevertheless, he held spiritualist belief about life continuing after death and in the ability of mediums to connect with the deceased in séances.[15] The First Spiritual Temple website says of Lodge that:

> Sir Oliver sought to bring together the transcendental world with the physical universe. He affirmed, with great conviction, that life is the supreme, enduring essence in the universe; that it fills the vast interstellar spaces; and the matter of which the physical world is composed is a particular condensation of ether for the purpose of manifesting life into a conscious, individual form.
>
> (www.fst.org/lodge.htm)

A hundred years after Lodge's less than illuminating musings, Edgar Dean Mitchell, the NASA astronaut who was the sixth person to walk on the Moon after piloting the Apollo 14 craft and who has science and engineering doctorate degrees from MIT, had a similar constellation of 'extra scientific' beliefs. Mitchell has claimed that, on his way back from the Moon, he had a Savikalpa Samadhi experience, during which his soul absorbed the fire of Spirit–Wisdom that 'roasts' or destroys the seeds of body-bound inclinations. After this experience, he conducted in-flight ESP experiments with his friends back home. These experiments were published in the *Journal of Parapsychology*. Mitchell believes a remote healer, Adam Dreamhealer, cured his kidney cancer over the telephone. He also believes in UFOs and interplanetary visitations and believes he has had personal encounters with these extraterrestrials.

There are hundreds of thousands, if not millions, of Lodges and Mitchells for whom first-rate science education seems to have little if any flow-over effect on the rest of their beliefs. This is a particular problem for those believing that science education should have beneficial impacts on students' personal life and for the advancement of culture more generally. This was the expectation of the Enlightenment philosophers and educators, it was John Dewey's hope, and it is the expectation of the American Association for the Advancement of Science (AAAS), which maintained that:

> The scientifically literate person is one who is aware that science, mathematics, and technology are interdependent human enterprises with strengths and limitations; understands key concepts and principles of science; is familiar with the natural world and recognises both its diversity and unity; *and uses scientific knowledge and scientific ways of thinking for individual and social purposes.*
>
> (AAAS 1989, p. 4; italics added)

In its *Benchmarks for Science Literacy*, the AAAS says that education has to: 'prepare students to make their way in the real world, a world in which problems abound – in the home, in the workplace, in the community, on the planet' (AAAS 1993, p. 282).

The unique contribution of the science programme to this more general, problem-solving and society-improving educational goal is the cultivation and refinement of scientific habits of mind. These are meant to 'flow on' from the laboratory bench to the home, workplace, community and planet. For the AAAS, the wider 'planetary' problems are not just material – they are social, cultural and ideological – but application of a 'scientific habit of mind' is necessary for solving these wider problems. They are not solved by listening to gurus, holding Ouija boards or consulting astrologers. A major problem is that scientific habits of mind are poorly cultivated in school science programmes.

The same hopes for flow-on effects energised Nehru's inclusion of the state's duty to promote 'scientific temper' in the first constitution of the independent India. However, 60 years later, despite enormous investment in, and spread of, science education, these expectations have not materialised. As two Indian scholars maintain:

> If one were to pick out three or four most important reasons for the country's backwardness or failure in many areas, the lack of scientific temper would be one of them.
>
> (Bhargava & Chakrabarti 2010, p. 277)

As will be shown in Chapter 2, such Enlightenment hopes depend upon science education embracing the history and philosophy of its subject; without such embrace, there is little chance that learning science will have positive personal, social and cultural effects beyond the classroom; indeed, the contrary. This recognition is one of the elements in the current rapprochement between science education and HPS. This is not to say that HPS-informed education is sufficient for the purpose, but, as Spinoza so wisely said, 'the best should not get in the way of the better'.

Critics of Science

Science has not been without its critics. In the seventeenth century, Giambattista Vico (1668–1744) turned his back on the new science of Galileo and the new mathematics of Descartes in favour of a return to 'ancient wisdom'. Subsequently, many other critics, including the literary Romantics, some religious traditions and various counter-cultural movements, have repeated Vico's stand.[16] Phenomenological philosophers such as Edmund Husserl (1859–1938) criticised the mathematisation of science inaugurated by Galileo because of its failure to grasp the experiential realities of the life world (Husserl 1954/1970). Postmodernist philosophers have attacked the universalist and realist assumptions of science. Prince Charles, the future King of England, has fulminated against Galileo and the modern science tradition he launched, saying that it is materialist, that it objectifies the world and that it is 'an affront to the world's sacred traditions'.[17] After criticising the two-century-old marriage of science and commerce, he opined:

> This imbalance, where mechanistic thinking is so predominant, goes back at least to Galileo's assertion that there is nothing in Nature but quantity and motion. This is the view that continues to frame the general perception of the way the world works and how we fit within the scheme of things. As a result, Nature has been completely objectified – 'She' has become an 'it' – and we are persuaded to concentrate on the material aspect of reality that fits within Galileo's scheme.

It is not just outsiders who criticise science. Glen Aikenhead, a senior Canadian educator and leading figure in international science-education research, has stated that, 'the social studies of science' reveal science as: 'mechanistic, materialist, reductionist, empirical, rational, decontextualised, mathematically idealised, communal, ideological, masculine, elitist, competitive, exploitive, impersonal, and violent' (Aikenhead 1997, p. 220).

It is imperative for science teachers to identify what is correct in these critiques, but also what is incorrect. If the claims of phenomenologists, postmodernists, Prince Charles and supposedly the social studies of science are accepted *in toto*, then the standard purposes and justifications of science teaching have to be abandoned, along with at least the compulsory teaching of science. Does anyone want children learning something that is exploitive, competitive, violent and destructive of comfortable worldviews? Clearly, the appraisal of these claims requires some knowledge of HPS, as this is precisely what the critics appeal to. The arguments of this book are that HPS can defend the core principles and practice of science, but also can contribute to the much-needed improvement and reform of science curricula and teaching.

Curriculum Developments

The HPS&ST programme is energised because of curriculum developments that, in the past few decades, have been instigated by numerous government and educational bodies. These will be documented in some detail in Chapter 3. Among these have been the AAAS in two of its very influential reports, *Project 2061* (AAAS 1989) and *The Liberal Art of Science* (AAAS 1990); the US NRC, with its *Next Generation Science Standards* (NRC 2013); the British National Curriculum Council (NCC 1988); the Science Council of Canada (SCC 1984); the Danish Science and Technology curriculum; and The Netherlands' PLON programme. In all of these cases, HPS is not simply another item of subject matter added to the science syllabus; what is proposed is the thesis of this book, namely more general incorporation of HPS themes into the content of curricula.

The AAAS provides a nice summation of the foregoing curricular initiatives when it says:

> Science courses should place science in its historical perspective. Liberally educated students – the science major and the non-major alike – should complete their science courses with an appreciation of science as part of an intellectual, social, and cultural tradition. . . . Science courses must convey these aspects of science by stressing its ethical, social, economic, and political dimensions.
>
> (AAAS 1989, p. 24)

It should be obvious that, for the realisation of the aims of all of these curricula, there needs to be HPS input into documents, teaching materials, assessment schemes, textbooks and teacher education.

Conclusion

Science and its associated technology are the defining features of the modern world; that they should be better understood is an educational truism. The inclusion of HPS in curricula, teacher education and classroom lessons does not, of course, provide all the answers to the problems of modern education – ultimately, these answers lie in the heart of culture, politics and the economic organisation of societies. However, HPS has a significant contribution to make to improving science teaching and learning and, consequently, personal and social flourishing. This contribution can be itemised as follows:

- HPS can humanise the sciences and connect them to personal, ethical, cultural and political concerns. There is evidence that this makes science and engineering programmes more attractive to the many students, and particularly girls, who currently reject them.
- HPS, particularly basic logical and analytic exercises – Does this conclusion follow from the premises? What do you mean by such and such? – can make classrooms more challenging, and enhance reasoning and critical thinking skills.
- HPS can contribute to the fuller understanding of scientific subject matter – it can help to overcome the 'sea of meaninglessness', as Joseph Novak once said, where formulae and equations are recited without knowledge of what they mean or to what they refer.
- HPS can improve teacher education by assisting teachers to develop a richer and more authentic understanding of science and its place in the intellectual and social scheme of things. This has a flow-on effect, as there is much evidence that teachers' epistemology, or views about the nature of science, affect how they teach, the message they convey to students and, ultimately, the epistemology of students.
- HPS can assist teachers in appreciating the learning difficulties of students, because it alerts them to the historic difficulties of scientific development and conceptual change. Galileo was 40 years of age before he formulated

the modern conception of acceleration; despite prolonged thought, he never worked out a correct theory for the tides. By historical studies, teachers can see what some of the intellectual and conceptual difficulties were in the early periods of scientific disciplines. This knowledge can assist with the organisation of the curriculum and the teaching of lessons.

- HPS can contribute to the clearer appraisal of many contemporary educational debates that engage science teachers and curriculum planners. Many of these debates – about constructivist teaching methods, multicultural science education, feminist critiques of science, issues about the relation between science and religion, environmental science, enquiry learning, science–technology–society curricula, teaching controversial issues such as evolution, and so forth – make claims and assumptions about the history and epistemology of science, or the nature of human knowledge and its production and validation. Without some grounding in HPS, teachers can be too easily carried along by fashionable ideas that, later, sadly, 'seemed good at the time', but that wreck educational and cultural havoc.

Notes

1 See at least: Mario Bunge (2000, 2003, 2011), Martin Carrier (2013), Hasok Chang (2011), Alberto Cordero (1992, 2009), Richard Grandy (1997), Rom Harré (1983), Gürol Irzik (2013, 2011 with Robert Nola, 2014 with Robert Nola), Peter Kosso (2009), Hugh Lacey (2009), Peter Machamer (1992), Martin Mahner (2012, 2014, 1996 with M. Bunge), Robert Nola (1997, 2003, 2005 with Gürol Irzik), Robert Pennock (2002), Cassandra Pinnick (2005, 2008), Demetris Portides (2007), Jürgen Renn (2013), Michael Ruse (1990), Harvey Siegel (1979, 1989, 1993, 1997, 2004), Peter Slezak (2000, 2014), Wallis Suchting (1992, 1995), Paul Thagard (2010 with S. Finlay, 2011) and Emma Tobin (2013).

2 See at least: Fabio Bevilacqua (1996 with E. Giannetto), William Brock (1989, 2014 with Edgar Jenkins), John Hedley Brooke (2010), Ricardo Lopes Coelho (2007, 2009), David Depew (2010), John Heilbron (1983), Mercé Izquierdo-Aymerich (2013), Helge Kragh (1992, 1998, 2014) and Cibelle Celestino Silva (2007).

3 See at least: *Hermeneutics and Science Education*, 1995, 4(2); *Religion and Science Education*, 1996, 5(2); *Philosophy and Constructivism in Science Education*, 1997, 6(1–2); *Galileo and Science Education*, 1999, 8(2); *Thomas Kuhn and Science Education*, 9(1–2); *Constructivism and Science Education*, 2000, 9(6); *Science Education and Positivism: A Re-evaluation*, 2004, 13(1–2); *Models in Science and in Science Education*, 2007, 16(7–8); *Feminism and Science Education*, 2008, 17(10); *Science, Worldviews and Education*, 2009, 18(6–7); *Darwinism and Education*, 2010, 19(4–5, 6–8); *Philosophical Considerations in the Teaching of Biology*, 2013, 22 (1–3); *Philosophical Considerations in the Teaching of Chemistry*, 2013, 22(7); *Mendel, Mendelism and Education*, 2015, 24; *Conceptual Change in Science and in Science Education*, 2014, 23.

4 Some useful discussions of the connection between history of science and philosophy of science can be found in Hacking (1992), Lakatos (1971), McMullin (1970, 1975), Shapere (1977) and Wartofsky (1976).

5 Some of the historical texts with introductions can be read in Matthews (1989).

6 There is a large literature on the theory and practice of liberal education. Sometimes, it is given the name 'general' or 'humanistic' education. Peters (1966, Chapters 1, 2) and Bantock (1981, Chapter 4) are useful introductions to these traditions.

7 The AAAS in its *Science for All Americans* lists ten episodes in history that have had major social and cultural impact in the West and beyond, and that should be appreciated by all citizens (Rutherford & Ahlgren 1990, Chapter 10).
8 The most visible and influential have been the NRC's *National Science Education Standards* (NRC 1996), *Inquiry and the National Science Education Standards* (NRC 2000), *America's Lab Report* (NRC 2006), *Taking Science to School* (NRC 2007), *A Framework for K-12 Science Education* (NRC 2012) and *Next Generation Science Standards* (NRC 2013); the AAAS's *Science for All Americans* (AAAS 1989), *The Liberal Art of Science* (AAAS 1990) and *Benchmarks for Science Literacy* (AAAS 1993).
9 The research literature on European science education reform, and especially the place of HPS in those reforms, is reviewed in Dibattista and Morgese (2014).
10 Miller's research is reviewed in Anelli (2011), Hobson (2008) and Trefil (2008, Chapter 6).
11 See, among others: DeBoer (2000), Laugksch (2000), Roberts (2007) and Shamos (1995).
12 On this, see: Chapman (1993), Feinstein (2011) and Wynne (2007).
13 The most sustained recent discussions of paranormal and pseudoscience belief are by Carl Sagan (1997) and Michael Shermer (1997). See also Mario Bunge (2011) and contributions to *Science & Education* 2011, 20(5–6), a thematic issue on Pseudoscience. A classic historical study of the subject was published 100 years ago by W.E.H. Lecky (Lecky 1914).
14 In 2014, folk were charged US$200 per hour to so sit, and it cost much the same for most other astro/psychic/out-of-world services in Sedona.
15 Oliver Lodge was just one of hundreds of prominent 'men of science' who embraced spiritualism and various other psychic movements in the late-nineteenth and early-twentieth centuries. The Society for Psychical Research has 2,710 letters written to Lodge by a credulous public. The former Catholic priest and professor of philosophy Joseph McCabe (1867–1955) wrote a convincing critique of Lodge's spiritualist–theological–philosophical edifice (McCabe 1914). Unfortunately, McCabe's voluminous publications in theology, philosophy, church history and popular science are now largely unknown, but see Cooke (2001).
16 A good account of 'Science and Its Critics' can be found in Passmore (1978), and in contributions to Gross *et al.* (1996) and Koertge (1998).
17 A lecture delivered at the Oxford University Centre for Islamic Studies in June 2010. See: www.princeofwales.gov.uk/media/speeches

References

AAAS (American Association for the Advancement of Science): 1989, *Project 2061: Science for All Americans*, AAAS, Washington, DC. Also published by Oxford University Press, 1990.

AAAS (American Association for the Advancement of Science): 1990, *The Liberal Art of Science: Agenda for Action*, AAAS, Washington, DC.

AAAS (American Association for the Advancement of Science): 1993, *Benchmarks for Science Literacy*, Oxford University Press, New York.

Aikenhead, G.S.: 1997, 'Towards a First Nations Cross-Cultural Science and Technology Curriculum', *Science Education* 81(2), 217–238.

Anelli, C.: 2011, 'Scientific Literacy: What Is It, Are We Teaching It, and Does It Matter?' *American Entomologist* 57(4), 235–243.

Ashby, C.M.: 2006, *Higher Education: Science, Technology, Engineering, and Mathematics Trends and the Role of Federal Programs*. U.S. Government Accountability Office, Washington, DC (Education Resources Information Center Document ED 491614).

Bantock, G.H.: 1981, *The Parochialism of the Present*, Routledge & Kegan Paul, London.

Bevilacqua, F. and Giannetto, E.: 1996, 'The History of Physics and European Physics Education', *Science & Education* 5(3), 235–246.

Bhargava, P.M. and Chakrabarti, C. (eds): 2010, *Devils and Science: A Collection of Articles on Scientific Temper*, National Book Trust, New Delhi, India.

Brock, W.H. and Jenkins, E.W.: 2014, 'Frederick W. Westaway and Science Education: An Endless Quest'. In M.R. Matthews (ed.) *International Handbook of Research in History, Philosophy and Science Teaching*, Springer, Dordrecht, The Netherlands, pp. 2359–2382.

Brock, W.H.: 1989, 'History of Science in British Schools: Past, Present and Future'. In M. Shortland and A. Warwick (eds) *Teaching the History of Science*, Oxford, Basil Blackwell, pp.30–41.

Brooke, J.H.: 2010, 'Darwin and Religion: Correcting the Caricatures', *Science & Education* 19(4–5), 391–405.

Brush S.G.: 1969, 'The Role of History in the Teaching of Physics', *The Physics Teacher* 7(5), 271–280.

Bunge, M.: 2000, 'Energy: Between Physics and Metaphysics', *Science & Education* 9(5), 457–461.

Bunge, M.: 2003, 'Twenty-Five Centuries of Quantum Physics: From Pythagoras to Us, and from Subjectivism to Realism', *Science & Education* 12(5–6), 445–466.

Bunge, M.: 2011, 'Knowledge: Genuine and Bogus', *Science & Education* 20(5–6), 411–438.

Carrier, M.: 2013, 'Values and Objectivity in Science: Value-Ladenness, Pluralism and the Epistemic Attitude', *Science & Education* 22(10), 2547–2568.

Chapman, B.: 1993, 'The Overselling of Science Education in the 1980s'. In R. Levinson (ed.) *Teaching Science*, Routledge, New York, pp. 192–207.

Chang, H.: 2011, 'How Historical Experiments Can Improve Scientific Knowledge and Science Education: The Cases of Boiling Water and Electrochemistry', *Science & Education* 20(3–4), 317–341.

Coelho, R.L.: 2007, 'The Law of Inertia: How Understanding Its History Can Improve Physics Teaching', *Science & Education* 16(9–10), 955–974.

Coelho, R.L.: 2009, 'On the Concept of Energy: How Understanding its History Can Improve Physics Teaching', *Science & Education* 18(8), 961–983.

Cohen, R.S.: 1964, 'Individuality and Common Purpose: The Philosophy of Science', *The Science Teacher* 31(4). Reprinted in *Science & Education* 3(4), 1994.

Collins, H.M. and Pinch, T.: 1992, *The Golem: What Everyone Should Know About Science*, Cambridge University Press, Cambridge, UK.

Cooke, B.: 2001, *Joseph McCabe and Rationalism*, Prometheus Books, Amherst, NY.

Cordero, A.: 1992, 'Science, Objectivity and Moral Values', *Science & Education* 1(1), 49–70.

Cordero, A.: 2009, 'Contemporary Science and Worldview-Making', *Science & Education* 18(6–7), 747–764.

Crombie, A.C.: 1981, 'Philosophical Presuppositions and the Shifting Interpretations of Galileo'. In J. Hintikka, D. Gruender and E. Agazzi (eds) *Theory Change, Ancient Axiomatics, and Galileo's Methodology*, Reidel, Boston, MA, pp. 271–286. Reproduced in A.C. Crombie, *Science, Optics and Music in Medieval and Early Modern Thought*, The Hambledon Press, London, 1990, pp. 345–362.

DeBoer, G.E.: 2000, 'Scientific Literacy: Another Look at Its Hisorical and Contemporary Meanings, and Its Relationship to Science Education Reform', *Journal of Research in Science Teaching* 37(6), 582–601.

Depew, D.J.: 2010, 'Darwinian Controversies: An Historiographical Recounting', *Science & Education* 19(4–5), 323–366.

Dibattista, L. and Morgese, F.: 2014, 'Incorporation of History and Philosophy of Science and Nature of Science Content in School and Teacher Education Programmes in Europe'. In M.R. Matthews (ed.) *International Handbook of Research in History, Philosophy and Science Teaching*, Springer, Dordrecht, The Netherlands, pp. 2083–2117.

EC (European Commission): 2004, *Europe Needs More Scientists! Increasing Human Resources for Science and Technology in Europe*, Brussels. Available at: http://ec.europa.eu/research/conferences/2004/sciprof/pdf/final_en.pdf

Elkana, Y.: 1970, 'Science, Philosophy of Science, and Science Teaching', *Educational Philosophy and Theory* 2, 15–35.

Ennis, R.H.: 1979, 'Research in Philosophy of Science Bearing on Science Education'. In P.D. Asquith and H.E. Kyburg (eds) *Current Research in Philosophy of Science*, PSA, East Lansing, MI, pp. 138–170.

Feigl, H.: 1955, 'Aims of Education for Our Age of Science: Reflections of a Logical Empiricist'. In N.B. Henry (ed.) *Modern Philosophies and Education: The Fifty-fourth Yearbook of the National Society for the Study of Education*, University of Chicago Press, Chicago, IL, pp. 304–341. Reprinted in *Science & Education* 13(1–2), 2004.

Feinstein, N.: 2011, 'Salvaging Science Literacy', *Science Education* 95, 168–185.

Frank, P.: 1947/1949, 'The Place of Philosophy of Science in the Curriculum of the Physics Student', *American Journal of Physics* 15 (3), 202–218. Reprinted in his *Modern Science and Philosophy*, Harvard University Press, Harvard, pp. 228–259.

Gallup Jr, G.H. and Newport, F.: 1991, 'Belief in Paranormal Phenomena Among Adult American', *Skeptical Inquirer* 15, 137–147.

Grandy, R.E.: 1997, 'Constructivism and Objectivity: Disentangling Metaphysics From Pedagogy', *Science & Education* 6(1–2), 43–53. Reprinted in M.R. Matthews (ed.) *Constructivism in Science Education: A Philosophical Examination*, Kluwer Academic Publishers, Dordrecht, The Netherlands, pp. 113–123.

Gross, P.R., Levitt, N. and Lewis, M.W. (eds): 1996, *The Flight from Science and Reason*, New York Academy of Sciences, New York (distributed by Johns Hopkins University Press, Baltimore, MD).

Hacking, I.: 1992, '"Style" for Historians and Philosophers', *Studies in History and Philosophy of Science* 23(1), 1–20.

Harré, R.: 1983, 'History & Philosophy of Science in the Pedagogical Process', in R.W. Home (ed.) *Science under Scrutiny*, Reidel, Dordrecht, The Netherlands, pp. 139–157.

Heilbron, J.L.: 1983, 'The Virtual Oscillator as a Guide to Physics Students Lost in Plato's Cave'. In F. Bevilacqua and P.J. Kennedy (eds) *Using History of Physics in Innovatory Physics Education*, Pavia, Italy, pp. 162–182. Reprinted in *Science & Education* 3(2), 1994, 177–188.

Heilbron, J.L.: 1987, 'Applied History of Science', *ISIS* 78, 552–563.

Hobson, A.: 2008, 'The Surprising Effectiveness of College Scientific Literacy Courses', *The Physics Teacher* 46, 404–406.

Holton, G.: 1975, 'Science, Science Teaching and Rationality'. In S. Hook, P. Kurtz and M. Todorovich (eds) *The Philosophy of the Curriculum*, Prometheus Books, Buffalo, NY, pp. 101–118.

Holton, G.: 1978, 'On the Educational Philosophy of the Project Physics Course'. In *The Scientific Imagination: Case Studies*, Cambridge University Press, Cambridge, UK, pp. 284–298.

Husserl, E.: 1954/1970, *The Crisis of European Sciences and Transcendental Phenomenology*, Northwestern University Press, Evanston, IL.

Irzik, G.: 2013, 'Introduction: Commercialization of Academic Science and a New Agenda for Science Education', *Science & Education* 22(10), 2375–2384.

Irzik, G. and Nola, R.: 2011, 'A Family Resemblance Approach to the Nature of Science for Science Education', *Science & Education* 20(7–8), 591–607.

Irzik, G. and Nola, R.: 2014, 'New Directions in Nature of Science Research'. In M.R. Matthews (ed.) *International Handbook of Research in History, Philosophy and Science Teaching*, Springer, Dordrecht, The Netherlands, pp. 999–1021.

Izquierdo-Aymerich, M.: 2013, 'School Chemistry: An Historical and Philosophical Approach', *Science & Education* 22(7), 1633–1653.

Kampourakis, K. (ed.): 2013, *The Philosophy of Biology: A Companion for Educators*, Springer, Dordrecht, The Netherlands.

Koertge, N. (ed.): 1998, *A House Built on Sand: Exposing Postmodern Myths about Science*, Oxford University Press, New York.

Koertge, N.: 1969, 'Towards an Integration of Content and Method in the Science Curriculum', *Curriculum Theory Network* 4, 26–43. Reprinted in *Science & Education* 5(4), 1996, 391–402 (with afterthoughts).

Kosso, P.: 2009, 'The Large-scale Structure of Scientific Method', *Science & Education* 18(1), 33–42.

Kragh, H.: 1992, 'A Sense of History: History of Science and the Teaching of Introductory Quantum Theory', *Science & Education* 1(4), 349–364.

Kragh, H.: 1998, 'Social Constructivism, the Gospel of Science and the Teaching of Physics', *Science & Education* 7(3), 231–243. Reprinted in M.R. Matthews (ed.) *Constructivism in Science Education: A Philosophical Examination*, Kluwer, Dordrecht, The Netherlands, pp. 125–137.

Kragh, H.: 2014, 'The Science of the Universe: Cosmology and Science Education'. In M.R. Matthews (ed.) *International Handbook of Research in History, Philosophy and Science Teaching*, Springer, Dordrecht, The Netherlands, pp. 643–665.

Lacey, H.: 2009, 'The Interplay of Scientific Activity, Worldviews and Value Outlooks', *Science & Education* 18(6–7), 839–860.

Lakatos, I.: 1971, 'History of Science and Its Rational Reconstructions'. In R.C. Buck and R.S. Cohen (eds) *Boston Studies in the Philosophy of Science* 8, pp. 91–135.

Lakatos, I.: 1978, 'History of Science and Its Rational Reconstructions'. In J. Worrall and G. Currie (eds) *The Methodology of Scientific Research Programmes: Volume I*, Cambridge University Press, Cambridge, UK, pp. 102–138 (originally 1971).

Laugksch, R.C.: 2000, 'Scientific Literacy: A Conceptual Overview', *Science Education* 84, 71–94.

Lecky, W.E.H.: 1914, *History of the Rise and Influence of the Spirit of Rationalism in Europe*, 2 volumes, D. Appleton, New York.

McCabe, J.: 1914, *The Religion of Sir Oliver Lodge*, Watts, London.

Machamer, P.: 1992, 'Philosophy of Science: An Overview for Educators'. In R.W. Bybee, J.D. Ellis, J.R. Giese and L. Parisi (eds) *Teaching About the History and Nature of Science and Technology: Background Papers*, BSCS/SSEC, Colorado Springs, pp. 9–18. Reprinted in *Science & Education* 7(1), 1998, 1–11.

McMullin, E.: 1970, 'The History and Philosophy of Science: A Taxonomy', *Minnesota Studies in the Philosophy of Science* 5, 12–67.

McMullin, E.: 1975, 'History and Philosophy of Science: a Marriage of Convenience?', *Boston Studies in the Philosophy of Science* 32, 515–531.

Mahner, M.: 2012, 'The Role of Metaphysical Naturalism in Science', *Science & Education* 21(10), 1437–1459.

Mahner, M.: 2014, 'Science, Religion, and Naturalism: Metaphysical and Methodological Incompatibilities'. In M.R. Matthews (ed.) *International Handbook of Research in History, Philosophy and Science Teaching*, Springer, Dordrecht, The Netherlands, pp. 1793–1835.

Mahner, M. and Bunge, M.: 1996, 'Is Religious Education Compatible With Science Education?' *Science & Education* 5(2), 101–123.

Martin, M.: 1971, 'The Use of Pseudo-Science in Science Education', *Science Education* 55, 53–56.

Martin, M.: 1972, *Concepts of Science Education: A Philosophical Analysis*, Scott, Foresman, New York (reprint, University Press of America, 1985).

Martin, M.: 1974, 'The Relevance of Philosophy of Science for Science Education', *Boston Studies in Philosophy of Science* 32, 293–300.

Martin, M.: 1986/1991, 'Science Education and Moral Education', *Journal of Moral Education* 15(2), 99–108. Reprinted in M.R. Matthews (ed.) *History, Philosophy and Science Teaching: Selected Readings*, OISE Press, Toronto, 1991, pp. 102–114.

Matthews, M.R. (ed.): 1989, *The Scientific Background to Modern Philosophy*, Hackett Publishing, Indianapolis, IN.

Matthews, M.R. (ed.): 1991, *History, Philosophy and Science Teaching: Selected Readings*, OISE Press, Toronto.
Matthews, M.R. (ed.): 2009, *Science, Worldviews and Education*, Springer, Dordrecht, The Netherlands.
Matthews, M.R. (ed.): 2014, *International Handbook of Research in History, Philosophy and Science Teaching*, 3 volumes, Springer, Dordrecht, The Netherlands.
Miller, J.D.: 1983, 'Scientific Literacy: A Conceptual and Empirical Review', *Daedalus* 112(2), 29–47.
Miller, J.D.: 1987, 'Scientific Literacy in the United States'. In E. David and M. O'Connor (eds) *Communicating Science to the Public*, John Wiley, London.
Miller, J.D.: 1992, *The Public Understanding of Science and Technology in the United States, 1990*, National Science Foundation, Washington, DC.
Miller, J.D.: 2007, 'Public understanding of science in Europe and the United States'. Paper presented at the 2007 annual meeting of AAAS.
Nagel, E.: 1969, 'Philosophy of Science and Educational Theory' *Studies in Philosophy and Education* 7(1), 16–27. Reprinted in J. Park (ed.), *Selected Readings in Philosophy of Education*, Macmillan, New York, 1974.
Nagel, E.: 1975, 'In Defense of Scientific Knowledge'. In S.Hook, P. Kurtz and M. Todorovich (eds) *The Philosophy of the Curriculum: The Need for General Education*, Prometheus Books, Buffalo, NY, pp. 119–126.
NCC (National Curriculum Council): 1988, *Science in the National Curriculum*, NCC, York, UK.
Nola, R.: 1997, 'Constructivism in Science and in Science Education: A Philosophical Critique', *Science & Education* 6(1–2), 55–83. Reproduced in M.R. Matthews (ed.), *Constructivism in Science Education: A Philosophical Debate*, Kluwer Academic Publishers, Dordrecht, The Netherlands, 1998, pp. 31–59.
Nola, R.: 2003, '"Naked Before Reality; Skinless Before the Absolute": A Critique of the Inaccessibility of Reality Argument in Constructivism', *Science & Education* 12(2), 131–166.
Nola, R. and Irzik, G.: 2005, *Philosophy, Science, Education and Culture*, Springer, Dordrecht, The Netherlands.
NRC (National Research Council): 1996, *National Science Education Standards*, National Academies Press, Washington, DC.
NRC (National Research Council): 2000, *Inquiry and the National Science Education Standards: A Guide for Teaching and Learning*, National Academies Press, Washington, DC.
NRC (National Research Council): 2006, *America's Lab Report: Investigations in High School Science*, National Academies Press, Washington, DC.
NRC (National Research Council): 2007, *Taking Science to School. Learning and Teaching Science in Grades K-8*, National Academies Press, Washington, DC.
NRC (National Research Council): 2012, *A Framework for K-12 Science Education: Practices, Crosscutting Concepts, and Core Ideas*, National Academies Press, Washington, DC.
NRC (National Research Council): 2013, *Next Generation Science Standards*, National Academies Press, Washington, DC.
Passmore, J.A.: 1978, *Science and Its Critics*, Rutgers University Press, New Brunswick, NJ.
Pennock, R.T.: 2002, 'Should Creationism be Taught in the Public Schools?', *Science & Education* 11(2), 111–133.
Peters, R.S.: 1966, *Ethics and Education*, George Allen & Unwin, London.
Pinnick, C.L.: 2005, 'The Failed Feminist Challenge to "Fundamental Epistemology"', *Science & Education* 14(2), 103–116.
Pinnick, C.L.: 2008, 'Science Education for Women: Situated Cognition, Feminist Standpoint Theory, and the Status of Women in Science', *Science & Education* 17(10), 1055–1063.

Portides, D.: 2007, 'The Relation Between Idealisation and Approximation in Scientific Model Construction', *Science & Education* 16(7–8), 699–724.
Quine, W.V.O.: 1960, *Word and Object*, MIT Press, Cambridge, MA.
Reichenbach, H.: 1938, *Experience and Prediction: An Analysis of the Foundations and the Structure of Knowledge*, University of Chicago Press, Chicago, IL.
Renn, J.: 2013, 'Einstein as a Missionary of Science', *Science & Education* 22(10), 2569–2591.
Roberts, D.A.: 2007, 'Scientific Literacy/Science Literacy'. In S.K. Abell and N.G. Lederman (eds) *Handbook of Research in Science Education*, Erlbaum, Mahwah, NJ, pp. 729–779.
Ruse, M.: 1990, 'Making Use of Creationism: A Case-study for the Philosophy of Science Classroom', *Studies in Philosophy and Education* 10(1), 81–92.
Rutherford, F.J. and Ahlgren, A.: 1990, *Science for All Americans*, Oxford University Press, New York.
Sagan, C.: 1997, *The Demon-Haunted World: Science as a Candle in the Dark*, Headline, London.
Scheffler, I.: 1973, 'Philosophy and the Curriculum'. In his *Reason and Teaching*, Bobbs-Merrill, Indianapolis, IN, pp. 31–41.
Science Council of Canada (SCC): 1984, *Science for Every Student: Educating Canadians for Tomorrow's World*, Report 36, SCC, Ottawam Canada.
Shamos, M.: 1995, *The Myth of Scientific Literacy*, Rutgers University Press, New Brunswick, NJ.
Shapere, D.: 1977, 'What Can the Theory of Knowledge Learn From the History of Knowledge?', *The Monist* LX(4), 488–508. Reproduced in his *Reason and the Search for Knowledge*, Reidel, Dordrecht, The Netherlands, pp. 182–202.
Shermer, M.: 1997, *Why People Believe Weird Things: Pseudoscience, Superstition, and Other Confusions of Our Time*, W.H. Freemand, New York.
Siegel, H.: 1979, 'On the Distortion of the History of Science in Science Education', *Science Education* 63, 111–118.
Siegel, H.: 1989, 'The Rationality of Science, Critical Thinking, and Science Education', *Synthese* 80(1), 9–42. Reprinted in M.R. Matthews (ed.) *History, Philosophy and Science Teaching: Selected Readings*, OISE Press, Toronto and Teachers College Press, New York, 1991.
Siegel, H.: 1993, 'Naturalized Philosophy of Science and Natural Science Education', *Science & Education* 2(1), 57–68.
Siegel, H.: 1997, 'Science Education: Multicultural and Universal', *Interchange* 28(2–3), 97–108.
Siegel, H.: 2004, 'The Bearing of Philosophy of Science on Science Education, and Vice Versa: The Case of Constructivism', *Studies in History and Philosophy of Science*, 35A, 185–198.
Silva, C.C.: 2007, 'The Role of Models and Analogies in the Electromagnetic Theory: A Historical Case Study', *Science & Education* 16(7–8), 835–848.
Slezak, P.: 2000, 'A Critique of Radical Social Constructivism'. In D.C. Phillips (ed.) *Constructivism in Education: 99th Yearbook of the National Society for the Study of Education*, NSSE, Chicago, IL, pp. 91–126.
Slezak, P.: 2014, 'Constructivism in Science Education'. In M.R. Matthews (ed.) *International Handbook of Research in History, Philosophy and Science Teaching*, Springer, Dordrecht, The Netherlands, pp. 1023–1055.
Suchting, W.A.: 1992, 'Constructivism Deconstructed', *Science & Education* 1(3), 223–254. Reprinted in M.R. Matthews (ed.) *Constructivism in Science Education: A Philosophical Examination*, Kluwer Academic Publishers, Dordrecht, The Netherlands, 1998, pp. 61–92.
Suchting, W.A.: 1995, 'The Nature of Scientific Thought', *Science & Education* 4(1), 1–22.
Suppe, F. (ed.): 1977, *The Structure of Scientific Theories*, University of Illinois Press, Urbana, IL.

Taylor, R.S. and Ferrari, M. (eds): 2011, *Epistemology and Science Education: Understanding the Evolution vs. Intelligent Design Controversy*. Routledge, New York.

Thagard, P.: 2011, 'Evolution, Creation, and the Philosophy of Science'. In R.S. Taylor and M. Ferrari (eds) *Epistemology and Science Education: Understanding the Evolution vs. Intelligent Design Controversy*, Routledge, New York, pp. 20–37.

Thagard, P. and Findlay, S.: 2010 'Getting to Darwin: Obstacles to Accepting Evolution by Natural Selection', *Science & Education* 19(6–8), 625–636.

Tobin, E.: 2013, 'Chemical Laws, Idealization and Approximation', *Science & Education* 22(7), 1581–1592.

Trefil, J.S.: 2008, *Why Science?* Teachers College Press, New York.

Wartofsky, M.W.: 1976, 'The Relation Between Philosophy of Science and History of Science'. In R.S. Cohen, P.K. Feyerabend and M.W. Wartofsky (eds) *Essays in Memory of Imre Lakatos*, Reidel, Dordrecht, The Netherlands, pp.717–738. (*Boston Studies in the Philosophy of Science* 39.) Republished in his *Models*, Reidel, Dordrecht, The Netherlands, 1979.

Whitehead, A.N.: 1947, 'Technical Education and Its Relation to Science and Literature'. In his *The Aims of Education and Other Essays*, Williams & Norgate, London, pp. 66–92.

Wynne, B.: 2007, 'Dazzled by the Mirage of Influence?' *Science, Technology & Human Values* 32(4), 491–503.

Chapter 2

The Enlightenment Tradition in Science Education

To better understand the reasons for contemporary advocacy of history and philosophy in science teaching, or for teaching the nature of science (NOS) as this is sometimes called, it is informative to go back to the origins of these concerns in the European Enlightenment. Although not co-extensive with the Enlightenment tradition in education, the HPS&ST programme shares many of that tradition's concerns for educational engagement by scientists, philosophers and historians in order to enlarge the domain of our knowledge of the natural and social worlds and to promote the betterment of culture and social life.

The European Enlightenment

The eighteenth-century Enlightenment philosophers – John Locke (1632–1704), Baruch Spinoza (1632–1677), Voltaire (1694–1778), Jean D'Alembert (1717–1783), Denis Diderot (1713–1784), Nicolas de Condorcet (1743–1794), Julien de la Mettrie (1709–1751), David Hume (1711–1776) and, a little later, Benjamin Franklin (1706–1790), Joseph Priestley (1733–1804), Thomas Jefferson (1743–1826) and Immanuel Kant (1724–1804) – were inspired by the dramatic achievements of the new science of the seventeenth century. The eighteenth-century Enlightenment was the fruit of the seventeenth-century scientific revolution. In Isaiah Berlin's estimation:

> The intellectual power, honesty, lucidity, courage and disinterested love of the truth of the most gifted thinkers of the eighteenth century remain to this day without parallel. Their age is one of the best and most hopeful episodes in the life of mankind.
>
> (Berlin 1956, p. 29)

With good reason, it was dubbed 'The Age of Reason'. Early-modern and Enlightenment philosophers believed in progress; they thought that, by the use of reason and following the methods of the outstandingly successful natural scientists (Galileo, Huygens, Newton), social life and structures could be made better, and that people could lead a happier and more fulfilling life. They varied in their enthusiasm for social reform (Newton famously opposed the admission of Roman Catholics into Cambridge University), and they varied a great deal in their conceptions of a good state; nevertheless, they did all share core commitments to free speech, free association, education, and the separation of church and state, including separation in law (blasphemy should not be a crime; other separations were harder won), education (the Church should not monopolise or control schooling) and state service (there should not be a religious test for state employment or promotion).

There was a wide spectrum of Enlightenment philosophers, and they held a diverse set of philosophical, religious and political views,[1] with there being a loose distinction between more moderate and more radical groupings.[2] Nevertheless, the following might be regarded as the central commitments of Enlightenment ideology:[3]

1 *Universalism*: All normal human beings share a similar nature and, consequently, are capable of acquiring knowledge and are equally subject to ethical considerations; universalism about law and human rights was a natural outgrowth of universalism about laws of nature and of scientific explanation.[4]
2 *Objectivity*: On matters of fact, whether particular or general, there is objective truth or falsity.
3 *Rationality*: Individuals are capable, in principle, of determining the truth or falsity of propositions concerning matters of fact.
4 *Empiricism*: Sensory evidence is required for the determination of matters of fact.
5 *Scientism*: The method of the new physical sciences needs to be followed in social, political, moral and religious investigations, in order to obtain knowledge in these fields.
6 *Anti-Revelationism*: The only sound method in theology is that of natural theology; knowledge of God is constrained to what can be reasoned from experience and the natural world. Reason thus judges putative revelations.
7 *Naturalism*: The only entities existing in the world, and hence capable of explaining events, are those revealed by science. This may or may not entail materialism. Myths and superstitions need to be rejected.
8 *Utilitarianism*: Ethical norms are to be formulated on the basis of their personal and social utility, not on the basis of revealed religion or their impact on one's afterlife, nor on any putative deontological grounds.
9 *Optimism*: Human beings and society can and should be improved, and this by the application of sound reasoning and right conduct.
10 *Independence*: Secular or religious authorities have no special means to determine truths about the world, ethics, politics or even religion; the claim of individual reason following right method is paramount over mere authoritarian pronouncements.

The historian Margaret Jacob has well expressed the contribution of the new science to the formation of early Enlightenment society in Europe:

> The creation of civil society – the zone of relatively free exchange that lies both between and outside the state and the domestic sphere – owes a debt to science. Experimental science requires voluntary associations and practices intended for verification by an independent audience, however gentlemanly or oligarchic its original composition.
>
> (Jacob 1998, p. 242)

The Enlightenment philosophers (and natural philosophers) believed that the method of the new science should be applied to the seemingly intractable social, political, religious, philosophical and cultural problems of the times, with hopefully something of the same success that was evident in its application to questions about the natural world. Jean Lerond d'Alembert's belief in the power of 'right thinking' was such that he thought that, if mathematicians were smuggled into Spain, the influence of their clear, rational thinking would spread until it undermined the Inquisition (Hankins 1985, p. 2). Here, the heart is clearly ruling the head. D'Alembert well captured the thought and enthusiasm of many eighteenth-century intellectuals when he wrote, in 1759:

> Our century is called . . . the century of philosophy par excellence. . . . The discovery and application of a new method of philosophizing, the kind of enthusiasm which accompanies discoveries, a certain exaltation of ideas which the spectacle of the universe produces in us – all these causes have brought about a lively fermentation of minds, spreading through nature in all directions like a river which has burst its dams.
>
> (Cassirer 1932/1951, pp. 3–4)

And there was much for clear thinkers to think about: seventeenth-century European society fell considerably short of the Garden of Eden. Among the horrific ills besetting society, at least the following warrant mention.[5] During the period of Galileo's most productive work, the terrible Thirty Years War (1618–1648) raged all over Europe – in German states, France, Italy, Spain, Portugal and The Netherlands; and was also fought out in the West Indies and in South America. It is widely accepted that between 15 and 20 per cent of the German population, Catholic and Protestant alike, were killed. The torture, burning and hanging of heretics by both Catholic and Protestant churches went on for centuries after the end of the overt religious wars; the Spanish Inquisition hanged its last heretic, a Deist schoolteacher, in 1834 (Burman 1984, p. 207). Along with religious wars and zealous inquisitions, witch crazes also engulfed Europe, with the worst bloodletting occurring in France, Switzerland, Germany and Scotland. In the Swiss canton of Vaud, in the 90 years between 1591 and 1680, 3,371 women were tried for witchcraft, and all were executed (Koenigsberger 1987, p. 136). The Salem witch trials took place in Massachusetts in 1692, 5 years after publication of Newton's *Principia*. As late as 1773, nearly 100 years after publication of Newton's *Principia*, the Presbyterian Church of Scotland reaffirmed its belief in witchcraft, but Catholic Spain has the distinction of being the last European country to burn a witch at the stake, this being in the early nineteenth century. And, as will be mentioned in Chapter 10, the unspeakable practice still goes on in Papua New Guinea, in Africa and doubtless many other traditional societies untouched by science and enlightened thinking.

Pleasingly, if some anachronism is allowed, there was a wide spectrum of people, including religious social reformers, who thought society might be better organised. Even the Spanish Inquisitor Alonzo Salazer de Frias had doubts about the prevalence, and even existence, of witches. In 1612 (2 years after Galileo's telescope observations), he wrote a report presciently saying:

> I also feel certain that, under present conditions, there is no need for fresh edicts or the prolongation of those existing, but rather that, in the diseased state of the public mind, every agitation of the matter is harmful and increases the evil. I deduce the importance of silence and reserve from the experience that there were neither witches nor bewitched until they were talked or written about.
>
> (Burman 1984, p. 182)[6]

However, it was the Enlightenment philosophers who consciously tried to understand the causes of, and then ameliorate, 'the diseased state of the public mind' – to use the Inquisitor's words.

In the last half-century there has been constant combat within education over the merits of the Enlightenment and of associated enlightened thinking. Numerous researchers – especially feminists, constructivists, postmodernists and multiculturalists – decry the Enlightenment, along with its values and philosophical assumptions. Michael Peters, the editor of an influential journal and book series, is a representative such voice. Peters, influenced by Foucault and other postmodernists, argues for a new philosophy of education that:

> will involve, most importantly, a reassessment of the 'philosophy of the subject' [person, not discipline], of subject-centred reason, as part of the project of modernity underpinning modern educational theory and the project of liberal mass schooling. . . . This line of philosophical investigation might question the way the modern 'subject of education' has been grounded in a European universalism and rationalism heavily buttressed by highly individualistic assumptions inherited from the Enlightenment grand narratives. Informed by a new awareness of the dangers of Western ethnocentrism and a critical understanding of difference and 'otherness' it would provide approaches to the constitution of subjectivity which recognise and redefine the relationship between representation and power at the levels of discourse and practice.
>
> (Peters 1995, pp. 327–328)

Whatever one makes of this claim, it is clear that HPS is germane to its appraisal. Such is the thesis of this book, as illustrated in a host of comparable debates. Without HPS, these educational claims simply produce heat, without much light.

The Enlightenment Tradition

Seventeenth-century England had witnessed the scientific triumphs of Boyle, Hooke, Newton and many lesser figures, along with the establishment of the Royal Society.[7] This new science spawned ideas and attitudes that were the foundation of the soon-to-flower European Enlightenment. In the early eighteenth century, England was the teacher of Europe. As one historian has written: 'in the 1730s and 1740s . . . virtually everything English was in demand in Europe. . . . Above all, Newton and Locke were almost everywhere eulogized and lionized' (Israel 2001, p. 515). The seventeenth-century scientific revolution was the seed that produced the eighteenth-century Enlightenment plant, with its philosophical, theological, political and educational fruit. The scientific accomplishments in mechanics, astronomy, horology, medicine and other fields are well known. These 'natural philosophy' endeavours were institutionalised with the establishment of the German Academy Leopoldina (1652), the Royal Society in England (1660) and the *Académie Royal des Sciences* in France (1666). Seeds do require, of course, nutrients and environs to grow; no one believes that Galileo's telescope or Newton's law of attraction produced by themselves the European Enlightenment, but, without the former, the latter would not have occurred when and where it did.

David Hume, in his *History of England*, wrote that Newton was 'the greatest and rarest genius that ever rose for the ornament and instruction of the species' (Hume 1754–1762/1828, Vol.IV, p. 434). This was, of course, one Englishman writing about another Englishman, but, nevertheless, Hume well expressed the general view of Newton's pre-eminence in seventeenth-century science. Newton famously said, in a letter to Robert Hooke (5 February 1676), 'If I have seen a little further it is by standing on the shoulders of Giants'. And there were many giants on whom to stand, including Galileo, Kepler and Huygens. His *Principia* (Newton 1713/1934) and *Opticks* (Newton 1730/1979) provided the foundation of modern science and the inspiration for the Enlightenment. Newton's self-styled 'under-labourer', John Locke, wrote five major Enlightenment texts in the decade after the publication of the *Principia* (Locke 1689/1924, 1689/1983, 1690/1960, 1693/1996) and *Concerning Education* (Locke 1693/1968).

Newton believed that there would be beneficial flow-on effects if the methods of the new science were applied to other fields. As he stated it: 'If natural philosophy in all its Parts, by pursuing this Method, shall at length be perfected, the Bounds of Moral Philosophy will be also enlarged' (Newton, 1730/1979, p. 405). He applied his scientific methods to historical questions, most notably in his persistent and detailed biblical studies and their extension to his massive, posthumously published 1728 study of *The Chronology of Ancient Kingdoms*. He studied the Bible assiduously in multiple translations, seeking evidence for authorship and for appraising different interpretations of texts; he had what he regarded as a scientific and critical approach to Biblical studies, church history and theology. These studies led him to believe that the doctrine of the Trinity was a Hellenistic corruption of early Church thought, but he was astute enough to keep this and other blasphemous beliefs private and unpublished. Newton wrote far more on biblical interpretation and theology than he ever did on natural philosophy. Unfortunately in those fields there were far fewer giants on whose shoulders he could stand.[8]

As will be mentioned in Chapter 7, within 50 years of Newton's heretical writing, Joseph Priestley applied the same methods to the same materials with the same results, but, unlike Newton, Priestley's philosophical and social convictions led him to make very public his unorthodox opinions. Fifty years before Mill's *On Liberty*, Priestley argued for the basic liberal position that free expression and public dispute were preconditions for the growth of all knowledge – scientific, religious, historical, political and everything else. Priestley's arguments for an 'Open Society' predated Karl Popper's by 150 years.

David Hume echoed Newton's expectation with the subtitle of his famous *Treatise of Human Nature*, which reads, *Being an Attempt to Introduce the Experimental Method of Reasoning into Moral Subjects* (Hume 1739/1888). The Marquis de Condorcet (1743–1794), a leading philosopher of the French Enlightenment, said in his 1782 acceptance speech at the French Academy that, 'the moral [social] sciences' would eventually 'follow the same methods, acquire an equally exact and precise language, attain the same degree of certainty' as the natural sciences (Condorcet 1976, p. 6).

In the circumstances of seventeenth-century Europe, it was not surprising that many with a reformist bent thought that Newton's scientific achievements might be replicated in fields outside natural philosophy if his approach and 'method' were applied more broadly. It was the hope of many that lessons from the new science might have flow-on effects for culture, society and personal life. It was the duty of education to promote this flow-on from academies to citizens.

All of the Enlightenment philosophers had a concern with education: they wanted enlightened ideas to fructify among citizens.[9] It is this educational or pedagogical commitment that underlies their writing and lecturing in the vernacular language, publishing books and pamphlets for wide readerships, engaging in very public debate in newspapers and periodicals, editing English and French encyclopedias, and so on. They also wrote explicitly on education, with works by Locke (1693/1996), Kant (1803/1899) and Rousseau (1762/1991) having great impact at the time and subsequently. The opening words of Locke's education treatise capture the Enlightenment's zeitgeist with its commitment to humanism, liberty, progress, the perfectability of individuals and society, and denial of all versions of pessimistic predestinationism:[10]

> I think I may say, that of all the Men we meet with, Nine Parts of Ten are what they are, Good or Evil, useful or not, by their Education. 'Tis that which makes the great difference in Mankind.
>
> (Locke 1693/1968, p. 114)

Joseph Priestley as Educator

The best and most striking Enlightenment precursor to modern 'Science for All' movements, and in particular a forerunner of the contemporary HPS&ST programme, is Joseph Priestley. He was born in Yorkshire in 1733 and died in Pennsylvania in 1804; his life spanned the core years of the European Enlightenment, in which he played a significant role. He was an enormously gifted person, a polymath, who made original and lasting contributions across a wide range of subjects. He wrote more than 200 books, pamphlets and articles, in history of science (specifically of electricity and optics), political theory, theology, biblical criticism, theory of language, philosophy of education and rhetoric, as well as authoring books and pamphlets on chemistry, for which he is now best known.[11] He was not just knowledgeable in many fields: there was an explicit interconnectedness to all his intellectual activity. For Priestley, knowledge was not compartmentalised: his epistemology (sensationalism) related to his ontology (materialism), and both related to his theology (Unitarianism) and to his psychology (Associationism), and these all bore upon his political and social theory (Liberalism). He was a consciously synoptic or systematic thinker: all components of knowledge (and life as a whole) had to relate consistently.

Priestley shared the Enlightenment conviction that a good education would benefit individuals and their societies. As he wrote in 'The Proper Objects of Education' (Priestley 1791):

> All great improvements in the state of society ever have been, and ever must be . . . the result of the most peaceable but assiduous endeavours in pursuing the slowest of all processes – that of enlightening the minds of men.

Although many advocated and wrote about better and more widespread education, Priestley was of the minority who practised what the Enlightenment preached: he had a lifelong engagement in schooling, teaching and learning. Priestley's educational views were part of his overall systematic position: his theology, philosophy, epistemology, psychology, social theory and science were all parts of a coherent whole. He was under-impressed with the state of English education, in particular education in science:

> I am sorry to have occasion to observe, that natural science is very little, if at all, the object of education in this country, in which many individuals have distinguished themselves so much by their application to it. And I would observe that, if we wish to lay a good foundation for a philosophical taste, and philosophical pursuits, persons should be accustomed to the sight of experiments, and processes, in early life. They should, more especially, be early initiated in the theory and practice of investigation, by which many of the old discoveries may be made to be really their own; on which account they will be much more valued by them.
>
> (Priestley 1790, p. xxix)

This is one of the first endorsements of enquiry teaching, and more specifically of historical-investigative teaching – following in the experimental footsteps of those who have gone before. This is, in part, why he wrote the first histories of optics (Priestley 1772)[12] and of electricity (Priestley, 1767/1775).[13] His assumption was that the habits and skills acquired in investigating nature – observing, hypothesising, seeking evidence for and against, experiments with controls – would flow on to the investigation of other matters: religion, revelation, politics, church history and so on. For Priestley, and a good many of the Enlightenment philosophers, science would be 'the means, under God, of extirpating all error and prejudice, and of putting an end to all undue and usurped authority in the business of religion, as well as of science' (Priestley 1775–1777, Vol.I, p. xiv).

Priestley had a good critical education at the Dissenting Academy at Daventry, where he was exposed to lively debate and argument on all subjects. The dissenting academies were a response by non-conformist clergy and laity to the Anglican Church's monopoly on English school and university education. Robert Merton has been one of many to draw attention to the role of these dissenting academies in fostering and promoting science in England (Merton 1938/1970, p. 119). One commentator has said:

> It is in Non-conformist England, the England excluded from the national universities, in industrial England with its new centres of population and civilisation that we must seek the institutions which gave birth to the utilitarian and scientific culture of the new era.
>
> (Halevy, quoted in Brooke 1987, p. 11)

Newton, at Cambridge, inspired them, but the Dissenters (and Catholics, Jews and atheists) were forbidden to enrol there. In contrast, 'Free Inquiry' was the entrenched motto of the dissenting academies.

In 1758, at age 25 years, Priestley took a pastor's position at Nantwich in Cheshire and, while there, he established a school with thirty boys and, in a separate room, six girls. He taught in the school for 3 years, 6 days a week, from 7a.m. to 4p.m., teaching Latin, Greek, English grammar and geography. In addition, he taught some natural philosophy and purchased an air pump and an electrical machine and instructed his pupils in their use. Thus, Priestley may well have been the first person to teach laboratory science to schoolchildren.

As well as some three decades of direct engagement in teaching, Priestley wrote a number of influential works on the theory and practice of education. His most famous work – *An Essay on a Course of Liberal Education for Civil and Active Life* (Priestley 1765/1965) – was written and published while he was teaching at Warrington Academy. It originally appeared as a pamphlet and then it became a twenty-five-page Prefix to his *Lectures on History and General Policy* (Priestley 1788). In this incarnation, it had sixteen printings and was translated into Dutch (1793) and French (1798). In the American edition of 1803, Priestley adds a note to the above text:

> Since this was written, which is near forty years ago, few persons have had more to do in the business of education than myself; and what I then planned in theory has been carried into execution by myself and others, with, I believe, universal approbation.
>
> (Passmore 1965, p. 289)

This theme of connecting theory to practice runs through all Priestley's work, including his opposition to Lavoisier's new oxygen theory. Although he is neither a harbinger of Marxism nor a premature Positivist, Priestley was always suspicious of theory that ran too far in front of practice, or removed itself too far from the facts of the matter; for him, to coin a later phrase, 'theory had to be proved in practice'.

As is common with the contemporary HPS&ST programme, Priestley advocated a coordinated curriculum, saying that: 'When subjects which have a connection are explained in a regular system, every article is placed where most light is reflected upon it from the neighbouring subjects' (ibid. p. 293), and also a structured and guided curriculum, saying that:

> The plainest things are discussed in the first place, and are made to serve as axioms, and the foundation of those which are treated of afterwards. Without this regular method of studying the elements of any science, it seems impossible ever to gain a clear and comprehensive view of it.
>
> (Ibid. p. 293)

Priestley contrasts favourably the learning and competence of a student instructed in this way with one who

> should only have considered the subject in a random manner, reading any treatise that might happen to fall in his way, or adopting his maxims from the company he might accidentally keep.
>
> (Ibid. p. 293)

One danger of unstructured instruction that he identifies is,

> being imposed upon by the interested views with which men very often both write and speak. For these are subjects on which almost every writer or speaker is to be suspected; so much has party and interest to do with everything relating to them.
>
> (Ibid. p. 293)

He advocates student discussion as part of the learning process, saying that:

> It is no wonder that many young gentlemen give but little attention to their present studies when they find that the subjects of them are never discussed in any sensible conversation to which they are ever admitted.
>
> (Ibid. p. 294)

Again connecting to themes in the current HPS&ST programme, Priestley reinforces his view that liberal education for civil and active life needs to promote the understanding of principles of subject matter, by saying:

> A man who has been used to go only in one beaten track and who has had no idea given him of any other. . . . Will be wholly at a loss when it happens that that track can no longer be used; while a person who has a general idea of the whole course of the country may be able to strike out another and perhaps a better road than the former.
>
> (Ibid. p. 295)

As a teacher at the Dissenting Academy at Warrington, Priestley insisted on students asking and answering questions; he promoted free engagement with all subjects, including Divinity; and he ensured that authorities on both sides of controversial issues be read and quoted. One of his Warrington students recalled that:

> At the conclusion of his lecture, he always encouraged his students to express their sentiments relative to the subject of it, and to urge any objections to what he had delivered, without reserve. It pleased him when anyone commenced such a conversation. . . . His object . . . was to encourage the students to examine and decide for themselves, uninfluenced by the sentiments of any other persons.
>
> (Rutt 1831–1832, Vol.1, p. 50. In Lindsay 1970, p. 15)

Priestley had some confidence that an educational regime such as he proposed and enacted would result in the betterment of society. He said:

> I cannot help flattering myself that were the studies I have here recommended generally introduced into places of liberal education, the consequences might be happy for this country in some future period.
>
> (Passmore 1965, p. 301)

This was the *reformist* Priestley. But, with reason, he was also regarded as a *revolutionary*. His understanding of the flow-on effects of scientific investigation and acquisition of its associated mental and character dispositions led him to proclaim from his Birmingham pulpit, in a sermon on 'The Importance and Extent of Free Inquiry':

> We are as it were, laying gunpowder, grain by grain, under the old building of error and superstition, which a single spark may hereafter inflame, so as to produce an instantaneous explosion; in consequence of which that edifice, the erection of which has been the work of ages, may be overturned in a moment and so effectually as that same foundation can never be built again.
>
> (Priestley 1785)

With Britain having just been defeated in the American Revolution (1775–1783), and with the first stirrings of the French Revolution (1787–1789) being felt in all European states and kingdoms, such words were not judicious; they led to his sobriquet 'Gunpowder Joe' and, in 1791, to an enraged 'King and Church' mob ransacking his home, library and laboratory and his flight from Yorkshire to America.[14]

Through Priestley's personal friendships with Benjamin Franklin, George Washington, John Adams and Thomas Jefferson, and the admiration they all had for him, there was a direct impact of Enlightenment ideas in late-colonial and early-independent US public life and education. Daniel Boorstin writes: 'Next to Paine, Priestley was the most vivid symbol of the cosmopolitan republican spirit; and while still abroad he had become a close collaborator of the Jeffersonians' (Boorstin 1948, p. 17).

The history of Enlightenment ideas and educational practice in the eighteenth and nineteenth centuries is complex and can here be passed over. Suffice to mention Ernst Mach, the Vienna Circle positivists and John Dewey, who are links in an educational chain connecting the present-day HPS&ST programme with its Enlightenment forbears.

Ernst Mach: Philosopher, Scientist, Educator

The first person to deal systematically with the contribution that HPS can make to science education was Ernst Mach (1838–1916), a major mid-nineteenth-century contributor to the Enlightenment tradition. Unfortunately, his contribution to science education has been almost entirely ignored in the English-speaking world.[15] This is a pity, because current trends in the practice and theory of science education are in many respects repeating Mach's century-old arguments concerning the purposes and aims of science teaching, the nature of understanding and the best ways to promote learning. Hopefully, this section will to some degree redo what was done 100 years ago in an obituary:

> It is Mach the educationalist whom we must here bring to the attention of our readers, particularly the younger ones, and not as someone who has passed on, but as a man whose seed is destined to put down ever further roots in physics teaching, and, with that, in all teaching about real things, and to fructify the whole spirit of this teaching.
>
> (Höfler 1916; trans. W.A. Suchting)

Mach was one of the great philosopher–scientists in the late nineteenth and early twentieth centuries. He was fluent in most European languages, an enthusiast of Greek and Latin classics, a physicist who made significant contributions to such diverse fields as electricity, gas dynamics, thermodynamics, optics, energy theory and mechanics, a historian and philosopher of science, a psychologist, Rector of Prague German University, a member of the Upper House in the Austrian Parliament and a writer of lucid prose.

He was a person of strong character and convictions, a socialist and outspoken liberal-humanist in the centre of the archconservative, Catholic Austro-Hungarian Empire. Einstein said of him that, 'he peered into the world with the inquisitive eyes of a carefree child taking delight in the understanding of relationships' (Hiebert 1976, p. xxi). Mach made scientific and philosophical contributions across the whole temporal span, from Darwin to Einstein. The first of Mach's 500 publications appeared in 1859, the year of Darwin's *The Origin of Species*; his last work was published 5 years after his death in 1921, the year of Einstein's *Relativity: The Special and General Theory*.[16]

Mach's Educational Contributions

Mach's understanding of science and philosophy bore upon his educational ideas. Mach was influenced by the ideas of the German philosopher–psychologist–educationalist Johann Friedrich Herbart. He applied Herbart's ideas in his first teaching assignment, 'Physics for Medical Students', and in the text he wrote arising from this course (*Compendium of Physics for Medical Students* 1863). Mach's concern here was with 'economy of thought', with getting across the general outline of the conceptual modes of physics, and with overcoming the compartmentalism of physics.

Psychology was a long-standing interest of Mach's. At 15 years of age, Mach had read Kant's *Prologomena* and signalled his subsequent positivist commitments – 'The superfluity of the role of the "thing-in-itself" suddenly dawned upon me' (Blackmore 1972, p. 11). His teaching was the occasion to unite pedagogical, psychological and scientific concerns. The first of his many science textbooks for school students, published in 1886, was widely used and went through several editions. Indeed, most of the major figures in European physics at the beginning of this century learned science from Mach's school texts. These texts provided a logical and historical introduction to science; they sought to present students with the 'most naive, simple, and classical observations and thoughts from which great scientists have built physics' (Pyenson 1993, p. 34). While at Prague German University, he taught courses on 'School Physics Teaching'. In 1887, Mach founded and co-edited the world's second-published science-education journal – *Zeitschrift für den Physikalischen und Chemischen Unterricht* (*Journal of Instruction in Physics and Chemistry*).[17] He contributed regularly to this journal until a stroke forced his retirement in 1898.

Mach did not write any systematic work on educational theory or practice; his ideas are scattered throughout his texts and journal articles. However, there are three lectures where he addressed pedagogical issues. One of these is perhaps his most systematic treatment of education in general and science education in particular – 'On Instruction in the Classics and the Mathematico-Physical Sciences' (Mach 1886/1986), translated in his *Popular Scientific Lectures*. His other chief pedagogical papers are 'On Instruction in Heat Theory' (1887) and 'On the Psychological and Logical Moment in Scientific Instruction' (1890), in Volumes 1 and 4, respectively, of his *Zeitschrift*.

As well as intellectual and practical interests in education, Mach had a notable Enlightenment-inspired political involvement in educational reform. The best of the Enlightenment thinkers connected thought to action. As Marx said, the point of philosophising was to change the world, not just think about the world. Mach addressed teacher organisations, spoke in the Austrian parliament on the need for school curricular change and was active in the struggles to transform the entrenched German gymnasium pattern of separating language and classics studies into separate schools from those for science and mathematics. Mach championed the creation of the new *Einheitsschule*, where integrated education in the humanities and the sciences could occur. There have been few scientists who have displayed such a wide-ranging interest in both formal (school) and informal (the reading public) education. Mach's relative neglect by English-speaking science educators is unfortunate.

Well-founded curricular and pedagogical proposals in school science are based upon two foundations: views about the nature and scope of science, and views about the nature and practice of education. There are, of course, other matters to be considered in drawing up curricula – political, social and psychological, to name just the obvious ones. But what one thinks, explicitly or implicitly, about the philosophy of science and about the philosophy of education will largely determine the form of the science curriculum promoted. Mach's suggestions for the conduct of science education stem, in part, from his theory of science and his Herbartian theory of education. Some of the major themes of Mach's philosophy of science (his view of the NOS) are the following:

- Scientific theory is an intellectual construction for economising thought and thereby conjoining experiences.
- Science is fallible; it does not provide absolute truths.
- Science is a historically conditioned intellectual activity.
- Scientific theory can only be understood if its historical development is understood.

Mach's educational ideas are fairly simple and uncontroversial; the HPS&ST programme can easily embrace them:

- Begin instruction with concrete materials and thoroughly familiarise students with the phenomena discussed.
- Aim for understanding and comprehension of the subject matter.
- Teach a little, but teach it well.
- Follow the historical order of development of a subject.
- Tailor teaching to the intellectual level and capacity of students.
- Address the philosophical questions that science entails and that gave rise to science.
- Show that, just as individual ideas can be improved, so also scientific ideas have constantly been, and will continue to be, overhauled and improved.
- Engage the mind of the learner.

Although a pre-eminent theorist and concerned with economy of thought in education, Mach firmly believed that abstractions in the science classroom should, as Hegel said of philosophy, take flight only at dusk:

> Young students should not be spoiled by premature abstraction, but should be made acquainted with their material from living pictures of it before they are made to work with it by purely ratiocinative methods.
>
> (Mach 1886/1986, p. 4)

A simple point, usually observed in its breach, as Arnold Arons has lamented:

> As physics teaching now stands, there is a serious imbalance in which there is an overabundance of numerical problems using formulae in canned and inflexible examples and a very great lack of phenomenological thinking and reasoning.
>
> (Arons 1988, p. 18)

Another of Mach's concerns was the tendency to overfill the curriculum. For him, the principal aims of education were to develop understanding, strengthen reason and promote imagination. A bloated curriculum counteracted these aims:

> I know nothing more terrible than the poor creatures who have learned too much. What they have acquired is a spider's web of thoughts too weak to furnish sure supports, but complicated enough to produce confusion.
>
> (Mach 1886/1986, p. 367)

One hundred years later, this lament is still being voiced about the US 'one mile wide and one inch deep' curriculum.

Mach believed in presenting science historically, or, as he put it, teaching should follow the genetic approach:

> every young student could come into living contact with and pursue to their ultimate logical consequences merely a few mathematical or scientific discoveries. Such selections would be mainly and naturally associated with selections from the great scientific classics. A few powerful and lucid ideas could thus be made to take root in the mind and receive thorough elaboration.
>
> (Mach 1886/1986, p. 368)

Mach's major textbooks on Mechanics (1883/1960), Heat (1869) and Optics (1922) all follow the genetic method of exposition. Mach realised that the logic of a subject was not necessarily the logic of its presentation – a point known to most schoolteachers, if not to administrators. The logic of a discipline and the logic of its pedagogy are not identical, as Mach's contemporary and fellow positivist Pierre Duhem also maintained:

> The legitimate, sure, and fruitful method of preparing a student to receive a physical hypothesis is the historical method . . . that is the best way, surely even

> the only way, to give those studying physics a correct and clear view of the very complex and living organisation of this science.
>
> (Duhem 1906/1954, p. 268)

The HPS&ST programme is not necessarily committed to such a historically structured pedagogy, but that is its first option: to walk in the footsteps, and appreciate the work, of the masters. The issue is discussed at some length in Chapter 4 (history in the curriculum), and Chapters 6 (the pendulum) and 7 (photosynthesis) outline such an approach.

The Positivist Tradition

The positivist Vienna Circle met initially as the Ernst Mach Circle; it saw itself as elaborating and advancing the Enlightenment programme of using scientific knowledge and thinking to improve society and culture. It was a social, cultural and philosophical movement, whose origins were with Comte, Spencer and Mach in the second half of the nineteenth century in Europe. As one commentator says: 'logical empiricism was stepped in the tradition of enlightenment thought and engaged in its continuation. Most importantly it was engaged in its renewal' (Uebel 1998, p. 418).[18] In slightly different guises (logical positivism, logical empiricism), it dominated Western philosophy through to the middle of the twentieth century.

For the past 50 years, positivism has been criticised from many sides, and it has been especially criticised and shunned in education, where being labelled 'a positivist' is akin to being labelled 'a terrorist' elsewhere. One prominent science educator writes that: 'as ideology [positivism] has led to the domination of class, race, gender and nature' (Tobin 1998, p. 196). Further, this pernicious influence has been operative for a very long time: 'The roots of positivism permeate science and science education and have done so since the birth of modern science and the time of Leonardo Da Vinci' (Tobin 1998, p. 209). Other educators have even less-kind things to say of positivism.

However, these criticisms are unfounded; as will be shown below, they trade on and perpetuate an image of 'village positivism'. Michael Friedman, with good reason, suggests that:

> As scholarly investigations of the past fifteen or twenty years into the origins of logical empiricism have increasingly revealed, such a simple-minded radically empiricist picture of this movement is seriously distorted. Our understanding of logical positivism and its intellectual significance must be fundamentally revised when we reinsert the positivists into their original intellectual context, that of the revolutionary scientific developments, together with the equally revolutionary philosophical developments, of their time. As a result, our understanding of the significance of the rise and fall of logical positivism for our own time also must be fundamentally revised.
>
> (Friedman 1999, p. xv)

It must certainly be revised for educational purposes, but, as will be a constant refrain in this book, educators are typically too busy with other things to keep abreast of philosophy. But if this is so, then one lesson is to be more modest and circumspect about philosophical (or historical, psychological, sociological or political) claims.

Positivists and the early logical empiricists belonged to the Enlightenment tradition. They believed in the possibility of progress across the board – in human life, medicine, social institutions and cultural components such as art, music, literature.[19] Many contributed to the brief Socialist Spring of 1920s Vienna. They recognised that this progress was entirely dependent on education, both formal (schools, universities, institutes) and informal (writing, newspapers, periodicals, radio and soapboxes in public parks and street corners). They all suffered with the rise of Nazism, and all fled or emigrated to Turkey, the UK, the US and other places.

The positivist philosophical programme took its canonical form when the term 'logical positivism' was used to designate the 1920s work of the Ernst Mach-inspired Vienna Circle of Moritz Schlick (1882–1936), Rudolf Carnap (1891–1970), Otto Neurath (1882–1945), Philipp Frank (1884–1966) and the circle's English populariser Alfred J. Ayer (1910–1989). The educational writings of two foundational positivists – Philipp Frank and Herbert Feigl – demonstrate that the populist and educationalist account of positivism is completely at odds with reality, and this raises the question of how such an enormous miscarriage could occur.

Philipp Frank

Philipp Frank was born in Vienna in 1884 and died in Cambridge, Massachusetts, in 1966. In 1907, he received his doctorate in theoretical physics at the University of Vienna, where he studied under Ludwig Boltzmann. Frank's first paper, published in 1907 at the age of 23 years – 'Experience and the Law of Causality' (Frank 1907/1949) – characterised his subsequent philosophical concern: namely prolonged and informed philosophical reflection on the structures, methodology and history of science. The meetings of the Vienna Circle that he instigated set the style of his subsequent intellectual career: there was a seriousness of purpose, coupled with a genuine open-mindedness towards different opinions and traditions:

> This apparent internal discrepancy [in the group] provided us, however, with a certain breadth of approach by which we were able to have helpful discussions with followers of various philosophical opinions. Among the participants in our discussions were, for instance, several advocates of Catholic philosophy. Some of them were Thomists, some were rather adherents of a romantic mysticism. Discussions about the Old and New Testaments, the Jewish Talmud, St. Augustine, and the medieval schoolmen were frequent in our group. Otto Neurath even enrolled for one year in the Divinity School . . . and won an award for the best

> paper on moral theology. This shows the high degree of our interest in the cultural background of philosophic theories and our belief in the necessity of an open mind which would enable us to discuss our problems with people of divergent opinions.
>
> (Frank 1949, pp. 1–2)

As for all Enlightenment figures, education was the crux of social reform. Peter Bergmann, the physicist, who in 1933 was an 18-year-old Berlin refugee from Nazism, recalled:

> In this overheated and jittery atmosphere there was one fatherly figure who represented all that was best at the University [of Prague], Philipp Frank. . . . He would encourage all of us students, and he gave us the feeling of a wide-open intellectual window, open to things that happened in and out of physics, and open to things that happened outside of the country as well. Philipp Frank saw to it that there was close contact with philosophy of science . . . with experimental physics . . . and with pure mathematics.
>
> (Blackmore *et al.* 2001, p. 69)

E.C. Kemble, a Harvard physicist, wrote of his colleague that:

> His was a gentle, unassuming spirit combined with a luminous mind and gifts of simplicity and humor that endeared him to all. He understood the nature of truth and the criteria that must be used to separate truth from mythology. He was a humanist as well as a scientist and philosopher . . . he had the patience, the perception and the wit to make profound truths intelligible to a wide public.
>
> (Frank 1931/1998, p. x)

It barely needs stating that these accounts of Frank's pedagogy are at odds with the popular educational view of positivists as dogmatic, overbearing, pupil-ignoring adherents of the 'banking' view of education. That these completely false claims are made is evidence for the thesis of this book, namely that knowledge of HPS can improve, to say the least, science-education debate and research.

Frank published two explicitly educational papers: 'Science Teaching and the Humanities' (Frank 1946/1949) and 'The Place of Philosophy of Science in the Curriculum of the Physics Student' (Frank 1947/1949). He regretted that the 'result of conventional science teaching has not been a critically minded type of scientist, but just the opposite' (Frank 1947/1949, p. 230). In part, this regret is because,

> the science student who has received the traditional purely, technical instruction in his field is extremely gullible when he is faced with pseudophilosophic and pseudoreligious interpretations that fill somehow the gap left by his science courses.
>
> (Frank 1947/1949, p. 230)

As a consequence:

> This failure prevents the science graduate playing in our cultural and public life the great part that is assigned to him by the ever-mounting technical importance of science to human society.
>
> (Frank 1947/1949, p. 231)

It is, of course, HPS that makes good these shortfalls, or rather, for Frank, just philosophy of science, because this indeed consists of two inseparable components, 'logico-empirical analysis' and 'socio-psychologic' analysis (Frank 1947/1949, p. 248). The first is conceptual or semantic analysis; the second is careful historical analysis. He says that, 'This analysis is the chief subject that we have to teach to science students in order to fill the gaps left by traditional science teaching' (Frank 1947/1949, p. 245).

Logico-empirical analysis of scientific theories consists primarily in, first, identifying purely logical statements; second, identifying observational statements; and third, specifying operational definitions whereby principles can be connected to observations (Frank 1947/1949, p. 243). The paper gives examples of such analyses of the Copernican controversy, Euclidean and non-Euclidean geometric systems, Newton's laws, relativity theory and quantum theory. Frank wants students to be able to decouple observational statements and statements that are deduced from these: 'For in all these fields the central problem is the relationship between sensory experience (often called fact finding), and the logical conclusions that can be drawn from it' (Frank 1947/1949, p. 234). He uses the Copernican controversy to illustrate his point:

> If we look, for example, at the treatment of the Copernican conflict in an average textbook of science, we notice immediately that the presentation is far from satisfactory. In almost every case, we are told that according to the testimony of our senses the sun seems to move around the earth. Then we are instructed that Copernicus has taught us to distrust this testimony and to look for truth in our reasoning rather than in our immediate sense experience.
>
> (Frank 1947/1949, p. 231)

Frank says that this account is mistaken and can be shown to be such by logico-empirical analysis:

> Actually our sense observation shows only that in the morning the distance between horizon and sun is increasing, but it does not tell us whether the sun is ascending or the horizon is descending.
>
> (Ibid.)

The statement that 'the sun is moving' is an elaboration of sensory evidence, not the sensory evidence; it is what Paul Feyerabend would later call a 'natural interpretation' of the sensory experience (Feyerabend 1975, Chapters 6–7). Frank is saying clearly that theory affects observation; the engaging philosophical task, and one empiricists are committed to, is to ascertain whether there is a level of observation statements that are not so affected. The conclusion Frank draws from the Copernican case is the same as he draws from most of the other examples he discusses, namely:

> By its failure to give an adequate presentation of this historic dispute our traditional physics teaching misses an opportunity to foster in the student an understanding of the relations between science, religion and government which is so helpful for his adjustment in our modern social life. With a good understanding of the Copernican and similar conflicts, the student of science would have even an inside track in the understanding of social and political problems. He would be put at least on an equal level with the student of the humanities.
>
> (Frank 1947/1949, p. 234)

Frank is an advocate of liberal education, affirming that a variety of subject matters should be mastered, and that, as much as possible, relations between the subjects should be brought out. He believes that humanities can be taught from within science, saying that:

> The student of science will get the habit of looking at social and religious problems from the interior of his own field and entering the domain of the humanities by a wide-open door . . . there is no better way to understand the philosophic basis of political and religious creeds than by their connection with science.
>
> (Frank 1946/1949, p. 281)

Frank's final claim in the foregoing will be examined further in Chapter 10.

Herbert Feigl

Herbert Feigl is the second classic positivist whose educational theory and practice are discordant with the popular image of positivism projected in educational writing. He was born in 1902 in Reichenberg, then in Austria–Hungary, a part of the Sudetenland that subsequently was incorporated into Czechoslovakia. He died in Minneapolis in 1988.[20] At age 16, he read an article on the theory of special relativity and set about trying, without success, to refute it. He said that the attempt resulted in him learning a lot of mathematics and physics. At age 20, he went to the University of Vienna to study philosophy with Moritz Schlick (and, additionally, to study mathematics, physics and psychology). He was a foundation member of the Vienna Circle, established by Schlick in 1924 as a weekly evening discussion group, and he remained a member of the Circle until his emigration to the US in 1930. In 1927, Feigl presented his doctoral thesis on 'Chance and Law: An Epistemological Investigation of Induction and Probability in the Natural Sciences'. In the US, he worked with Percy Bridgman at Harvard on the foundations of physics, including the theory of operational definitions of theoretical terms. In 1940, he was appointed professor of philosophy at the

University of Minnesota; in 1953, he established the Minnesota Center for the Philosophy of Science, a centre that would make a significant contribution to the articulation and spread of logical empiricist philosophy in the US and worldwide, especially through contributions to the many volumes of *Minnesota Studies in Philosophy of Science*.

Feigl published one explicitly educational paper: 'Aims of Education for Our Age of Science: Reflections of a Logical Empiricist' (Feigl 1955). The paper was a contribution to *The Fifty-fourth Yearbook of the National Society for the Study of Education*, which dealt with 'Modern Philosophies and Education'. It included contributions from Thomists (Jacques Maritain), Liberal Christians (Theodore Greene), Marxists (Robert Cohen) and others. Against fatalistic or mechanically deterministic views of human freedom, Feigl regards promotion of individual autonomy as the prime educational achievement.[21]

> As long as education promotes the formation of intelligence and character in a manner that allows for free learning, rational choices, and critical reflection, human beings so educated will have an excellent opportunity for being masters of their own activities and achievements.
>
> (Feigl 1955, p. 322)

This is almost, and not accidently, a verbatim repetition of the opening sentences of Kant's 1784 'What is Enlightenment?':

> Enlightenment is man's release from his self-incurred tutelage. Tutelage is man's inability to make use of this understanding without direction from another. Self-incurred is this tutelage when its cause lies not in a lack of reason but in lack of resolution and courage to use it without direction from another. 'Have courage to use your own reason!' – that is the motto of the enlightenment.
>
> (Kant 1784/2003, p. 54)

Not surprisingly, Feigl advocates teaching science in a historically and philosophically informed manner, saying:

> It is my impression that the teaching of science could be made ever so much more attractive, enjoyable, and generally profitable by the sort of approach that is more frequently practised in the arts and the humanities. The dull and dry-as-dust science courses can be replaced by an exciting intellectual adventure if the students are permitted to see the scientific enterprise in broader perspective. Preoccupation with the purely practical values of applied science has overshadowed the intellectual and cultural values of the quest for knowledge.
>
> (Feigl 1955, p. 337)

And, further, he embraces the orthodox liberal education position wherein: 'training in the sciences and in the scientific attitude should, of course, be combined with studies in history, literature, and the arts' (Feigl 1955, p. 338). As important as science is, it is not the only thing that Feigl treasures:

> I consider truly great music the supreme achievement of the human spirit . . . I am inclined to think that music expresses (even more than poetry) what is inexpressible in cognitive and especially in scientific language.
>
> (Cohen 1981, p. 5)

Feigl has a robust account of values and recognises that they are an intrinsic part of education; that they mould and direct educational processes and are crucial to the establishment of educational aims. Feigl has an even more robust account of rationality and its place in education. He believes that the classical Aristotelian conception of man as rational animal 'may still be a good beginning' (Feigl 1955, p. 335), and then explicates the idea for education, stressing that rationality covers at least six virtues of thought and conduct:

- clarity of thought (the meaningful use of language and avoidance of gratuitous perplexities);
- consistency of reasoning (conformity with the principles of formal logic);
- reliability of knowledge claims (wherever the evidence is too weak, belief should be withheld);
- objectivity of knowledge claims (knowledge claims should be testable by anyone sufficiently equipped with intelligence and competence);
- rationality of purposive behaviour (maximum positive outcomes are to be gained at the cost of minimum negative outcomes); and
- moral rationality (adherence to principles of justice, equity or impartiality, and abstention from coercion and violence in the settlement of conflicts of interest (Feigl 1955, pp. 335–336ff.). Rationality is not just an intellectual virtue, it is connected intimately with conduct, or at least with dispositions towards rational conduct.[22]

The Myth and the Reality of Positivism

There is a disjunction between the faults of positivism as commonly adumbrated by science educators and the principles and practice of science education advocated by at least two foundational positivists – Philipp Frank and Herbert Feigl. A source of confusion in educational writing on positivism is that, among educators, positivism is routinely taken to mean materialism or realism. These were metaphysical positions that were explicitly rejected by Postitivism; their rejection defined Machian positivism and the philosophy of the later logical positivists. Eventually, with good reasons, as enunciated by Karl Popper (1902–1994), the phenomenal metaphysics of positivism was abandoned on account of its being inconsistent with the pursuit of science. However, the educational theory of Frank and Feigl can survive the demise of their metaphysics.

The following table summarises the disjunction:

Educational Myths About Positivism	*Frank and Feigl's Position*
Positivism regards scientific knowledge as secure and privileged	They are committed to the fallibility of science. Science is not privileged by anything (revelation, metaphysics, intuition) outside its own methodology and established knowledge
Positivism does not recognise the theoretical dependence of observation	They recognise the theory dependence of observation but try to identify and isolate the dependence
Positivism regards scientific knowledge as a codification of sense data	They would consider the reduction of science to the codification of sense data to be a completely bizarre idea
Positivism promotes unquestioning textbook learning	They reject unquestioning and unreflective teaching and learning
Positivism is tied to a behaviourist psychology	They reject behaviourist reduction of mind to behaviour, and also reject treating mind as a theoretical 'fiction'. They support the scientific study of mind, while rejecting dualist views
Positivism regards knowledge as a commodity	They would be appalled at the image
Positivism believes scientific knowledge can easily be transmitted	They maintain the opposite
Positivism is blind to the effect of culture on the generation of scientific knowledge	They explicitly say that effects of culture need to be recognised, but maintain that such recognition does not in itself compromise the truth of knowledge claims
Positivism regards scientific knowledge as devoid of history and removed from society	They devoted most of their intellectual life to showing the exact opposite
Positivism ignores the value dimension of science	They explicitly and in detail address the value dimensions of science
Positivism is divorced from, or indifferent to, action for the improvement of society and culture	They supported progressive, left-wing social justice causes, as did Ernst Mach and most of the Vienna Circle members
Positivism is the dominant ideology of Western society	They could only wish that science was the dominant ideology for understanding nature and society
Positivism adheres to the Enlightenment tradition	They agree

In summary, the educational views of Frank and Feigl (and the canonical positivists) are the same as those of the HPS&ST programme and the same as held in the Anglo-American liberal education and European *Bildung* traditions. The stark disjunction between myth and reality above is partly explained by the common generalisation to all positivism and logical empiricism of its 'mature' variant. George Reisch writes:

> Logical empiricism was originally a project that self-consciously sought engagement not only with science but with progressive social and cultural developments (both in Europe of the 1920s and in North America of the 1930s and '40s). In the space of about ten years, however, from roughly 1949 to 1959, it became the scrupulously non-political project in applied logic and semantics that most philosophers today associate with the name 'logical empiricism' or 'logical positivism'.
>
> (Reisch 2005, p. xi)

Undoubtedly, the mature variant of positivism, with its insistence on a phenomenal epistemology and its rejection as 'metaphysical' of all attempts to seek an underlying, non-visible, mechanism for phenomena, did give rise to and sustain behaviourism in psychology, and the extension of this ideology to education. All of this is another instance where psychologists, social scientists and educators became uncritical enthusiasts for an unsound philosophy of science.[23]

John Dewey

Two centuries after Priestley, John Dewey (1859–1952) repeated the core position of the Enlightenment education tradition when, in 1910, he wrote:

> Scientific method is not just a method which it has been found profitable to pursue in this or that abstruse subject for purely technical reasons. It represents the only method of thinking that has proved fruitful in any subject.
>
> (Dewey 1910, p. 127)

He elaborated this conviction in his justly famous 1916 *Democracy and Education*:

> Our predilection for premature acceptance and assertion, our aversion to suspended judgment, are signs that we tend naturally to cut short the process of testing. We are satisfied with superficial and immediate short-visioned applications. . . . Science represents the safeguard of the race against these natural propensities and the evils which flow from them. . . . It is artificial (an acquired art), not spontaneous; learned, not native. To this fact is due the unique, the invaluable place of science in education.
>
> (Dewey 1916/1966, p. 189)

And, in 1938, in his contribution to the *International Encyclopedia of Unified Science*, he further elaborated, saying:

> In short, the scientific attitude as here conceived is a quality that is manifested in any walk of life. What, then, is it? On its negative side, it is freedom from control by routine, prejudice, dogma, unexamined tradition, sheer self-interest. Positively, it is the will to inquire, to examine, to discriminate, to draw conclusions only on the basis of evidence after taking pains to gather all available evidence. It is the intention to reach beliefs, and to test those that are entertained, on the basis of observed fact, recognizing also that facts are without meaning save as they point to ideas.
>
> (Dewey 1938, p. 31)

As an applied and engaged philosopher, Dewey laments 'the spirit in which the sciences are often taught' (Dewey 1938, p. 36) and despairs that 'scientific subjects are taught very largely as bodies of subject matter rather than a method of universal attack and approach' (ibid.).[24] Concerning the educational and cultural isolation of science, he warns that, 'There are powerful special interests which strive in any case to keep science isolated so that the common life may be immune from its influence' (Dewey 1938, p. 37). The latter is developed to a pitch in all authoritarian regimes.

Dewey here lays out a number of themes that will recur through this book:

- that science involves some modicum of sophisticated testing, not just 'look and see' appraisals; science is experimental;
- that scientific thinking and procedures are not natural, they do not spontaneously unfold as maturation proceeds: science needs to be taught;
- that scientific thinking – 'habits of mind' or 'scientific temper' – needs to be applied outside the laboratory and is the way to analyse and address social and cultural problems, as well as natural and environmental problems.

As will be seen, there is debate on each of these claims, and many educators reject some or all of them.

Spread of Science Education and Enlightenment Ideas

From the eighteenth century, modern science moved beyond Western Europe, becoming in the next two centuries the orthodox, universal science of the present day.[25] Hundreds of thousands of men and women from almost every nation, creed, caste and class contribute to science, and millions of students study the subject.

The early spread of science was connected to voyages of discovery and the imperialist programmes of various European powers. This brought the science of Galileo, Huygens, Newton, Leibniz and others to all parts of the globe.[26] Enlightenment ideas unevenly accompanied modern science and the European merchants, colonisers and missionaries who spread over the globe. Although the technological power of science was generally desired by people with whom European powers came into contact, Enlightenment values were often unwelcome. Enlightenment ideas of freedom of speech and association, the rule of law, open questioning of authority and the legitimacy of hereditary powers, the separation of church and state, along with the decriminalising of heretical beliefs and immoral behaviours, the provision of universal access to education, human rights and so on, were by no means widely welcomed in Europe, even though there were public champions of enlightenment and liberalism, and they were generally even less welcome in the European colonies.

In Europe, there have been two centuries of struggle over the 'enlightenment of education'. This struggle was largely between the Churches (Roman Catholic and Protestant) and Enlightenment advocates who were seeking to fashion more liberal, secular, democratic and educated societies. In the mid nineteenth century, Robert Owen founded a secular school at his New Lanark socialist–industrial complex; this was a lonely bloom and opposed by both government and church. The first UK government grant for education (£20,000) was in 1833, and all of the money went for church schools. In Spain, Portugal, Italy, Poland and elsewhere, the Roman Catholic Church directly controlled education and, in a number of these countries, did so up to the end of the twentieth century. Pope Pius IX, in his anti-Modernist 1864 *Syllabus of Errors*, condemned all those who affirmed that:

> The best theory of civil society requires that popular schools open to children of every class of the people, and, generally, all public institutes intended for instruction in letters and philosophical sciences and for carrying on the education of youth, should be freed from all ecclesiastical authority, control and interference, and should be fully subjected to the civil and political power at the pleasure of the rulers, and according to the standard of the prevalent opinions of the age.

This core Enlightenment belief was listed as Error 47 out of the identified 80 errors of 'Modernism'.[27] In Europe, Latin America and the European colonies, the Church's influence in education (hiring of teachers, setting curricula, denying the application of Darwinism to human origins, compulsory religious instruction, etc.) continued into the late twentieth century. Of course, the Churches and missionaries were usually the only groups interested in education in the colonies. Without their devotion and effort, there would have been little, if any, education of girls, and nearly all leaders of liberation struggles in Africa and Asia acknowledge a debt to their education in missionary schools.

Nevertheless, modern science and Enlightenment-influenced education did spread from Europe. The example of Turkey and the education reforms initiated by Mustafa Kemal (Atatürk) in the 1920s is especially interesting, but too complex to elaborate here. Also complex, but more possible to describe briefly, is the case of Nehru's education reforms in India in the 1950s. This will be elaborated a little, as it brings into sharp relief some of the themes of this book.

India

India is a clear case where the introduction of Enlightenment-informed science education to a country and its subsequent contested history provides material for philosophical, political and educational analysis. In the nineteenth century, Britain, with the energetic championing of Charles Trevelyan (1807–1886), brought English education, including science classes (and cricket), to India for the Indian elite, but it assuredly did not have an Enlightenment purpose or agenda.[28] One century later, with independence, there was a huge change, at least in ideology. The constitution of the newly independent India has many claims to international attention, but one is unique. Article 51A(h) of the Constitution of India states as a Fundamental Duty of the state: 'To develop the scientific temper, humanism and the spirit of enquiry and reform.' Benjamin Franklin, Thomas Jefferson, Joseph Priestley, Ernst Mach, John Dewey, the positivists, most national Associations for the Advancement of Science and national Science Teachers Associations could only dream of the inclusion of a duty to develop the 'scientific outlook' or 'scientific habit of mind' or 'scientific sensibility' in a national constitution! How did the provision get there, what has it achieved, and what intellectual, educational and political controversy has it occasioned? These are all illuminating questions.

The term 'scientific temper' and the Indian Scientific Temper programme have their origins in the convictions of Pandit Jawaharlal Nehru (1889–1964) who, in 1947, was installed as the first president of independent India. Nehru's convictions were articulated in his *The Discovery of India* (Nehru 1946/1981). The convictions owe a good deal to the science degree he completed at Cambridge (1907–1910) at a time when Enlightenment, liberal and socialist currents coursed through the university corridors and wider English society.[29] Nehru praised, in distinctly Enlightenment terms:

> The adventurous and yet critical temper of science, the search for truth and new knowledge, the refusal to accept anything without testing and trial, the capacity to change previous conclusions in the face of new evidence, the reliance on observed fact and not on pre-conceived theory, the hard discipline of the mind.
>
> (Nehru 1946/1981, p. 36)

Importantly, and often overlooked, Nehru's convictions were supported and strengthened by Bhimrao Ramji Ambedkar (1891–1956), who was his political collaborator, independent India's first Law Minister, writer-in-chief of the Constitution and forceful and relentless opponent of the Indian caste system and the associated Hindu beliefs that underwrite it.[30] John Dewey had an enormous influence on him while he was a postgraduate student at Columbia

University; everything about Dewey's philosophy and social programme resonated with his own views and experiences (Mukherjee 2009). Ambedkar and Nehru saw that the deeply entrenched ills, backwardness, irrationality and inequities of Indian society and culture could not be legislated away: education, specifically science education, was needed to change outlooks and orientations. The Congress government reaffirmed its commitment to modern science, technology and scientific temper in its 1958 Science Policy Resolution.

Predictably, this change did not happen on a large scale. India embraced, as did China and many other newly modernised societies, technical and industrial science education and science policies – nuclear energy, massive dams, the Green Revolution, agribusiness, manufacturing industries, world-leading institutes of technology, scientific research centres and much more were created – but day-to-day life in villages, towns, cities and even in the elite universities did not much change. In July 1981, the Nehru Centre in Delhi published another Scientific Temper Statement, signed by many prominent scientists and intellectuals (Haksar *et al.* 1981), that it hoped would reposition scientific temper as a national educational and cultural priority. The statement was brief, affirming:

> Scientific temper . . . leads to the realization that events occur as a result of the interplay of understandable and describable natural and social forces and not because someone, however great, so ordained them.
>
> (in Nanda 2003, p. 210)

As will be seen in Chapter 10, this is a statement of both methodological and ontological naturalism.

No sooner had this restatement of Enlightenment-sourced conviction been published than it was swamped by a tsunami of post-Kuhnian, post-colonial, postmodern, multicultural, and traditional-knowledge-affirming critics. Proponents of scientific temper were, as might be expected, labelled Positivists. The critics of the scientific temper declarations appealed to the supposed findings of current science studies, with the names of Kuhn, Feyerabend, Marcuse and Latour recurring in publications. Ashis Nandy published a widely discussed 'Counter-statement on Humanistic Temper' in 1981 that decried the 'obscene and amoral logic of science', and that informed readers that, 'science is no less informed by culture and society than any other human effort' (Nandy 1981, in Nanda 2003, p. 212).

All of this was expanded in a book, *The Intimate Enemy*, that drew on 'support from seventy-five years of work in history, philosophy and sociology of science' (Nandy 1983). Nandy and fellow critics wanted to 'decolonise the Indian mind'. Countless other books, anthologies, articles and seminars all proclaimed and defended the same counter-Enlightenment theses: The supposed truths and universality of Western science are an illusion; there is no scientific method or privileged outlook; orthodox science is just a tool of the colonial oppressors; native Vedic science and Indian technologies are on an intellectual and practical par with modern science; Hindu culture should be valued over alien and oppressive culture; and so on. All, or some, of these theses are propounded by senior figures in numerous international science-education journals.

In India, the 'Science Wars' had real and serious social consequences. Tragically, Dr Narendra Dabholkar was murdered in the street in August 2013 because of his campaign for passage of an anti-superstition bill in the Maharashtra state parliament. When the Hindu nationalist Bharatiya Janata Party first came to national power in Delhi (1998) and to the governing benches in many states. Notoriously, astrology was welcomed into many universities to take its place (and money) alongside astronomy departments; likewise for various traditional medical practices. Scientific temper as a curriculum aim disappeared from numerous school programmes.[31]

Thirty years after the Nehru Centre's Scientific Temper Statement and consequent debate, the matter was reactivated by scientists and intellectuals in 2011 in *The Palampur Declaration*, which recounted the original Nehruvian hopes, repeated the core arguments of the 1981 declaration and went on to say:

> During the past 30 years there has been a marked increase in public display of religious and sectarian identities, ascendance of irrational cults, glorification of obscurantist practices, religiosity and wielding of religious symbols. This has provided the ideological basis for, at times, brutal unscientific actions in both public and personal domains. Discrimination based on caste, gender and ethnic identities, perpetuated on the basis of irrational beliefs and superstitions are still widely prevalent, and are a blot on our society.
>
> (Various 2011, p. 2)

It identifies the distinction between knowledge and information, saying that India has lots of the latter, but education should aspire to the former.

> Modern education is the strongest determinant of scientific information, knowledge and attitude. It is true that over the years the scientific information base in the country has enlarged, but it will be far from reality to assume that this information is getting transformed into knowledge and thereby bringing a change in attitude. Unfortunately, our education system is still not sufficiently evolved to inculcate Scientific Temper in young minds.
>
> (Various 2011, p. 5)

These Indian scientists, politicans and educators must be dismayed when they read of the more prosaic aspirations for science education held by some intellectuals in the 'advanced' world: namely, that such education enables citizens to know 'where in the oven to put a soufflé, [and] lower one's energy bills' (Collins & Pinch, 1992, p. 150). It would be nice if it did this, but the serious cultural and educational issue is whether it should do more, whether a 'flow-on' effect should be the test of a competent science education?

Conclusion

These debates in India and elsewhere clearly demonstrate the ways in which education is embedded in philosophy, politics, economics and religion. The Enlightenment tradition in science education has always recognised this and has advocated the acquisition of historical and philosophical knowledge, so as to enable teachers and administrators to better understand their own scientific tradition and and to more fruitfully engage with their wider social, cultural and historic traditions. The HPS&ST programme is a broad church of which the Enlightenment tradition is one congregation. Some proponents of technical education appeal to HPS for its betterment, all proponents of liberal education are clear about the necessity of HPS (indeed, of the history and philosophy of any subject being taught), but the Enlightenment tradition puts HPS on centre stage, because of the value it places on science and on the scientific outlook and mentality required by science. From Locke and Priestley to Mach, Dewey and the positivists, the Enlightenment tradition has provided a developing body of method, argument and analysis that can illuminate contemporary social, cultural and educational disputes.

Neil Postman, co-author of the educational classic *Teaching as a Subversive Activity* (Postman & Weingartner 1969) recently published a book with the engaging title *Building a Bridge to the Eighteenth Century: How the Past Can Improve Our Future* (Postman 1999). After laying out a familiar litany of modern social and cultural ills, he writes:

> With this in mind, I suggest that we turn our attention to the eighteenth century. It is there, I think, that we may find ideas that offer a humane direction to the future, ideas that we can carry with confidence and dignity across the bridge to the twenty-first century.
>
> (Postman 1999, p. 17)

Echoing Kant and all intelligent commentators, Postman recognises that it is the spirit, outlook and methods that define the historic Enlightenment. Many would endorse his view:

> Let us not turn to the eighteenth century in order to copy the institutions she fashioned for herself but in order that we may better understand what suits us. Let us look there for instruction rather than models. Let us adopt the principles rather than the details.
>
> (Postman 1999, p. 17)

Notes

1 There is a huge literature on the Enlightenment and its history. See at least: Dupré (2004), Gay (1970), Grayling (2007, 2009), Himmelfarb (2004), Israel (2001), Pagden (2013) and Porter (2000).
2 On this distinction, see especially Israel (2001).
3 For the delineation of these characteristics, see Shimony (1997).
4 A complex matter, but see at least Hunt (2007).

5 See Munck (1990) for a brief introduction to seventeenth-century life and accounts of its serfdom, slavery, feudalism, despotism, ignorance, kingly and ecclesiastical control of publication and speech, epidemics of smallpox and plague, enforced religion, witchcrazes, entrenched superstitions, and much more.
6 Four hundred years later, as will be documented in Chapter 12, some sociologists of scientific knowledge will openly affirm that 'talking about' creates and brings into existence the entity talked about. De Frias appears to have common-sense doubt about this idealist and confused ontological manœuvre.
7 The classic discussion of the scientific revolution in England is Merton's 1938 *Science, Technology and Society in Seventeenth Century England* (Merton 1938/1970).
8 On Newton's application of scientific method to biblical questions, see Buchwald and Feingold (2012).
9 For discussion of the main figures and guides to literature on 'Education and the Enlightenment', see Parry (2007) and Schmitter *et al.* (2003).
10 For the Enlightenment context of Locke's educational theory, see Tarcov (1989).
11 More will be written of Priestley in Chapter 7. The most definitive studies of Priestley are the two biographical volumes of Robert Schofield (1997, 2004), with the latter containing a full bibliographic listing of Priestley's many books, pamphlets and articles.
12 For the next 150 years, this was the only English-language history of optics.
13 This authoritative work led to productive correspondence with Franklin, Volta and many others at the birth of electrical science.
14 For background and more general treatment of the anti-Jacobin riots of the 1790s, see Thompson (1963/1980, pp. 111–130).
15 John Bradley, the English chemist and educator, organised his chemistry instruction on Machian principles (Bradley 1963–1968) and wrote a useful book on Mach's philosophy of science (Bradley 1971). Mach the educator is discussed in Matthews (1990). The most comprehensive and best-documented discussion of the subject is Siemsen (2014).
16 An excellent documentary source of Mach's staggering influence in science, philosophy and beyond is Blackmore *et al.* (2001).
17 The first such journal was *Zeitschrift für mathematischen und naturwissenschaflichen Unterricht*, which began publication in 1870. It was edited by J.C.V. Hoffmann, a secondary schoolteacher in the Saxony mining town of Freiberg (thanks to Kathryn Olesko for this information).
18 A standard history is von Mises (1951); some classic texts are in Ayer (1959); informed contemporary appraisals are in Parrini *et al.* (2003). See also Thomas Uebel's informative overview of the 'Vienna Circle'in the web-based *Stanford Encyclopedia of Philosophy* (Uebel 2012).
19 The positivist hoped-for connection of science and philosophy with other disciplines and with educational, political and cultural endeavours can be seen in chapters of the *International Encyclopedia of Unified Science*, Vol.1, edited by Neurath *et al.* (1938). Bertrand Russell and John Dewey were both invited to contribute explicitly educational pieces to this volume. See Reisch (2005, Chapters 1, 2) and Uebel (1998).
20 Sources of biographical information are Feigl's own informal life story (Feigl 1974/1981) and Paul Feyerabend's Introduction to the *Festschrift* for Feigl (Feyerabend 1966).
21 For further elaboration of autonomy as an educational goal, see Dearden (1975).
22 The topic of rationality and education is much written upon; a good starting point for the arguments and literature is Siegel (1997).
23 B.F. Skinner is explicit in his intellectual debt to logical empiricism. The whole complex of behaviourism, positivism and philosophy is examined by many, but see: Mackenzie (1977), Scriven (1956) and Smith (1986).
24 For a review of literature on Dewey and science education, particularly his conception of historical and philosophical elements in science education, see Johnston (2014).
25 See contributions to Porter and Teich (1992).

26 The expansion of European science, its connection to imperialism and its reception in the colonies is a separate study, but see, at least, Pyenson (1993).

27 The eightieth and final error was belief that: 'The Roman Pontiff can, and ought to, reconcile himself, and come to terms with progress, liberalism and modern civilization.' The entire Syllabus is now on the web at www.papalencyclicals.net

28 Trevelyan returned to England and oversaw the British occupation of Ireland. Notoriously, commenting on the starvation and epidemic deaths of 1 million people in 1846–1851, he said that it was 'the will of God'. This is a sad comment on the common failure of science education to have 'flow-on' effects for general intelligence or social understanding. For some, however, science education is not meant to have such 'flow-on' effects.

29 The Fabian Society had been founded in 1884, and the British Labour Party was founded in 1900; the writings and speeches of Cambridge philosopher Bertrand Russell and, a little later, Cambridge chemist J.D. Bernal had wide audiences.

30 Ambedkar deserves to be much better known. Among the first 'untouchables' to have a college education, he earned doctorates in economics (Columbia University) and law (University of London). Despite being India's Chief Law Minister, while travelling, Ambedkar could not sleep in hotels, as this would 'pollute' them, and guests would leave and new ones never enter. Although the hallowed Ghandi tried to ameliorate some of the worst features of the iniquitous Hindu caste system, he never rejected the concept. Ambedkar campaigned against it all of his life. On this, see Jaffrelot (2005).

31 On these themes, see Mahanti (2013), Nanda (2003, Chapter 8), Sarukkai (2014) and contributions to Bhargava and Chakrabarti (2010).

References

Arons, A.B.: 1988, 'Historical and Philosophical Perspectives Attainable in Introductory Physics Courses', *Educational Philosophy and Theory* 20(2), 13–23.

Ayer, A.J. (ed.): 1959, *Logical Positivism*, The Free Press, New York.

Berlin, I. (ed.): 1956, *The Age of Enlightenment: The Eighteenth Century Philosophers*, Mentor Books, New York.

Bhargava, P.M. and Chakrabarti, C.: 2010, *Angels, Devil and Science: Collection of Articles on Scientific Temper*, National Book Trust, New Delhi.

Blackmore, J.T.: 1972, *Ernst Mach: His Work, Life and Influence*, University of California Press, Berkeley, CA.

Blackmore, J.T., Itagaki, R. and Tanaka, S. (eds): 2001, *Ernst Mach's Vienna 1895–1930*, Kluwer Academic Publishers, Dordrecht, The Netherlands.

Boorstin, D.J.: 1948, *The Lost World of Thomas Jefferson*, Beacon Press, Boston, MA.

Bradley, J.: 1963–1968, 'A Scheme for the Teaching of Chemistry by the Historical Method', *School Science Review* 44, 549–553; 45, 364–368; 46, 126–133; 47, 65–71, 702–710; 48, 467–474; 49, 142–150; 454–460.

Bradley, J.: 1971, *Mach's Philosophy of Science*, Athlone Press of the University of London, London.

Brooke, J.H.: 1987, 'Joseph Priestley (1733–1804) and William Whewell (1794–1866): Apologists and Historians of Science. A Tale of Two Stereotypes'. In R.G.W. Anderson and C. Lawrence (eds) *Science, Medicine and Dissent: Joseph Priestley (1733–1804)*, Wellcome Trust & Science Museum, London, pp. 11–27.

Buchwald, J. and Feingold, M.: 2012, *Newton and the Origin of Civilisation*, Princeton University Press, Princeton, NJ.

Burman, E.: 1984, *The Inquisition: The Hammer of Heresy*, Dorset Books, Wellingborough, UK.

Cassirer, E.: 1932/1951, *The Philosophy of the Enlightenment* (trans. Fritz C.A. Koelln and James P. Pettegrove), Princeton University Press, Princeton, NJ.

Cohen, R.S. (ed.): 1981, *Inquiries and Provocations: Selected Writings of Herbert Feigl 1929–1974*, Reidel, Dordrecht, The Netherlands.

Collins, H.M. and Pinch, T.: 1992, *The Golem: What Everyone Should Know About Science*, Cambridge University Press, Cambridge, UK.
Condorcet, N.: 1976, *Selected Writings*, K.M. Baker (ed.), Bobbs-Merrill, Indianapolis, IN.
Dearden, R.F.: 1975, 'Autonomy as an Educational Ideal I'. In S.C Brown (ed.) *Philosophers Discuss Education*, Macmillan, London, pp. 3–18.
Dewey, J.: 1910, 'Science as Subject-Matter and as Method', *Science* 31, 121–127. Reproduced in *Science & Education*, 1995, 4(4), 391–398.
Dewey, J.: 1916/1966, *Democracy and Education*, Macmillan, New York.
Dewey, J.: 1938, 'Unity of Science as a Social Problem'. In O. Neurath, R. Carnap and C.W. Morris (eds) *International Encyclopedia of Unified Science*, Vol.1, pp. 29–38.
Duhem, P.: 1906/1954, *The Aim and Structure of Physical Theory* (trans. P.P. Wiener), Princeton University Press, Princeton, NJ.
Dupré, L.: 2004, *The Enlightenment and the Intellectual Foundations of Modern Culture*, Yale University Press, New Haven, CT.
Feigl, H.: 1955, 'Aims of Education for Our Age of Science: Reflections of a Logical Empiricist'. In N.B. Henry (ed.) *Modern Philosophies and Education: The Fifty-fourth Yearbook of the National Society for the Study of Education*, University of Chicago Press, Chicago, IL, pp. 304–341. Reprinted in *Science & Education* 13(1–2), 2004.
Feigl, H.: 1974/1981, 'No Pot of Message'. In R.S. Cohen (ed.) *Inquiries and Provocations: Selected Writings 1929–1974*, Reidel, Dordrecht, The Netherlands, pp.1–20.
Feyerabend, P.K.: 1966, 'Herbert Feigl: A Biographical Sketch'. In P.K. Feyerabend and G. Maxwell (eds) *Mind, Matter, and Method: Essays in Philosophy of Science and Science in Honor of Herbert Feigl*, University of Minnesota Press, Minneapolis, MN, pp. 3–13.
Feyerabend, P.K.: 1975, *Against Method*, New Left Books, London.
Frank, P.: 1907/1949, 'Experience and the Law of Causality'. In his *Between Physics and Philosophy*, Harvard University Press, Cambridge, MA, pp. 53–60.
Frank, P.: 1931/1998, *The Law of Causality and Its Limits*, R.S. Cohen (ed.), Kluwer Academic Publishers, Dordrecht, The Netherlands.
Frank, P.: 1946/1949, 'Science Teaching and the Humanities', *Etc.: A Review of General Semantics* 4(3). In his *Modern Science and Its Philosophy*, Harvard University Press, Cambridge, MA (1949), pp. 260–285.
Frank, P.: 1947/1949, 'The Place of Philosophy of Science in the Curriculum of the Physics Student', *American Journal of Physics* 15(3), 202–218. Reprinted in his *Modern Science and Philosophy*, Harvard University Press, Cambridge, MA, pp. 228–259.
Frank, P.: 1949, 'Introduction: Historical Background'. In his *Modern Science and Its Philosophy*, Harvard University Press, Cambridge, MA, pp. 1–52.
Friedman, M.: 1999, *Reconsidering Logical Positivism*, Cambridge University Press, New York.
Gay, P.: 1970, *The Enlightenment: An Interpretation*, 2 volumes, Weidedfeld & Nicolson, London.
Grayling, A.C.: 2007, *Towards the Light: The Story of the Struggles for Liberty & Rights That Made the Modern West*, Bloomsbury, London.
Grayling, A.C.: 2009, *Liberty in the Age of Terror: A Defence of Civil Society and Enlightenment Values*, Bloomsbury, London.
Haksar, P.N. *et al.*: 1981, *A Statement on Scientific Temper*, Nehru Centre, Bombay.
Hankins, T.L.: 1985, *Science and the Enlightenment*, Cambridge University Press, Cambridge, UK.
Hiebert, E.N.: 1976, 'Introduction'. In E. Mach *Knowledge and Error*, Reidel, Dordrecht, The Netherlands (orig. 1905).
Himmelfarb, G.: 2004, *The Roads to Modernity. The British, French, and American Enlightenments*, Alfred A. Knopf, New York.
Höfler, A.: 1916, 'Ernst Mach: Obituary', *Zeitschrift fur den Physikalischen und Chemischen Unterricht* 29(2).
Hume, D.: 1739/1888, *A Treatise of Human Nature: Being an Attempt to Introduce the Experimental Method of Reasoning into Moral Subjects*, Clarendon Press, Oxford, UK.

Hume, D.: 1754–1762/1828, *The History of England: From the Invasion of Julius Caesar to the Revolution in 1688*, 4 volumes, Bennett & Walton, Philadelphia, PA.
Hunt, L.: 2007, *Inventing Human Rights: A History*, W.W. Norton, New York.
Israel, J.: 2001, *Radical Enlightenment: Philosophy and the Making of Modernity 1650–1750*, Oxford University Press, Oxford, UK.
Jacob, M.C.: 1998, 'Reflections on Bruno Latour's Version of the Seventeenth Century'. In N. Koertge (ed.) *A House Built on Sand: Exposing Postmodernist Myths About Science*, Oxford University Press, New York, pp. 240–254.
Jaffrelot, C.: 2005, *Ambedkar and Untouchability: Fighting the Indian Caste System*, Columbia University Press, New York.
Johnston, J.S.: 2014, 'John Dewey and Science Education'. In M.R. Matthews (ed.) *International Handbook of Research in History, Philosophy and Science Teaching*, Springer, Dordrecht, The Netherlands, pp. 2409–2432.
Kant, I.: 1784/2003, 'What is Enlightenment?' In P. Hyland (ed.) *The Enlightenment: A Sourcebook and Reader*, Routledge, London, pp. 54–58.
Kant, I.: 1803/1899, *Kant on Education* (trans. A. Churton), Kegan Paul, London.
Koenigsberger, H.G.: 1987, *Early Modern Europe 1500–1789*, Longman, London.
Lindsay, J.: 1970, 'Introduction'. In *Autobiography of Joseph Priestley*, Adams & Dart, Bath, pp. 11–66.
Locke, J.: 1689/1924, An Essay Concerning Human Understanding, abridged and edited by A.S. Pringle-Pattison, Clarendon Press, Oxford, UK.
Locke, J.: 1689/1983, *A Letter Concerning Toleration*, J. Tully (ed.), Hackett Publishing, Indianapolis, IN.
Locke, J.: 1690/1960, *Two Treatises of Government*. Introduction and Notes by Peter Laslett, Cambridge University Press, Cambridge, UK.
Locke, J.: 1693/1968, Some Thoughts Concerning Education. In J.L. Axtell (ed.) *The Educational Writings of John Locke*, Cambridge University Press, Cambridge, UK, pp. 114–325.
Locke, J.: 1693/1996, *Some Thoughts Concerning Education & Of the Conduct of the Understanding*, R.W. Grant and N. Tarcov (eds), Hackett Publishing, Indianapolis, IN.
Mach, E.: 1883/1960, *The Science of Mechanics*, Open Court Publishing, LaSalle, IL.
Mach, E.: 1886/1986, 'On Instruction in the Classics and the Sciences'. In his *Popular Scientific Lectures*, Open Court Publishing, LaSalle, IL, pp. 338–374.
Mackenzie, B.D.: 1977, *Behaviourism and the Limits of Scientific Method*, Humanities Press, Atlantic Highlands, NJ.
Mahanti, S.: 2013, 'A Perspective on Scientific Temper in India', *Journal of Scientific Temper* 1, 46–62.
Matthews, M.R.: 1990, 'Ernst Mach and Contemporary Science Education Reforms', *International Journal of Science Education* 12(3), 317–325.
Merton, R.K.: 1938/1970, *Science, Technology and Society in Seventeenth Century England*, Harper & Row, New York.
Mukherjee, A.P.: 2009, 'B.R. Ambedkar, John Dewey, and the Meaning of Democracy', *New Literary History* 40(2), 345–370.
Munck, T.: 1990, *Seventeenth Century Europe: State, Conflict and the Social Order in Europe 1598–1700*, Macmillan, London.
Nanda, M.: 2003, *Prophets Facing Backward. Postmodern Critiques of Science and Hindu Nationalism in India*, Rutgers University Press, New Brunswick, NJ.
Nandy, A.: 1983, *The Intimate Enemy*, Oxford University Press, New Delhi.
Nehru, J.L.: 1946/1981, *The Discovery of India*, Oxford University Press, New Delhi.
Neurath, O., Carnap, R. and Morris, C. (eds): 1938, *International Encyclopedia of Unified Science*, Vol.1, University of Chicago Press, Chicago, IL.
Newton, I.: 1713/1934, *Principia Mathematica*, 2nd edn (trans. Florian Cajori), University of California Press, Berkeley, CA (1st edition, 1687).
Newton, I.: 1730/1979, *Opticks*, 4th edn, I.B. Cohen and D.H.D. Roller (eds), Dover, New York (1st edition, 1704).

Pagden, A.: 2013, *The Enlightenment and Why It still Matters*, Oxford University Press, Oxford, UK.
Parrini, P., Salmon, W. and Salmon, M. (eds): 2003, *Logical Empiricism: Historical and Contemporary Perspectives*, University of Pittsburgh Press, Pittsburgh, PA.
Parry, G.: 2007, 'Education and the Reproduction of the Enlightenment'. In M. Fitzpatrick, P. Jones, C. Knellwolf and I. McCalman (eds) *The Enlightenment World*, Routledge, London, pp. 217–233.
Passmore, J.A. (ed.): 1965, *Priestley's Writings on Philosophy, Science and Politics*, Collier Macmillan, London.
Peters, M.: 1995, 'Philosophy and Education "After" Wittgenstein'. In P. Smeyers and J.D. Marshall (eds) *Philosophy and Education: Accepting Wittgenstein's Challenge*, Kluwer Academic Publishers, Dordrecht, The Netherlands, pp. 189–328.
Porter, R.: 2000, *The Enlightenment: Britain and the Creation of the Modern World*, Penguin, London.
Porter, R. and Teich, M. (eds): 1992, *The Scientific Revolution in National Context*, Cambridge University Press, Cambridge, UK.
Postman, N.: 1999, *Building a Bridge to the 18th Century: How the Past Can Improve Our Future*, Alfred A. Knopf, New York.
Postman, N. and Weingartner, C.: 1969, *Teaching as a Subversive Activity*, Dell Publishing, New York.
Priestley, J.: 1765/1965, *An Essay on a Course of Liberal Education for Civil and Active Life*. In J.A. Passmore (ed.) *Priestley's Writings on Philosophy, Science and Politics*, Collier Macmillan, London, pp. 285–304.
Priestley, J.: 1767/1775, *The History and Present State of Electricity, with Original Experiments*, 2nd edn, J. Dodsley, J. Johnson and T. Cadell, London; 3rd edition, 1775, reprinted Johnson Reprint Corporation, New York, 1966, with Introduction by Robert E. Schofield.
Priestley, J.: 1772, *The History and Present State of the Discoveries Relating to Vision, Light, and Colours*, 2 volumes, London.
Priestley, J.: 1775–1777, *Experiments and Observations on Different Kinds of Air*, 2nd edn, 3 volumes, J. Johnson, London. Sections of the work have been published by the Alembic Club with the title *The Discovery of Oxygen*, Edinburgh, 1961.
Priestley, J.: 1785, *The Importance and Extent of Free Inquiry in Matters of Religion*, to which is added *The Present State of Free Inquiry in this Country*, J. Johnson, Birmingham, UK. In Rutt, *Collected Works*, Vol.15, pp. 70–82.
Priestley, J.: 1788, *Lectures on History and General Policy to Which is Prefixed, An Essasy on the Course of Liberal Educatioin for Civil and Active Life*, P. Byrne, Dublin.
Priestley, J.: 1790, *Experiments and Observations on Different Kinds of Air, and Other Branches of Natural Philosophy, Connected with the Subject. Being the Former Six Volumes Abridged and Methodized*, 3 volumes, J. Johnson, Birmingham, UK.
Priestley, J.: 1791, 'The Proper Objects of Education'. In J.T. Rutt (ed.) *The Theological and Miscellaneous Works of Joseph Priestley*, Vol.15, pp. 420–440.
Pyenson, L.: 1993, 'The Ideology of Western Rationality: History of Science and the European Civilizing Mission', *Science & Education* 2(4), 329–344.
Reisch, G.A.: 2005, *How the Cold War Transformed Philosophy of Science. To the Icy Slopes of Logic*, Cambridge University Press, New York.
Rousseau, J.J.: 1762/1991, *Emile, or On Education* (trans. Allan Bloom), Penguin, Harmondsworth, UK.
Rutt, J.T.: 1831–1832, *The Life and Correspondence of Joseph Priestley*, 2 volumes, London.
Sarukkai, S.: 2014, 'Indian Experiences with Science: Considerations for History, Philosophy and Science Education'. In M.R. Matthews (ed.) *International Handbook of Research in History, Philosophy and Science Teaching*, Springer, Dordrecht, The Netherlands, pp. 1693–1720.

Schmitter, A.M., Tarcov, N. and Donner, W.: 2003, 'Enlightenment Liberalism'. In R. Curren (ed.) *A Companion to the Philosophy of Education*, Blackwell Publishing, Malden, MA, pp. 73–93.

Schofield, R.E.: 1997, *The Enlightenment of Joseph Priestley: A Study of His Life and Work from 1733 to 1773*, Penn State Press, University Park, PA.

Schofield, R.E.: 2004, *The Enlightened Joseph Priestley: A Study of His Life and Work from 1773 to 1804*, Penn State Press, University Park, PA.

Scriven, M.: 1956, 'A Study of Radical Behaviourism', *Minnesota Studies in the Philosophy of Science*, 1, 88–130.

Shimony, A.: 1997, 'Presidential Address: Some Historical and Philosophical Reflections on Science and Enlightenment'. In L. Darden (ed.) *Proceedings of the 1996 PSA Meeting*, S1–14.

Siegel, H.: 1997, *Rationality Redeemed? Further Dialogues on an Educational Ideal*, Routledge, New York.

Siemsen, H.: 2014, 'Ernst Mach: A Genetic Introduction to His Educational Theory and Pedagogy'. In M.R. Matthews (ed.) *International Handbook of Research in History, Philosophy and Science Teaching*, Springer, Dordrecht, The Netherlands, pp. 2329–2357.

Smith, L.D.: 1986, *Behaviorism and Logical Positivism*, Stanford University Press, Stanford, CA.

Tarcov, N.: 1989, *Locke's Education for Liberty*, University of Chicago Press, Chicago, IL.

Thompson, E.P.: 1963/1980, *The Making of the English Working Class*, Penguin, Harmondsworth, UK.

Tobin, K.: 1998, 'Sociocultural Perspectives on the Teaching and Learning of Science'. In M. Larochelle, N. Bednarz and J. Garrision (eds) *Constructivism and Education*, Cambridge University Press, Cambridge, UK, pp. 195–212.

Uebel, T.E.: 1998, 'Enlightenment and the Vienna Circle's Scientific World-Conception'. In A.O. Rorty (ed.) *Philosophers on Education: New Historical Perspectives*, Routledge, New York, pp. 418–438.

Uebel, T.E.: 2012, 'Vienna Circle', *The Stanford Encyclopedia of Philosophy* (Summer 2012 edn), Edward N. Zalta (ed.), available at: http://plato.stanford.edu/archives/sum2012/entries/vienna-circle

Various: 2011, *Scientific Temper Statement Revisited: The Palampur Declaration*. Available at: http://st.niscair.res.in/scientific-temper-statement-revisited

von Mises, R.: 1951, *Positivism*, Harvard University Press, Cambridge, MA.

Chapter 3

Historical and Current Developments in Science Curricula

In order to appraise the value of HPS to science teaching, it is useful to be aware of the history and diversity of school science curricula, and of the major debates that have occurred in efforts to improve classroom instruction. This and the following two chapters will outline the development of school science with a view to understanding present claims for contextual or liberal pedagogy and curricula. The fact of diversity and change prompts questions about the justification of different curricular orientations, and about the degree to which change is driven by educational versus other considerations.[1]

Natural Philosophy in the Curriculum

Science, then called 'natural philosophy', was introduced into schools, the few that there were, in the middle of the eighteenth century. Its introduction was not universally lauded. Theology, the classics and humanities were regarded as appropriate subjects for the elite, and basic literacy, numeracy and religion, along with simple trade and domestic skills, were thought appropriate for the masses. In the nineteenth century, Thomas Huxley, Henry Armstrong and Thomas Percy Nunn in England, John Dewey in the United States, Ernst Mach and Johann Friedrich Herbert in Germany and, earlier, the mathematician de Condorcet in France were some who championed a popular presence for science education.[2] No sooner was science included in the curriculum than debate began about its contents, objectives, teaching methods and clientele. The clientele debate revolved around what is now called the 'Science for All' issue: whether science should be the same for all students, or whether there should be different programmes depending upon whether students were proceeding with university studies or terminating their education at the end of school, or simply having zero interest in studying science.

In Britain, a practical approach was widespread. Science was a servant of the Industrial Revolution, and this was reflected in educational endeavours (Uglow 2002). A noteworthy text was James Ferguson's *Natural Philosophy* (1750), which went through many editions, was revised in 1806 by Sir David Brewster and was published in America in 1806. Brewster's introduction says, 'The chief object of Mr. Ferguson's labours was to give a familiar view of physical science and to render it accessible to those who are not accustomed to mathematical investigation' (Woodhull, 1910, p. 18). Brewster went on to say that, 'No book upon the same subject has been so generally read, and so widely circulated, among all ranks of the community'. Sixty-two pages of the text were devoted to machines, and forty pages to pumps. This applied, technical, everyday emphasis was repeated in other widely used texts, such as the twenty-two editions of R.G. Parker's *The School Compendium of Experimental Philosophy* (1837), the seventy-three editions of J.L. Comstock's *System of Natural Philosophy* (1846), and J.W. Draper's *Natural Philosophy for Schools* (1847).

Draper stated what was to be a long-standing dilemma in the teaching of science when he said:

> There are two different methods in which Natural Philosophy is now taught: (1) as an experimental science; (2) as a branch of mathematics. I believe that the proper course is to teach physical science experimentally first.
>
> (Woodhull 1910, p. 21)

US colleges and British universities did not agree. Natural philosophy disappeared from American schools around 1872, to be replaced by high-school physics and texts that increasingly were filled with algebra and mathematical formulae, in which diagrams of common machines were replaced by abstract line drawings. Along with the new texts came the long-standing problem of the over-stuffed curricula. The New York State Department of Education issued its *Topical Syllabus in Physics* in 1905; it contained 260 topics, which, for a course of 120 hours, meant a new topic each half-hour of class time (Mann 1912, p. 66). This was the harbinger of the long-lamented US preference for 'mile-wide and inch-deep science curricula'.

Not all agreed that the new science teaching was an improvement on the old, and, at the end of the nineteenth century, many, gathered under the banner of 'The New Movement in Physics Teaching', advocated a return to the applied, experimental focus of the old natural philosophy courses and texts.[3] A part of this advocacy was for the teaching of the principles of science in science programmes, and it was reasonably held that a topic every 30 minutes, discussed in essentially a foreign language, was not conducive to children learning the principles of science. An example of the sort of science that the new movement opposed was the setting of questions such as, 'A force of 5,000 dynes acts for 10 seconds on a mass of 250 grams; what momentum is imparted to the body?', without students knowing experientially what a force of 5,000 dynes meant in everyday life. Could such a force, for instance, knock an adult down? Is it sufficient to move an orange on a table? (Mann 1912, p. 89).

US Science Education to the 1950s

There have been three competing traditions in US science education up to the present time: theoretical, stressing the conceptual structure of the disciplines;

applied, stressing the science and workings of everyday things; and liberal or contextual, stressing the historical development and cultural implications of science. These traditions have, of course, not been exclusive; like many borders, they are porous.

A significant trend in the development of science education up to the 1950s was the increasing recognition of the practical, vocational, social and humanitarian aspects of science, and the inclusion of these aspects in the curriculum. In many respects, this was a return to the past – a swing of the educational pendulum. Biology teaching, for instance, became less theoretical over this period (Hurd 1961, Rosenthal 1985). One teacher in 1909 complained that school biology texts were so encyclopaedic and theoretical that they were more appropriate for doctoral exams. After observing a class, the teacher wondered what meaning 'oogonia', 'antheridia' and 'oospore' conveyed to students (Rosenthal 1985). During the first half of the twentieth century, in response to a multitude of pressures – among them the Progressive Education Society, business and industrial demands, environmental problems, demographic changes and health concerns – school biology increasingly diverged from university biology. Finley wrote a 1926 text that stressed the 'practical, ecological, economic, human welfare aspects of biology'. He observed that generally 'the aim of biology teaching . . . changed from "biology for the sake of biology" to "biology in relation to human welfare" ' (Rosenthal 1985, p. 105). A review of these developments is aptly titled: 'Emergence of the Biology Curriculum: A Science of Life or a Science of Living' (Rosenthal & Bybee 1987).[4]

World War Two gave further impetus to practical biology: disease prevention, hygiene and agriculture were all part of the practical applications that guided course design. Columbia Teachers College developed a curriculum that stressed the 'content and methods of science in dealing with personal and social issues that have been raised largely as a result of advances in science'. The aim was to give a 'clearer understanding of society [and] of the social function of science' (Layton & Powers 1949). This concern with making science personally relevant can be seen in a report of the Consumer Education Society of the National Association of Secondary Principals. This report, *The Place of Science in the Education of the Consumer*, was published in 1945 by the National Science Teachers Association. It urged that science teaching should focus on knowledge that helps consumers purchase wisely and on procedures useful in the solution of consumer problems (Hurd 1961, p. 85).

It was not only biology that developed more practical concerns: physics texts up to the mid 1950s were also concerned with applied questions and gave everyday illustrations of physical principles. As Douglas Roberts (1982) has pointed out, it was common for the chapters on electricity to discuss the workings of the telephone, the electric iron, home circuits and fuses and everyday electrical appliances; the chapters on liquids dealt with town water systems, hydraulic brakes and other such matters.

There were predictable tensions in this applied science tradition. Some stressed applications at the personal level – hygiene, consumer decisions, planting gardens, hobbies and so on; others responded to the demands of business for vocational skills and stressed the social applications of science (Callahan 1962). Others stressed understanding of the interaction of society and science. Present-day science–technology–society (STS) programmes are in the same tradition as these interwar applied science courses.

The applied tradition was criticised from two sides: on the right, so to speak, were advocates of teaching the theoretical, disciplinary structure of science, and on the left were advocates of the humanistic, cultural aspects of science. The Union of American Biological Societies criticised the tendency to teach biology, not as a science, but as 'a way to pleasing hobbies, and a series of practical technologies' (Rosenthal 1985, p. 109). It championed specialist, disciplinary courses. This call was echoed in the 1947 report of the AAAS titled, 'The Present Effectiveness of our Schools in the Training of Scientists'. It stated:

> The report is based on the premise that our people should take such steps as may be necessary to ensure (1) enough competent scientists to do whatever job may be ahead, and (2) a voting public that understands and supports the scientists' role in defense and in the design for better living.
>
> (In Klopfer & Champagne 1990, p. 137)

In contrast, the Harvard Committee (1945) advocated a science programme in which, 'the facts of science must be learned in another context, cultural, historical, and philosophical'. The committee produced a manifesto for liberal science education. It claimed:

> Science instruction in general education should be characterized mainly by broad integrative elements – the comparison of scientific with other modes of thought, the comparison and contrast of the individual sciences with one another, the relations of science with its own past and with general human history, and of science with problems of human society. These are areas in which science can make a lasting contribution to the general education of all students. . . . Below the college level, virtually all science teaching should be devoted to general education.
>
> (Conant 1945, pp. 155–156)

In 1944, the National Educational Association issued a report, *Education for All American Youth*, that proposed a liberal approach to the sciences for precollege programmes. In addition to knowledge of specific subject matters, science, by the tenth grade, should introduce students to the role of science in human progress, to the scientific view of the world and of man, to the history of science and an imaginative association with the great scientists and their major experiments (Hurd 1961, p. 83). Clarence Faust, speaking at a 1958 national conference of presidential science advisers held at Yale University, stressed this contextual approach:

> What American life most needs, is a new respect for intelligence, for intellectual achievement, for the life of the mind, for books and for learning, for basic science and for philosophic wisdom . . . education cannot realize its promise if it is viewed merely as a means to individual advancement, social achievement, and national power . . . we need wisdom, not merely power . . . a commitment to the basic function of education.
>
> (Elbers & Duncan 1959, p. 178)

Thus, at the time of the mid 1950s *Sputnik* crisis, at least three competing views about the nature, purposes and emphases of school science can be identified:

1 a practical, technical, applied emphasis;
2 a liberal, generalist, humanistic emphasis;
3 a specialist, theoretical, disciplinary emphasis.

These are akin to what Elliot Eisner (1979) calls 'curricular orientations'. Roberts (1982), in his survey of numerous science curricula, identified seven 'curriculum emphases'. The above three correspond, approximately, with his 'everyday coping', 'the self as explainer' and 'correct explanations'. Neither Roberts's distinctions, nor the above tripartite divisions, are meant to be mutually exclusive. Curricula that stress one usually include something of the others. What is in contention between the views is the general orientation of the science programme and the goals that it seeks to achieve.

The serious educational issue is to identify the grounds for these curricular choices and then to justify the choice. Are there educational and philosophical grounds for the decisions, or does a society's curriculum just take the shape of the last political, economic or special interest group's foot that trod upon it? It is obvious that efforts to justify choices will lead to philosophy of education; justifications will need to appeal to the aims of education, to what is required for individual growth and flourishing, and to political understandings about the mutual relationship of individuals to their society.[5]

National Science Foundation Curricula (1950s–1960s)

In the early 1950s, American academics, scientists and professional associations, with physicists at the forefront, led agitation for the reform of US science education. These groups were concerned about the decline of science and mathematics in schools. In the 40 years between 1910 and 1950, the number of non-academic subjects (cooking, typing, driving and so on) in US schools increased from 8 to 215, separate physics and chemistry courses were amalgamated into general science, and algebra became part of general mathematics.[6]

On 4 October 1957, the Soviet *Sputnik* went into orbit, and its shock waves swept across the US political and educational landscape. Dianne Ravitch commented:

> The Soviet launch . . . promptly ended the debate that had raged for several years about the quality of American education. Those who had argued since the late 1940s that American schools were not rigorous enough and that life adjustment education had cheapened intellectual values felt vindicated, and as one historian later wrote, 'a shocked and humbled nation embarked on a bitter orgy of pedagogical soul-searching'.
>
> (In DeBoer 1991, p. 146)

Sputnik brought the claims of reformers of science education to national prominence. The launch triggered a flurry of legislation, the principal one being the 1957 National Defense Education Act, which gave $94 million for science education in the 3 years from 1958 to 1961, and a further $600 million in the years from 1961 to 1975. Conferences and meetings occurred across the country. A representative one was the above-mentioned Yale conference, sponsored by the President's Committee on Scientists and Engineers (Elbers & Duncan 1959).

The National Science Foundation (NSF) was instrumental in the transformation of school science into proto-university science, a process sometimes called the professionalisation of school science. The NSF's first school curriculum grant was for $1,725 in 1954; its 1956 grant to the Physical Science Study Committee (PSSC) was $300,000. The National Defense Act transformed this meagre level of funding and subsequently transformed US science education. In 1957, the NSF said that its curriculum projects:

> Seek to respond to the concern, often expressed by scientists and educators, over failure of instructional programs in primary and secondary schools to arouse motivating interest in, and understanding of, the scientific disciplines. General agreement prevails that much of the science taught in schools today does not reflect the current state of knowledge nor does it necessarily represent the best possible choice of materials for instructional purposes.
>
> (Crane 1976, pp. 56–57)

In 1956, Jerrold Zacharias,[7] a physicist at MIT, used a small grant from the National Science Foundation to set up the PSSC. This was a case of 'from small grants, big projects grow', especially when fuelled by a national *Sputnik* fear. This committee produced the PSSC *Physics* text, which was eventually to be used by millions of students in the US and throughout the world. With its multiple translations, it was the most utilised science textbook in history.[8] It was the MacDonald's or Coca-Cola of education. The Spanish and Portuguese translations, along with scholarships to bring Latin American teachers to the US for training, shaped the form of Latin American physics teaching for decades. The intention of PSSC physics was to focus upon the conceptual structure of physics and teach the subject as a discipline: applied material was almost totally absent from the text. As Zacharias stated:

> One should always design a curriculum by picking out the end point and working back. . . . We should like them [students] to understand the whole notion of quantization, the whole notion of particles and waves . . . working backwards, we said it was necessary, clearly, to understand the electrical nature of matter . . . and then of course working back, Newtonian mechanics. It is also necessary to know why one believes Newtonian mechanics. One believes Newtonian mechanics because of celestial mechanics, not because of blocks of wood on tables.
>
> (Zacharias 1964, p. 67)

With this curriculum theory or educational compass, it is easy to understand that teachers and classes sometimes never got back to blocks of wood on tables or, as will be mentioned later, 'where to put a soufflé in the oven'; the allocated time ran out. As will be illustrated in Chapter 4, air pressure, for instance, is not mentioned in the PSSC index; it is discussed in the chapter on 'The Nature of Gases', and the chapter proceeds entirely without mention of barometers or steam engines, the former making their first appearance in the notes to the chapter. And, as will be shown in Chapter 6, Zacharias commends beginning to teach pendulum matters with a coupled pendulum and trusting that the class will want to understand the simple pendulum, but even this latter understanding will not encompass the multitude of applied uses of the pendulum.[9]

The NSF put scientists firmly in the saddle of curriculum reform, teachers were at best stable hands, and the education faculty rarely got as far as the stable. The PSSC project epitomised 'top–down' curriculum development; its maxim was 'Make physics teacher-proof'. Zacharias stated that PSSC physics must 'have the materials in a form which is refractory, which cannot be changed easily' (Zacharias 1964, p. 69). In a 1962 explanation of its policies, the NSF said that, 'Projects are directed by college-level scientists, and grants are made to institutions of higher learning and professional scientific societies. Emphasis is placed on subject matter rather than pedagogy' (Klopfer & Champagne 1990, p. 139).

Trialing of projects did not always have the significance that the policy gave it. One teacher who participated said:

> My own experience with that process suggests the results of classroom tryouts had little effect on subsequent versions. Scientists were usually hesitant to accept the criticism of their 'science' from school teachers unless very convincing substantiating data were provided.
>
> (Welch 1979, p. 288)

The NSF supported the explosion of 'alphabet curricula' in the late 1950s and early 1960s. The first curriculum to be widely used was MIT's PSSC. Then followed the Chemical Bond Approach (CBA), Biological Sciences Curriculum Study (BSCS), Chemical Education Materials (CHEMS), Earth Science Curriculum Project (ESCP), Introductory Physical Science (IPS), Project Physics and a host of others. By 1975, the NSF supported twenty-eight science curriculum reform projects. A number of these were directed at the elementary school: Elementary Science Study (ESS), Science Curriculum Improvement Study and Science – A Process Approach (SAPA). During the boom period, millions of students studied these NSF-supported curricula: PSSC (1 million in 1956–1960), CHEMS (1 million in 1959–1963), BSCS (10 million in 1959–1990), IPS (1 million in 1963–1972), ESS (1 million in 1961–1971) and SAPA (1 million in 1963–1974). These constituted the major league of curricula. In 1976–1977, it was estimated that 19 million students were using the new curriculum materials; this number represented 43 per cent of the school population.[10]

Most of the NSF-funded projects neglected practical and technological applications of science. One review said:

> There is little or nothing of STS in currently available textbooks. Our group reviewed a number of widely used textbooks . . . and found virtually no references to technology in general, or to our eight specific areas of concern. In fact, we found fewer references to technology than in textbooks of twenty years ago. The books have become more theoretical, more abstract with fewer practical applications. They appear to have evolved in a context where science education is considered the domain of an 'elite' group of students.
>
> (Piel 1981, p. 106)

The success of the Russian *Sputnik*, along with the vocal demands of science professionals, created enormous legislative and commercial pressure to use school science as a means of preparing students for tertiary science studies. In the 30 years between 1957 and 1987, the practical and the liberal curriculum emphases progressively gave way to the academic, or professional, model of curriculum design. And, as to be expected, these took on what their writers saw as the settled orthodoxy about scientific method. An inductive-empiricist view of science, for instance, dominated the curricula reforms of the 1960s. This can be seen in representative documents, such as the 1966 *Education and the Spirit of Science*, published by the Education Policies Commission. There, it is stated that, in science: 'generalizations are induced from discrete bits of information gathered through observation conducted as accurately as the circumstances permit', and that science seeks for 'verification' of its claims (Education Policies Commission 1966, p. 18). Just a little bit of HPS input could have corrected this glaring mistake: science does not proceed by induction, nor does it seek to verify its claims; more modestly, it seeks to confirm them.

Two important exceptions to the general ahistorical, professional curricula supported by the NSF were the Harvard Project Physics course and the Yellow version of the BSCS high-school biology course. Another small-scale example of a historical–philosophical science programme was the Klopfer and Cooley 'Use of Case Histories in the Development of Student Understanding of Science and Scientists'. These case histories were consciously aimed at replicating the well-established Harvard Case Studies in Experimental Science, used successfully at the college level. One review of the utilisation of the case-study approach said that, 'the method is definitely effective in increasing student understanding of science and scientists when used in biology, chemistry, and physics classes in high schools' (Klopfer & Cooley 1963, p. 46).[11]

Appraisal of the NSF Reforms

By the mid 1970s, after 20 years of energetic involvement, and $1.5 billion in financial support, the NSF withdrew from school curriculum development. In 1975, federal funding for the NSF's curriculum developments was below what it had been in 1959. The times had changed: the Soviet threat had receded, the US had its man on the Moon, school enrolments were falling, and there was a state and local authority backlash against the de facto introduction of a national curriculum – such federal interference was (and still is) a matter of grave concern to the more than 16,000 fiercely independent local school boards in the US.

Numerous studies were done on the effectiveness of the massive federal intervention. Among the more prominent was that of Helgeson, Blosser and Howe, which reviewed all research appearing between 1955 and 1975 (Helgeson *et al.* 1977), and Project Synthesis, directed by Norris Harms, which scrutinised hundreds of studies (Harms & Yager 1981). These studies found that the curricular reforms were only partially successful in meeting their own objectives and in fulfilling the hopes that government and society held for them. In 1979, the original director of the PSSC project lamented that the curriculum reform movement was suffering a 'deadening sense of frustration and near defeat'. To this proponent, it was a time of 'despair and confusion' (Jackson 1983, p. 152).

Fifty years later, when school science reform is once more on the political and educational agenda, it is timely to know how much of this failure and confusion was due to the curriculum materials, how much to teacher inadequacies, how much to implementation and logistic failures, how much to general anti-intellectual or anti-scientific cultural factors, and how much to a residue factor of faulty learning theory and inadequate views of scientific method that the schemes incorporated. It may be, however, that there are no overall answers to the question; perhaps the reasons for failure may be localised, varying from curriculum to curriculum, from school district to school district or even school to school.

One respected physics teacher, researcher, textbook writer and curriculum planner, Arnold Arons, has drawn attention to the fact that 'curricular material, however skilful and imaginative, cannot "teach themselves"' (Arons

1983, p. 117). He believed that, 'a substantial body of interesting, imaginative, and educationally sound material was developed' in the NSF-sponsored curricula. He attributes the failures to two causes: first, inadequate logistic support for schoolteachers; second, and more importantly, the inadequate training of teachers.

The first factor covers such commonplace things as the absence of laboratory assistants in schools and of money for equipment or films, little free time to set up experiments and maintain displays, and minimum study-leave provisions. The second factor covers such things as lack of knowledge of subject matter, failure to appreciate the psychological requirements for science learning – particularly the need for experience and familiarity with reality to precede theory and concepts – poor in-service courses, where teachers were 'given more of the same rapidly paced, irrelevant, and unintelligible college courses that had had no visible intellectual effect in the past' (Arons 1983, p. 120), and the failure of science teachers to appreciate and convey the rich intellectual and cultural import of their subject. Science was taught as a rhetoric of conclusions, to use Schwab's term, and the fluid nature of scientific enquiry and conclusions was seldom apparent.

Other studies support Arons's reluctance to blame the NSF curricula. Wayne Welch concludes that, 'when compared to teacher effectiveness, student ability, time on task, and the many other things that influence learning, curriculum does not appear to be an important factor'. He cites studies that show that only 5 per cent of the variance in student achievement was due to curriculum/non-curriculum treatments. Welch reports that his Project Physics team 'eventually concluded that 5% was an acceptable return on our investment since we could seldom find greater curricular impact on the students' (Welch 1979, p. 301).

One way of looking at these results is that, although curriculum is important, it is not important by itself: the mere change of curriculum, without change of teacher education, assessment tasks, resources and support, is not going to have any dramatic effect on student engagement, interest and learning of science, or of any other subject. It is of little use to set up high-powered curriculum committees that devise curricula that are then sent in the mail to schools. A curriculum without appropriate texts, examinations, teacher commitments and systematic support is like a car without petrol – it looks nice, but doesn't go anywhere. What many have said is that results such as Welch's and analyses such as Arons's, point to the fundamental importance of teachers – their knowledge, enthusiasm, attitudes, educational philosophy and views about their subject, science – for successful teaching.

Current US Curricula Reforms

By the early 1980s, it was apparent to all that there was a second-generation crisis in US science education; it was variously labelled 'the science literacy crisis' or 'the flight from science' (Bishop 1989). Despite all the money and effort that had been expended since *Sputnik*, the bulk of American high-school graduates and citizens had minimal scientific understanding. A few knew a great deal; the vast majority knew very little. This state of affairs had been documented in countless research articles and government reports. But what brought it to popular attention in the US, and galvanised the government to action, was the publication in 1983 of *A Nation at Risk* (NCEE, 1983). Its conclusion was stark: 'the educational foundations of our society are presently being eroded by a rising tide of mediocrity that threatens our very future as a nation and as a people'.[12] It expressed a particular concern about the abysmal state of the scientific and mathematical knowledge of high-school graduates.

In the 5 years after the publication of *A Nation At Risk*, more than 300 reports documented the sorry state of US education. In 1983, twenty bills were put before Congress designed to offer solutions to the national crisis in science education. These bills and reports all urged the adoption of 'scientific and technology literacy for all' (Mansell 1976) as the goal of school science instruction.[13] *Science for All* has been adopted as a goal for science education, not just in the US (Rutherford & Ahlgren 1990), but in the UK, Canada, Australia, New Zealand and most other countries.

Of course, it should not be thought that the crisis was entirely one of curriculum, or of instruction, or that the schools should be able to counteract major cultural, social or economic forces. The fact that the US has 500 lawyers for each engineer, whereas Japan has 500 engineers for each lawyer, is not something that schools can control; nor can schools influence the massive disparity in salaries paid to lawyers, fund managers and accountants, in contrast to science teachers or engineers. Further, schools have little effect on a mass culture that is anti-intellectual and operates at the sound-bite level of analysis. Boyer, in his influential 1983 report, *High School*, drew attention to this:

> After visiting schools from coast to coast, we are left with the distinct impression that high schools lack a clear and vital mission. They are unable to find common purposes or establish educational priorities that are widely shared. They seem unable to put it all together. The institution is adrift.
>
> (Boyer 1983, p. 63)

An enriched understanding of science, its methods, achievements and cultural interactions, in other words of HPS, can contribute a little to the 'clear and vital mission' of which Boyer wrote, as can some articulation of a philosophy of education that can guide classroom, curricular and organisational decisions. Such recourse to HPS and philosophy of education has been taken up in various national science-education reform proposals, from the AAAS's *Project 2061* of the late 1980s through the National Science Education Standards and now the *Next Generation Science Standards*.[14] The outcomes at each of these stages could only be improved by the engagement of historians and philosophers with educators.

Project 2061

In 1985, the AAAS established an extensive national study called *Project 2061*[15] to stimulate and promote an overhaul of science education in schools (its brief included mathematics, technology and social science, along with natural sciences). Recognising that, in the US, educational decisions are made by thousands of different entities, including 16,000 separate school districts, and federal and state courts constantly mandate, and then reverse, major programmes, the project realistically said of itself that, '*Project 2061* constitutes, of course, only one of many efforts to chart new directions in science, mathematics, and technology education' (AAAS 1989, p. 155). The explicit recourse to history and philosophy in this project is noteworthy.

The first report, *Science for All Americans*, was published in 1989 (AAAS 1989, Rutherford & Ahlgren 1990). It advocates the achievement of scientific literacy by all American high-school students. Its proposals were based on the belief that:

> The scientifically literate person is one who is aware that science, mathematics, and technology are interdependent human enterprises with strengths and limitations; understands key concepts and principles of science; is familiar with the natural world and recognises both its diversity and unity; and uses scientific knowledge and scientific ways of thinking for individual and social purposes.
>
> (AAAS 1989, p. 4)

As explained in Chapter 2, the final clause – 'uses scientific knowledge and scientific ways of thinking for individual and social purposes' – links *Project 2061* to the Enlightenment tradition in science education.

The report has a chapter on philosophy of science and another on history of science. These are among twelve chapters that range over topics such as mathematics, technology, the physical world, the living environment, the human organism, human society and the designed world. The history and philosophy chapters are encouraging to those who advocate the inclusion of HPS in the school science curricula.

Philosophical Commitments

All science curricula contain views about the NOS: images of science that influence what is included in the curriculum, how material is taught, and how the curriculum is assessed. The image of science held by curriculum framers sets the tone of the curriculum, and the image of science held by teachers influences how the curriculum is taught and assessed. When spelled out, these images of science become statements about the NOS, or more narrowly about the epistemology of science.

Project 2061's view of the NOS can be found in its Chapter 1, titled 'The Nature of Science', where there are discussions on objectivity, the mutability of science, the demarcation dispute about how science is distinguished from non-science, evidence and how it relates to theory appraisal, scientific method as logic and as imagination, explanation and prediction, ethics, social policy and the social organisation of science. These themes are intended to be developed in science courses; it stresses that the themes are to be developed within the subject matter of science, and not treated as 'add-ons'. The following philosophical theses are advocated in Chapter 1 of *Science for All Americans*. These could be regarded as items in the AAAS 'nature of science' list, or, as will be argued in Chapter 11, more expansively they can be understood as 'features of science' (FOS):

1 *Realism*: There is an existing material world apart from, and independent of, human experiences and knowledge. This ontological position is in contrast to varieties of idealism that maintain that, either there is no world outside human experience, or that such a world, and human experience, is all ideational. The report says that, 'Science assumes that the universe is . . . a vast single system in which the basic rules are everywhere the same' (AAAS 1989, p. 25). Realism is only committed to the existence of an external world. The claim that its laws are everywhere the same is an elaboration of the basic realist position. To what extent the 'basic rules' are assumed to be everywhere the same, and to what extent they are discovered to be the same, is a moot point even among realists.
2 *Fallibilism*: Humans can have knowledge of the world, even though such knowledge is imperfect, and reliable comparisons can be made between competing theories or opinions. Fallibilism is an epistemological position that is opposed, on the one hand, to relativism, which holds that no reliable comparison can be made between competing views, and, on the other hand, to absolutism, which holds that current theory constitutes absolute, unimprovable knowledge. The report says that, 'Scientists assume that even if there is no way to secure complete and absolute truth, increasingly accurate approximations can be made to account for the world and how it works' (AAAS 1989, p. 26). The notion of 'approximate truth' is much debated, with many philosophers preferring to simply speak of better, or more progressive, theories.
3 *Durability*: Science characteristically does not just abandon its central ideas. The simple falsificationist picture of scientists examining and rejecting ideas in some sort of quality-control process does not hold up. The report says, 'The modification of ideas, rather than their outright rejection, is the norm in science, as powerful constructs tend to survive and grow more precise' (AAAS 1989, p. 26). The philosopher Otto Neurath first gave picturesque expression to this view when he spoke of the correction of scientific theory as the fixing of a leaking boat at sea: the entire hull is not taken out; rather, planks are examined and replaced one at a time. Willard van Orman Quine gave wide currency to the image:

> We are like sailors who on the open sea must reconstruct their ship but are never able to start afresh from the bottom. Where a beam is taken away a

> new one must at once be put there, and for this the rest of the ship is used as support. In this way, by using the old beams and driftwood the ship can be shaped entirely anew, but only by gradual reconstruction.
>
> (Quine 1960, p. 3)

Imre Lakatos formalised this conception with his idea of science as a series of research programmes, with hard-core commitments that were very resistant to change, and protective belt commitments that changed to accommodate discordant or falsifying data (Lakatos 1970).

4 *Rationalism*: The report holds to a modified form of rationalism, saying that,

> sooner or later scientific arguments must conform to the principles of logical reasoning – that is, to testing the validity of arguments by applying certain criteria of inference, demonstration, and common sense.
>
> (AAAS 1989, p. 27)

The old view was that science was always rational in its deliberations among competing views, theories or research programmes. As a result of research in history, philosophy and sociology of science, this old view has been modified, and the roles of personal and external interest have been recognised in the short-term resolution of disputes; the long-term resolution is not so easily accounted for by interests. The Christian churches finally accepted the Copernican solar system, and Soviet agriculture finally accepted Mendelian genetics. This topic will be further developed in Chapter 5, in the section 'Sociological Challenges to the Rationality of Science'.

5 *Antimethodism*: Although rationalist in its justification of scientific theory, the report rejects the idea that there is a single method of scientific discovery, saying that, 'There simply is no fixed set of steps that scientists always follow, no one path that leads them unerringly to scientific knowledge' (AAAS 1989, p. 26). The report stresses the creative dimension of science, saying,

> Scientific concepts do not emerge automatically from data or from any amount of analysis alone. This aspect is often overlooked in schools. Inventing hypotheses or theories about how the world works and then figuring out how they can be put to the test of reality is as creative as writing poetry, composing music, or designing sky-scrapers.
>
> (AAAS 1989, p. 27)

6 *Demarcationism*: Science can nevertheless be separated from non-scientific endeavours. This is a contentious and debated matter that lay at the heart of the 1981 creationist trial, when creation scientists were arguing that their activity was every bit as scientific as mainstream science, and so they

ought to have a place in the school science curriculum. Whether creation science falls inside or outside the divide is one question, that there is a divide is another, and the report is unambiguous about it, saying: 'There are, however, certain features of science that give it a distinctive character as a mode of inquiry' (AAAS 1989, p. 26).[16]

7 *Predictability*: The report says:

> It is not enough for scientific theories to fit only the observations that are already known. Theories should also fit additional observations that were not used in formulating the theories in the first place; that is, theories should have predictive power.
>
> (AAAS 1989, p. 28)

A part of the distinctiveness of science is its concern with predicting phenomena and having the results count. There are problems with this idea, and it is known that testing is not a simple matter; yet there is reasonable agreement on one aspect, namely that good scientific theories should uncover phenomena not currently known. They cannot merely keep accounting for what other theories bring to light, or what common sense has already ascertained.

8 *Objectivity*: It is recognised that science is a far more human activity than it was once conceived to be. Francis Bacon's (1561–1626) Idols of the Mind have persisted long after he urged their eradication in 1620. But the report, although recognising this human face of science, nevertheless maintains that science at its best tries to correct for, and rise above, subjective interests in the determination of truth. It says:

> Scientific evidence can be biased in how the data are interpreted, in the recording or reporting of the data, or even in the choice of what data to consider in the first place. Scientists' nationality, sex, ethnic origin, age, political convictions, and so on may incline them to look for or emphasize one or another kind of evidence or interpretation . . . but scientists want to know the possible sources of bias and how bias is likely to influence evidence.
>
> (AAAS 1989, p. 28)

The possibility of objectivity in science has been challenged by some feminists, some constructivists and most philosophical postmodernists. The issue is further discussed in Chapter 5 of this book.

9 *Moderate externalism and interests*: The attempt to eliminate subjectivity and interest from the determination of truth claims is not the same as saying that various interests should not influence what spheres of knowledge science should investigate. Whether research is conducted on space travel or cheaper public transport, on nuclear energy or solar energy, on chemical insecticide development or biological controls will be a function of personal, social and commercial interests. Science does not

proceed in a political vacuum; most countries draw up lists of national priority areas and will only release public funds for scientific research in these areas. Being on or off the list is a political matter. The report recognises that:

> As a social activity, science inevitably reflects social values and viewpoints . . . The direction of scientific research is affected by informal influences within the culture of science itself, such as prevailing opinion on what questions are most interesting or what methods of investigation are most likely to be fruitful. . . . Funding agencies influence the direction of science by virtue of the decisions they make on which research to support.
>
> (AAAS 1989, p. 29)

When decisions about truth or otherwise are made in order to serve the interests of funding or political bodies, then science has moved from moderate to complete externalism. Although some sociologists of science argue the latter view, it is rejected in the report. The question is investigated in Chapter 5 of this book in the section 'Ethics, Values and Science Education'.

10 *Ethics*: The report recognises that scientists do not determine the ethical values of society; it rejects a triumphal or scientistic 'leave it all to the scientists' view, but it does show how scientific work is crucial to informed ethical deliberations. It says:

> Nor do scientists have the means to settle issues concerning good and evil, although they can sometimes contribute to the discussion of such issues by identifying the likely consequences of particular actions.
>
> (AAAS 1989, p. 26)

The question is also investigated in Chapter 5 in the section 'Ethics, Values and Science Education'.

From the foregoing sketch, it is clear that *Project 2061*'s image of science is informed by current history, philosophy and sociology of science. As the document is meant to be a curriculum framework, and not an academic treatise, it does not contain detailed arguments for the theses advanced. However, as the project intends that local bodies will reflect on and respond to the document, then these theses will have to be more fully developed at the local level. On just about every point listed above, philosophers, historians and sociologists of science will be aware of a body of contending literature. Nevertheless, the document is a valuable starting point for reflection, and it clearly requires that teachers and decision-makers be comfortable with philosophising about science. The latter is going to be done better, the more there is engagement between the education and philosophy communities.[17]

Historical Perspectives

Curriculum proposals usually do say something about philosophy of science, although not as explicitly as *Project 2061*. This project is distinctive in the place it gives to the history of science in school science teaching. In introducing Chapter 10 on 'Historical Perspectives', the report says:

> The emphasis here is on ten accounts of significant discoveries and changes that exemplify the evolution and impact of scientific knowledge: the planetary earth, universal gravitation, relativity, geologic time, plate tectonics, the conservation of matter, radioactivity and nuclear fission, the evolution of species, the nature of disease, and the Industrial Revolution.
>
> (AAAS 1989, p. 111)

Project 2061 says of these that, 'although other choices may be equally valid, these clearly fit our dual criteria of exemplifying historical themes and having cultural significance'.

Project 2061 advances two types of argument for bringing history into school science, both of which are of interest to philosophers of science and to educators. The first is that:

> Generalizations about how the scientific enterprise operates would be empty without concrete examples. Consider for example, the proposition that new ideas are limited by the context in which they are conceived; are often rejected by the scientific establishment; sometimes spring from unexpected findings; and usually grow slowly, through contributions from many different investigators. Without historical examples, these generalizations would be no more than slogans, however well they might be remembered.
>
> (AAAS 1989, p. 111)

The second reason for bringing the history of science into science classrooms is that:

> some episodes in the history of the scientific endeavor are of surpassing significance to our cultural heritage. Such episodes certainly include Galileo's role in changing our perception of our place in the universe; Newton's demonstration that the same laws apply to motion in the heavens and on earth; Darwin's long observations of the variety and relatedness of life forms that led to his postulating a mechanism for how they came about; Lyell's careful documentation of the unbelievable age of the earth; and Pasteur's identification of infectious disease with tiny organisms that could be seen only with a microscope. These stories stand among the milestones of the development of all thought in Western civilization.
>
> (AAAS 1989, p. 111)

These comments underline the unfortunate fact that the history of science has fallen between academic stools. Arguably the greatest achievement of

Western civilisation, and that which has undoubtedly been responsible in large part for the shape of world history, is usually not dealt with in school (or university) history departments, because it is thought too technical or difficult, and it is not dealt with in science departments because it is thought irrelevant. Bringing HPS into science programmes can in part rectify this situation. It can spur cooperation between school history and science departments. It can assist the integrative goals of education.

Project 2061 devotes one and a half pages to Galileo and his achievement in physics – 'Displacing the Earth from the Center of the Universe'. It is an informed treatment of the complexities of astronomical evidence at the time, the role of sense perception in Aristotelian science, the status of mathematical models in ancient astronomy, the tradition of realist versus instrumentalist interpretations of scientific theory, the interplay of metaphysics and physics at the beginning of the scientific revolution, the function of technology in the establishment of the new science, Galileo's use of rhetorical argument to establish his position, and the complex role of theological considerations in the evaluation of Galilean science. It deals with other historical episodes, providing material for teachers, programmers and curriculum developers to consider.

There are philosophical and educational problems with these recommendations. As in its first chapter on philosophy, it could be asked: Whose nature of science is going to be taught? So, with its historical chapter, it could be asked: Whose history of science is to be taught? Whigs, internalists, externalists, idealists, Marxists – all have different accounts of the major episodes that *Project 2061* commends to teachers. There is a great deal of unresolved controversy in the history of science: after nearly 400 years, the intellectual dust has still not settled on the trial of Galileo – as can be evidenced in the recent Vatican collection on the matter (Poupard 1987) and other studies of the episode.[18] The report does recognise that it deals with milestones in the development of Western civilisation, but in multicultural classrooms there may be need for other milestones to be recognised and investigated. Some other specifically educational problems created by the inclusion of history of science in the science curriculum will be discussed in the following chapter.

Habits of Mind

Science for All Americans includes a final chapter on 'Habits of Mind', which deals with values, attitudes, communication, reasoning, manipulation skills and so on. This topic raises the perennial issue of whether being scientific is meant to extend beyond the laboratory? As outlined in Chapter 2, the Enlightenment tradition hoped that learning to be scientific in investigation of nature would have a natural flow-over effect on how people thought about social and cultural problems. This expectation was voiced by John Dewey, and it was enshrined in the Indian Constitution's requirement that the state promote 'Scientific Temper'. As with *Project 2061*'s philosophical and historical sections, this final chapter is yet another occasion for educators to work with historians, philosophers and psychologists in fleshing out and defending the attributes of a 'scientific habit of mind'.[19] Beyond this task, the chapter moves discussion towards a very old question, namely: What are the qualities of an educated person? Education is far more than instruction in science, or even the sum of instruction in various disciplines. Teachers need to provide such instruction, but they need to recognise that the goals of education are far wider, and they have to contribute to these wider goals within the teaching of their own subjects.

National Science Education Standards

Following the AAAS report, in the US, the first ever National Science Education Standards were published by the National Research Council in 1996 (NRC 1996). They recognise the centrality of philosophical and historical knowledge in the teaching of science, maintaining, for instance, that students should learn how:

- science contributes to culture (NRC 1996, p. 21);
- technology and science are closely related: a single problem has both scientific and technological aspects (NRC 1996, p. 24);
- curriculum will often integrate topics from different subject-matter areas . . . and from different school subjects – such as science and mathematics, science and language arts, or science and history (NRC 1996, p. 23);
- scientific literacy also includes understanding the nature of science, the scientific enterprise and the role of science in society and personal life (NRC 1996, p. 21);
- effective teachers of science possess broad knowledge of all disciplines and a deep understanding of the disciplines they teach (NRC 1996, p. 60);
- tracing the history of science can show how difficult it was for scientific innovators to break through the accepted ideas of their time to reach conclusions that we currently take for granted (NRC 1996, p. 171);
- progress in science and technology can be affected by social issues and challenges (NRC 1996, p. 199);
- if teachers of mathematics use scientific examples and methods, understanding in both disciplines will be enhanced (NRC 1996, p. 218).

These aspirations for science classrooms cannot be achieved without teachers who care about HPS and have some competence in it. A position paper of the US Association for the Education of Teachers in Science, the professional association of those who prepare science teachers, has recognised this in its recommendation: '*Standard 1d*: The beginning science teacher educator should possess levels of understanding of the philosophy, sociology, and history of science exceeding that specified in the [US] reform documents' (Lederman *et al.* 1997, p. 236).

Next Generation Science Standards

For the past 3 years in the US, a new national science-education standards document, called the *Next Generation Science Standards* (NGSS) (NRC 2012, 2013), has been progressively developed.[20] As the NGSS say:

> The impetus for this project grew from the recognition that, although the existing national documents on science content for grades K-12 (developed in the early to mid-1990s) were an important step in strengthening science education, there is much room for improvement. Not only has science progressed, but the education community has learned important lessons from 10 years of implementing standards-based education, and there is a new and growing body of research on learning and teaching in science that can inform a revision of the standards and revitalize science education.
>
> (NRC 2012 p. ix)

The NGSS incorporate and build on the 'existing national documents', but a novel feature is the conscious effort to connect science learning to engineering and to scientific practices, and to make it progressive and cumulative from the beginning of elementary school. These are seen as its differentia from 'the existing national documents'. As previously mentioned, after two decades of consultation and trialling, the US National Research Council, in its much anticipated *A Framework for K-12 Science Education*, writes:

> Epistemic knowledge is knowledge of the constructs and values that are intrinsic to science. Students need to understand what is meant, for example, by an observation, a hypothesis, an inference, a model, a theory, or a claim and be able to distinguish among them.
>
> (NRC 2012, p. 79)

These sentences need only be read for us to realise that HPS is required for their realisation in classrooms and curricula. If students need to know and understand what is meant by 'an observation, a hypothesis, an inference, a model, a theory, or a claim and be able to distinguish among them', then surely teachers must know and be able to promote interest in these topics. If so, the immediate question is, where do they acquire such knowledge? Where does HPS come into the programme of pre-service or in-service teacher education? This question will be returned to in Chapter 12.

British Science Curricular Reform

Natural philosophy entered British schools, such as they were, in the mid eighteenth century.[21] By the middle of the nineteenth century, the 'science of everyday things' was common in primary schools (Jenkins 1979). The work of the Reverends Charles Mayo and Richard Dawes was influential. Not surprisingly, given the widespread enthusiasm for Paley's *The Evidences*, much of this science of everyday things, and nature study, was used to promote religious perspectives such as creation, design and providence. However, this religious uplifting did not save science from those who thought that the lower classes were being dangerously over-educated and becoming far too critical. The Revised Curriculum Code of 1862 basically removed all science from state-funded primary schools. The Clarendon Commission, in 1864, supported the importance of classical studies, but it also lamented the absence of scientific studies in the education of the upper classes. In 1867, the British Association for the Advancement of Science (BAAS) threw its influence behind efforts to have science reinstated and reconstituted in the curriculum. Thomas Huxley, in his influential address, 'A Liberal Education; and Where to Find it' (Huxley 1868/1964), given at the opening of the South London Working Men's College, focused attention on the importance of science to education and ridiculed contemporary curricula that excluded science. The philosopher C.E.M. Joad typifies the circumstance Huxley railed against:

> I left my public school in 1910, an intelligent young barbarian. . . . My acquaintance with the physical sciences was confined to their smells. I had never been in a laboratory; I did not know what an element was or a compound. Of biology I was no less ignorant. I knew vaguely that the first Chapter of Genesis was not quite true, but I did not know why. Evolution was only a name to me and I had never heard of Darwin.
>
> (Joad 1935, p. 9)

Henry Armstrong and the Heuristic Method

At the turn of the century, Henry Armstrong (1848–1937),[22] professor of chemistry at Imperial College, London, led a crusade against the dry, verbal, didactic pedagogy that then prevailed in science classrooms, indeed all classrooms. Armstrong said of this scholastic approach that:

> I have no hesitation in saying that at the present day the so-called science taught in most schools, especially that which is demanded by examiners, is not only worthless, but positively detrimental.
>
> (Armstrong 1903, p. 170)

In contrast to these didactic methods, Armstrong advocated the heuristic method (or what might loosely be called the discovery method), which he characterised thus:

> Heuristic methods of teaching are methods which involve our placing students as far as possible in the attitude of the discoverer – methods which involve their finding out instead of merely being told about things. It should not be necessary to justify such a policy in education . . . discovery and invention are divine prerogatives, in some sense granted to all, meant for daily usage and that it is

> consequently of importance that we be taught the rules of the game of discovery and learn to play it skilfully.
>
> (Armstrong 1903, p. 236)

Armstrong's views ought not to be identified with the extreme 'discovery *ex nihilo*' view or the 'Robinson Crusoe' view, advocated by some enthusiasts of discovery learning. To place students as far as possible in the attitude of the discoverer does mean that students have to have some stock of concepts, of techniques, of instruments, of calculating abilities and so on – the things that the discoverer surely starts out with. Armstrong said:

> It is needless to say that young scholars cannot be expected to find out everything themselves; but the facts must always be presented to them so that the process by which results are obtained is made sufficiently clear as well as the methods by which any conclusion based on the facts are deduced.
>
> (Armstrong 1903, p. 255)

In contemporary terms, for Armstrong, discovery has to be guided by teachers; teachers need to transmit concepts, methods and methodologies to students; and the more and better guidance by teachers, the more and better learning by students.

Armstrong also believed that the heuristic method should be historical. In this, he acknowledges an 1884 paper of Meiklejohn, who had said:

> This view has its historical side; and it will be found that the best way, the truest method, that the individual can follow is the path of research that has been taken and followed by whole races in past times.
>
> (Armstrong 1903, p. 237)

So, for Armstrong, the discovery method was something that stressed pupil activity and individual reasoning, but this was in a context created by the teacher, and this context was designed to follow the historical path of the development of science.

Armstrong's crusade had mixed results: some victories, many defeats, some converts, many unmoved. Nevertheless, he started a tradition in British science education that has emphasised enquiry teaching, historical study, pupil activity and investigation. The fortunes of this tradition have fluctuated during the past century. At different times and with different people, different aspects of Armstrong's ideas have been emphasised: enquiry learning can be ahistorical; practical work can be didactic and reduced merely to the following of cookbook recipes; historical study can be just a sweetener for technocratic science. Edgar Jenkins surmises that:

> Despite the virtual eclipse of the heuristic method of teaching science, many of Armstrong's ideas were to continue to influence school science education. An emphasis on practical experimental teaching and a belief in the importance of

> learning by doing became established features, and science curriculum reform, a generation after his death in 1937, was to incoprporate Armstrong's view that science could best contribute to liberal education by initiating pupils into its greatest professional mystery, its method.
>
> (Jenkins 1979, p. 52)

Between the Wars

Between the wars, there were some significant contributions to both the theory and practice of HPS-informed science education. Three individuals stand out: Federick Westaway (1864–1946), E.J. Holmyard (1891–1959) and John Bradley.

Frederick Westaway was one of 'His Majesty's Inspectors of Schools' in the UK in the 1920s and also authored substantial books on history of science and philosophy of science.[23] In a widely used teacher training textbook, he wrote that a successful science teacher is one who:

> knows his own subject . . . is widely read in other branches of science . . . knows how to teach . . . is able to express himself lucidly . . . is skilful in manipulation . . . is resourceful both at the demonstration table and in the laboratory . . . is a logician to his finger-tips . . . is something of a philosopher . . . is so far an historian that he can sit down with a crowd of [students] and talk to them about the personal equations, the lives, and the work of such geniuses as Galileo, Newton, Faraday and Darwin. More than this he is an enthusiast, full of faith in his own particular work.
>
> (Westaway 1929, p. 3)

One wonders whether 90 years of educational research and debate have added significantly to this account of a good science teacher.

E.J. Holmyard was another influential figure in between-the-wars English science education. As with Westaway, he combined being a science teacher and prolific textbook writer with making his own contributions to the history of science.[24] He argued that:

> The historical method is not, I believe, one of several equally good alternative schemes of teaching chemistry in schools: it is the only method which will effectively produce all the results at which it is at once our privilege and duty to aim.
>
> (Holmyard 1924, p. 229)

Elsewhere, he argued against a merely 'utilitarian standpoint' in the teaching of science, saying that science needed to be regarded as the greatest of the humanities (Holmyard 1922).

John Bradley, at the University of Hull, was one of the finest exponents of Armstrong's heurism.[25] He had a passionate commitment to teaching chemistry in a manner that allowed students to fall in love with it:

> This falling in love with chemistry is the Real Right Thing about learning chemistry; and it is the only item of educational psychology which the teacher of chemistry needs to know.
>
> (Bradley 1964, p. 364)

He was an admirer of Ernst Mach and wrote an important book on Mach's philosophy of science (Bradley 1971). Bradley endorsed Mach's instrumentalist view of theory, insisting that theoretical discussion be X-rated, and that children not be exposed to it until the final school years. He colourfully said: 'The young people of this country come hopefully to school asking for the bread of experience; we give them the stones of atomic models' (Bradley 1964, p. 366).

Bradley built an introductory chemistry course around the celebrated 'copper problem' (oxidation), where students begin with heating copper and noticing that it puts on two coats – a scarlet inner one and a black outer one. From there, the course takes off, with students suggesting reasons for this, testing them, asking whether copper gains or loses weight on heating and why, devising ways to heat copper without air, the investigation of oxygen, reduction problems and so on. All of this is very low-technology teaching. He was a resolute opponent of the 'Post-Sputnik NSF Education' that swept the US, and most of the rest of the Western world, after 1957 and that still has a commanding presence. He wrote:

> By returning from the far country [US] with its painted Jezebels of atomic models to the homeland and pure gospel of Armstrong, the teaching of chemistry could be immensely improved without the expenditure of a penny. Indeed money could be saved, because sulphuric acid is cheaper than models of models of models.
>
> (Bradley 1964, p. 366)

Nuffield Science

By the 1960s, disquiet was being expressed at English science achievement levels and participation rates. The major response to this was the Nuffield science courses. Like the NSF courses, a number of the Nuffield courses advocated discovery learning and the enquiry method of teaching. The Nuffield schemes were developed at the time of the *Plowden Report* (1967), which recommended child-centred teaching for British primary schools. The Nuffield schemes resurrected the enquiry portion of the Armstrong tradition, while largely neglecting the historical dimension.

As with the NSF courses, the Nuffield courses held an inductivist view of scientific method (Stevens 1978). This is seen in the Physics Year 4 Teachers Guide where, discussing Newton's Second Law, the advice is given that:

> Students should be left on their own to draw conclusions from their graphs. It is much less valuable, though much quicker for the teacher to impose a well-taught conclusion. What the pupils find out for themselves from the slopes of these

> graphs (without ever being told to look at the slopes) will remain in their minds as one of their great discoveries in physics – particularly if we can tell them that they are finding out part of the story of Newton's great Laws of Motion.
>
> (Harris & Taylor 1983, p. 285)

The Physics Year 3 Guide says: 'what [students] need are simple general instructions, where to look but not what to look at' (Harris & Taylor 1983, p. 278).

The Nuffield courses dominated British school science teaching in the 1960s and 1970s. As in the US, the idea was to produce 'little scientists' by having students engage in scientific discovery. Some of the problems with the approach surfaced very early. The Association for Science Education (ASE), in its 1963 *Training of Graduate Science Teachers*, stressed the obvious problem of teachers who did not understand, or have an interest in, the nature of science itself. Of graduate teachers, it said: 'Many behave and think scientifically as a result of their training but they lack an understanding of the basic nature and aims of science' (ASE 1963, p. 13).

Contextual and STS Science

In the decade after its adoption, voices were increasingly raised against the Nuffield approach. The sociologist Michael Young observed, in 1976, that, 'Despite a decade of unprecedented investment in curriculum innovation, school science displays many of the manifestations of a continuing "crisis"' (Young 1976, p. 47). The ASE, in its 1979 report, *Alternatives for Science Education*, advocated science education for all students to the age of 16 years, saying that such a curriculum should 'incorporate a reasonable balance between the specialist and generalist aspects of science' and should 'reflect science as a cultural activity'. In a later report, *Education through Science* (1981), the teaching of science as a cultural activity was spelled out as:

> the more generalized pursuit of scientific knowledge and culture that takes account of the history, philosophy and social implications of scientific activities, and therefore leads to an understanding of the contribution science and technology make to society and the world of ideas.
>
> (ASE 1981)

The ASE recognised the importance of HPS in its own 'Science in Society' project, which includes a reader on the subject.

English National Curriculum

The British Education Reform Act (1988) provided for the establishment in England (but not Scotland, Wales or Northern Ireland) of a national school curriculum to replace the variety of university entrance curricula, local education authority curricula and other courses of study and examinations that had characterised British secondary education. The National Curriculum Council recommended that science constitute 20 per cent of the curriculum for all students aged from 5 to 16 years. Its first report, *Science in the National Curriculum*, was produced in 1988 (NCC 1988) and revised in 1991 (NCC 1991). In the original 1988 document, the importance given to HPS and the detail with which HPS goals were spelled out were exemplary – an HPS advocate's educational dream come true.[26]

A study of educational and political struggles over the place of history and philosophy in the national curriculum is a rewarding if sobering exercise for all advocates of HPS in science programmes.[27] A significant feature of the science curriculum was that about 5 per cent of it was devoted to HPS. This was the last of seventeen 'attainment targets' in the first report. The NCC, at the beginning of its first report, draws attention to this field. It says that the curriculum,

> is concerned with the nature of science, its history and the nature of scientific evidence. Council recognises that this aspect has not enjoyed a traditional place in science education in schools. . . . Since this target may be relatively unfamiliar to teachers, several examples have been given for each level to illustrate the area of study.
>
> (NCC 1988, p. 21)

The committee elaborates its intentions when it says of Attainment Target 17 that:

> pupils should develop their knowledge and understanding of the ways in which scientific ideas change through time and how the nature of these ideas and the uses to which they are put are affected by the social, moral, spiritual and cultural contexts in which they are developed.
>
> (NCC 1988, p. 113)

Although, in the first report, the history and philosophy of science were singled out as a separate attainment target, the expectation was that the themes identified would be taught as they arose in the context of the other attainment targets; this is made explicit in the second report.

Concerning the programme of study for 11–14-year-olds, the first report said that students should, through their investigations of the life of a famous scientist and/or the development of an important idea in science, be given the opportunity to:

- study the ideas and theories used in other times to explain natural phenomena;
- relate these ideas and theories to present scientific and technological understanding and knowledge;
- compare these ideas and theories with their own emerging understanding and relate them to available evidence.

For 14–16-year-olds, the committee recommends that pupils continue the course of study outlined above, but, in addition, they should also:

- distinguish between claims and arguments based on scientific data and evidence and those that are not;
- consider how the development of a particular scientific idea or theory relates to its historical and cultural, including the spiritual and moral, context;
- study examples of scientific controversies and the ways in which scientific ideas have changed (NCC 1988, p. 113).

Beyond providing a programme of study, the NCC report also itemised expected competence levels. The report says pupils should at:

- Level 4, be able to describe the story of some scientific advance, for example, in the context of medicine, agriculture, industry or engineering, describing the new ideas and investigation or invention and the life and times of the principal scientist involved.
- Level 7, be able to give a historical account of a change in accepted theory or explanation and demonstrate an understanding of its effects on people's lives – physically, socially, spiritually, morally; for example, understanding the ecological balance and the greater concern for our environment; or, the observations of the motion of Jupiter's moons and Galileo's dispute with the Church.
- Level 10, be able to demonstrate an understanding of the differences in scientific opinion on some topic, either from the past or the present, drawn from studying the relevant literature (NCC 1988, pp. 114–115).

The NCC said of these levels of attainment that they are 'pitched both to be realistic and challenging across the whole ability range' (NCC 1988, p. 117). There is no doubt that they were challenging, and not just for students. How realistic they were would depend in large measure on the HPS knowledge and interest of teachers, and thus on how much HPS figured in their own scholarly formation. How realistic also depends on the demands of the examination system – if something is not examined, it will not be taught.[28] The first revision of the NCC eliminated Attainment Target 17 by collapsing it into Attainment Target 1, which dealt with practical work and investigations in science.

The subsequent history of the National Curriculum, especially the chequered career of 'NOS', is detailed and discussed by Edgar Jenkins (2013) and also by James Donnelly, who remarks that:

> I have suggested that the curricular emphases associated with the nature of science represents potentially more than a collection of disparate issues. . . . These domains represent the main possibilities for a reform of the science curriculum which is other than cosmetic or instrumental. Yet, without some sense of coherence and

> underlying educational purpose they are likely to retain that marginality which their unruly educational provenance has promoted.
>
> (Donnelly 2001, p. 193)

The importance of being clear about educational purpose, or philosophy of science education, is a recurrent theme in all serious science-education debate; unfortunately, it is not something that teachers are prepared well for, and they are less and less prepared as foundation courses are stripped from teacher education programmes.

The Perspectives on Science Course

The most recent concerted UK effort to teach NOS material is the new optional Upper Level *Perspectives on Science* course for England and Wales (Swinbank & Taylor 2007). The course has four parts:

- Part 1: Researching the history of science;
- Part 2: Discussing ethical issues in science;
- Part 3: Thinking philosophically about science;
- Part 4: Carrying out a research project.

The textbook for this course, on its opening page, says:

> *Perspectives on Science* is designed to help you address historical, ethical and philosophical questions relating to science. It won't provide easy answers, but it will help you to develop skills of research and argument, to analyse what other people say and write, to clarify your own thinking and to make a case for your own point of view.
>
> (Swinbank & Taylor 2007, p.vii)

The Philosophy section begins with sixteen pages outlining fairly standard matters in philosophy of science – NOS, induction, falsifiablity, paradigms, revolutions, truth, realism, relativism, etc. Importantly, the book then introduces the subject of 'Growing your own philosophy of science' by saying:

> Having learned something about some of the central ideas and questions within the philosophy of science, you are now in a position to evaluate the viewpoints of some scientists who were asked to describe how they viewed science. The aim here is to use these ideas as a springboard to develop and support your own thinking.
>
> (Swinbank & Taylor 2007, p. 149)

This is a key element of all competent proposals for the inclusion of HPS into classrooms, curricula and teacher education: the HPS is not to be learned catechism-like, but is to be utilised, developed and evaluated.

Beyond 2000

In the United Kingdom, a group of prominent science educators, reflecting on Britain's National Curriculum and the most appropriate form of science education for the new millennium, wrote a report with ten recommendations, the sixth of which said that:

> The science curriculum should provide young people with an understanding of some key-ideas-about science, that is, ideas about the ways in which reliable knowledge of the natural world has been, and is being, obtained.
>
> (Millar & Osborne 1998, p. 20)

In elaborating this recommendation, the writers say that pupils should also become familiar with stories about the development of important ideas in science that illustrate the following general ideas:

- that scientific explanations 'go beyond' the available data and do not simply 'emerge' from it but involve creative insights (e.g. Lavoisier and Priestley's efforts to understand combustion);
- that many scientific explanations are in the form of 'models' of what we think may be happening, on a level which is not directly observable;
- that new ideas often meet opposition from other individuals and groups, sometimes because of wider social, political or religious commitments (e.g. Copernicus and Galileo and the Solar System);
- that any reported scientific findings, or proposed explanations, must withstand critical scrutiny by other scientists working in the same field, before being accepted as scientific knowledge (e.g. Pasteur's work on immunisation) (Millar & Osborne 1998, pp. 21–22).

Science–Technology–Society Curricula

In looking at the contributions of HPS to curriculum development and debate in science education, it is useful to identify the vicissitudes of STS curricula. There are points of connection between the two traditions that ought to have been more fully explored and utilised.[29]

Contemporary (post-1980s) STS education had its origins in the failures of the discipline-based curricular reforms of the 1960s. Many saw the flight from science as a demand for more useful and relevant science courses: courses that would capture the attention of students and give them some understanding of the myriad technical devices that they lived among and of the transport and manufacturing technology that drove modern economies. In this regard, however, STS courses adopted one of the chief tenets of the progressivism of the 1930s – make education relevant to the lives of students; they continue the tradition of the science of everyday life that was common in the US between the world wars, and the science of common things that was prevalent in the UK between the wars.

The educational STS movement drew intellectual inspiration and support from the academic field of STS studies that, from the 1970s, had began to differentiate itself from orthodox HPS studies in universities, that formed its own academic society – Society for the Social Studies of Science (4S) – and that had its own 'in-house' journal – *Social Studies of Science*.[30] The international spread of STS curricula and research was reflected in the creation of the International Organisation of Science and Technology Education (IOSTE) in the late 1970s.[31]

Rodger Bybee (1985, 1993), Paul DeHart Hurd (1985) and Robert Yager (1993, 1996) were three prominent US advocates of STS-informed science education. The US National Science Teachers Association (NSTA) endorsed the STS orientation to science in its 1971 statement, *School Science Education for the 1970s* (NSTA 1971). The 1985 NSTA Yearbook dealt with the rationale and content of such STS programmes (Bybee 1985), and the NSTA publication *The Science, Technology, Society Movement* (Yager 1993) reviewed their implementation. STS made significant inroads in Canadian provincial science eduction programmes, with Glen Aikenhead being perhaps its most prominate advocate (Aikenhead 1994, 2000).[32]

In England, STS education was championed by, among others, John Lewis and Joan Solomon and found expression in the SATIS (Science and Technology in Society) and SISCON (Science in a Social Context) courses. The UK ASE funded two curriculum projects, the 1981 'Science and Society' course (ASE 1981), and the 1983 'Science in its Social Context' (SISCON) course (Solomon 1985). The latter course was influenced by successful university STS programmes, of which an exemplary textbook was John Ziman's *Teaching and Learning About Science and Society* (Ziman 1980).

An energised contemporary manifestation of the STS tradition, but one that consciously goes 'beyond STS', is the Socio-Scientific Issues (SSI) movement.[33] This movement recognises that philosophical discussion of ethics and politics cannot be avoided – the core concern of social justice cannot be elaborated without engagement with moral theory, political theory and economic theory. The educational task is to see that such discussion avoids indoctrination and manipulation of students to have them commit to whatever the teacher believes is a good cause; it needs also to rise above slogans and immersion in whatever the current cultural fashion might be. These caveats are aided by teachers having a better grounding in historical and philosophical subject matter. Even cursory thinking on the difference between education and indoctrination is of great value for SSI teaching.

Clearly, the STS and HPS orientations to science education should be complementary: pure science and applied science have gone together; technology and technology-related social issues have a history and they clearly have philosophical and cultural dimensions; both considerations need to be incorporated in good educational programmes. A 1990 Alberta, Canada, departmental guide to STS education, *Unifying the Goals of Science Education* (Alberta Education 1990), laid out one of the basic continuing tensions in STS education, namely how much can and should STS education be conducted independently of HPS education? The Alberta guide made explicit a commitment to teaching about the nature of science, insisting that this 'includes teaching the concepts that philosophers of science have developed to describe the nature of the scientific endeavour and the origins, limits and nature of scientific knowledge'. Often, the classroom practice of STS falls far short of the curriculum HPS rhetoric.

Enquiry Teaching and Discovery Learning

The US, British and most other curricular reforms from the 1960s to the present aimed at more than just specifying content areas or laying down topics to be taught; they were also concerned to develop scientific attitudes and methods among students. Reformers wanted students to become scientific, not just learn science. To this end, 'enquiry' or 'discovery learning' was a prominent feature of the US NSF and the UK Nuffield reforms, one advocate saying, 'All of modern science curriculum developments stress teaching science as inquiry' (Sund & Trowbridge 1967, p. 22).[34] The appraisal of enquiry learning is interesting for the light that it sheds on a number of important educational and philosophical matters – concept acquisition, social dependence of learning and so forth – but it is especially interesting for those promoting HPS in science education.[35]

Enquiry in the 1960s

In the 1960s, the enquiry or discovery approach to teaching and learning was separately advocated by two very prominent theorists – Joseph Schwab, a University of Chicago educationalist involved with the BSCS project,[36] and Jerome S. Bruner, a Harvard cognitive psychologist. Schwab's first publication on the subject was in 1958; he elaborated upon it in 1960 in what was to become a classic of enquiry theory (Schwab 1960). Bruner was director of a working party of thirty-five that the National Academy of Sciences convened in the summer of 1959 at Woods Hole on Cape Cod, Massachusetts, to investigate the rash of new curricula and to see whether basic principles of learning and curriculum construction could be elucidated.[37]

Bruner's main contribution was to bring to educational discussion and research the 'cognitive turn' that was taking place in psychology. He was in large part responsible for initiating this turn with his 1956 *The Study of Thinking*. At the time, psychology had been taken over by behaviourism, and his own Harvard department was 'locked in a standoff between Skinner's operant conditioning and Steven's psychophysics' (Bruner 1983, p. 122). Whatever the advances of the cognitive turn were in psychology departments, it was much slower in departments of education, where some such departments in the 1960s made successful pigeon training a precondition for the award of the doctorate in education. Joseph Novak (1977) outlines the lingering grip of behaviourism on educational psychology. Bruner also brought a concern with classrooms, teachers and educational practices, at a time when educational psychologists preferred to think of 'learning theory' in terms of rats, pigeons, stimulii and reinforcement schedules, and of human learning only in experimental situations. In this context, he introduced the cognitive, human-centred ideas of Jean Piaget to the group. He also stressed the importance of 'structure' for learning. This was connected with his idea of the 'generativeness' of knowledge:

> 'Learning' is, most often, figuring out how to use what you already know in order to go beyond what you currently think. There are many ways of doing that. Some are more intuitive; others are formally derivative. But they all depend on knowing something 'structural' about what you are contemplating – how it is put together. Knowing how something is put together is worth a thousand facts about it. It permits you to go beyond it.
>
> (Bruner 1983, p. 183)

There is an ambiguity here between the material object of knowledge and the theoretical object of knowledge. The structures of disciplines that Bruner and Schwab elevate to the forefront of science learning are structures in the theoretical objects of science: the structure of interrelating definitions and concepts contained in Newton's *Principia*, the structure of geometry as contained in Euclid's *Elements*, the structure of evolutionary theory in Darwin's *Origin*, the structure of Brönsted's acid/base theory or of plate tectonic theory. Once these structures are grasped, then distant theorems can be derived from axioms, and predictions can be made about likely intervening species or the acidity of new chlorides and so on. But these are not the objects contemplated by neophytes: they contemplate material objects, such as falling stones, triangles or a range of flora. There are two very different senses of 'structure' being used here: the structure of objects and processes, and the structure of disciplines. The structure of a leaf is one thing; the structure of photosynthesis theory is quite another, and two very different modes of contemplation and manipulation are required for the different objects: one is turned around in the hand; the other is turned over in the mind. Aristides Baltas well noted this:

> The concept, say, of a 'material point with a determinate mass' does not constitute the common essence of apples, planets and projectiles. It is rather a concept of the conceptual scheme of physics which is produced together with the other concepts of this system and which is, precisely, attributed to such real objects so that their movement may be accounted for by this system as a whole.
>
> (Baltas 1988, p. 216)

Bruner's 1961 *Harvard Educational Review* article, 'The Act of Discovery', popularised discovery learning. With its popularisation came its inevitable distortion. In 1966, Bruner wrote a follow-up essay, 'Some Elements of Discovery', distancing himself from the educational excesses touted in the name of discovery learning. He confided that, 'I am not sure any more what discovery is' (Bruner 1974, p. 84), and complained that, 'Discovery was being treated by some educators as if it were valuable in and of itself, no matter what it was a discovery of or in whose service' (Bruner 1974, p. 15). This was a crucial recognition that should have been heeded by all naive advocates of 'science is fun' or students being 'scientists for a day'.

Discovery learning aimed to promote thinking and reasoning skills and independent research. In the words of one advocate:

> It gives students more opportunities to think and learn how to think critically. As inquirers, students learn to be independent, to compare, to analyze, to synthesize knowledge, and to develop their mental and creative faculties.
>
> (Sund & Trowbridge 1967, p. 22)

More specifically, discovery learning was welcomed as a way for students to grasp the nature of scientific enquiry. Students would learn about the nature of scientific discovery and reasoning by themselves, participating in enquiry. Jerome Bruner, in his classic *The Process of Education*, optimistically writes that, 'The schoolboy learning physics *is* a physicist' (Bruner 1960, p. 14), and his emphasis on 'is' seems to take degree out of the claim and places it in the already swollen ranks of educational hyperbole – is someone learning the piano to be called a pianist? Many other texts and curricula took up this theme of students doing enquiry and so being 'scientists for a day' (Harlen 1996).

Much was written on the theory and practice of discovery learning. Numerous conferences on the subject were convened, with one conference in particular drawing together many prominent critics of the programme (Shulman & Keislar 1966). The convenors of this conference pointed to a major problem:

> Examination of both the exhaustive reviews of the literature and deliberations of the conference lead to an inescapable conclusion: The question as stated is not amenable to research solutions because the implied experimental treatment, the discovery method, is far too ambiguous and imprecise to be used meaningfully in an experimental investigation.
>
> (Shulman & Keislar 1966, p. 191)

One major study did document better performance for US students in the NSF curricula, compared with extant curricula (Shymansky *et al.* 1990), but this study only dealt with curricula, not with teaching methods. It may have been that the curricula were better, the teachers of it were more innovative and enthusiastic, the support materials were richer, and so on. The performance may well have had nothing to do with supposed enquiry methods. This big study is strangely silent on this vital matter. Fifty years later, this is the very same problem that bedevils efforts to evaluate contemporary constructivist and enquiry-based pedagogy. More generally, poor design, Hawthorne effects and lack of control groups, or even just controls, are all good reasons for examining carefully any conclusions drawn from educational research (Shavelson & Towne 2002).

At one level, the enquiry approach was very attractive. For students to discover by experiment and manipulation what materials are attracted to a magnet and what materials are not, what makes one hour-glass empty quickly while another empties slowly, rather than being told this by a teacher, is an advance on rote learning or textbook learning. Bruner said that discovery methods were preferable because they promoted an increase in intellectual potency, they involved a shift from extrinsic to intrinsic rewards, they taught the heuristics of discovering, and they were an aid to memory processing (Bruner 1961).

However, the theoretical promise of enquiry teaching was not always fulfilled. One extensive review of the American enquiry-based programmes and curricula of the 1960s concluded: 'In spite of new curricula, better trained teachers, and improved facilities and equipment, the optimistic expectations for students becoming inquirers have seldom been fulfilled' (Welch *et al.* 1981, p. 33).

The reviewers said that the problem lay in large part with teachers:

> Science was something teachers took in college, but it was not something they experienced as a process of inquiry. . . . The values associated with speculative, critical thinking were often ignored and sometimes ridiculed.
>
> (Welch *et al.* 1981, pp. 38–39)

Other reviewers documented the positive effects that participation in NSF teacher programmes had on student performance in those teachers' classes (Shymansky *et al.* 1990). How much this was due to a Hawthorne effect, how much to the teacher being a more enthusiastic and capable teacher, was not teased out.

Philosophers and psychologists in the 1960s and 1970s detailed many flaws in the fundamental assumptions of enquiry teaching; they indicated that the problems of discovery learning, the gap between promise and fulfilment, were not just 'practical' ones that could be overcome with money or more resources: they were theoretical ones.[38] These are the same problems that recur with contemporary constructivism and enquiry learning. One critic located the flaw with discovery learning in the image of science, or the epistemology of science, that infused the curriculum:

> A basic flaw in the process is the apparent assumption that science is a sort of commonsensical activity, and that the appropriate 'skills' are the primary ingredients in doing productive work. There seems to be no explicit recognition of the powerful role of the conceptual frames of reference within which scientists and children operate and to which they are firmly bound. These general views of the physical world demand careful nurture . . . by a variety of means.
>
> (J. Myron Atkin, in Glass 1970, p. 20)

David Ausubel, who, with Bruner, was largely responsible for bringing the 'cognitive turn' to education (Ausubel 1968), was sceptical of the whole NSF curricular endeavour and its underpinnings. Of the endeavour, he observed that: 'Despite their frequent espousal of discovery principles, the various curriculum projects have failed thus far to yield any research evidence in support of the discovery method' (Ausubel 1964/1969, p. 110). Of the theoretical underpinnings, he observed that:

> Actually, a moment's reflection should convince anyone that most of what he really knows and meaningfully understands, consists of insights discovered by others which have been communicated to him in meaningful fashion.
>
> (Ausubel 1964/1969, p. 98)

A persistent problem was the very meaning of 'discovery' in discovery learning. There is an inescapably epistemological aspect of the concept. Not every propositional belief that students entertain or come to, or agree upon, deserves to be called a 'discovery', just those beliefs or claims that are true. Someone might come to the belief 'that 2 + 2 = 5' or 'that Moscow is the capital of Finland', but these cannot be discoveries, because they are both false beliefs. And, as these judgements about truth and falsity are public, not private, it cannot be individuals alone or groups in isolation that made discoveries. Not all groups that come together and generate sound are making music. Not all combinations of sounds are music; just sounds competently judged to be so are music. To say that all sound is music means some other term will need be introduced, 'sound +', to pick out what we mean by music. Sound is necessary for music, but not sufficient. In practice, it is going to be some external authority that sifts the convictions from the discoveries. Eventually, making sounds is not good enough for learning music; likewise with enquiring and being scientific.

Further, there is an inescapably process aspect to the 'discovery' concept; a discovery means that the individual or the group has to make the discovery. There is an element of engagement and intention required. And, in education, there needs be some element of rationality or at least good reason-giving involved in the process. If a student does not understand the proposition or claim they enunciate, then they have not discovered it, even if it is true; they may have learned it, but not discovered it. Similarly, if they do understand the proposition, but have been forced to believe or enunciate it, then they cannot be said to have discovered it. If a student discovers that '2 + 2 = 4' and, when asked why, says that '4' is their favourite number, then, although the answer is correct, it has not been a discovery in the educational sense. In education, 'discovery' is closer to 'finding out'. There is a process connected to legitimate discovery, and elucidating this process is a philosophical endeavour, whether done by philosophers or not.

Contemporary Enquiry Programmes

In the US, the *National Science Education Standards* are explicitly committed to enquiry teaching, with the title of its teachers' guide being: *Inquiry and the National Science Eduation Standards: A Guide for Teaching and Learning* (NRC 2000). There, it is confidently stated that:

> Inquiry is a central part of the teaching standards. The standards say, for example, that teachers of science 'plan an "inquiry-based" science program', 'focus and support inquiries' and 'encourage and model the skills of scientific inquiry'.
>
> (NRC 2000, p. 21)

Many make the link between enquiry and constructivism:

> Hence, the current teaching standards in the US call for teachers to embrace a social constructivist view of learning and teaching in which science is described as a way of knowing about natural phenomena and science teaching as facilitation of student learning through science inquiry . . . In particular, the reform emphasizes teacher education by promoting social constructivist teaching approaches.
>
> (Kang 2008, p. 478)

All European Union science and mathematics education 'renewal' projects are premised on the efficacy of enquiry teaching, or enquiry-based science education (EBSE) as it is labelled. Currently, there are at least half-a-dozen major EBSE projects being supported, with nearly €100 million of EU funds. Among them are the following projects: PRIMUS, Pollen, Sinus-Transfer and Fibonacci, which together involve more than twenty countries, hundreds and hundreds of schools, hundreds of teachers, and multiple thousands of students. The Fibonacci project alone involves sixty universities, 3,000 teachers and 45,000 students.[39] The 2007 Rocard Report aggregated the rationales and programmes of these enquiry-based projects. It requested, and received, a 'not unreasonable budget of 60 million euros over 6 years' to pursue its policy (among others):

> Improvements in science education should be brought about through the new forms of pedagogy: The introduction of the inquiry-based approaches in schools and the development of teachers' networks should actively be promoted and supported.
>
> (Rocard *et al.* 2007, p. 17)

In an editorial in *Science* magazine, it was reported that the above €60 million was just 'seed money', and that:

> In all four nations, the 'science as inquiry' pedagogy encourages students (ages 5 to 16) to develop a sense of wonder, observation, and logical reasoning. Because of their interactions with scientists, as well as new assessment and professional

development methods, teachers gain increased confidence and a better understanding of science as a process.

(Léna 2009, p. 501)

We are not told to what degree the laudable 'increased confidence and better understanding' flows over into more enquiry teaching; nor are we told how much the latter, of itself, is responsible for whatever increase there might be in student enrolment in science or students' learning. This requires fine-grained and controlled educational research, which is too seldom conducted.

The PRIMAS project (www.primas-project.eu), involving twelve European countries – Cyprus, Denmark, Germany, Hungary, Malta, The Netherlands, Norway, Romania, Slovakia, Spain, Switzerland and the UK – says in its published report:

> Fourteen universities from twelve different countries are working together to further promote the implementation and use of inquiry-based learning in mathematics and science. PRIMAS provides materials for direct use in class and for professional development.
>
> (PRIMAS 2013, p. 1)

Revealingly, the PRIMAS website says that:

> A common misunderstanding is to confuse IBL [inquiry-based learning] with doing experiments or some practical work in the classroom. If the knowledge needed to conduct the experiment is provided by the teacher or by the task as a kind of cookbook recipe, the experiment can hardly be called inquiry-based. The degree of inquiry depends on the openness of the situation as well as on the distribution of responsibilities between the teacher and the pupils.
>
> (www.primas-project.eu/artikel/en/1302/What+exactly+does+inquiry-based+learning+mean/view.do?lang=en)

That is, it is committed to the view that the 'less guidance, the better the enquiry'. This, then, skirts around the question of whether better enquiry (so defined) results in better learning; overwhelmingly, the evidence is that it does not: learning is a function of guidance, and, the more guidance that is given, the more learning will occur.

Neither the current US nor European enquiry programmes pay much attention to the critics of 50 years ago. In the 200 glossy pages and 150 references of the US *Teachers Guide*, there is no mention of any of the above-listed educational, psychological or philosophical critiques of discovery learning and enquiry teaching; no mention of Bruner, Ausubel, Strike, Dearden or any other of the major critics or of their lines of criticism. Nor is there mention of the extensive contemporary critiques of 'minimally guided instruction' (Kirschner *et al.* 2006, Mayer 2004). These latter will be taken up in Chapter 8, when the efficacy of constructivist teaching methods is discussed; at present, it suffices to repeat the conclusion drawn in one major study:

> Pure discovery did not work in the 1960s, it did not work in the 1970s, and it did not work in the 1980s, so after these three strikes, there is little reason to believe that pure discovery will somehow work today.
>
> (Mayer 2004, p. 19)

It surely is a mistake to concentrate on pedagogy as the solution to acknowledged problems of science education. The problems are many, as are the solutions. Seeking the Holy Grail in pure discovery or minimally guided enquiry sets teachers and everyone else off down the wrong path. As a blanket ideal, it presents an unattainable standard for teachers, and it has no input at all to make on the subject matter of the curriculum, on what enquiry will be about.

James Rutherford, early in the 1960s, made a prescient observation that went largely unheeded at the time, and continues to go unheeded:

> Science teachers must come to know just how inquiry is in fact conducted in the sciences. Until science teachers have acquired a rather thorough grounding in the history and philosophy of the sciences they teach, this kind of understanding will elude them, in which event not much progress toward the teaching of science as inquiry can be expected.
>
> (Rutherford 1964, p. 84)

This book endorses Rutherford's claim that teachers' familiarity with HPS is essential for the improvement of science teaching. Such familiarity would have enabled past and present teachers to avoid much of the naivety associated with the claims of discovery learning – naive and false views such as: that scientific method is inductive, that observation does not depend upon conceptual understanding, and that messing about with real objects can reveal the structure of the scientific theories that apply to those objects.

Conclusion

With lessons learned from the curriculum reforms of the 1960s and from the science-education crisis of the 1980s, and with better comprehension of how children learn science, curricular projects in most of the world are at present attempting to embody the following ideas:

- Less content should be taught, but it should be taught and evaluated in a way that encourages understanding and comprehension rather than memorisation and rote leaning.
- Some of the connections between science, technology and society need to be appreciated. This is independent of whether full STS programmes are taught.

- The cultural dimensions of science, its history and philosophy, its moral and religious implications, need to be appreciated; a science course should entail some learning about science, as well as learning of science.
- Curriculum change will only be effective if it is accompanied by widespread systematic changes involving teacher education or re-education programmes, funding, assessment schemes and texts.

These ideas are not without their problems and internal tensions. It is probably the case that the curriculum documents overestimate the amount and sophistication of HPS material that can be conveyed in a school programme that is devoted principally to the teaching of science. The topics mentioned in *Project 2061* and in the first version of the British National Curriculum are very complex, and teachers and students need to realise that there are few simple answers. It may be that an interest in the questions and some appreciation of the complexity are as much as can realistically be conveyed to most students, given the demands of the syllabus and of other subjects. Although a modest ambition, it is nevertheless an important one.

Curriculum development and classroom teaching need to be cognisant of the psychological preparedness of students. Mature scientific thinking requires formal thought processes in the Piagetian scheme of sensorimotor, preoperational, concrete operational and formal operational modes of thinking. These thought processes are late in developing. In the US, it is estimated that most first-year college students have not reached the stage of formal operational thought. One US study showed that fewer than 6 per cent of 17-year-olds can solve simple algebra problems (Cromer 1993, p. 26). In the UK, it is estimated that fewer than 20 per cent of 16-year-olds in comprehensive schools are in the formal stage of reasoning (Black & Lucas 1993, p. 34). This is a powerful reason for science teaching to initially be as phenomenological and concrete as possible, and for explicit attention to be paid to the promotion of formal and abstract reasoning.

Teachers concerned with the HPS dimensions of science need also to be sensitive to the realities of psychological development. Teachers and curriculum framers need to seek in the HPS sphere the equivalent of phenomenological material. Things should be kept simple, concrete and focused. The big questions – What is the nature of science? How does science relate to religion? Is knowledge of the world truly possible? Is science just a social product? – should be approached very slowly. Meaningful discussion of these questions requires sophisticated thinking and a good stock of basic information about particular parts of the history of science and the philosophy of science.

Premature attention to the big questions in HPS can cheapen and devalue intellectual activity. This happens enough in society, with sound-bite-level analyses of complex economic, political and ethical questions the norm. Depressingly, it happens enough in schools too, where children are encouraged to give opinions about all sorts of issues, independently of any knowledge of them. This educational practice systematically devalues knowledge acquisition and sustained thought. Good HPS-informed science teaching can counteract these narcissistic practices by showing that things are more complex than they appear.

Notes

1 George DeBoer (1991) provides a comprehensive history of leading ideas in science education. Other accounts of the history of science education are Glass (1970), Hodson (1987), Rudolph (2002, 2008) and Waring (1979). Good sources for charting the history of US science education are the occasional yearbooks published by the National Society for the Study of Education. These include the third, *Nature Study* (1904), the thirty-first, *A Program for Teaching Science* (1932), the forty-sixth, *Science Education in American Schools* (1947) and the fifty-ninth, *Rethinking Science Education* (1960).
2 Some of this seminal debate can be read in Armstrong (1903), Dewey (1910), Huxley (1885/1964), Mach (1886/1986) and Nunn (1907).
3 Accounts of this turn-of-the-century debate can be read in Mann (1912) and Woodhull (1910).
4 The titles of other commonly used texts were: *Applied Biology*, Bigelow & Bigelow, 1911; *A Civic Biology*, Hunter, 1914; *Practical Biology*, Smallwood, Reveley, & Bailey, 1916; *Civic Biology*, Hodge & Dawson, 1918; *Biology and Human Welfare*, Peabody & Hunt, 1924.
5 For a comprehensive review of literature on the engagement of science education with philosophy of education, see Schulz (2014).
6 The history of this 1950s crisis, and its attendant educational reforms, can be found in numerous sources. Some useful ones are: DeBoer (1991), Jackson (1983), Klopfer and Champagne (1990), Raizen (1991), Rudolph (2002) and Welch (1979).
7 Jerome Bruner says of Zacharias that, 'it was Zack more than anybody else who converted *Sputnik* shock into the curriculum reform movement that it became rather than taking some other form' (Bruner 1983, p. 180).
8 In 1963, at the IUPAP International Conference on Physics Teaching, held in Rio de Janeiro, Zacharias reported that there had been translations into Spanish, Portuguese, Hebrew, Japanese, Turkish, Thai, Swedish, Danish, French and Norwegian, with more pending (Zacharias 1964, p. 68).
9 A discussion of the educational theory behind PSSC can be found in Rudolph (2002, Chapter 5).
10 Expository papers on each of these curricula are gathered in Andersen (1969, section 7).
11 Another review of these case studies can be found in Klopfer (1964).
12 The NCEE publication received wide media coverage. Articles appeared with headings such as 'Can American Schools Produce Scientifically Literate High School Graduates?' and 'Can Democracy Survive Scientific Illiteracy?' (Bauer 1992, p. 1).
13 Gerald Holton, a member of the National Commission for Excellence in Education (NCEE), which prepared the report, has provided an account of its disturbing contents that document the 'tide of mediocrity' in US education and its recommendations for turning the tide (Holton 1986).
14 William McComas (2014) provides a detailed history and commentary on these developments.
15 The project began in 1985, which was the year of the Halley's Comet visitation, and 2061 was the year of its next visitation. AAAS realistically thought that science education could be improved over that time span.
16 On the demarcation dispute in philosophy of science, see at least: Laudan (1983), Mahner (2007) and Pennock (2011).

17 The Biological Sciences Curriculum Study published a collection of background papers on HPS for teachers that can assist this engagement; the paper of Peter Machamer (1992) on 'Philosophy of Science: An Overview for Educators' is particularly useful.
18 See at least Fantoli (1994, Chapter 7), Finocchiaro (1989, 2005), McMullin (2005) and Redondi (1988).
19 For some discussion on this matter, see Gauld (2005) and Good (2005).
20 The 320-page draft is available free from the National Academies Press website; it is titled *A Framework for K-12 Science Education*. Background studies for the NGSS are in NRC (2007).
21 It was not until the Education Act of 1870 that state-funded primary education appeared, and not until the Acts of 1902 and 1903 that state secondary schools appeared. On the history of British science education, see at least: Brock (1989, 1996), Jenkins (1979, 2013), Layton (1973) and Taylor and Hunt (2014).
22 Accounts of the life, times and achievements of Armstrong can be found in Brock (1973) and Jenkins (1979). A collection of twenty-four of his own articles and addresses were published as *The Teaching of Scientific Method and Other Papers on Education* (Armstrong 1903).
23 An extensive account of the life, writings and achievement of Westaway can be found in Brock and Jenkins (2014).
24 Details of his life, work and achievements can be found in Jenkins (2014).
25 Bradley's views are presented in numerous *The School Science Review* articles over a span of 40 years. His first journal article was in 1933. He elaborated his approach to molecular chemistry in a series of articles in 1957 Vol.39, 1958 Vol.39, 1959 Vol.40 and 1961 Vol.42. His famous 'The Copper Problem', where he defends Armstrong's heurism against Nuffield science, is elaborated in eight articles in *The School Science Review*: 1963, 1964, 1964, 1965, 1966, 1967, 1967 and 1968. G. van Praagh's *Chemistry by Discovery* (1949) is another example of Armstrong's heurism.
26 In a reminder that HPS endorsements in curricula need to be elaborated and appraised, the historian Stephen Pumfrey published a detailed critique of the HPS assumptions in the NCC documents (Pumfrey 1991).
27 The unpleasant politics of the NCC revisions are described by Duncan Graham, the first chairperson of the NCC (Graham 1993). There are serious questions about the degree to which the clear HPS attainments of the 1988 report can be met within the guidelines of the second and subsequent NCC reports. The political origins and educational implications of the 1988 Act are discussed in Flude and Hammer (1990) and Taylor and Hunt (2014).
28 Martin Monk and Jonathan Osborne (1997) shed light on these considerations.
29 A comprehensive account of the commonalities in the STS and HPS traditions can be found in Vesterinen *et al.* (2014).
30 Studies in STS were not new; they had long been part of Marxist-influenced histories of science, where the work of Bernal (1939) and Hogben (1940) is exemplary. John Ziman's work dealt with many of the core philosophical and disciplinary issues of STS (Ziman 1968, 1980, 1994).
31 A recent study of the effectiveness of STS education is Bennett *et al.* (2007).
32 Canadian STS/HPS initiatives are reviewed in Metz (2014).
33 Discussion can be found in Sadler (2011) and Zeidler and Sadler (2008).
34 Among the vast literature on the enquiry approach of the 1960s, the following are particularly useful: Ausubel (1964/1969), Bruner (1961), Rutherford (1964), Schwab (1960) and contributions to Shulman and Keislar (1966).
35 These aspects of enquiry teaching have been extensively canvassed in Kelly (2014).
36 For the role of HPS in Schwab's work, see DeBoer (2014).
37 Bruner wrote a chairperson's report on the Woods Hole conference, which was published as *The Process of Education* (Bruner 1960). This immediately became an international bestseller, being translated into nineteen languages, and was described in the *New York Herald Tribune* as a 'classic, comparable in its philosophical centrality

and humane concreteness to Dewey's essays on education'. A more personal account of the conference can be found in Bruner's autobiography *In Search of Mind* (1983, pp. 181–188).

38 For discussion of some of the fundamental problems of enquiry learning that emerged from these early efforts, see: Atkinson and Delamont (1977), Ausubel (1964/1969), Dearden (1967), Harris and Taylor (1983), Herron (1971), Strike (1975), Wellington (1981) and contributions to Shulman and Keislar (1966). For a contemporary review of the field, see Kelly (2014).

39 There are websites for each project, and all are under the mantle of the European Union.

References

AAAS (American Association for the Advancement of Science): 1989, *Project 2061: Science for All Americans*, AAAS, Washington, DC. Also published by Oxford University Press, 1990.

Aikenhead, G.S.: 1994, 'What is STS Teaching?' In J. Solomon and G. Aikenhead (eds) *STS Education: International Perspectives on Reform*, Teachers College Press, New York, pp. 47–59.

Aikenhead, G.S.: 2000, 'Renegotiating the Culture of School Science'. In R. Millar and J. Osborne (eds) *Improving Science Education*, Open University Press, Philadelphia, PA, pp. 245–264.

Alberta Education: 1990, *Unifying the Goals of Science Education*, Curriculum Support Branch, Edmonton, Canada.

Andersen, H.O. (ed.): 1969, *Readings in Science Education for the Secondary School*, Macmillan, New York.

Armstrong, H.E.: 1903, *The Teaching of Scientific Method and Other Papers on Education*, Macmillan, London.

Arons, A.B.: 1983, 'Achieving Wider Scientific Literacy', *Daedalus* 112(2), 91–122.

ASE (Association for Science Education): 1963, *Training of Graduate Science Teachers*, ASE, Hatfield, UK.

ASE (Association for Science Education): 1979, *Alternatives for Science Education*, ASE, Hatfield, UK.

ASE (Association for Science Education): 1981, *Education Through Science*, ASE, Hatfield, UK.

Atkinson, P. and Delamont, S.: 1977, 'Mock-ups & Cock-ups'. In M. Hammersley and P.Woods (eds) *The Process of Schooling*, London, pp. 87–108.

Ausubel, D.P.: 1964/1969, 'Some Psychological Aspects of the Structure of Knowledge'. In S. Elam (ed.) *Education and the Structure of Knowledge*, Rand McNally, Chicago, IL.

Ausubel, D.P.: 1968, *Educational Psychology: A Cognitive View*, Holt, Rinehart & Winston, New York (2nd edn, 1978, with Novak & Hanesion).

Baltas, A.: 1988, 'On the Structure of Physics as a Science'. In D. Batens and J.P. van Bendegens (eds) *Theory and Experiment*, Reidel, Dordrecht, The Netherlands, pp. 207–225.

Bauer, H.H.: 1992, *Scientific Literacy and the Myth of the Scientific Method*, University of Illinois Press, Urbana, IL.

Bennett, J., Hogarth, S. and Lubben, F.: 2007, 'Bringing Science to Life: A Synthesis of the Research Evidence on the Effects of Context-based and STS Approaches to Science Teaching', *Science Education* 91(3), 347–370.

Bernal, J.D.: 1939, *The Social Function of Science*, Routledge & Kegan Paul, London.

Bishop, J.: 1989, 'Scientific Illiteracy: Causes, Costs, and Cures'. In A.B. Champagne, B.E. Lovitts and B.J. Callinger (eds) *This Year in School Science 1989. Scientific Literacy*, American Association for the Advancement of Science, Washington, DC, pp. 41–88.

Black, P.J. and Lucas, A.M. (eds): 1993, *Children's Informal Ideas in Science*, Routledge, New York.

Boyer, E.L.: 1983, *High School: A Report on Secondary Education in America*, Harper & Row, New York.
Bradley, J.: 1963–1968, 'A Scheme for the Teaching of Chemistry by the Historical Method', *School Science Review* 44, 549–553; 45, 364–368; 46, 126–133; 47, 65–71, 702–710; 48, 467–474; 49, 142–150; 454–460.
Bradley, J.: 1964, 'Chemistry II: The Copper Problem', *School Science Review* 45, 364–368.
Bradley, J.: 1971, *Mach's Philosophy of Science*, Athlone Press of the University of London, London.
Brock, W.H. and Jenkins, E.W.: 2014, 'Frederick W. Westaway and Science Education: An Endless Quest'. In M.R. Matthews (ed.) *International Handbook of Research in History, Philosophy and Science Teaching*, Springer, Dordrecht, The Netherlands, pp. 2359–2382.
Brock, W.H.: 1973, *H.E. Armstrong and the Teaching of Science 1880–1930*, Cambridge University Press, Cambridge, UK.
Brock, W.H.: 1989, 'History of Science in British Schools: Past, Present and Future'. In M. Shortland and A. Warwick (eds) *Teaching the History of Science*, Basil Blackwell, Oxford, UK, pp. 30–41.
Brock, W.H.: 1996, *Science for All: Studies in the History of Victorian Science and Education*, Variorum Press, Aldershot, UK.
Bruner, J.S.: 1960, *The Process of Education*, Random House, New York.
Bruner, J.S.: 1961, 'The Act of Discovery', *Harvard Educational Review* 31, 21–32. Reprinted in R.C. Anderson and D.P. Ausubel (eds) *Readings in the Psychology of Cognition*, Holt, Rhinehart and Winston, New York, 1965.
Bruner, J.S.: 1974, 'Some Elements of Discovery'. In his *Relevance of Education*, Penguin, Harmondsworth, UK, pp. 84–97. Originally published in L. Shulman and E. Keislar (eds) *Learning by Discovery*, Rand McNally, Chicago, IL, 1966.
Bruner, J.S.: 1983, *In Search of Mind: Essays in Autobiography*, Harper & Row, New York.
Bybee, R.W.: 1993, *Reforming Science Education: Social Perspectives and Personal Reflections*, Teachers College Press, New York.
Bybee, R.W. (ed.): 1985, *Science, Technology, Society, Yearbook of the National Science Teachers Association*, NSTA, Washington, DC.
Callahan, R.E.: 1962, *Education and the Cult of Efficiency*, University of Chicago Press, Chicago, IL.
Conant, J.B.: 1945, *General Education in a Free Society: Report of the Harvard Committee*, Harvard University Press, Cambridge, MA.
Crane, L.T.: 1976, *The National Science Foundation & Pre-College Science Education: 1950–1975*, US Government Printing Office, Washington, DC.
Cromer, A.: 1993, *Uncommon Sense: The Heretical Nature of Science*, Oxford University Press, New York.
Dearden, R.F.: 1967, 'Instruction and Learning by Discovery'. In R.S. Peters (ed.) *The Concept of Education*, Routledge & Kegan Paul, London, pp. 135–155.
DeBoer, G.E.: 1991, *A History of Ideas in Science Education*, Teachers College Press, New York.
DeBoer, G.E.: 2014, 'Joseph Schwab: His Work and His Legacy'. In M.R. Matthews (ed.) *International Handbook of Research in History, Philosophy and Science Teaching*, Springer, Dordrecht, The Netherlands, pp. 2433–2458.
Dewey, J.: 1910, 'Science as Subject-Matter and as Method', *Science* 31, 121–127. Reproduced in *Science & Education*, 1995, 4(4), 391–398.
Donnelly, J.F.: 2001, 'Contested Terrain or Unified Project? "The Nature of Science" in the National Curriculum for England and Wales', *International Journal of Science Education* 23(2), 181–195.
Education Policies Commission: 1966, *Education and the Spirit of Science*, National Education Association, Washington, DC.
Eisner, E.: 1979, *The Educational Imagination: On the Design and Evaluation of School Programs*, Macmillan, New York.

Elbers, G.W. and Duncan, P. (eds): 1959, *The Scientific Revolution: Challenge and Promise*, Public Affairs Press, Washington, DC.
Fantoli, A.: 1994, *Galileo: For Copernicanism and for the Church* (trans. G.V. Coyne), Vatican Observatory Publications, Vatican City (distributed by University of Notre Dame Press).
Finocchiaro, M.A.: 1989, *The Galileo Affair: A Documentary History*, University of California Press, Berkeley, CA.
Finocchiaro, M.A.: 2005, *Retrying Galileo: 1633–1992*, University of California Press, Berkeley, CA.
Flude, M. and Hammer, M.: 1990, *The Education Reform Act 1988*, Falmer Press, Basingstoke, UK.
Gauld, C.F.: 2005, 'Habits of Mind, Scholarship and Decision-Making in Science and Religion', *Science & Education* 14(3–5), 291–308.
Glass, B.: 1970, *The Timely and the Timeless: The Interrelations of Science Education and Society*, Basic Books, New York.
Good, R.G.: 2005, *Scientific and Religious Habits of Mind*, Peter Lang, New York.
Graham, D.: 1993, *A Lesson for Us All*, Routledge, London.
Harlen, W.: 1996, *The Teaching of Science in Primary Schools*, David Fulton, London.
Harms, N.C. and Yager, R.E.: 1981, *What Research Says to the Science Teacher*, Vol.3, NSTA, Washington, DC.
Harris, D. and Taylor, M.: 1983, 'Discovery Learning in School Science: The Myth & the Reality', *Journal of Curriculum Studies* 15, 277–289.
Helgeson, S.L., Blosser, P.E. and Howe, R.W.: 1977, *The Status of Pre-College Science, Mathematics, and Social Science Education: 1955–1975*, US Government Printing Office, Washington.
Herron, M.D.: 1971, 'The Nature of Scientific Inquiry', *School Review* 79, 170–212.
Hodson, D.: 1987, 'Social Control As a Factor in Science Curriculum Change', *International Journal of Science Education* 9, 529–540.
Hogben, L.: 1940, *Science for the Citizen*, 2nd edn, George, Allen & Unwin, London (1st edition, 1938).
Holmyard, E.J.: 1922, *Inorganic Chemistry: A Textbook for Schools and Colleges*, Edward Arnold, London.
Holmyard, E.J.: 1924, 'The Historical Method of Teaching Chemistry', *School Science Review* 20(5), 227–233.
Holton, G.: 1986, '"A Nation At Risk" Revisited'. In his *The Advancement of Science and Its Burdens*, Cambridge University Press, Cambridge, UK, pp. 253–278.
Hurd, P.D.: 1961, *Biological Education in American Secondary Schools 1890–1960*, American Institute of Biological Science, Washington, DC.
Hurd, P.D.: 1985, 'A Rationale for a Science, Technology, and Society Theme in Science Education'. In R.W. Bybee (ed.) *Science, Technology, Society, Yearbook of the National Science Teachers Association*, NSTA, Washington, DC, pp. 94–101.
Huxley, T.H.: 1868/1964, 'A Liberal Education; and Where to Find It'. In his *Science and Education*, Appleton, New York, 1897 (orig. 1885). Reprinted with Introduction by C. Winick, Citadel Press, New York, 1964, pp. 72–100.
Huxley, T.H.: 1885/1964, *Science and Education*, The Citadel Press, New York.
Jackson, P.W.: 1983, 'The Reform of Science Education: A Cautionary Tale', *Daedalus* 112(2), 143–166.
Jenkins, E.W.: 1979, *From Armstrong to Nuffield*, John Murray, London.
Jenkins, E.W.: 2013, 'The "Nature of Science" in the School Curriculum: The Great Survivor', *Journal of Curriculum Studies* 45(2), 132–151.
Jenkins, E.W.: 2014, 'E.J. Holmyard and the Historical Approach to Science Teaching'. In M.R. Matthews (ed.) *International Handbook of Research in History, Philosophy and Science Teaching*, Springer, Dordrecht, The Netherlands, pp. 2383–2408.
Joad, C.E.M.: 1935, *The Book of Joad: A Belligerent Autobiography*, Faber & Faber, London.

Kang, N.H.: 2008, 'Learning to Teach Science: Personal Epistemologies, Teaching Goals, and Practices of Teaching', *Teaching and Teacher Education* 24, 478–498.
Kelly, G.J.: 2014, 'Inquiry Teaching and Learning: Philosophical Considerations'. In M.R. Matthews (ed.) *International Handbook of Research in History, Philosophy and Science Teaching*, Springer, Dordrecht, The Netherlands, pp. 1363–1380.
Kirschner, P., Sweller, J. and Clark, R.E.: 2006, 'Why Minimally Guided Learning Does Not Work: An Analysis of the Failure of Discovery Learning, Problem-Based Learning, Experiential Learning and Inquiry-Based Learning', *Educational Psychologist* 41(2), 75–96.
Klopfer, L.E.: 1964, 'The Use of Case Histories in Science Teaching', *School Science and Mathematics*, November, 660–666. In H.O. Andersen (ed.) *Readings in Science Education for the Secondary School*, Macmillan, New York, 1969, pp. 226–233.
Klopfer, L.E. and Champagne, A.B.: 1990, 'Ghosts of Crisis Past', *Science Education* 74(2), 133–154.
Klopfer, L.E. and Cooley, W.W.: 1963, 'Effectiveness of the History of Science Cases for High Schools in the Development of Student Understanding of Science and Scientists', *Journal of Research in Science Teaching* 1, 35–47.
Lakatos, I.: 1970, 'Falsification and the Methodology of Scientific Research Programmes'. In I. Lakatos and A. Musgrave (eds) *Criticism and the Growth of Knowledge*, Cambridge University Press, Cambridge, UK, pp. 91–196.
Laudan, L.: 1983, 'The Demise of the Demarcation Problem'. In R.S. Cohen and L. Laudan (eds) *Physics, Philosophy and Psychoanalysis*, Reidel, Dordrecht, The Netherlands, pp. 111–127.
Layton, A.D. and Powers, S.R.: 1949, *New Directions in Science Teaching*, McGraw-Hill, New York.
Layton, D.: 1973, *Science for the People. The Origins of the School Science Curriculum in England*, George Allen & Unwin, London.
Lederman, N.G., Kuerbis, P.J., Loving, C.C., Ramey-Gassert, L., Roychoudhury, A. and Spector, B.S.: 1997, 'Professional Knowledge Standards for Science Teacher Educators', *Journal of Science Teacher Education* 8(4), 233–240.
Léna, P.: 2009, 'Editorial: Europe Rethinks Education', *Science* 324, 501.
McComas, W.F.: 2014, 'Nature of Science in the Science Curriculum and in Teacher Education Programmes in the United States'. In M.R. Matthews (ed.) *International Handbook of Research in History, Philosophy and Science Teaching*, Springer, Dordrecht, The Netherlands, pp. 1993–2023.
Mach, E.: 1886/1986, 'On Instruction in the Classics and the Sciences'. In his *Popular Scientific Lectures*, Open Court Publishing, LaSalle, IL, pp. 338–374.
Machamer, P.: 1992, 'Philosophy of Science: An Overview for Educators'. In R.W. Bybee, J.D. Ellis, J.R. Giese and L. Parisi (eds) *Teaching About the History and Nature of Science and Technology: Background Papers*, BSCS/SSEC, Colorado Springs, pp. 9–18. Reprinted in *Science & Education* 7(1), 1998, 1–11.
McMullin, E. (ed.): 2005, *The Church and Galileo*, University of Notre Dame Press, Notre Dame, IN.
Mahner, M.: 2007. 'Demarcating Science from Pseudoscience'. In T. Kuipers (ed.) *Handbook of the Philosophy of Science: General Philosophy of Science – Focal Issue*, Elsevier, Amsterdam, pp. 515–575.
Mann, C.R.: 1912, *The Teaching of Physics for Purposes of General Education*, Macmillan, New York.
Mansell, A.E.: 1976, 'Science for All', *School Science Review* 57, 579–585.
Mayer, R.E.: 2004, 'Should There be a Three-Strikes Rule Against Pure Discovery Learning? The Case for Guided Methods of Instruction', *American Psychologist* 59(1), 14–19.
Metz, D.: 2014, 'The History and Philosophy of Science in Science Curricula and Teacher Education in Canada'. In M.R. Matthews (ed.) *International Handbook of Research in History, Philosophy and Science Teaching*, Springer, Dordrecht, The Netherlands, pp. 2025–2043.

Millar, R. and Osborne, J.: 1998, *Beyond 2000: Science Education for the Future*, School of Education, King's College, London.
Monk, M. and Osborne, J.: 1997, 'Placing the History and Philosophy of Science on the Curriculum: A Model for the Development of Pedagogy', *Science Education* 81(4), 405–424.
Novak, J.D.: 1977, *A Theory of Education*, Cornell University Press, Ithaca, NY (paperback edn, 1986).
NCC (National Curriculum Council): 1988, *Science in the National Curriculum*, NCC, York, UK.
NCC (National Curriculum Council): 1991, *Science for Ages 5 to 16*, DES, London.
NCEE (National Commission on Excellence in Education): 1983, *A Nation At Risk: The Imperative for Education Reform*, US Department of Education, Washington, DC.
NRC (National Research Council): 1996, *National Science Education Standards*, National Academies Press, Washington, DC.
NRC (National Research Council): 2000, *Inquiry and the National Science Education Standards: A Guide for Teaching and Learning*, National Academies Press, Washington, DC.
NRC (National Research Council): 2007, *Taking Science to School. Learning and Teaching Science in Grades K-8*, National Academies Press, Washington, DC.
NRC (National Research Council): 2012, *A Framework for K-12 Science Education: Practices, Crosscutting Concepts, and Core Ideas*, National Academies Press, Washington, DC.
NRC (National Research Council): 2013, *Next Generation Science Standards*, National Academies Press, Washington, DC.
NSTA (National Science Teachers Association): 1971, *School Science Education for the '70s*, NSTA, Washington, DC.
Nunn, T.P.: 1907, *The Aims and Achievements of the Scientific Method*, Macmillan, New York.
Pennock, R.T.: 2011, 'Can't Philosophers Tell the Difference Between Science and Religion? Demarcation Revisited', *Synthese* 178(2), 177–206.
Piel, E.J.: 1981, 'Interaction of Science, Technology, and Society in Secondary Schools'. In N.C. Harms and R.E. Yager (eds) *What Research Says to the Science Teacher*, Vol.3, NSTA, Washington, DC, pp. 94–112.
Poupard, P. (ed.): 1987, *Galileo Galilei: Toward a Resolution of 350 Years of Debate – 1633–1983*, Duquesne University Press, Pittsburgh, PA.
Praagh, G. van: 1949, *Chemistry by Discovery*, Murray, London.
PRIMAS: 2013, *Inquiry-Based Learning in Maths and Science*, European Union, Freiburg, Germany.
Pumfrey, S.: 1991, 'History of Science in the British National Science Curriculum: A Critical Review of Resources and Their Aims', *British Journal for the History of Science* 24, 61–78.
Quine, W.V.O.: 1960, *Word and Object*, MIT Press, Cambridge, MA.
Raizen, S.A.: 1991, 'The Reform of Science Education in the U.S.A.: Déjà Vu or De Nova', *Studies in Science Education* 19, 1–41.
Redondi, P.: 1988, *Galileo Heretic*, Allen Lane, London.
Roberts, D.A.: 1982, 'Developing the Concept of "Curriculum Emphases" in Science Education', *Science Education* 66, 243–260.
Rocard, M., Osermely, P., Jorde, D., Lenzen, D. and Walberg-Henniksson, H.: 2007, *Science Education Now: A Renewed Pedagogy for the Future of Europe*, European Commission, Brussels.
Rosenthal, D.B. and Bybee, R.W.: 1987, 'Emergence of the Biology Curriculum: A Science of Life or a Science of Living?' In T.S. Popkewitz (ed.) *The Formation of the School Subjects: The Struggle For Creating an Amercan Institution*, Falmer Press, New York, pp. 123–144.

Rosenthal, D.B.: 1985, 'Biology Education in a Social and Moral Context'. In R.W. Bybee (ed.) *Science, Technology, Society, Yearbook of the National Science Teachers Association*, NSTA, Washington, DC, pp. 102–116.

Rudolph, J.L.: 2002, *Scientists in the Classroom: The Cold War Reconstruction of American Science Education*, Palgrave, New York.

Rudolph, J.L.: 2008, 'Historical Writing on Science Education: A View of the Landscape', *Studies in Science Education* 44(1), 63–82.

Rutherford, F.J.: 1964, 'The Role of Inquiry in Science Teaching', *Journal of Research in Science Teaching* 2, 80–84. Reprinted in W.D. Romey (ed.) *Inquiry Techniques for Teaching Science*, Prentice Hall, Englewood Cliffs, NJ, 1968, pp. 264–270.

Rutherford, F.J. and Ahlgren, A.: 1990, *Science for All Americans*, Oxford University Press, New York.

Sadler, T.D. (ed.): 2011, *Socio-scientific Issues in the Classroom: Teaching, Learning and Research*, Springer, Dordrecht, The Netherlands.

Schulz, R.M.: 2014, 'Philosophy of Education and Science Education: A Vital but Underdeveloped Relationship'. In M.R. Matthews (ed.) *International Handbook of Research in History, Philosophy and Science Teaching*, Springer, Dordrecht, The Netherlands, pp. 1259–1315.

Schwab, J.J.: 1960, 'The Teaching of Science as Enquiry'. In J.J. Schwab and P. Brandwein (eds) *The Teaching of Science*, Harvard University Press, Cambridge, MA, pp. 1–103.

Shulman, L.S. and Keislar, E.R. (eds): 1966, *Learning by Discovery: A Critical Appraisal*, Rand McNally, Chicago, IL.

Shavelson, R.J. and Towne, L. (eds): 2002, *Scientific Research in Education*, National Academy Press, Washington, DC.

Shymansky, J.A., Hedges, L.V. and Woodworth, G.: 1990, 'A Reassessment of the Effects of Inquiry-Based Science Curricula of the 60's on Student Performance', *Journal of Research in Science Teaching* 27(2), 127–144.

Solomon, J.: 1985, 'Science in a Social Context: Details of a British High School Course', in R.W. Bybee (ed.) *Science, Technology, Society, Yearbook of the National Science Teachers Association*, NSTA, Washington, DC, pp. 144–157.

Stevens, P: 1978, 'On the Nuffield Philosophy of Science', *Journal of Philosophy of Education* 12, 99–111.

Strike, K.A.: 1975, 'The Logic of Learning by Discovery', *Review of Educational Research* 45, 461–483.

Sund, R.B. and Trowbridge, L.W. (eds): 1967, *Teaching Science by Inquiry*, Charles Merrill, Columbus, OH.

Swinbank, E. and Taylor, J. (eds): 2007, *Perspectives on Science: The History, Philosophy and Ethics of Science*, Heinemann, Harlow, UK.

Taylor, J.L. and Hunt, A.: 2014, 'History and Philosophy of Science and the Teaching of Science in England'. In M.R. Matthews (ed.) *International Handbook of Research in History, Philosophy and Science Teaching*, Springer, Dordrecht, The Netherlands, pp. 2045–2081.

Uglow, J.: 2002, *The Lunar Men: Five Friends Whose Curiosity Changed the World*, Faber & Faber, London.

Vesterinen, V.-M., Manassero-Mas, M.-A. and Vázquez-Alonso, A.: 2014, 'History, Philosophy and Sociology of Science and Science-Technology-Society Traditions in Science Education: Continuities and Discontinuities'. In M.R. Matthews (ed.) *International Handbook of Research in History, Philosophy and Science Teaching*, Springer, Dordrecht, The Netherlands, pp. 1895–1925.

Waring, M.: 1979, *Social Pressures & Curriculum Innovation: A Study of the Nuffield Foundation Science Teaching Project*, Methuen, London.

Welch, W.W.: 1979, 'Twenty Years of Science Education Development: A Look Back', *Review of Research in Education* 7, 282–306.

Welch, W.W., Klopfer, L., Aikenhead, G. and Robinson, J.: 1981, 'The Role of Inquiry in Science Education: Analysis and Recommendations', *Science Education* 65(1), 33–50.

Wellington, J.J.: 1981, 'What's Supposed to Happen, Sir? – Some Problems with Discovery Learning', *School Science Review* 63(222), 167–173.
Westaway, F.W.: 1929, *Science Teaching*, Blackie and Son, London.
Woodhull, J.F.: 1910, 'The Teaching of Physical Science', *Teachers College Record* 11(1), 1–82.
Yager, R.E. (ed.): 1993, *The Science, Technology, Society Movement*, NSTA, Washington, DC.
Yager, R.E. (ed.): 1996, *Science/Technology/Society as Reform in Science Education*, SUNY Press, Albany, NY.
Young, M.F.D.: 1976, 'The Schooling of Science'. In G. Whitty and M.F.D. Young (eds) *Explorations in the Politics of School Knowledge*, Nafferton Books, Driffield, UK, pp. 47–61.
Zacharias, J.R.: 1964, 'Curriculum Reform in the USA'. In S.C. Brown, N. Clarke and J. Tiomno (eds) *Why Teach Physics: International Conference on Physics in General Education*, MIT Press, Cambridge MA, pp. 66–70.
Zeidler, D.L. and Sadler, T.D. (eds): 2008, 'Social and Ethical Issues in Science Education', Special issue of *Science & Education* 17(8–9).
Ziman, J.: 1968, *Public Knowledge: The Social Dimension of Science*, Cambridge University Press, Cambridge, UK.
Ziman, J.: 1980, *Teaching and Learning about Science and Society*, Cambridge University Press, Cambridge.
Ziman, J.: 1994, 'The Rationale of STS Education Is in the Approach'. In J. Solomon and G. Aikenhead (eds) *STS Education: International Perspectives on Reform*, Teachers College Press, New York, pp. 21–31.

Chapter 4

History of Science in the Curriculum and in Classrooms

In the middle of the nineteenth century, the Duke of Argyll, in his presidential address to the BAAS, stated that 'what we want in the teaching of the young, is, not so much mere results, as the methods and above all, the history of science' (Jenkins 1990, p. 274). The Duke's exhortation has been more ignored than followed, but there has been a minority tradition in science education that has attempted to bring something of the history of science into science instruction. Leo Klopfer, long active on this task in the US, made the following melancholy observation about this tradition:

> Proposals for weaving the history and nature of science into the teaching of science in schools and colleges have a history of more than sixty years. Over this long period, various kinds of instructional materials which entwine science and the history of science were produced. The historical accounts, lessons, or units usually served to convey a philosophy of science in which educators believed at the time. Their philosophy of science identified ideas about the nature of science which they wished students to understand or appreciate. These ideas anchored a web, and the strands of science content and science history formed the web's pattern. Yet each of these webs was fragile; they rarely persisted for very long and left little trace on the science education landscape.
>
> (Klopfer 1992, p. 105)

This chapter will outline the changing fortune of the history of science in science curricula and will illustrate some of the arguments for the inclusion of history in science programmes. To illustrate these arguments, it will contrast historical with 'professional' or 'technical' approaches to the teaching of air pressure. Finally, it will consider and reject some arguments that have been raised by scientists and historians against the inclusion of history in the science curriculum.

Reasons for History

At different times and places, there have been appeals to the following reasons for including a historical component in science programmes:

1 History promotes the better comprehension of scientific concepts and methods.
2 Historical approaches connect the development of individual thinking with the development of scientific ideas.
3 History of science is intrinsically worthwhile. Important episodes in the history of science and culture should be familiar to all students.
4 History is necessary to understand the nature of science.
5 History, by examing the life and times of individual scientists, humanises the subject matter of science, making it less abstract and more engaging for students.
6 History allows connections to be made within topics and disciplines of science, as well as with other academic disciplines; history displays the integrative and interdependent nature of human achievements.

History Promotes Conceptual Comprehension

Underlying the first argument is the belief that well-founded understanding is necessarily historical. The importance of history for the proper understanding of social institutions, such as political parties and churches, or of social customs and mores, such as marriage rites and associated laws, is widely appreciated. It is less well recognised that the same considerations apply to understanding the intellectual products of science. Ernst Mach's view was that: 'Historical investigation not only promotes the understanding of that which now is, but also brings new possibilities before us' (Mach 1883/1960, p. 316). For Mach, the perspective of history allows people generally, and scientists in particular, to locate themselves in a tradition of thought, and to see how their concepts and the intellectual frameworks that give them meaning are historically conditioned. Thus, the historical perspective encourages the having of new ideas, and novel conceptualisations. He recognised that the same disposition could be developed by classical studies:

> A person who has read and understood the Greek and Roman authors has felt and experienced more than one who is restricted to the impressions of the present. He sees how men placed in different circumstances judge quite differently of the same things from what we do today. His own judgments will be rendered thus more independent.
>
> (Mach 1886/1986, p. 347)

In support of Mach's view, Albert Einstein writes, in his autobiographical essay, about the grip that the mechanical worldview had upon all scientists of his generation, including Maxwell and Hertz. He says that, 'It was Ernst Mach who, in his *History of Mechanics*, shook this dogmatic faith; this book exercised a profound influence upon me in this regard' (Schilpp 1951, p. 21).

The importance of a historical perspective for understanding has generally been more widely recognised in Continental thought than it has in British and American writing. Ludwik Fleck,[1] in a book that was instrumental in the development of Thomas Kuhn's philosophy of science, succinctly states this view as follows:

> There can be no ahistorical understanding, that is to say an understanding separated from history, just as there can be no asocial act of understanding performed by an isolated researcher.
>
> (Fleck 1935/1979, in Sibum 1988, p. 139)

More recently, Ernst Mayr, in the opening pages of his *The Growth of Biological Thought*, commends historical study to scientists in these terms:

> I feel that the study of the history of a field is the best way of acquiring an understanding of its concepts. Only by going over the hard way by which these concepts were worked out – by learning all the earlier wrong assumptions that had to be refuted one by one, in other words by learning all past mistakes – can one hope to acquire a really thorough and sound understanding. In science one learns not only by one's own mistakes but by the history of the mistakes of others.
>
> (Mayr 1982, p. 20)

Conceptual Change in Individuals and in Science

The second argument holds that, not only does a historical perspective allow students to situate their concepts and conceptual schemes on the larger canvas of intellectual systems and the history of scientific ideas, but also, historical presentation is grounded in certain psychological realities about the development of individual cognition. Ernst Mach was a strong advocate of this genetic method. This argument claims that the development of individual cognition in some way naturally mirrors the development of species cognition. Hegel was perhaps the first to enunciate this idea; Herbert Spencer followed him. In England, the chemist and textbook writer J.C. Hogg wrote, in his 1938 text, that:

> The historic development is a logical approach. The slow progress of the early centuries was owing to a lack of knowledge, to poor technique and to unmethodical attack. But these are precisely the difficulties of the beginner in chemistry. There is a bond of sympathy between the beginner and the pioneer.
>
> (Hogg 1938, p. vii)

Fifty years later James Wandersee (1985) suggested ways in which knowledge of the historical development of a discipline can assist teachers in anticipating and understanding the difficulties that contemporary students have with learning subjects. The history can also suggest questions and experiments that promote appropriate conceptual change in students. There are approximately 3,000 published studies on children's misconceptions in science; the information on the resistance of science learning to science instruction is overwhelming (Duit 2009). Knowledge of the slow and difficult path traversed in the historical development of particular sciences can assist teachers planning the organisation of a programme, the choice of experiments and activities, their responses to classroom questions and puzzles, the 'redoing' of original experiments and reliving of historical interpretations and debate about the experiments.

There have been numerous studies on the utilisation of history and historical resources (papers, experiments, life stories) in science teaching. Table 4.1 lists just some representative research.

History of Science is Intrinsically Worthwhile

This third argument has not been advanced as much as it needs to be. To its credit, *Project 2061* does advance the argument and lists the following ten episodes that should be known and appreciated by all students who have had a good education:

> The emphasis here is on ten accounts of significant discoveries and changes that exemplify the evolution and impact of scientific knowledge: the planetary earth,

Table 4.1 Utilising History of Science in Pedagogy: Some Studies

Topic	*Studies*
Science (general)	Cohen (1950), Conant (1947), Finocchiaro (1980), Kindi (2005), Klopfer (1992), Kokkotas *et al.* (2011), Lennox & Kampourakis (2013), Stinner *et al.* (2003), Wandersee (1985)
Physics (general)	Brush (1969), Hong & Lin-Siegler (2012), Jung (1983), Seroglou & Koumaras (2001)
Optics	Andreou & Raftopoulos (2011), Galili (2014), Galili & Hazan (2001), Kipnis (1992), Mihas & Andreadis (2005)
Oxygen and combustion	Cartwright (2004), Pumfrey (1987)
Genetics	Burian (2013), El-Hani *et al.* (2014), Jamieson & Radick (2013)
Electricity	Binnie (2001), Leone (2014), Sibum (1988)
Chemistry	Chamizo (2007), Chang (2010), Kauffman (1989), Padilla & Furio-Mas (2008)
Quantum theory	Garritz (2013), Greca & Friere (2014), Kragh (1992)
Evolution	Gauld (1992), Jensen & Finley (1995), Kampourakis (2013)
Mathematics	Fauvel (1990), Gulikers & Blom (2001), Panagiotou (2011)
Thermodynamics	Besson (2014), Cotignola *et al.* (2002), De Berg (2008)
Relativity	Levrini (2014), Villani & Arruda (1998)
Mechanics	Besson (2013), Coelho (2007, 2009), Gauld (1998), Kalman (2009), Schecker (1992)

> universal gravitation, relativity, geologic time, plate tectonics, the conservation of matter, radioactivity and nuclear fission, the evolution of species, the nature of disease, and the Industrial Revolution.
>
> (AAAS 1989, p. 111)

Others can easily be added – genetics and heredity leap out, as does the connection of early modern science to the Enlightenment. Unfortunately, most countries allow students to complete history courses without any knowledge of major scientific, mathematical and technical achievements, which constitute some of the most important episodes in the development of civilisation. If as much history time were devoted to the scientific revolution as to political revolutions, to Mendel and genetics as to generals, to the development of time-keeping as to the development of constitutions – then the overall education of society would be considerably advanced, and the 'two-cultures' gap lamented by C.P. Snow would be less apparent (Snow 1963).

History is Required to Understand the Nature of Science

The fourth argument has also been advocated in detail by *Project 2061* and discussed in the previous chapter. It was also elaborated in earlier discussion on the linkage between history of science and philosophy of science. This book endorses the qualified position that historicised philosophy of science is required for educational purposes.

History Humanises the Subject Matter of Science

This fifth argument has often been advanced in reaction to widespread abuse of science, and in reaction to authoritarian teaching practices sometimes associated with naive understandings of science. Some historical study can counteract the revulsion for science and technology felt by many witnesses of high-tech wars, sonar-guided whale kills, napalm attacks and so on. The lives and times of the great and not-so-great scientists are usually full of interesting and appealing incidents and issues that students can read about, debate and re-enact. This might go against the 'no heroes' school of history writing, but members of the public vote with their wallets. It is not accidental that biographies of scientists have got on to 'bestseller' lists and are published in multiple languages: Galileo (Heilbron 2010), Newton (Westfall 1980), Harrison (Sobel 1994), Galton (Gillham 2001), Darwin (Desmond & Moore 1991), Planck (Heilbron 1986), Einstein (Pais 1982), Curie (Pflaum 1989) and Bohr (Pais 1991), to just name the more obvious.

The Dava Sobel case is illustrative. Academics had extensively researched the 'longitude problem' and published articles and books that barely sold (Matthews 2000b, Chapter 7). Sobel, the journalist, attended one Harvard conference on the subject and went away, did some work of her own and published a multi-translated international bestseller on the subject (Sobel

1994). This brought a history of science to the widest possible audience. There are lessons to be learned by both historians and educators concerning the popularisation of science (Gascoigne 2007).

The use of role-play and drama, from elementary-school level through to senior years, has been very successful. Students who may not remember much about Planck's constant can remember that, as director of the Kaiser Wilhelm Institute during the Third Reich, he was faced with the painful 'The Dilemmas of an Upright Man', to use the subtitle of John Heilbron's masterful biography (Heilbron 1986). Science teachers can combine with history teachers and drama teachers to enact something of the circumstances and dilemmas facing Planck and give some dramatic expression to his resolution of them.

History is a way of putting a face on what otherwise is just foreign terminology. Boyle's law, Ohm's law, Newton's laws, Hooke's law, Curie's discoveries, Mach bands, Planck's constant, units of measurement such as volts and ohms, and so on. James Wandersee has successfully incorporated historical vignettes into science programmes. These can be made as sophisticated as the class and resources allow (Wandersee & Roach 1998).

History Promotes Curriculum Connections

The sixth argument, that history integrates the sciences and other disciplines, has been the backbone of liberal approaches to the teaching of science. The integrative function of history was recognised by Percy Nunn, James Conant, Gerald Holton and others. It was one of the central planks of the Harvard Committee Report, *General Education in a Free Society* (Conant 1945), and was prominent in the Harvard Project Physics programme (Holton 1978, 2003). Science has developed in conjunction with mathematics, philosophy, technology, theology, commerce, art and literature. In turn, it has affected each of these fields. History allows science programmes to reveal to students something of this rich tapestry and engender their appreciation of the interconnectiveness of human intellectual and practical endeavours.

Galileo's physics was dependent upon Euclidean geometry and the then just translated mechanical analyses of Archimedes (brought to Italy by those fleeing the Turkish invasion of Constantinople). It was also dependent upon technological advances, lens grinding and the telescope being the most obvious. His philosophy allowed him to first understand, and then to break from, central Aristotelian concepts that constrained the physics of those around him. His theological views also freed him to investigate the heavens and to experiment with falling objects. Even music had a role to play, as in the timing of rolling bodies. And, of course, patronage, commerce and communications all contributed to Galileo's achievements. In turn, his new, mathematical, experimental physics had an enormous effect on further physics, philosophy, technology, commerce, mathematics and theology.

The same rich pattern of influence and effect can be seen in the achievements of Newton, Darwin and Einstein, to name just the most obvious. A historical approach to science allows students to connect the learning of specific scientific topics with their learning of mathematics, literature, political history, theology, geography, philosophy, art and so on. When the richness of science's history is appreciated, then collaboration between science teachers and the teachers of other subjects can fruitfully be encouraged, and engaging examples can thus be given to students.[2]

History in US Science Curricula: The Conant Legacy

In the US, as Leo Klopfer remarked, 'Proposals for weaving the history and nature of science into the teaching of science in schools and colleges have a history of more than 60 years' (Klopfer 1992, p. 105).[3] In the 1920s, some chemists, following Holmyard's example in the UK, advocated a historical approach in science instruction, and a number of historically oriented texts were written.[4]

After World War Two, the generalist or contextual approach to science teaching gained momentum. The dominant influence here was the president of Harvard University, James B. Conant, whose case-study approach to science education was widely adopted. He developed this while in charge of undergraduate general education at Harvard and popularised it in a widely distributed government report, *General Education in a Free Society* (Conant 1945), and paperback bestsellers (Conant 1947, 1951). His two-volume *Harvard Case Histories in Experimental Science* (Conant 1948) became a popular university textbook. The *General Education* report proposed that:

> Science instruction in general education should be characterized mainly by broad integrative elements – the comparison of scientific with other modes of thought, the comparison and contrast of the individual sciences with one another, the relations of science with its own past and with general human history, and of science with problems of human society.
>
> (Conant 1945, p. 155)

It then identifies crucial features of science, its abstractness and tradition-dependence, that make learning difficult and that, as Mach noted, history can render more intelligible:

> The facts of science and the experience of the laboratory no longer can stand by themselves; they no longer represent simple, spontaneous, and practical elements directly related to the everyday life of the student. As they become further removed from his experience, more subtle, more abstract, the facts must be learned in another context, cultural, historical and philosophical. Only such broader perspectives can give point and lasting value to scientific information and experience for the general student.
>
> (Conant 1945, p. 155)

Conant's influence cannot be overestimated (Hershberg 1993). Thomas Kuhn, in the Preface of his first book, *The Copernican Revolution*, which arose from his lectures in Harvard's General Education programme, says that:

> Work with him [Conant] first persuaded me that historical study could yield a new sort of understanding of the structure and function of scientific research. Without my own Copernican revolution, which he fathered, neither this book nor my other essays in the history of science would have been written.
>
> (Kuhn 1957, p. xi)

For good or bad, Kuhn's personal transformation caused a massive transformation of professional history, philosophy and sociology of science – indeed, of many other academic fields. For the past half-century, the ambiguous and oft-misunderstood idea of 'paradigm' and its associated relativist epistemology and idealist ontology has been a loose cannon on the scholarly deck.[5]

Gerald Holton makes a similar admission of debt. Holton was subsequently instrumental in developing, in the early 1960s – with Stephen Brush, Fletcher Watson, James Rutherford and others – the Harvard Project Physics course for secondary schools. Holton produced a number of substantial defences of the liberal view of science education (Holton 1975, 1978) and wrote a college-physics text embodying historical and philosophical themes (Holton 1952).

The then-young physics graduate I. Bernard Cohen worked with Conant on his Yale Invitation Lectures, subsequently published as *On Understanding Science: An Historical Approach* (Conant 1947). This hugely popular book, among other things, argued that the history of science was indispensable for the understanding of science. Cohen then worked with Conant's Harvard Committee that produced the above-mentioned 1945 report, the famous 'red book'. Additionally, Cohen wrote a substantial essay, in 1950, on the importance of history for the teaching of science.[6] After the war, Conant organised a series of conferences of teachers of chemistry and physics, plus historians of science. One outcome of these conferences was the collection *Science in General Education* (McGrath 1948).

The success of Conant's Harvard Case Studies in college courses and the example of Joseph Schwab's historical text-based science course at the University of Chicago (Schwab 1950) prompted Leo Klopfer, then at the University of Chicago, to emulate the approach in the teaching of secondary science. His rationale was presented in articles with Fletcher Watson, who was later to work on Harvard Project Physics (Klopfer & Watson 1957). One of their major concerns was to increase students' understanding of the scientific enterprise and of its interactions with society. Their disquiet concerned students' poor grasp of what would shortly be labelled 'scientific literacy' – a term introduced by Hurd (1958) and Fitzpatrick (1960). They saw historical studies as a way of expanding and enriching students' understanding of science. Klopfer later said that scientific literacy encompasses five components:

- knowledge of significant science facts, concepts, principles and theories;
- the ability to apply relevant science knowledge in situations of everyday life;
- an understanding of general ideas about the organisation of the scientific enterprise, the important interactions of science, technology and society, and the characteristics of scientists;
- the ability to utilise the processes of scientific enquiry and an understanding of the nature of scientific enquiry;
- the possession of informed attitudes and interests related to science (Klopfer 1990, p. 3).

Klopfer and Watson produced a course of *History of Science Cases for Schools* (HOSC) (Klopfer 1969b).[7] Each of eight cases was presented in a separate booklet containing the historical narrative, quotations from scientists' original papers, pertinent student experiments and exercises, marginal notes and questions, and space for students to write answers to questions. Teachers' guides and supplementary material were also produced. The experimental version was tested and evaluated in 108 classes, with encouraging results:

> The [HOSC] method is definitely effective in increasing student understanding of science and scientists when used in biology, chemistry, and physics classes in high schools . . . moreover . . . they achieve these significant gains in understanding of science and scientists with little or no concomitant loss of achievement in the usual content of high school science courses.
>
> (Klopfer & Cooley 1963, p. 46)

After this success, the individual case studies were produced over a number of years and published by Science Research Associates in Chicago (Klopfer 1964–1966); a version was also published by Wadsworth, San Francisco (Klopfer 1969b). Despite their initial success, they seem to have been one of the webs that, as Klopfer said, 'rarely persisted for very long and left little trace on the science education landscape'.

The liberal or generalist programme was endorsed by the National Society for the Study of Education in its fifty-ninth yearbook, where it was advised that:

> A student should learn something about the character of scientific knowledge, how it has developed, and how it is used. He must see that knowledge has a certain dynamic quality and that it is quite likely to shift in meaning and status with time.
>
> (NSSE 1960)

However, as the yearbook was being written, the curricular and social times were changing. The National Science Foundation was formed in 1950 and made its first grant for the development of a high-school science curriculum in 1956. This was to the PSSC at the Massachusetts Institute of

Technology (MIT), whose draft text was published at the same time that the Soviet *Sputnik* was launched. There quickly followed the spate of NSF-funded curricular projects previously discussed, in which historical, technological and cultural matters were ignored. The emphasis was upon the mastery of science content in its most theoretical form. The NSF had a professional or technical or disciplinary approach to school science, in contrast to the generalist or humanistic or contextual approach recommended in the NSSE 1960 yearbook. The NSF's credo is stated in *Policies for Science Education*, prepared in 1960 by the Science Manpower Project at Teachers College, Columbia University:

> Let us note that [education] is the basic factor upon which an adequate science manpower supply depends. We must have improved science-education programs in the schools. . . . Then and only then, will we secure a flow of new scientific and technological personnel adequate to meet the present and projected needs of our culture.
>
> (Fitzpatrick 1960, p. 195)

These science curriculum reforms of the early 1960s proceeded without the participation of either historians or philosophers of science. There were two prominent exceptions, the Harvard Project Physics course and the Yellow version of the BSCS high-school biology. Less prominent were the Klopfer and Cooley case studies for high school developed in the period from 1956 to 1960.

The BSCS text was informed by the ideas of the previously mentioned Chicago University biologist–philosopher–educationalist J.J. Schwab. He wrote an influential essay on 'The Nature of Scientific Knowledge as Related to Liberal Education' (Schwab 1949), and he vigorously promoted the Deweyean idea of 'science as enquiry'.[8] Schwab wrote the *Teachers' Handbook* for the BSCS curriculum, in which he advocated the historical approach, saying that,

> the essence of teaching of science as enquiry would be to show some of the conclusions of science in the framework of the way they arise and are tested . . . [it] would also include a fair treatment of the doubts and incompleteness of science.
>
> (Schwab 1963, p. 41)

History is also advocated because it 'concerns man and events rather than conceptions in themselves. There is a human side to enquiry' (Schwab 1963, p. 42).

History in British Science Curricula

In Britain, there has also been a long, if uneven, tradition of incorporating the history of science in science education.[9] The BAAS, at its 1917 conference, repeated the call made by the Duke of Argyll at its 1855 conference. The association said that the history of science 'supplied a solvent of that artificial barrier between literary studies and science which the school timetable sets up' (Jenkins 1990, p. 274). The influential government report of 1918 (*Natural Science in Education*, known as the Thompson Report, after its chair, J.J. Thompson) also saw a creative role for history:

> It is desirable . . . to introduce into the teaching some account of the main achievements of science and of the methods by which they have been obtained. There should be more of the spirit, and less of the valley of dry bones . . . One way of doing this is by lessons on the history of science.
>
> (Brock 1989, p. 31)

The report went on to say that: 'some knowledge of the history and philosophy of science should form part of the intellectual equipment of every science teacher in a secondary school'. These recommendations were included in the 'Science for All' curriculum that was developed in the immediate post-war years (Mansell 1976).

Percy Nunn, the philosopher of science, Richard Gregory and other historically minded educationalists argued the case for history in the interwar years. They were influenced by the Hegelian, Spencerian and Herbartian idea that the development of individual thinking in some sense replicates the historical development of human thinking, a view later popularised by Piaget's genetic epistemology (Kitchener 1986). Popular science textbooks incorporating these ideas were written by E.J. Holmyard (1924, 1925), J.A. Cochrane and J.R. Partington. Holmyard's *Elementary Chemistry* (Holmyard 1925) sold over half a million copies between 1925 and 1960.

The British *School Science Review* welcomed the founding of the first professional journals in the history of science – *Annals of Science* (1936) and *Ambix* (1937) – with the comment that schoolteachers,

> knew from experience the value of historical details in arousing and maintaining interest and in meeting the criticism that science is unhuman . . . [the journal] ought to be placed in every school library.
>
> (Sherratt 1983, p. 421)

In the 1920s and 1930s, special courses on the history of science were offered to science teachers in teacher training colleges, and, beginning in 1921, a Master's degree on the subject was offered at University College, London.

After World War Two, history gradually diminished in importance. It was a small part of the Nuffield O-level course, but, generally, the experiential Nuffield courses ignored the historical, social and cultural dimensions of science. A number of examining boards ran separate courses in the history of science, but, by the 1980s, the number of candidates presenting had dwindled dramatically. Prior to the National Curriculum, the history of science found some place in the Nuffield programmes and in the SISCON and SATIS courses introduced in the early 1980s.

This decline in the contextual dimension of school science was of concern to the ASE, which, in a number of its reports, urged the incorporation of more historical and philosophical material into the science curriculum (ASE 1979, 1981).[10] Its 1979 *Alternatives for Science Education* maps three approaches to science education, all of which emphasise HPS. Its 1981 report recommended:

> Teaching science as a cultural activity: the more generalised pursuit of scientific knowledge and culture that takes account of the history, philosophy and social implications of scientific activities, and therefore leads to an understanding of the contribution science and technology make to society and the world of ideas.
>
> (ASE 1981)

In a recurring theme in all such efforts all around the world, the ASE, as early as its 1963 report, *Training of Graduate Science Teachers* (ASE 1963), recognised that teachers were not adequately prepared to teach this contextual science. Then, as now, specific efforts needed to be made to incorporate HPS into pre-service and in-service programmes for teachers.

Teaching About Air Pressure

The best way to appreciate the contrasts between professional and contextual approaches to teaching science is to examine the different ways that specific topics are taught using the two approaches or orientations. Air pressure, a central topic in most elementary- and high-school science courses, serves as a good example, but others could be chosen.

Historical–Humanistic Approach

There have been good historical treatments of air pressure in science programmes. Air pressure and Boyle's vacuum pump were the subject of the first of Conant's 1957 Harvard Case Studies. One of three physics units in the nine Klopfer case studies (Klopfer 1969b) is on air pressure (Case 6). The unit comprises a collection of texts, extracts, activities, slides, hardware and experiments. The case study combines the story of the overthrow, in the seventeenth century, of the ancient Aristotelian doctrine that nature abhors a vacuum with the application of hydrostatic principles to explain the phenomena associated with atmospheric pressure. The pioneer work of Torricelli with the barometer included the idea that the mercury column standing at a height of about 30 inches above the level of mercury in a dish was balanced by the weight of the 'sea of air' pressing on the surface of the mercury (Klopfer & Cooley 1961, p. 10).

Case Six contains material on Galileo's incorrect account of why the lift pump can only bring water up 34 feet – his idea was that, if any longer than this, the column would break under its own weight. The case asks students to hold up a length of chewed gum and see what its critical (non-breaking) length is, and asks them whether, by analogy, it is possible that a similar situation will occur in a long column of water. The case also has material on Pascal's Law and recommends the building of a simple hydraulic press to illustrate the principles. Among other benefits, the case allows students to see that great scientists such as Galileo get things wrong and persist in erroneous beliefs. This is even more apparent in Galileo's commitment to a completely false account of the tides, a subject with which he occupied the final day of his 1633 *Dialogue* and which he believed provided the best argument for the Copernican worldview.[11] History shows the fallibility of science and scientists, as well as the the triumphs – something usefully learned by students.

Each of Klopfer's case studies has objectives listed under three headings:

1 information about science subject matter and the narrative of the case;
2 understanding of science concepts and principles;
3 understanding of ideas concerning science and scientists.

The objectives that it lists under (3) for the unit on air pressure are instructive. It is said that, after studying the unit, students should understand the following ideas concerning science and scientists:

- the meanings and functions of scientific hypotheses, principles and theories, and their interconnections;
- the difference between science and applied science or technology;
- the dynamic interaction between ideas and experiments, between thinking and doing, in scientific work;
- that a chain of reasoning, which often involves many assumptions, connects a theory with hypotheses that can actually be tested by experiments and observations;
- that factors involved in the establishment of a scientific theory or concept include experimental evidence, the personal convictions of participating scientists and the theory's usefulness;
- that scientific explanations of natural phenomena are given in terms of accepted laws and principles;
- that scientists are individuals possessing a wide range of personal characteristics and abilities;
- that science is an international activity;
- the nature and functions of scientific societies;
- that progress in science is, in part, dependent upon the existing state of technology and on other factors outside science itself;
- that free communication among scientists through journals, books, meetings and personal correspondence is essential to the development of science;
- that new observations have a trigger effect: they shake up established concepts and lead to new hypotheses and new experiments;

- that new apparatus and new techniques are important in making possible new experiments and the exploration of new ideas.

This list could double as a suitable statement of the objectives of any course in HPS; it is also a list that would be at home in any 'science studies' programme. It is of some note that this sophisticated and nuanced characterisation of science was first written in 1961 – that is, prior to the publication of Thomas Kuhn's *Structure of Scientific Revolutions*, which gave wide exposure to such views. All three individuals were indebted to Conant and their engagement in the Harvard General Education programme.

The HOSC materials aimed at providing an education about science as well as an education in science, and with what are often called 'intangible' outcomes. Of forty-seven teachers participating in one review of the HOSC materials, 64 per cent said that their students gained intangible benefits that were not measured by tests. Some teachers commented as follows:

- The students obtained a new feeling for the meaning of science.
- Discussion and opinions of class members played a larger part than normal . . . critical evaluation of science and scientists in our society was encouraged.
- Students gained a feeling of being part of a great adventure.

(Klopfer & Cooley 1961, p. 128)

These unmeasured intangibles are as important as high-stakes, standardised-test outcomes, in part because they frequently last much longer and guide subsequent engagements with science.

Air pressure is a ready-made field for integrating history of science into science teaching. There is a natural progression and parallelism between the evolving ideas and investigations of students and the historical story. Thirty years after the Klopfer case study, Joan Solomon, in the UK, also wishing to incorporate historical and social themes into school science, wrote a booklet on the same subject, titled *The Big Squeeze*, for the UK-based ASE (Solomon 1989). It promotes understanding of air pressure by traversing the ancient Egyptians and Greeks; medieval pumps and bagpipes; Galileo's ideas; his student Torricelli's famous experiment with a tube of water to create a vacuum, and the diverse interpretations advanced to account for the 'space' above the water column; Torricelli's mercury barometer; Pascal's experiment of taking a barometer up a mountain and recording the changes in height of mercury supported and thus suggesting that air pressure is the result of the weight of air above the mercury; von Guericke's Magdeburg hemispheres; and, finally, Boyle's vacuum pump and his speculations about the 'springiness of air'.

With good teaching, students can easily be led through this sequence of concepts and experiments. A nice sequence of such lessons can be seen in Börje Ekstig (1990). A second-year University of New South Wales science student, training to be a secondary teacher, wrote the following after reading Ekstig's article:

> I am a student who did not do physics for the Higher School Certificate and only did half a year of university physics before dropping out after failing the mid-year exam. I have heard myself say many times that I dislike immensely and cannot do physics. After reading this article I wonder at my negative attitude. Basically I have never given the subject much of a chance, but on the other hand I have never heard it or read it presented in such an interesting and relevant way ... The thing that amazed me was that I actually understood ... Because my exposure to physics generally left me confused and I was convinced that it was beyond me.
>
> (private communication to author)

This is a pleasing testimony for the pedagogical worth of history in professional science courses. With knowledge of the history of the science of air pressure, students can be engaged with the following problems or investigations:

- First, students can conjecture about whether there is anything in air or whether it is essentially empty. After thinking about tests of their conjectures, they can be shown that air is difficult to compress – an empty test tube pushed into water shows this. If the same test tube is filled with water and then raised out of the beaker, we see the barometer situation.
- Second, students can be asked whether there would be a limit to the length of the water column supported in the test tube, and why the column is supported. Holding a clear plastic garden hose clamped at one end and placing the other end in a bucket of water and then suspending it from a building provides an answer to this question.
- Third, students can conjecture whether a heavier liquid would have a less high column supported, and what the predicted height of a mercury column would be.
- Fourth, the creation of a vacuum in a cylinder and the subsequent pulling of a piston into it can be shown, and thus the basis of Newcomen's steam engine can be demonstrated.

With judicious use of assignments, experiments and essays, a great many of the objectives of the HOSC unit on air pressure and the more general objectives of a contextual science programme can be met. The interplay of science with philosophy on the one hand, and with technology on the other, can be beautifully seen: The Aristotelian doctrine of 'nature abhors a vacuum' can be appreciated; the influence of this on scientists as prominent as Galileo can be seen; the efforts to support this philosophical and scientific doctrine in the light of Pascal's and Torricelli's demonstrations of its seeming falsity can be outlined; and, by students making their own primitive steam engines (versions of Newcomen's cooling-induced vacuum engine), or just pistons and cylinders, the technical difficulties in the advancement of the science of air pressure can be appreciated.

Professional–Disciplinary Approach

A standard professional approach to the topic of air pressure can be found in the PSSC *Physics* text (1960), which was the first NSF-funded high-school science programme; it was published in numerous languages and has been used by millions of students throughout the world. PSSC contrasts markedly with the above historical treatment found in the Harvard Case Studies, the HOSC, Project Physics and British materials. It is noteworthy that, in the thirty-four chapters of the text, not one is devoted to air pressure, nor is it mentioned in the index of approximately 1,000 entries. Without mentioning air pressure, its treatment of the subject begins with Boyle's Law and a model of colliding molecules in a chamber. The discussion of this law assumes the existence of air pressure; however, all developments up to Boyle are ignored. There is no mention of Torricelli or Pascal, much less are Aristotle and the *horror vacui* doctrine mentioned. Boyle's Law is explained using the mole concept, and it is stated as:

> At a given temperature the pressure exerted by a gas is proportional to the number of molecules divided by the volume they occupy.
>
> $P = K \times N/V$, where K is the proportionality factor

Notably absent from the PSSC discussion is any mention of technology or the applications of the science of air pressure. Although the expected change in the P–V relation is discussed for rarefied atmospheres, there is no mention of a barometer in the chapter; barometers are relegated to end-of-chapter exercises. Children can study the gas laws in the PSSC physics programme without their connections to barometers and weather changes being mentioned or explained. Similarly, water pumps, steam engines and all other technological uses of air pressure are omitted. The momentous connection of science with technology and its dramatic effect on the transformation of economic and social life are entirely omitted from PSSC physics. Not just PSSC, but many of the other reform projects of the early 1960s removed applied aspects of science from their programmes. One reviewer of the 1960s reforms has said:

> The first major changes in all the NSF supported curriculum reform of the '60s was removing all technology and presenting pure science 'in a way it is known to the scientist'. It is only recently that many are proclaiming the fallacy of such efforts.
>
> (Yager & Penick 1987, p. 53)

Metaphysics and Physics in the Science of Air Pressure

A teacher's interest in history will influence how much a class learns from discussing and re-enacting the historical progression that led to the contemporary understanding of air pressure.

Aristotle on Air

Aristotle was one of the earliest contributors to the philosophical/scientific investigations of air. He regarded air as one of the five fundamental elements; air was all of the one kind, it was not a mixture of different components. This was one of the great 'epistemological obstacles'[12] that needed to be overcome by Joseph Priestley and the early pneumatic chemists of the seventeenth century, whose investigations led them to the conclusion that air was a composite of gases (a matter that will be detailed in Chapter 7). Importantly, Aristotle, for philosophical and empirical reasons, denied that a vacuum could exist in nature. He advanced his arguments for the *horror vacui* against the atomists, for whom the existence of a void between atoms was philosophically fundamental. Aristotle's chief arguments against the possibility of a void are contained in his *Physics* Book IV (reproduced in Matthews 1989). The historian Ernest Moody says of this text that:

> It was, in a very definite sense, the cradle of mediaeval mechanics. And for Galileo . . . this text was a constant point of departure. Not only in this Pisan dialogue, but in the great *Discorsi* of Galileo's maturity, it was as a criticism of this Aristotelian text that he developed his dynamic theory of the motion of heavy bodies.
>
> (Moody 1951, p. 175)

Aristotle argued, reasonably enough given everyday experience, that the velocity (V) of a moving body varied directly as the force applied (F) and inversely as the resistance of the medium (R) through which it moved. (Think of pushing a car along a smooth road and then through sand.) That is:

$$V = K \times F/R$$

In a vacuum, R would be zero, and a body once pushed would move with infinite speed. Thus, its time of movement between two points, A and B, would be zero seconds, and thus it could not be said to move but would dissolve at A and be recreated instanteously at B. Thus, Aristotle's conclusion was that, in a vacuum, there could be no motion. However, as motion can be seen everywhere, then a vacuum is impossible. This basic belief that in nature there could be no vacuum dominated and constrained physics for over 1,000 years, and Galileo struggled with it at the beginning of his philosophical/scientific investigations.

One enduring lesson from the consideration of these early Greek speculations on air pressure is the awareness of the way in which Aristotelian science and philosophy are rooted in the experience of the everyday world. The modern Aristotelian, Mortimer J. Adler, recognised this when he observed, in his Introduction to *Aristotle For Everybody*, that:

> In an effort to understand nature, society, and man, Aristotle began where everyone should begin – with what he already knew in the light of his ordinary, commonplace experience. Beginning there, his thinking used notions that all of us possess, not because we are taught them in school, but because they are the common stock of human thought about anything and everything.
>
> (Adler 1978, p. xi)

The ancient world was familiar with all the phenomena that introductory students can see and experience. The siphon was used to drain fluid; the pipette or clepsydra was used to transfer fluid; and the drinking straw was, of course, used. The move from all of this familiar experience to the belief that nature abhors a vacuum was very easy. Aristotle appealed to this common experience and then added certain logical arguments about motion and place, and the outcome was the long-lasting and powerful doctrine of *horror vacui*.

A similar train of argument was used in discussing another aspect of the question: Does air have weight? The ancients, through observation of windmills, sails, balloons made of animal bladders and so on, realised that space contained something, namely air; it was not empty. Yet it did not appear to weigh anything; indeed, it seemed to have a negative weight – it did not press down, but rather it seemed to go upwards. Aristotle held that things in their proper place had no weight, or *gravitas* (as the Latin speakers would say). Stones and matter had *gravitas*, because they were trying to press down into the centre of the earth, their natural home; air had no *gravitas*, but in contrast had *levitas*, because its natural tendency was to go upwards to the sky, its natural place. Thus, the claim that air had weight, in the sense that stones had weight, would invalidate an important plank in Aristotle's philosophy.

This same move from everyday experience to scientific and philosophical doctrines can also be seen in Aristotelian theories of motion, astronomy, biology and much else. If this point can be appreciated, then students are in a position to grasp the most important feature of the scientific revolution of Galileo and Newton – the reinterpretation of everyday experience and certainties in the formulation of their new sciences; the move from Aristotle's explanation of the unfamiliar in terms of the familiar, to Newton's explanation of the familiar (falling bodies) in terms of the unfamiliar (inertia).

Seventeenth-Century Debates

Galileo, from at least 1614, was confident that he had experimentally demonstrated that, contra Aristotle, air had weight (Drake 1978, p. 231). He persisted in his belief that the weight of air had nothing to do with the limit to the height of water in siphons (Drake 1978, p. 314). His associate and student, Evangelista Torricelli (1608–1647), made the initial steps towards proper understanding of air pressue and its explanation in interpreting barometers.

Torricelli's 1643 experiment (repeated in France in 1646 by the child prodigy Blaise Pascal (1623–1662)) used an inverted closed tube of mercury placed in an open dish of mercury.[13] The mercury fell a distance from the top of the tube, but always stayed at about 30 inches or 76 cm above the level of mercury in the dish (see Figure 4.1). This demonstration concentrated the minds of philosophers and scientists (to use an anchronistic distinction): the long-standing philosophical dispute about a vacuum seemed to be settled by a simple experiment. This same experiment can also engage the minds of contemporary students.

Students can be shown Torricelli's apparatus – or now, in mercury-free laboratories, some variant of it – and with or without guidance can see that there are two questions that need an answer:

1 What holds the column of mercury up?
2 Is there anything in the space above the mercury in the column?

These are separate questions, although they are often merged. At the time of Pascal and Torricelli, a few said the answer to the first question was that air pressure was forcing down on the surface of the mercury in the dish. Others, denying that air had weight, had other accounts of how the column of mercury

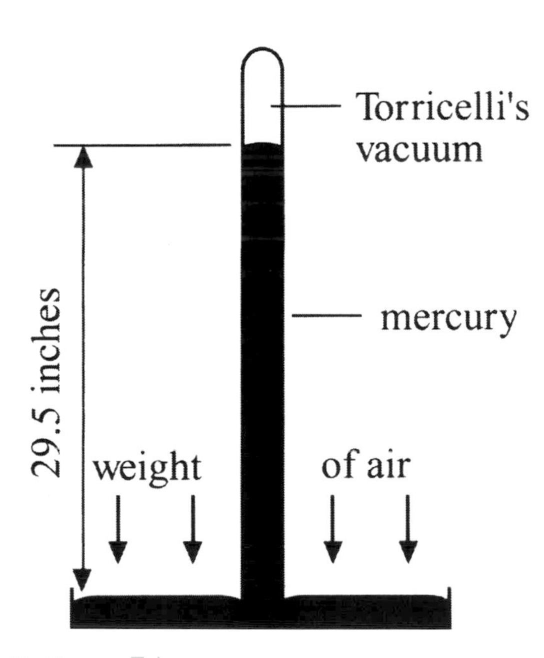

Figure 4.1 Torricelli's Vacuum Tube

was supported. In answer to the second question, many said that there was no vacuum; fewer said that there was a vacuum. In the middle of the seventeenth century, all four possible answers had adherents.

The above questions can be reformulated in the following matrix (following Dijksterhuis 1961/1986, p. 445), with representative adherents to different answers included.

	Yes	*No*
Does air pressure hold up the mercury?	Descartes Pascal	All Aristotelians Roberval
Is there a vacuum above the mercury?	Roberval Boyle	All Aristotelians Descartes Galileo Hobbes

The resolution of these competing views depended upon logical, technical and experimental considerations. Some who denied air pressure and the vacuum said that what caused the column to sink was the generation of vapours or spirits from the liquid. To test this view, Pascal took wine and water and asked his opponents what would sink further in Torricelli's tube. His audience reasoned that, as wine was more volatile, it would vaporise more and, thus, sink further down the tube than water. When the experiment was done, it was seen that water sank further than wine. Pascal knew this would happen, because it was heavier than wine. So the spirits hypothesis had either to be abandoned or reworked. Others who denied the vacuum said that a small amount of air had been left behind in the tube. Pascal took tubes of different diameters and established that what was constant was the height of the column, not the volume of the space, as the rival hypothesis would have it. Again, the rival hypothesis had to be abandoned or reworked. Students can be led through these options – by questions, by debates, by reading source materials – and can thus learn something of the process of scientific argument and hypothesis testing, at the same time as they learn about air pressure. Such historical introductions and re-enactments of actual scientific disputes provide students with an invaluable window on to scientific argument.

This sequence illustrates the difference between simple observation and scientific experiment, the relationship of theory to the construction of experiments, and the less than straightforward reassessment of theory in the light of disconfirming experimental results. For instance, the *horror vacui* doctrine could be reconciled with the aberrant experimental results by a simple twist: it could have been said that the varying degrees of fall of the mercury column established the degree to which nature abhors a vacuum. Nature's abhorrence is not absolute, but relative to the substance at hand. Nature is prepared to pull mercury up a certain amount, and other liquids up by differing amounts, in its effort to avoid a vacuum. So, the column height, rather than a measure of the pressure of air forcing down on the dish, was a measure of the degree to which nature abhors a vacuum. To combat this move requires that purely ad hoc hypotheses be ignored, and more formally ruled out, in science.

Pascal, in his best-known experiment, had a barometer taken up the Puy-de-Dôme mountain in 1648, confirming that the height of the column decreased, the higher it was taken. He thought that this was because, the higher up the mountain, the less air was pushing down on the surface of the mercury. Students can be encouraged to imagine this experimental test and may, in places, with very tall buildings or roads reaching to high elevations, have the opportunity to conduct it. Pascal's brother-in-law conducted the experiment, and the results were as predicted. He left a barometer at the base of the mountain to see that its level did not change during the day, a control that students might be encouraged to think about and, by doing so, learn the importance of controlled experiment in science (and even in educational research). The results were wonderfully consistent with the air-pressure hypothesis, and, indeed, Pascal looked upon it as an *experimentum crucis* between the two doctrines.

It would seem that, against fundamental tenets of Aristotelian philosophy, the existence of a vacuum had been demonstrated, and at the same time it had been demonstrated that air has weight. But, as in life, so also in science and philosophy, things were not always simple. The Aristotelian anti-vacuum theorists could easily say that Pascal's experiment simply showed that, as we go higher towards the heavens, nature's abhorrence of a vacuum diminishes. Thomas Hobbes (1588–1679) and René Descartes (1596–1650), who were both firm opponents of Aristotelianism, acknowledged all the experimental evidence presented by Pascal and Roberval, yet denied their conclusion that it established the existence of a vacuum. Thomas Hobbes, for instance, said of Torricelli's supposedly definitive proof of the vacuum that:

> If the force with which the quicksilver descends be great enough . . . it will make the air penetrate the quicksilver in the vessel, and go up into the cylinder to fill the place which they [vacuists] thought was left empty.
>
> (Shapin & Schaffer 1985, p. 89)

Hobbes and Descartes, respectively, thought that the void was filled with aetherial substance and with subtle matter.[14] When students are reminded that a vacuum entails zero or very low pressure, and further that liquids vaporise and then boil at low pressures, they might see that the subtle-matter view of the space above the mercury is not entirely without merit. With a vacuum above, surely the mercury will vaporise.

Methodological Lessons

At just about every stage in the foregoing development, the basic scientific move from phenomenal evidence to invisible mechanisms has to be made.

Students can easily repeat the processes of evidence collection, conjecture and testing, and thus appreciate the nature of scientific hypotheses and their appraisal.[15] By doing this, they can learn more about the nature of science.

Following the early history of air-pressure investigation reveals the difficulties and complexities of: describing the evidence in a manner acceptable to all competing theories; formulating empirically testable hypotheses; testing hypotheses; designing and conducting experiments with adequate controls; refuting hypotheses given contrary evidence; and the rescuing of hypotheses despite this evidence. Discussion of this history can lead to such central, and much written upon, philosophical questions as the possibility of crucial experiments (vide the Duhem–Quine thesis), the difference between ad hoc and justified alteration of theories in the light of contradictory evidence (vide Popper and Lakatos), and the role of metaphysics in the maintenance of scientific theories (vide Burtt and Buchdahl). Issues in the sociology of science can also be canvassed in this style of instruction. The dependence of one researcher on another can be seen – Torricelli upon Galileo, Boyle upon von Guericke – and then the consequent importance of open communication and truthfulness for science can be appreciated.

Most students' initial understanding of the testing of a scientific theory is the naïve hypothetico-deductive method:

> Theory (*T*) implies Observation (*O*)
> *O* occurs
> Therefore *T* is confirmed

or

> *O* does not occur
> Therefore *T* is falsified

The foregoing has shown that this simple view needs to be elaborated to take into account that it is the theory along with statements of initial conditions (*C*) that constitute the test situation. Thus we have:

> *T* and *C* together imply *O*
> If not *O*,
> Then, not *T*, or not *C*

But this is still too simple, because the test also embodies assumptions about the reliability and validity of the test apparatus and measuring instruments (*I*). Thus we have:

> *T* and *C* and *I* together imply *O*
> If not *O*,
> Then, not *T*, or not *C*, or not *I*

When metaphysics (M) is assumed in a scientific experiment, we have this situation:

> M and T and C and I together imply O
> If not O,
> Then, not M, or not T, or not C, or not I

This is the argument structure that will be elaborated in Chapter 6, concerning debate about the shape of the Earth.

The educational objective of critical thinking and careful reasoning can be realised with this historical approach, because it engages the student's mind. A teacher well versed in the history of the topic can identify when students are making the same intellectual moves as previous scientists, and can encourage the reconsideration of earlier debates. This allows an appreciation of both the achievements and mistakes of earlier scientists, and perhaps some empathy with them.[16]

Opposition to History

The inclusion of history in science programmes has been opposed from two sides: from historians who see history in science lessons either as poor history or as downright fabrication of history in support of current scientific ideology, and from scientists who see it as taking up valuable time that could be devoted to science proper, and who see it as possibly eroding the student's conviction that their hard effort is uncovering the truth about the world. In 1970, at an MIT conference sponsored by the International Commission on Physics Education (Brush & King 1972), arguments from both sides were advanced.

Pseudo-history

Martin Klein's argument was basically that teachers of science who select and use historical materials do so to further contemporary scientific or pedagogical purposes, that such selection is contrary to the canons of good history, and, thus, 'in trying to teach physics by means of its history, or at least with the help of its history, we run a real risk of doing an injustice to the physics or to its history – or to both' (Klein 1972, p. 12). He quotes, approvingly, Arthur O. Lovejoy's caution that:

> The more a historian has his eye on the problems which history has generated in the present, or has his inquiry shaped by the philosophic or scientific conceptual material of the period in which he writes, the worse historian he is likely to be.
>
> (Klein 1972, p. 13)

The result of this partial or selective approach,

> is almost inevitably bad history, in the sense that the student gets no idea of the problems that really concerned past physicists, the contexts within which they worked, or the arguments that did or did not convince their contemporaries to accept new ideas.
>
> (Klein 1972, p. 13)

Further, Klein suggested that there was a basic difference in the very enterprises of science and history that makes their marriage most improbable and, where it does occur, makes the union short and stormy:

> One reason it is difficult to make the history of physics serve the needs of physics teaching is an essential difference in the outlooks of physicist and historian . . . it is so hard to imagine combining the rich complexity of fact, which the historian strives for, with the sharply defined simple insight that the physicist seeks.
>
> (Klein 1972, p. 16)

In support of this view of the historical enterprise and why it should be kept out of science classrooms, he repeats Herbert Butterfield's injunction that,

> When he describes the past the historian has to recapture the richness of the moments . . . and far from sweeping them away, he piles up the concrete, the particular, the personal.
>
> (Klein 1972, p. 16)

He also mentions Otto Neugebauer, the historian of classical science, who, like Butterfield, believed that the historian's role was to recover the complexity of the past. Neugebauer, in his *The Exact Sciences in Antiquity* (1969), wrote:

> I do not consider it as the goal of historical writing to condense the complexity of historical processes into some kind of 'digest' or 'synthesis'. On the contrary, I see the main purpose of historical studies in the unfolding of the stupendous wealth of phenomena which are connected with any phase of human history and thus to counteract the natural tendency toward oversimplification and philosophical constructions which are the faithful companions of ignorance.
>
> (Neugebauer 1969, p. 208)

Klein's conclusion was that, if good science teaching is historically informed, then it will be informed by bad history. He prefers no history to bad history.

Quasi-history

M.A.B. Whitaker pushed these claims further in a two-part article titled 'History and Quasi-history in Physics Education' (Whitaker 1979). Like Klein, his concern was to identify the prevalent fabrication of history to suit, not just pedagogical ends, but the ends of scientific ideology, or the view of science held by the writer. These cases abound in textbooks.

One that has been much discussed is the widespread account of Einstein's postulation of the photon, following the perceived contradiction between the photoelectric effect and the wave theory of light. The photoelectric effect is apparent in the creation of current between plates in a vacuum when light shines upon one plate. According to the standard textbook account (PSSC 1960, p. 596), anomalous aspects of the photoelectric effect – such as the energy of the emitted electrons not depending upon the intensity of the light and the threshold frequency levels for producing the effect independently of the intensity of the light – were known by the end of the nineteenth century to be a problem for the orthodox wave theory of light. In the orthodox theory, the intensity of light was a measure of the energy of light, so that light of any frequency, provided it was intense enough, should be able to produce the photoelectric effect. This does not happen.

The standard account says that Einstein's 1905 photon theory of light, with its Planck-inspired formula of $E = hf$ (energy of a photon equals a constant times the frequency of the light ray), was put forward as a brilliant solution to the anomalies and the harbinger of a new period in the physics of radiation. The old battle between particle and wave theories of light had been resolved in favour of a compromise view that saw light waves as coming in packages and, hence, being particle-like. This account reinforces the public and scientific image of Einstein, it accords with the hypothetico-deductive model of scientific theory, it emphasises the rationality of science and it demonstrates the progressiveness of scientific work. In other words, there is nothing in the standard account to disturb the rational, methodical and inevitable picture of scientific progress commonly held both by scientists and the public. The only problem with the account is that the actual history was nowhere near as straightforward as this.

For many years, respectable scientists such as Lenard, Thomson and Lorentz put forward accounts of the photoelectric effect that focused on within-the-atom structures and behaviour (resonance effects triggered by the light), rather than properties of the light beam.[17] These could account for the effect just as well as Einstein's peculiar hypothesis. Planck, the originator of the quantum theory, rejected Einstein's 'wave package' or photon interpretation in his 1912 book on heat radiation. Robert Millikan, who was to receive a Nobel Prize for his confirmation of Einstein's 1905 hypothesis, says in his autobiography:

> I think it is correct to say that the Einstein view of light pulses, or as we now call them, photons, had practically no convinced adherents prior to about 1915 . . . Nor in those earlier stages was even Einstein's advocacy vigorous or definite.
>
> (Millikan 1950, p. 67)

The mixed, indeed lukewarm, reception accorded Einstein's hypothesis is evidenced by the fact that he did not receive the Nobel Prize for his paper until 1921, some 16 years after its original publication. This suggests some slowness in the process of rational conversion of the scientific community.

Even when adherents began to appear, they were adherents to Einstein's equation, not to his physical interpretation of the equation – a big difference. Millikan had written in 1916 that:

> Despite . . . the apparently complete success of the Einstein equation, the physical theory on which it was designed . . . is found so untenable that Einstein himself, I believe, no longer holds to it.
>
> (Pais 1982, p. 380)

The extremely popular Halliday and Resnick physics text makes much of Millikan's 1916 experimental confirmation of Einstein's photon theory, saying that his experiments 'verified Einstein's ideas in every detail'. But again, the experiments did not confirm Einstein's theory, only his equation. And even the confirmation of his equation (linking the energy of the emitted electron to the frequency of bombarding light) was far from unequivocal. A series of experimental physicists interpreted Einstein's data as showing that the energy varied as frequency squared, or as frequency to the two-thirds power, or even that it had no connection with frequency (Kragh 1986, p. 74). Millikan's data were open to a variety of mathematical interpretations apart from the one he chose – energy varied as frequency – and this mathematical equation did not carry its physical interpretation upon its sleeve; it did not prove that light travelled in little bundles.

Klein's and Whitaker's charges about the inaccuracy and bias of a good deal of history in science texts are certainly proven. The reason for these inaccuracies is an interesting question, and its answer would reveal much about the ideology of science education and the function of textbooks. Whitaker says of quasi-history that it is the,

> result of the large numbers of books by authors who have felt the need to enliven their account of [these episodes] with a little historical background, but have in fact rewritten the history so that it fits in step by step with the physics.
>
> (Whitaker 1979, p. 109)

He does not see such history as arising from a conscious effort to support an author's vision of science:

> I do not assume that writers of quasi-history necessarily have any philosophical intent, even subconsciously. I see quasi-history more often merely as a result of a rather misguided desire for order and logic, as a convenience in teaching and learning.
>
> (Whitaker 1979, p. 239)

Whitaker traces the mistakes of quasi-history to a neglect of the 'public and social nature of science'.

Quasi-history is not just Klein's pseudo-history, or simplified history, where mistakes of omission are likely to occur, or where the story might fall short of the lofty standard of 'the truth, the whole truth, and nothing but the truth'; rather, in quasi-history, we have manufactured history masquerading as genuine history. This is akin to Lakatos's 'rational reconstructions' of history (Lakatos 1971), but for Lakatos the historical story was plainly labelled as a 'rational reconstruction'. Historical figures are painted in the hues of the current methodological orthodoxy. Galileo has been a fine example of the treatment: he appears as an experimentalist in empiricist texts, as an instrumentalist in other texts, and as a rationalist in still others. He has become, as will be documented in Chapter 6, a man for all philosophical seasons. Where quasi-history is substituted for history, the power of history to inform the present is nullified. If the historian rigidly selects and interprets his material according to a prior philosophical position, it is difficult, if not impossible, for these reconstructed data to feed back into the proper assessment of the philosophical position.

Historical, philosophical and scientific writing on the decades-long delay in final acceptance of Alfred Wegener's plate-tectonic mechanism for his 1912 'Continental Drift' hypothesis well illustrates these problems of quasi-history that Whitaker points out.[18] The common accounts are that the hypothesis was finally endorsed because plate tectonics was either able to make novel predictions that were vindicated or plate tectonics unified a large body of hitherto disparate data, and, conversely, the theory was rejected because it left more data unexplained than explained. One commentator has noted that:

> However, a careful reading of the historical record fails to substantiate any decision of rejection or acceptance being made on the basis of any of these reasons. A problem, I would argue, is that philosophers of science start with some preconceived general criteria for theory acceptance, and their reading of the history of the development of that theory selects the events that validate their preconception.
>
> (Ryan 1992, p. 71)

Confidence Destroying

The third type of criticism brought against the introduction of history of science into science courses is that it saps the neophyte scientific spirit. At the MIT conference, the historian Harold Burstyn elaborated Klein's problem in terms of the different outlooks of students, rather than the different outlooks of teachers or professional historians and scientists. Burstyn cautions that:

> There is a lot of evidence (including my own experience in teaching history of science to science students) that science students and students of other subjects have different outlooks on the world. To phrase it pejoratively, the science students are looking for the 'right' answers, they are 'convergent' rather than 'divergent' thinkers. The problem Klein is getting at is this: Can you in fact use the historical materials, whose hallmark is their complexity, their diffuseness and imprecision, in the teaching of people who are interested in getting right answers, and who, if

> they are successful, can't be diverted from this quest as we historians might want to divert them? Isn't history therefore somewhat subversive of the aims of physics pedagogy?
>
> (Brush & King 1972, p. 26)

This charge was earlier made by Thomas Kuhn, in a 1959 address to a conference on scientific creativity (Kuhn 1959). Kuhn repeated the charge in the first (1962) and second (1970) editions of his immensely popular *The Structure of Scientific Revolutions*. In his conference address, he drew attention to the fact that:

> The single most striking feature of this [science] education is that, to an extent wholly unknown in other fields, it is conducted entirely through textbooks. Typically, undergraduate and graduate students of chemistry, physics, astronomy, geology, or biology acquire the substance of their fields from books written especially for students.
>
> (Kuhn 1959, p. 228)

He noted that science students are not encouraged to read the historical classics of their fields, 'works in which they might discover other ways of regarding the problems discussed in their textbooks' (Kuhn 1959, p. 229). All of this produces a rigorous training in convergent thought, and Kuhn maintains that the sciences 'could not have achieved their present state or status without it' (Kuhn 1959, p. 228) – a position altogether at odds with Ernst Mach's view of the kind of pedagogy required for the advancement of science. Kuhn justifies this training by its results: not just the production of good convergent thinkers, but also the production of a smaller group of innovators and creative scientists who would not have been able to be innovative unless they were thoroughly seeped in the orthodox thought of their discipline. These ideas provided the title both for his conference address and his subsequent book, *The Essential Tension* (1971/1977). Kuhn elaborated these ideas in his *The Structure of Scientific Revolutions*, where he says that, in a science classroom, the history of science should be distorted, and earlier scientists should be portrayed as working upon the same set of problems that modern scientists work upon, in order that the apprentice scientist should feel himself part of a successful truth-seeking tradition (Kuhn 1970, p. 138).[19]

Stephen Brush developed the Kuhn charge further in his 'Should the History of Science be Rated X?' (Brush 1974). Here it was suggested, tongue in cheek, that history of science could be a bad influence on students, because it undercuts the certainties of scientistic dogma seen as necessary for maintaining the enthusiasm of apprentices on a difficult task. He warned teachers that,

> the teacher who wants to indoctrinate his students in the traditional role of the scientist as a neutral fact finder should not use historical materials of the kind now being prepared by historians of science: they will not serve his purposes.

Defence of History

The Klein–Whitaker–Kuhn charges are serious but not fatal; their main concerns can be addressed without ejecting history from science courses. The charges will be briefly restated and then commented upon.

Charge I: Science and history are very different intellectual enterprises, because the former looks for simplicity and ignores extraneous circumstances, whereas the latter celebrates and seeks complexity; thus, there are two antagonistic mental outlooks to be cultivated if history is brought into the the science classroom.

First, if this charge is true, is it such an unfortunate thing? The cultivation of different mental outlooks should be an educational goal. A good school curriculum is one that encourages a range of perspectives and ways of dealing with problems; thus, students are required to study mathematics, literature, art, history, science and perhaps morals, civics and religion. The problem seems to be that there are different habits of mind being cultivated within the one classroom, but even this should not be a problem. The English teacher sometimes encourages creativity and free expression, at other times rote learning and disciplined thought, at still other times empathic understanding and moral reasoning. And, of course, the English teacher needs to provide historical and political contexts to the literature being examined. Can Dickens' novels be appreciated without some knowledge of nineteenth-century English society and its economic sinews? Can Orwell's novels be understood without knowledge of twentieth-century totalitarianism of both the Left and the Right? These different outlooks are not regarded as disruptive to the English teacher's overall task of developing literate students. The science teacher should be no more worried than the English teacher by such heterogeneity in a lesson programme. Further, we have myriad examples of successful cross-disciplinary programmes that avoid the putative pitfalls and achieve some of the objectives of a liberal education.

It is not just history that brings intellectual schizophrenia to the science classroom. Increasingly, morals and politics are regarded as legitimate and, indeed, necessary components of science education. This is most clear in the numerous STS and socio-scientific-issues courses, where moral/political issues, such as pollution, alternative energy sources, conservation, sustainability and so on, are used as themes around which the science course is developed. Such courses require that students think in moral and political as well as scientific ways within the one class. But, apart from STS courses, the English National Curriculum, *Project 2061* and other mainstream curriculum developments also require of science students that they consider their subject from a variety of perspectives. The argument proffered against history would also rule out of the science classroom these other considerations. However, there seem to be no empirical grounds for so doing, apart from lack of time in a crowded syllabus, and there are good pedagogical grounds for including the wider considerations.

Second, are the differences between a scientific and a historical approach as great as claimed? At one level, Klein's account of history as seeking complexity and putting nothing aside is simply wrong: all historical writing has to be selective. It is true that Klein's empiricist, fact-finding account of history has often been proposed. The eminent nineteenth-century historian Ranke proposed that the task of history was 'simply to show how it was'. This is plainly rhetorical or silly, or both. The criticisms levelled against it by E.H. Carr are well known (Carr 1964).

The simple point is that history cannot tell everything; it has to be selective. A history of railway development in England will legitimately ignore developments in the theatre; it will focus upon matters related to railways, but there is a superabundance of such matters – patronage, arrival and departures of trains, architecture of platforms, the work force and its costs, railway meals, orders for steel and so on – and selection needs to occur. A historian is not an archivist: the latter's job may be to file away all the timetables, meal menus, order books and so on (even this has to involve a sense of what is likely to be useful). The historian is supposed to select and, further, make something of the historical record. To say this, and to oppose simple empiricist views of history, is not to endorse extreme postmodernist accounts that maintain that history is just all construction, that there are no facts of the matter to ascertain.

Detail of correct dates, a concern with uncovering all the relevant correspondence, examining changes between editions and other such scholarly endeavour can be of the utmost importance, provided some objective is in mind, and provided some principle of inclusion/exclusion is operative. The scientist does leave aside the colour, texture and composition of a falling ball and replaces all this richness with a simple point mass; historians also have to leave aside some of the richness of historical episodes and seek for some essentials that are pertinent to the story they wish to tell. In this sense, their discipline is not so different from science. A scholarly article might concentrate upon the trees, but, in classrooms and student texts, there should not be such attention paid to trees that the forest can no longer been seen.

Charge II: Inevitably, the history used in science courses is pseudo-history in virtue of its being in the service of science instruction.

This claim is a variant of the first and need not deter a science teacher. Its apparent strength lies in a confusion between writing history and using history in science classes. With some notable exceptions, such as F.W. Westaway and E.J. Holmyard, a science teacher is not a historian. There may be problems with writing history in order to serve ulterior ends, if this results in the distortion of history. Writing for a purpose need not result in pseudo-history. Be this as it may, a science teacher is explicitly using history for pedagogical purposes, and his or her use of history is to be judged on criteria different from those of a practising historian: the two activities are very different.

It needs to be remembered that science teaching is not historical research: they are different activities, with different purposes and different criteria of success and authenticity. Standards of sophistication required for historical research are misplaced when applied to science pedagogy. In pedagogy, the subject matter needs to be simplified. This is as true of history of science as it is of economics or of science itself. The pedagogical task is to produce a simplified history that illuminates the subject matter and promotes student interest in it, yet is not a caricature of the historical events. The simplification will be relevant to the age group being taught and the overall curriculum being presented. The history can become more complex as the educational situation demands. To criticise elementary-school teachers for hagiography is to misunderstand what they are doing, namely trying to interest students in important figures in the history of science; to criticise secondary-school teachers for simplifying the history of genetics is again to misunderstand what they are doing, namely trying to teach about genetics in a way that is interesting and comprehensible to adolescents. The pedagogical art is to simplify subject matter, and historical stories, in such a way that the inevitable approximations and distortions are educationally benign, not pernicious.

This art results in what Lee Shulman usefully called 'pedagogical content knowledge' (PCK) (Shulman 1986, 1987) and what is commonly labelled as 'didactical transposition' in European didactic traditions.[20] It is the everyday classroom practice whereby good teachers make formal professional knowledge into teachable school knowledge. As Shulman says, PCK requires the subject specialist to know 'the most useful forms of analogies, illustrations, examples, explanations, and demonstrations – in a word, the ways of representing and formulating the subject in order to make it comprehensible to others' (Shulman 1986, p. 6). This seems sensible and laudatory; nevertheless, Shulman's view has been criticised by educators having a more constructivist and postmodernist orientation, because it is 'informed by an essentially objectivist epistemology' and it 'focuses primarily on the skills and knowledge that the teacher possesses, rather than on the process of learning' (Banks *et al.* 2005, p. 333).

Mindful of these supposed problems, history is one element that can usefully contribute to PCK. Helge Kragh well expressed this defence of history:

> In an educational context, history will necessarily have to be incorporated in a pragmatic, more or less edited way. There is nothing illegitimate in such pragmatic use of historical data so long as it does not serve ideological purposes or violate knowledge of what actually happened.
>
> (Kragh 1992, p. 360)

Charge III: It is likely that the history used in science courses will be quasi-history, because of the purposes and limitations of the science teacher.

First, as has been said, there is a great deal of truth in this claim, and as such it serves as a timely caution to those advocating the use of history. The problem of 'revisionist' history is notorious in the political realm and often destroys the mind-expanding purpose of school history. We know that official Soviet histories of the Communist Party are historically worthless, the official history itself changing with each change in party leadership. After August

1991, all such histories are being consigned to the dust heap; and, across the entire ex-Soviet bloc, school history curricula have been rewritten. Despite law suits brought by the courageous, non-postmodern historian Saburo Ienaga, the official Japanese school history texts have rewritten the history of the Pacific War: the period is largely omitted, and, where it is mentioned, it is in terms of Japan's efforts to encourage Asian economic growth. Many American histories, of the 'Opening of the West', of the conquest of Mexican territories, of the 1905 Spanish–US war, of the Vietnam War, of labour history and so forth, are themselves driven by ideology and distort the historical record. Notoriously, there are as many histories of the Middle East as there are national, religious and commercial interests. Communist party histories of China dealing with Mao, the Cultural Revolution, Tibet, and Tiananmen Square are as corrupted as their Soviet counterparts.

Given the importance and status of science and its accomplishments, it is not surprising that various political and ideological groups should write histories of science showing their own group as the champions and as responsible for the achievements of science. The Nazis wrote Aryan histories of science that demonstrated that Jewish scientists either did poor research or stole good ideas from Germans (Beyerchen 1977). The Soviet Union produced its own ideological version of the history of science (Graham 1973). In the history of warfare between the Church and science, both sides produced histories appropriate to their case. Sometimes, the distortions are conscious; other times less so. Many have claimed that the monumental histories of science of Duhem and the case studies of Poincaré are both influenced and, some would say, compromised by their Catholicism (Nye 1975, Paul 1979). Undoubtedly, myths and ideologies abound in histories of science, just as they do in political, social and religious histories;[21] it is salutary for everyone to recognise this and to tread warily in classrooms. Howard Zinn, the US historian, well expressed the point:

> By the time I began teaching and writing, I had no illusions about 'objectivity', if that meant avoiding a point of view. I knew that a historian (or a journalist, or anyone telling a story) was forced to choose, out of an infinite number of facts, what to present, what to omit. And that decision inevitably would reflect, whether consciously or not, the interests of the historian.
>
> (Zinn 1999, p. 657)

Charge IV: Good historical study is corrosive of scientific commitment.

This is an empirical claim for which the evidence is slight. The author has taught 'History and Philosophy for Science Teachers' courses for many years, without seeing any such deleterious results. In fact, comments such as 'teachers are hungry for this information', 'I never realised that Galileo did such things', 'this makes me want to teach science better' are commonplace. The experience of Einstein, when given Mach's *The Science of Mechanics* by his friend Besso, might be more typical: exposure to history enlivened Einstein's commitment to science. Certainly, for the history to make sense, a body of scientific knowledge and technique has to be mastered, but there is no evidence that this mastery is impeded or threatened by historical study. On the contrary, the extensive research done on the subject-matter mastery of the hundreds of thousands of students who studied the Harvard Project Physics materials in the 1970s is impressive and contradicts the pessimistic claim of Kuhn. Likewise, the much more restricted evidence from the Klopfer and Cooley high-school case studies suggests that history enlivens student interest in, and understanding of, science. Independent of the effectiveness claim, there are serious educational issues involved in trading putative student commitment to science for historical truthfulness about science (Siegel 1979). This merges very quickly into the issue of indoctrination in education.

The History of Science and the Psychology of Learning

Apart from all the foregoing curricular and pedagogical considerations in appraising the role of history of science in science classrooms, there is an important theoretical issue, namely the putative common cognitive mechanisms involved in transformations in the history of science and in the conceptual development of children. Jean Piaget famously connected the history of science with his psychological account of accommodation and assimilation in the maturing mind of individuals; he advanced his own version of the 'ontogeny recapitulates phylogeny' thesis.[22] In Piaget's words: 'The fundamental hypothesis of Genetic Epistemology is that there is a parallelism between the progress made in logical and rational organisation of knowledge and the corresponding formative psychological processes' (Piaget 1970, p. 13).

In turn, Piaget's research was launched into the HPS community by Thomas Kuhn's remark in the 1962 Preface of the first edition of his *The Structure of Scientific Revolutions*, where he says:

> A footnote encountered by chance led me to the experiments by which Jean Piaget has illuminated both the various worlds of the growing child and the process of transition from one to the next.
>
> (Kuhn 1970, p. vi)

A decade later, Kuhn reaffirmed his debt to Piaget:

> Part of what I know about how to ask questions of dead scientists has been learned by examining Piaget's interrogations of living children . . . it was Piaget's children from whom I had learned to understand Aristotle's physics.
>
> (Kuhn 1971/1977, p. 21)

The linkage was reinforced for Kuhn when Alexandre Koyré told him that, 'It was Aristotle's physics that had taught him to understand Piaget's children' (ibid.).

The recapitulation thesis has been exhaustively researched and debated. There have been thousands of studies on children's thinking about nature and astronomical processes, their reasoning, their concept acquisition, mental maturation, epistemology and 'scientific ideas'.[23] Not surprisingly, a common thread in all of these studies is the recognition that cognition is social, the 'I think' is dependent upon the 'we think'; our thoughts and concepts are expressed in language and for this we require degrees of participation in a community. Abstraction theories of concept acquisition fail because they are circular; we cannot, Robinson Crusoe-like, abstract 'hard' from experience of a number of hard things, because, along with hardness, there are always other properties. The concept has to be given to us.

If one study might be identified as the major link between the body of Piagetian/Kuhnian research on conceptual change and science pedagogy, it would be Posner *et al.*'s (1982) 'Accommodation of a Scientific Conception: Toward a Theory of Conceptual Change'. This enormously cited study draws upon the accounts of scientific theory change given by Kuhn, Toulmin and Lakatos. They propose that, for individual conceptual change or learning to take place, four conditions must be met:

1 There must be dissatisfaction with current conceptions.
2 The proposed replacement conception must be intelligible.
3 The new conception must be initially plausible.
4 The new conception must offer solutions to old problems and to novel ones; it must suggest the possibility of a fruitful research programme.

The study, along with others, sparked thousands of classroom and laboratory conceptual-change interventions and researches.[24]

A decade later, Strike and Posner pointed out something that was being overlooked by many researchers in the field, namely that their original conceptual change theory was: 'largely an epistemological theory, not a psychological theory . . . it is rooted in a conception of the kinds of things that count as good reasons' (Strike & Posner 1992, p. 150). Their original theory is concerned with the 'formation of rational belief' (p. 152); it does not 'describe the typical workings of student minds or any laws of learning' (p. 155). Their theory of individual learning is dependent upon the historical and philosophical analyses of scientific change provided by Thomas Kuhn, Imre Lakatos and Stephen Toulmin. Once Strike and Posner focus on 'rational' conceptual change, then clearly philosophy enters the psychological picture.[25] This point goes back at least to Aristotle (Jastrzebski 2012) and was made by Ned Block in the Preface to his important anthology on *Philosophy of Psychology*:

> It is increasingly clear that progress in philosophy of mind is greatly facilitated by knowledge of many areas of psychology and also that progress in psychology is facilitated by knowledge of philosophy. [. . .] A host of crucial issues do not 'belong' to either philosophy or psychology, but rather fall equally well in both

> disciplines because they reflect the traditional concerns of both fields. The problems will yield only to philosophically sophisticated psychologists or to psychologically sophisticated philosophers.
>
> (Block 1980, v)

This, of course, supports the thesis of this book, that there are many theoretical issues in science education that require the input of HPS for their elaboration and resolution.

Conclusion

Science has been enormously influential in shaping the material, technical, religious and cultural dimensions of the modern world, and in turn it has been shaped by these societal aspects. Modern science is one of the major accomplishments of the human race. We are seeing something of the constitution of the largest and smallest bodies in the world around us and understanding more and more about our own bodies, brains, health, and more about our environment and the other species with which we share it. The professional purpose of science education is to introduce students into the conceptual and procedural realms of science. It has been argued that history of science facilitates this introduction. But science education also has a wider purpose, which is to help students learn about science – its changing methods, its forms of organisation, its methods of proof, its interrelationships with the rest of culture and so forth. It has been argued that this requires contextual and historical approaches to science teaching.

The integrative function of history is perhaps its fundamental value to science education. History allows seemingly unrelated topics within a science discipline to be connected to each other – Einstein's analysis of Brownian motion to confirm the atomic hypothesis with Brown's attempts to prove Vitalism in biology, and maybe even Brown's botantical work in the early exploration of Australia. History also connects topics across the scientific disciplines – unravelling of the DNA code connected geology, crystallography, chemistry and molecular biology. Historical study shows the interconnections between different realms of knowledge – mathematics, philosophy, theology and physics all had parts to play in the development of, for instance, Newtonian mechanics. Darwinian theory depended upon advances in geology, botany, chemistry, zoology, philosophy, theology and genetics. Finally, history allows some appreciation of the interconnections of realms of academic knowledge with economic, societal and cultural factors. Darwinian evolutionary theory was affected by, and in turn affected, religion, literature, political theory and educational practice. Historical presentation can weave all sorts of seemingly separate topics into strands within disciplines and connect the strands into an intellectual tapestry. Having students develop such a picture is a central concern of liberal education. The cultural significance of science education is, in part, fulfilled to the extent that it contributes to students having a picture of the interconnectedness of human achievement (Suchting 1994).

As with most educational matters, teachers are the key to successful historical teaching of science. Teachers need to be interested in, and to an appropriate degree trained in, history. If they are so prepared and resourced, then, in numerous formal and informal, planned and unplanned ways, history will contribute to the professional and cultural tasks of science education; if they are not, then merely legislating for history, or including it in the curriculum, will have little effect. As has often been said, good teachers can rescue the worst curriculum, and bad teachers can kill the best.

Notes

1 For the writings and arguments of Fleck, see the collection of essays in Cohen and Schnelle (1986).
2 This 'integrative' function of history will be developed below in separate chapters on pendulum motion (Chapter 6) and photosynthesis (Chapter 7).
3 Reviews of the chequered career of history in US science education can be found in Brush (1989) and Klopfer (1969a, 1992). Kauffman (1989) examines specifically the use of history in teaching chemistry.
4 See the articles of Sammis (1932), Oppe (1936) and Jaffe (1938, 1955) and the much-reprinted text of Jaffe (1942).
5 For early critical accounts of Kuhn's theory, see: Gutting (1980), Lakatos and Musgrave (1970), Shapere (1964) and Shimony (1976). These did not slow the almost out-of-control enthusiasm across the academy for all things Kuhnian. For 'Kuhn and Education', see Fuller (2000), Kindi (2005) and Matthews (2000b, 2004).
6 This was a 1950 address to the American Association of Physics Teachers – 'A Sense of History in Science' (Cohen 1950). After obtaining his PhD in the history of science, the second such degree awarded in the US, Cohen taught in the general science course and wrote his own best seller, *The Birth of a New Physics* (Cohen 1961), for the PSSC school physics committee.
7 The first edition contained eight cases: three in biology – The Sexuality of Plants, Frogs and Batteries, Cells of Life; two in chemistry – Discovery of Bromine, Chemistry of Fixed Air; and three in physics – Fraunhofer Lines, Speed of Light, Air Pressure.
8 Joseph Schwab was long associated with the University of Chicago and was imbued with its 'great books' tradition. He had, independently of Kuhn and contemporaneously with him, enunciated a distinction between 'fluid' and 'stable' periods of scientific enquiry, which parallels Kuhn's better known distinction between 'revolutionary' and 'normal' science (Siegel 1978). Selections of his articles are in Ford and Pugno (1964) and Westbury and Wilkof (1978). His work and achievements are reviewed in DeBoer (2014).
9 The philosopher and historian William Whewell was one of the first to advocate the contributions of the history of science to education more generally (Whewell 1855). This tradition has been well documented by Bill Brock (1989), Edgar Jenkins (1979, 1990) and W.J. Sherratt (1983).
10 These reports were the subject of much debate and controversy, with some labelling them 'Alternatives to Science Education' (Jenkins 1998).
11 For a discussion of Galileo's erroneous theory of the tides, see Brown (1976) and Shea (1970).
12 This is the expression coined by Gaston Bachelard (Bachelard 1934/1984) to indentify deep-seated conceptual barriers to scientific investigations. These categories blocked completely some lines of investigation and shaped the form of others. The notion was elaborated and utilised by Louis Althusser (Althusser 1969).
13 The torr, the unit used in vacuum measurement, and the pascal, the international pressure unit, perpetuate their names in the present day.

14 For a discussion of the sixteenth-century controversy about a void, see Schmitt (1967).
15 For non-realists, as will be elaborated in Chapter 9, the postulation of invisible mechanisms is *tout court*; they are just shorthand ways, or a convenience, for connecting phenomenal regularities.
16 The history of the science of air pressure is complex, and experts disagree on various aspects of it. A useful beginning is Middleton (1964). Shapin and Schaffer (1985) have provided an extensive case study of the interactions of science and philosophy in the debate between Hobbes and Boyle over the latter's famous air-pump experiments.
17 See Kragh (1992) for an account of these.
18 On this episode, see Dolphin and Dodick (2014).
19 On Kuhn's views of science education, see Andersen (2000), Kindi (2005), Matthews (2004), Siegel (1979) and contributions to Matthews (2000a).
20 The expression was introduced in 1975 by the sociologist Michel Verret and elaborated in 1985 by Yves Chevallard, in his book *La Transposition Didactique*.
21 See Chapter 10 of Kragh (1987) for a review of such influences.
22 Piaget's position is stated most fully in Piaget and Garcia (1989). See Franco and Colinvaux-de-Dominguez (1992) and contributions to Strauss (1988).
23 Among the better-known contributions are: Carey (2009), Gopnik (1996), Kitcher (1988) and Vosniadou (2013). This tradition of research is reviewed in Dunst and Levine (2014).
24 For some of the literature see: diSessa and Sherin (1998), Limón and Mason (2002), Nersessian (1989, 2003) and West and Pines (1985).
25 To understate the problem, Strike and Posner do not fully engage with the problem of using Kuhn to identify rational conceptual change, this was the very thing that Kuhn, in his 'purple passages', denied the possibility of.

References

AAAS (American Association for the Advancement of Science): 1989, *Project 2061: Science for All Americans*, AAAS, Washington, DC. Also published by Oxford University Press, 1990.

Adler, M.J.: 1978, *Aristotle for Everybody*, Macmillan, New York.

Althusser, L.: 1969, *For Marx*, Penguin, Harmondsworth, UK.

Andersen, H.: 2000, 'Learning by Ostension: Thomas Kuhn on Science Education', *Science & Education* 9(1–2), 91–106.

Andreou, C. and Raftopoulos, A.: 2011, 'Lessons from the History of the Concept of the Ray for Teaching Geometrical Optics', *Science & Education* 20(10), 1007–1037.

ASE (Association for Science Education): 1963, *Training of Graduate Science Teachers*, ASE, Hatfield, UK.

ASE (Association for Science Education): 1979, *Alternatives for Science Education*, ASE, Hatfield, UK.

ASE (Association for Science Education): 1981, *Education Through Science*, ASE, Hatfield, UK.

Bachelard, G.: 1934/1984, *The New Scientific Spirit*, Beacon Books, Boston, MA.

Banks, F., Leach, J. and Moon, B.: 2005, 'Extract From New Understandings of Teachers' Pedagogic Knowledge 1', *The Curriculum Journal* 16(3), 331–340.

Besson, U.: 2013, 'Historical Scientific Models and Theories as Resources for Learning and Teaching: The Case of Friction', *Science & Education* 22(5), 1001–1042.

Besson, U.: 2014, 'Teaching About Thermal Phenomena and Thermodynamics: The Contribution of the History and Philosophy of Science'. In M.R. Matthews (ed.) *International Handbook of Research in History, Philosophy and Science Teaching*, Springer, Dordrecht, The Netherlands, pp. 245–283.

Beyerchen, A.D.: 1977, *Scientists Under Hitler: Politics and the Physics Community in the Third Reich*, Yale University Press, New Haven, CT.

Binnie, A.: 2001, 'Using the History of Electricity and Magnetism to Enhance Teaching', *Science & Education* 10(4), 379–389.

Block, N.J. (ed.): 1980, *Readings in Philosophy of Psychology*, Vol.1, Harvard University Press, Cambridge, MA.

Brock, W.H.: 1989, 'History of Science in British Schools: Past, Present and Future'. In M. Shortland and A. Warwick (eds) *Teaching the History of Science*, Basil Blackwell, Oxford, UK, pp.30–41.

Brown, H.I.: 1976, 'Galileo, the Elements, and the Tides', *Studies in History and Philosophy of Science* 7(4), 337–351.

Brush S.G.: 1969, 'The Role of History in the Teaching of Physics', *The Physics Teacher* 7(5), 271–280.

Brush, S.G.: 1989, 'History of Science and Science Education', *Interchange* 20(2), 60–70.

Brush, S.G. and King, A.L.Y. (eds): 1972, *History in the Teaching of Physics*, University Press of New England, Hanover, NH.

Brush, S.G.: 1974, 'Should the History of Science be Rated X?' *Science* 183, 1164–1172.

Burian, R.M.: 2013, 'On Gene Concepts and Teaching Genetics: Episodes From Classical Genetics', *Science & Education* 22(2), 325–344.

Carey, S.: 2009, *The Origin of Concepts*, Oxford University Press, Oxford, UK.

Carr, E.H.: 1964, *What Is History?* Penguin, Harmondsworth, UK.

Cartwright, J.: 2004, *The Discovery of Oxygen: Student Guide*, Department of Chemistry, University of Chester, UK.

Chamizo, J.A.: 2007, 'Teaching Modern Chemistry Through Recurrent Historical Teaching Models', *Science & Education* 16(2), 197–216.

Chang, H.: 2010, 'How Historical Experiments Can Improve Scientific Knowledge and Science Education: The Cases of Boiling Water and Electrochemistry', *Science & Education* 20(3–4), 317–341.

Coelho, R.L.: 2007, 'The Law of Inertia: How Understanding Its History Can Improve Physics Teaching', *Science & Education* 16(9–10), 955–974.

Coelho, R.L.: 2009, 'On the Concept of Energy: How Understanding Its History Can Improve Physics Teaching', *Science & Education* 18(8), 961–983.

Cohen, I.B.: 1950, 'A Sense of History in Science', *American Journal of Physics* 18, 343–359. Reprinted in *Science & Education* 2(3), 1993, 251–277.

Cohen, I.B.: 1961, *The Birth of a New Physics*, Heineman, London.

Cohen, R.S. and Schnelle, T.: 1986, *Cognition and Fact: Materials on Ludwick Fleck*, Reidel, Dordrecht, The Netherlands.

Conant, J.B.: 1945, *General Education in a Free Society: Report of the Harvard Committee*, Harvard University Press, Cambridge, MA.

Conant, J.B.: 1947, *On Understanding Science*, Yale University Press, New Haven, CT.

Conant, J.B. (ed.): 1948, *Harvard Case Histories in Experimental Science*, 2 volumes, Harvard University Press, Cambridge, MA.

Conant, J.B.: 1951, *Science and Common Sense*, Yale University Press, New Haven, CT.

Cotignola, M.I., Bordogna, C., Punte, G. and Cappannini, O.M.: 2002, 'Difficulties in Learning Thermodynamic Concepts: Are They Linked to the Historical Development of this Field?' *Science & Education* 11, 279–291.

De Berg, K.C.: 2008, 'The Concepts of Heat and Temperature: The Problem of Determining the Content for the Construction of an Historical Case Study which Is Sensitive to Nature of Science Issues and Teaching-Learning Issues', *Science & Education* 17, 75–114.

DeBoer, G.E.: 2014, 'Joseph Schwab: His Work and His Legacy'. In M.R. Matthews (ed.) *International Handbook of Research in History, Philosophy and Science Teaching*, Springer, Dordrecht, The Netherlands, pp. 2433–2458.

Desmond, A. and Moore, J.: 1992, *Darwin: The Life of a Tormented Evolutionist*, Penguin Books, London.

Dijksterhuis, E.J.: 1961/1986, *The Mechanization of the World Picture*, Princeton University Press, Princeton, NJ.
diSessa, A.A. and Sherin, B.L.: 1998, 'What Changes in Conceptual Change?', *International Journal of Science Education* 20(10), 1155–1191.
Dolphin, G. and Dodick, J.: 2014 'Teaching Controversies in Earth Science: The Role of History and Philosophy of Science'. In M.R. Matthews (ed.) *International Handbook of Research in History, Philosophy and Science Teaching*, Springer, Dordrecht, The Netherlands, pp. 553–599.
Drake, S.: 1978, *Galileo at Work*, University of Chicago Press, Chicago, IL.
Duit, R.: 2009, *Bibliography – STCSE*. Available at: www.ipn.uni-kiel.de/aktuell/stcse/stcse.html
Dunst, B. and Levine, A.: 2014, 'Conceptual Change: Analogies Great and Small, and the Quest for Coherence'. In M.R. Matthews (ed.) *International Handbook of Research in History, Philosophy and Science Teaching*, Springer, Dordrecht, The Netherlands, pp. 1345–1361.
Ekstig, B.: 1990, 'Teaching Guided by the History of Science: The Discovery of Atmospheric Pressure'. In M.R. Matthews (ed.) *History, Philosophy, and Science Teaching: Selected Readings*, OISE Press, Toronto, pp. 213–217.
El-Hani, C.N. *et al.*: 2014, 'The Contribution of History and Philosophy to the Problem of Hybrid Views about Genes in Genetics Teaching'. In M.R. Matthews (ed.) *International Handbook of Research in History, Philosophy and Science Teaching*, Springer, Dordrecht, The Netherlands, pp. 469–520.
Fauvel, J. (ed.): 1990, *History in the Mathematics Classroom*, The Mathematical Association, Leicester, UK.
Finocchiaro, M.A.: 1980, 'A Symposium on the Use of the History of Science in the Science Curriculum', *Journal of College Science Teaching* 10(1), 14–33.
Fitzpatrick, F. (ed.): 1960, *Policies for Science Education*, Teachers College, Columbia University, New York.
Fleck, L.: 1935/1979, *Genesis and Development of a Scientific Fact*, T.J. Trenn and R.K. Merton (eds), University of Chicago Press, Chicago, IL.
Ford, G.W. and Pugno, L. (eds): 1964, *The Structure of Knowledge and the Curriculum*, Rand McNally, Chicago, IL.
Franco, C. and Colinvaux-de-Dominguez, D.: 1992, 'Genetic Epistemology, History of Science, and Science Education', *Science & Education* 1(3), 255–272.
Fuller, S.: 2000, 'From Conant's Education Strategy to Kuhn's Research Strategy', *Science & Education* 9(1–2), 21–37.
Galili, I.: 2014, 'Teaching Optics: A Historico-Philosophical Perspective'. In M.R. Matthews (ed.) *International Handbook of Research in History, Philosophy and Science Teaching*, Springer, Dordrecht, The Netherlands, pp. 97–128.
Galili, I. and Hazan, A.: 2001, 'The Effect of a History-Based Course in Optics on Students' Views about Science', *Science & Education* 10(1–2), 7–32.
Garritz, A.: 2013, 'Teaching the Philosophical Interpretations of Quantum Mechanics and Quantum Chemistry through Controversies', *Science & Education* 22(7), 1787–1807.
Gascoigne, J.: 2007, 'Getting a Fix', *Isis* 98(4), 769–778.
Gauld, C.F.: 1992, 'Wilberforce, Huxley and the Use of History in Teaching About Evolution', *The American Biology Teacher* 54(7), 406–410.
Gauld, C.F.: 1998, 'Solutions to the Problem of Impact in the 17th and 18th Centuries and Teaching Newton's Third Law Today', *Science & Education* 7(1), 49–67.
Gillham, N.W.: 2001, *A Life of Sir Francis Galton: From African Exploration to the Birth of Eugenics*, Oxford University Press, Oxford, UK.
Gopnik, A.: 1996, 'The Scientist as Child', *Philosopy of Science* 63(4), 485–514.
Graham, L.R.: 1973, *Science and Philosophy in the Soviet Union*, Alfred A. Knopf, New York.

Greca, I.M. and Friere Jr, O.: 2014, 'Meeting the Challenge: Quantum Physics in Introductory Physics Courses'. In M.R. Matthews (ed.) *International Handbook of Research in History, Philosophy and Science Teaching*, Springer, Dordrecht, The Netherlands, pp. 183–209.

Gulikers, I. and Blom, K.: 2001, 'A Historical Angle: A Survey of Recent Literature on the Use and Value of History in Geometrical Education', *Educational Studies in Mathematics* 47, 223–258.

Gutting, G. (ed.): 1980, *Paradigms and Revolutions: Applications and Appraisals of Thomas Kuhn's Philosophy of Science*, University of Notre Dame Press, Notre Dame, IN.

Heilbron, J.L.: 1986, *The Dilemmas of an Upright Man: Max Planck as Spokesman for German Science*, University of California Press, Berkeley, CA.

Heilbron, J.L.: 2010, *Galileo*, Oxford University Press, Oxford, UK.

Hershberg, J.G.: 1993, *James B. Conant: Harvard to Hiroshima and the Making of the Nuclear Age*, Knopf, New York.

Hogg, J.C.: 1938, *Introduction to Chemistry*, Oxford University Press, New York.

Holmyard, E.J.: 1924, *The Teaching of Science*, Bell, London.

Holmyard, E.J.: 1925, *An Elementary Chemistry*, Edward Arnold, London.

Holton, G.: 1952, *Introduction to Concepts and Theories in Physical Science*, Princeton University Press, Princeton, NJ (2nd edn, revised with S.G. Brush, 1985, 3rd edition *Physics the Human Adventure* 2001).

Holton, G.: 1975, 'Science, Science Teaching and Rationality'. In S. Hook, P. Kurtz and M. Todorovich (eds) *The Philosophy of the Curriculum*, Prometheus Books, Buffalo, NY, pp. 101–118.

Holton, G.: 1978, 'On the Educational Philosophy of the Project Physics Course'. In his *The Scientific Imagination: Case Studies*, Cambridge University Press, Cambridge, UK, pp. 284–298.

Holton, G.: 2003, 'The Project Physics Course: Then and Now', *Science & Education* 12(8), 779–786.

Hong, H.-Y. and Lin-Siegler, X.: 2012, 'How Learning About Scientists' Struggles Influences Students' Interest and Learning in Physics', *Journal of Educational Psychology* 104, 469–484.

Hurd, P.D.: 1958, 'Science Literacy: Its Meaning for American Schools', *Educational Leadership* 16, 13–16.

Jaffe, B.: 1938, 'The History of Chemistry and Its Place in the Teaching of Chemistry', *Journal of Chemical Education* 15, 383–389.

Jaffe, B.: 1942, *New World of Chemistry*, Silver Burdett, New York. Revised edns 1947, 1952, 1955, 1959 and 1964.

Jaffe, B.: 1955, 'Using the History of Chemistry in Our Teaching', *Journal of Chemical Education* 32, 183–185.

Jamieson A. and Radick G.: 2013, 'Putting Mendel in His Place: How Curriculum Reform in Genetics and Counterfactual History of Science Can Work Together'. In K. Kampourakis (ed.) *The Philosophy of Biology: a Companion for Educators*, Springer, Dordrecht, The Netherlands, pp. 577–595.

Jastrzebski, A.: 2012, 'Towards a Better Understanding of the Philosophy of Psychology', *History and Philosophy of Psychology* 14(1), 13–33.

Jenkins, E.W.: 1979, *From Armstrong to Nuffield*, John Murray, London.

Jenkins, E.W.: 1990, 'History of Science in Schools: Retrospect and Prospect in the UK', *International Journal of Science Education* 12(3), 274–281. Reprinted in M.R. Matthews (ed.) *History, Philosophy and Science Teaching: Selected Readings*, OISE Press, Toronto, 1991, pp. 33–42.

Jenkins, E.W.: 1998, 'The Association for Science Education and the Struggle to Establish a Policy for School Science in England and Wales, 1976–81', *History of Education* 27(4), 441–459.

Jensen, M.S. and Finley, F.N.: 1995, 'Teaching Evolution Using Historical Arguments in a Conceptual Change Strategy', *Science Education* 79(2), 147–166.

Jung, W.: 1983, 'Toward Preparing Students for Change: A Critical Discussion of the Contribution of the History of Physics to Physics Teaching'. In F. Bevilacqua and P.J. Kennedy (eds) *Using History of Physics in Innovatory Physics Education*, Pavia University, Italy, pp. 6–57. Reprinted in *Science & Education* 1994, 3(2), 99–130.

Kalman, C.S.: 2009, 'A Role for Experiment in Using the Law of Inertia to Explain the Nature of Science: A Comment on Lopes Coelho', *Science & Education* 18(1), 25–31.

Kampourakis, K.: 2013, 'Teaching About Adaptation: Why Evolutionary History Matters', *Science & Education* 22(2), 173–188.

Kauffman, G.B.: 1989, 'History in the Chemistry Curriculum', *Interchange* 20(2), 81–94. Reprinted in M.R. Matthews (ed.) *History, Philosophy and Science Teaching: Selected Readings*, OISE Press, Toronto, 1991, pp. 185–200.

Kindi, V.: 2005, 'Should Science Teaching Involve the History of Science? An Assessment of Kuhn's View', *Science & Education* 14(7–8), 721–731.

Kipnis, N.: 1992, *Rediscovering Optics*, BENA Press, Minneapolis, MN.

Kitchener, R.F.: 1986, *Piaget's Theory of Knowledge: Genetic Epistemology and Scientific Reason*, Yale University Press, New Haven, CT.

Kitcher, P.: 1988, 'The Child as Parent of the Scientist', *Mind and Language* 3(3), 217–228.

Klein, M.J.: 1972, 'Use and Abuse of Historical Teaching in Physics'. In S.G. Brush and A.L. King (eds) *History in the Teaching of Physics*, University Press of New England, Hanover, NH, pp. 12–18.

Klopfer, L.E.: 1964–1966, *History of Science Cases*, Science Research Associates, Chicago, IL.

Klopfer, L.E.: 1969a, 'The Teaching of Science and the History of Science', *Journal of Research in Science Teaching* 6, 87–95.

Klopfer, L.E.: 1969b, *Case Histories and Science Education*, Wadsworth Publishing, San Francisco, CA.

Klopfer, L.E.: 1990, 'Scientific Literacy', in *The International Encyclopedia of Curriculum*, Pergamon Press, Oxford, UK.

Klopfer, L.E.: 1992, 'An Historical Perspective on the History and Nature of Science in School Science Programs'. In R. Bybee, J.D. Ellis, J.R. Giese and L. Parisi (eds) *Teaching About the History and Nature of Science and Technology: Background Papers*, BSCS/SSEC, Colorado Springs, pp. 105–130.

Klopfer, L.E. and Cooley, W.W.: 1961, *The Use of Case Histories in the Development of Student Understanding of Science and Scientists*, Graduate School of Education, Harvard University, Cambridge, MA.

Klopfer, L.E. and Cooley, W.W.: 1963, 'Effectiveness of the History of Science Cases for High Schools in the Development of Student Understanding of Science and Scientists', *Journal of Research in Science Teaching* 1, 35–47.

Klopfer, L.E. and Watson, F.G.: 1957, *Historical Materials and High School Science Teaching*, *The Science Teacher* 24, 264–265, 292–293.

Kokkotas, P.V., Malamitsa, K.S. and Rizaki, A.A. (eds): 2011, *Adapting Historical Knowledge Production to the Classroom*, Sense Publishers, Rotterdam, The Netherlands.

Kragh, H.: 1986, 'Physics and History: Noble Lies or Immoral Truths?'. In P.V. Thomsen (ed.) *Science Education and the History of Physics*, University of Aarhus, Denmark, pp. 70–76.

Kragh, H.: 1987, *An Introduction to the Historiography of Science*, Cambridge University Press, Cambridge, UK.

Kragh, H.: 1992, 'A Sense of History: History of Science and the Teaching of Introductory Quantum Theory', *Science & Education* 1(4), 349–364.

Kuhn, T.S.: 1957, *The Copernican Revolution*, Random House, New York.

Kuhn, T.S.: 1959, 'The Essential Tension: Tradition and Innovation in Scientific Research', *The Third University of Utah Research Conference on the Identification of Scientific Talent*, University of Utah Press, Salt Lake City. Reprinted in his *The Essential Tension*, University of Chicago Press, Chicago, IL, pp. 225–239.

Kuhn, T.S.: 1970, *The Structure of Scientific Revolutions*, 2nd edn, Chicago University Press, Chicago, IL (1st edition, 1962).

Kuhn, T.S.: 1971/1977, 'Concepts of Cause in the Development of Physics'. In his *The Essential Tension: Selected Studies in Scientific Tradition and Change*, University of Chicago Press, Chicago, IL, pp. 21–30.

Lakatos, I.: 1971, 'History of Science and Its Rational Reconstructions'. In R.C. Buck and R.S. Cohen (eds) *Boston Studies in the Philosophy of Science* 8, pp. 91–135.

Lakatos, I. and Musgrave, A. (eds): 1970, *Criticism and the Growth of Knowledge*, Cambridge University Press, Cambridge, UK.

Lennox, J.G. and Kampourakis, K.: 2013, 'Biological Teleology: The Need for History'. In K. Kampourakis (ed.) *The Philosophy of Biology: a Companion for Educators*, Springer, Dordrecht, The Netherlands, pp. 421–454.

Leone, M.: 2014, 'History of Physics as a Tool to Detect the Conceptual Difficulties Experienced by Students: The Case of Simple Electric Circuits in Primary Education', *Science & Education* 23(4), 923–953.

Levrini, O.: 2014, 'The Role of History and Philosophy in Research on the Teaching and Learning of Relativity'. In M.R. Matthews (ed.) *International Handbook of Research in History, Philosophy and Science Teaching*, Springer, Dordrecht, The Netherlands, pp. 157–181.

Limón, M. and Mason, L. (eds): 2002, *Reconsidering Conceptual Change: Issues in Theory and Practice*, Kluwer Academic Publishers, Nowell, MA.

Mach, E.: 1883/1960, *The Science of Mechanics*, Open Court Publishing, LaSalle, IL.

Mach, E.: 1886/1986, 'On Instruction in the Classics and the Sciences'. In his *Popular Scientific Lectures*, Open Court Publishing, LaSalle, IL, pp. 338–374.

McGrath, E. (ed.): 1948, *Science in General Education*, W.C. Brown, Dubuque, IA.

Mansell, A.E.: 1976, 'Science for All', *School Science Review* 57, 579–585.

Matthews, M.R. (ed.): 1989, *The Scientific Background to Modern Philosophy*, Hackett Publishing Company, Indianapolis, IN.

Matthews, M.R. (ed.): 2000a, 'Thomas Kuhn and Science Education', thematic issue, *Science & Education* 9(1–2).

Matthews, M.R.: 2000b, *Time for Science Education: How Teaching the History and Philosophy of Pendulum Motion Can Contribute to Science Literacy*, Kluwer Academic Publishers, New York.

Matthews, M.R.: 2004, 'Thomas Kuhn and Science Education: What Lessons Can be Learnt?' *Science Education* 88(1), 90–118.

Mayr, E.: 1982, *The Growth of Biological Thought*, Harvard University Press, Cambridge, MA.

Middleton, W.E.K.: 1964, *A History of the Barometer*, Johns Hopkins University Press, Baltimore, MD.

Mihas, P. and Andreadis, P.: 2005, 'A Historical Approach to the Teaching of the Linear Propagation of Light, Shadows and Pinhole Cameras', *Science & Education* 14(7–8), 675–697.

Millikan, R.A.: 1950, *Autobiography*, Prentice-Hall, New York.

Moody, E.A.: 1951, 'Galileo and Avempace: The Dynamics of the Leaning Tower Experiment', *Journal of the History of Ideas* 12, 163–193, 375–422. Reprinted in his *Studies in Medieval Philosophy, Science and Logic*, University of California Press, Berkeley, CA, 1975, pp. 203–286.

Nersessian, N.J.: 1989, 'Conceptual Change in Science and in Science Education', *Synthese* 80(1), 163–184. Reprinted in M.R. Matthews (ed.) *History, Philosophy and Science Teaching: Selected Readings*, OISE Press, Toronto, 1991.

Nersessian, N.J.: 2003, 'Kuhn, Conceptual Change and Cognitive Science'. In T. Nickles (ed.) *Thomas Kuhn*, Cambridge University Press, Cambridge, UK, pp. 178–211.
Neugebauer, O.: 1969, *The Exact Sciences in Antiquity*, 2nd edn, Dover, New York.
NSSE (National Society for the Study of Education): 1960, *Rethinking Science Education. 59th Yearbook*, University of Chicago Press, Chicago, IL.
Nye, M.J.: 1975, 'The Moral Freedom of Man and the Determinism of Nature: The Catholic Synthesis of Science and History in the *Revue des Questions Scientifiques*', *British Journal for the History of Science* 8, 274–292.
Oppe, G: 1936, 'The Use of Chemical History in the High School', *Journal of Chemical Education* 13, 412–414.
Padilla, K. and Furio-Mas, C.: 2008, 'The Importance of History and Philosophy of Science in Correcting Distorted Views of "Amount of Substance" and "Mole" Concepts in Chemistry Teaching', *Science & Education* 17(4), 403–424.
Pais, A.: 1982, *Subtle is the Lord: The Science and Life of Albert Einstein*, Oxford University Press, New York.
Pais, A.: 1991, *Neils Bohr's Times, in Physics, Philosophy, and Polity*, Clarendon Press, Oxford, UK.
Panagiotou, E.V.: 2011, 'Using History to Teach Mathematics: The Case of Logarithms', *Science & Education* 20(1), 1–35.
Paul, H.: 1979, *The Edge of Contingency: French Catholic Reaction to Scientific Change from Darwin to Duhem*, University of Florida Press, Gainesville.
Pflaum, R.: 1989, *Grand Obsession: Madame Curie and Her World*, Doubleday, New York.
Piaget, J.: 1970, *Genetic Epistemology*, Columbia University Press, New York.
Piaget, J. and Garcia, R.: 1989, *Psychogenesis and the History of Science*, Columbia University Press, New York.
Posner, G.J., Strike, K.A., Hewson, P.W. and Gertzog, W.A.: 1982, 'Accommodation of a Scientific Conception: Toward a Theory of Conceptual Change', *Science Education* 66(2), 211–227.
PSSC (Physical Science Study Committee): 1960, *Physics*, D.C. Heath, Boston, MA.
Pumfrey, S.: 1987, 'The Concept of Oxygen: Using History of Science in Science Teaching'. In M. Shortland and A. Warwick (eds) *Teaching the History of Science*, Basil Blackwell, Oxford, UK, pp. 142–155.
Ryan, J.: 1992, 'Finding Generalizable Strategies in Scientific Theory Debates'. In S.P. Norris (ed.) *The Generalizability of Critical Thinking: Multiple Perspectives on an Educational Ideal*, Teachers College Press, New York, pp. 66–79.
Sammis, J.H.: 1932, 'A Plan for Introducing Biographical Material into Science Courses', *Journal of Chemical Education* 9, 900–902.
Schecker, H.: 1992, 'The Paradigmatic Change in Mechanics: Implications of Historical Processes on Physics Education', *Science & Education* 1(1), 71–76.
Schilpp, P.A. (ed.): 1951, *Albert Einstein: Philosopher–Scientist*, 2nd edn, Tudor, New York.
Schmitt, C.B.: 1967, 'Experimental Evidence For and Against a Void: The Sixteenth-Century Arguments', *Isis* 58, 352–366.
Schwab, J.J.: 1949, 'The Nature of Scientific Knowledge as Related to Liberal Education', *Journal of General Education* 3, 245–266. Reprinted in I. Westbury and N.J. Wilkof (eds) *Joseph J. Schwab: Science, Curriclum, and Liberal Education*, University of Chicago Press, Chicago, IL, 1978.
Schwab, J.J.: 1950, 'The Natural Sciences: The Three Year Programme'. In University of Chicago Faculty, *The Idea and Practice of General Education*, University of Chicago Press, Chicago.
Schwab, J.J.: 1963, *Biology Teacher's Handbook*, Wiley, New York.
Seroglou, F. and Koumaras, P.: 2001, 'The Contribution of the History of Physics in Physics Education: A Review', *Science & Education* 10 (1–2), 153–172.
Shapere, D.: 1964, 'The Structure of Scientific Revolutions', *Philosophical Review* 73, 383–394.

Shapin, S. and Schaffer, S.: 1985, *Leviathan and the Air-Pump: Hobbes, Boyle, and the Experimental Life*, Princeton University Press, Princeton, NJ.
Shea, W.R.: 1970, 'Galileo's Claim to Fame: The Proof that the Earth Moves From the Evidence of the Tides', *British Journal for the History of Science* 5, 111–127.
Sherratt, W.J.: 1983, 'History of Science in the Science Curriculum: An Historical Perspective', *School Science Review* 64, 225–236, 418–424.
Shimony, A.: 1976, 'Comments on Two Epistemological Theses of Thomas Kuhn'. In R.S. Cohen, P.K. Feyerabend and M.W. Wartofsky (eds) *Essays in Memory of Imre Lakatos*, Reidel, Dordrecht, The Netherlands, pp.569–588.
Shulman, L.S.: 1986, 'Those Who Understand: Knowledge Growth in Teaching', *Educational Researcher* 15(2), 4–14.
Shulman, L.S.: 1987, 'Knowledge and Teaching: Foundations of the New Reform', *Harvard Educational Review* 57(1), 1–22.
Sibum, H.O.: 1988, 'The Beginning of Electricity: Social and Scientific Origins and Experimental Setups'. In C. Blondel and P. Brouzeng (eds) *Science Education and the History of Physics*, Université Paris-Sud, Paris, pp. 139–146.
Siegel, H.: 1978, 'Kuhn and Schwab on Science Texts and the Goals of Science Education', *Educational Theory* 28, 302–309.
Siegel, H.: 1979, 'On the Distortion of the History of Science in Science Education', *Science Education* 63, 111–118.
Snow, C.P.: 1963, *The Two Cultures: A Second Look*, Cambridge University Press, Cambridge, UK.
Sobel, D.: 1994, *Longitude: The True Story of a Lone Genius Who Solved the Greatest Scientific Problem of His Time*, Walker Publishing, New York.
Solomon, J.: 1989, *The Big Squeeze*, Association for Science Education, Hatfield, UK.
Stinner, A., McMillan, B., Metz, D., Jilek, J. and Klassen, S.: 2003, The Renewal of Case Studies in Science Education', *Science & Education* 12, 617–643.
Strauss, S. (ed.): 1988, *Ontogeny, Philogeny and Historical Development*, Ablex, Norwood, NJ.
Strike, K.A. and Posner, G.J.: 1992, 'A Revisionist Theory of Conceptual Change'. In R. Duschl and R. Hamilton (eds) *Philosophy of Science, Cognitive Psychology, and Educational Theory and Practice*, State University of New York Press, Albany, NY, pp. 147–176.
Suchting, W.A.: 1994, 'Notes on the Cultural Significance of the Sciences', *Science & Education* 3(1), 1–56.
Villani, A. and Arruda, S.M.: 1998, 'Special Theory of Relativity, Conceptual Change and History of Science', *Science & Education* 7(1), 85–100.
Vosniadou, S.: 2013, 'Conceptual Change in Learning and Instruction: The Framework Theory Approach'. In S. Vosniadou (ed.) *The International Handbook of Conceptual Change*, 2nd edn, Routledge, New York, pp. 11–30.
Wandersee, J.H.: 1985, 'Can the History of Science Help Science Educators Anticipate Students' Misconceptions?', *Journal of Research in Science Teaching* 23(7), 581–597.
Wandersee, J.H. and Roach, L.M.: 1998, 'Interactive Historical Vignettes'. In J.J. Mintzes, J.H. Wandersee and J.D.Novak (eds) *Teaching Science for Understanding. A Human Constructivist View*, Academic Press, San Diego, CA, pp. 281–306.
West, L.H.T. and Pines, A.L. (eds): 1985, *Cognitive Structure and Conceptual Change*, Academic Press, New York.
Westbury, I. and Wilkof, N.J. (eds): 1978, *Joseph J. Schwab: Science, Curriculum, and Liberal Education*, University of Chicago Press, Chicago, IL.
Westfall, R.S.: 1980, *Never at Rest: A Biography of Isaac Newton*, Cambridge University Press, Cambridge, UK.
Whewell, W.: 1855, 'On the Influence of the History of Science Upon Intellectual Education'. In *Lectures on Education Delivered at the Royal Institution on Great Britain*, J.W.Parker, London.

Whitaker, M.A.B.: 1979, 'History & Quasi-history in Physics Education Pts I, II', *Physics Education* 14, 108–112, 239–242.
Yager, R.E. and Penick, J.E.: 1987, 'Resolving the Crisis in Science Education: Understanding Before Resolution', *Science Education* 71(1), 49–55.
Zinn, H.: 1999, *A People's History of the United States: 1492–Present*, 2nd edn, Harper Collins, New York.

Chapter 5

Philosophy in Science and in Science Classrooms

Whenever science is taught, philosophy is taught. Messages are conveyed, explicitly or implicitly, about epistemology, ontology, ethics, plausible reasoning, argumentation and other philosophical topics, including religion and aesthetics. In this chapter, ways in which philosophy's presence in science education has been and can be made more explicit will be examined. It will be argued that, by making philosophy more explicit, the goals of good technical science education can be advanced – students will understand the subject better and be more proficient at scientific reasoning – and, at the same time, something of the more general cultural and epistemological dimension of science can be conveyed. This consideration can be extended to the teaching of history, mathematics, geography, economics and most school subjects: they all have philosophical dimensions, and these are best made clear and explicitly engaged with in an informed manner.

Science and Philosophy

The separation of science education from philosophy results in a distorted science education. From the ancient Greeks to the present, science has been interwoven with philosophy: science, metaphysics, logic and epistemology have been inseparable. Most of the great scientists – Democritus, Aristotle, Copernicus, Galileo, Descartes, Newton, Leibnitz, Boyle, Faraday, Darwin, Mach, Einstein, Planck, Heisenberg, Schrödinger – were at the same time philosophers. In the early twentieth century, the German theologian Adolf von Harnack commented that, 'People complain that our generation has no philosophers. Quite unjustly: it is merely that today's philosophers sit in another department, their names are Planck and Einstein' (Scheibe 2000, p. 31).

Despite revolutions, paradigm changes, commercialisation and much else, contemporary science, and more especially contemporary school science, is continuous with the new science of Galileo and Newton and prompts the same range of philosophical questions: science and philosophy continue to go hand in hand.[1] Peter Bergmann expressed this point when he said that he learned from Einstein that, 'the theoretical physicist is . . . a philosopher in workingman's clothes' (Bergmann 1949, p. v, quoted in Shimony 1983, p. 209).[2] One commentator on the work of Niels Bohr remarked that:

> For Bohr, the new theory [quantum theory] was not only a wonderful piece of physics; it was also a philosophical treasure chamber which contained, in a new form, just those thoughts he had dreamed about in his early youth.
>
> (Petersen 1985, p. 300)

Most of the major physicists of the nineteenth and twentieth centuries wrote books on philosophy and the engaging overlaps between science and philosophy.[3] Many less-well-known physicists also wrote such books, teasing out relations between their scientific work and the ontology, epistemology and ethics that it presupposed and for which it had implications.[4] And not just physicists: many chemists and biologists have made contributions to this genre.[5]

This is not, of course, to say that all these good scientists wrote good philosophy or drew sound conclusions from their experience in science: some did; others did not.[6] The point is not that the scientists had sound philosophy, it is rather that they all philosophised; they all reflected on their discipline and their activity, and they saw that such reflection bore upon the big and small questions of philosophy. This fact supports the contention that philosophy is inescapable in good science;[7] it should also suggest that philosophy is inescapable in good science education.

The Oxford philosopher and historian R.G. Collingwood (1889–1943), in his landmark study *The Idea of Nature*, wrote on the history of the mutual interdependence of science and philosophy and commented that:

> The detailed study of natural fact is commonly called natural science, or for short simply science; the reflection on principles, whether those of natural science or of any other department of thought or action, is commonly called philosophy . . . but the two things are so closely related that natural science cannot go on for long without philosophy beginning; and that philosophy reacts on the science out of which it has grown by giving it in future a new firmness and consistency arising out of the scientist's new consciousness of the principles on which he has been working.
>
> (Collingwood 1945, p. 2)

He goes on to write that:

> For this reason it cannot be well that natural science should be assigned exclusively to one class of persons called scientists and philosophy to another class called philosophers. A man who has never reflected on the principles of his work has not achieved a grown-up man's attitude towards it; a scientist who has never philosophized about his science can never be more than a second-hand, imitative, journeyman scientist.
>
> (Collingwood 1945, p. 2)

What Collingwood says about the requirement of 'reflecting upon principles' being necessary for the practice of good science can equally be said for the practice of good science teaching. Liberal education promotes just such deeper reflection and the quest to understand the meaning of basic concepts, laws or methodologies for any discipline (mathematics, history, economics, theology) being taught, including science.

Metaphysical issues naturally emerge from the subject matter of science. Historical studies portray the interdependence of science and metaphysics. The Galilean/Aristotelian controversy over final causation, the Galilean/Keplerian controversy over the lunar theory of tides, the Newtonian/Cartesian argument over action at a distance, the Newtonian/Berkelian argument over the existence of absolute space and time, the Newtonian/Huygensian–Fresnelian argument over the particulate theory of light, the Darwinian/Paleyian argument over design and natural selection, the Mach/Planck argument over the realistic interpretation of atomic theory, the Einstein/Copenhagen dispute over the deterministic interpretation of quantum theory – all bring to the fore metaphysical issues. Metaphysics is pervasive in science.[8]

As has been mentioned, Galileo was an outstanding example of the scientist–philosopher. He made substantial philosophical contributions in a variety of areas: in ontology, with his distinction of primary and secondary qualities; in epistemology, both with his criticism of authority as an arbiter of knowledge claims and with his subordination of sensory evidence to mathematical reason; in methodology, with his development of the mathematical–experimental method; and in metaphysics, with his critique of the Aristotelian causal categories and rejection of teleology as an explanatory principle. It is, thus, unfortunate that, despite his important contributions to the subject, and despite his acknowledged influence on just about all philosophers of the seventeenth century and such subsequent philosophers as Kant and Husserl, Galileo makes scant appearence in most histories of philosophy, and most science texts ignore his philosophical interests and contributions.

Science has always been conducted within the context of the philosophical ideas of the time. This is to be expected. Scientists think, write and talk with the language and conceptual tools available to them; more generally, people who form opinions are themselves formed in specific intellectual circumstances, and their opinions are constrained by these circumstances. Newton said that he was able to see further than others because he stood upon the shoulders of giants: without Copernicus, Kepler and Galileo, not to mention Euclid's geometry, there would not have been a unified theory of terrestrial and celestial mechanics. A scientist's understanding and approach to the world is formed by his or her education and milieu, and this milieu is pervaded by the philosophies of the period. From an objectivist point of view, what these claims are pointing to is the fact that science is a system of concepts, definitions, methodologies, results, instruments and professional organisations created and developed by individuals, but it predates the individual who comes to learn it and to work within it. Inasmuch as the former embodies philosophical suppositions, then the work of the scientist will be shaped by philosophy.[9]

The connection of science with philosophy, broadly understood, is promoted in much popular, bestseller scientific literature. This literature conveys, with sometimes more, and other times less, understanding, the basic idea that science affects, and is affected by, other disciplines – philosophy, psychology, theology, mathematics – and, more generally, the worldviews of a culture. The widespread impact of current books is comparable to that enjoyed in the interwar period by Arthur Eddington's (1882–1944) *The Nature of the Physical World* (1928/1978), J.D. Bernal's (1901–1971) *The Social Function of Science* (1939) and James Jeans' (1877–1946) *Physics and Philosophy* (1943/1981). All of these were enormously influential books, affecting thinking and outlooks across the academy and well beyond. Jeans, in the Preface to his book, wrote:

> The aim of the present book is very simply stated; it is to discuss . . . that borderland territory between physics and philosophy which used to seem so dull, but suddenly became so interesting and important through recent developments of theoretical physics. . . . The new interest extends far beyond the technical problems of physics and philosophy to questions which touch human life very closely.
>
> (Jeans 1943/1981, p. i)

Philosophy in the Science Classroom: The Law of Inertia

Science teachers do not have to 'bring philosophy into the classroom from outside'; it is already inside. At a most basic level, any text or scientific discussion will contain terms such as 'law', 'theory', 'model', 'explanation', 'cause,' 'truth,' 'knowledge', 'hypothesis', 'confirmation', 'observation', 'evidence', 'idealisation', 'time', 'space', 'fields', 'species', 'proof', 'evidence', 'mass' and so on. Philosophy begins when students and teachers slow down the science lesson and ask what these terms mean and what the conditions are for their correct use. Students and teachers can be encouraged to ask the philosopher's standard questions – What do you mean by . . .? and, how do you know . . .? – of all these concepts. Such introductory philosophical analysis allows greater appreciation of the distinct empirical and conceptual issues involved when, for instance, Boyle's Law, Dalton's model or Darwin's theory is discussed. It also promotes critical and reflective thinking more generally. Such analysis can be as sophisticated as the classroom occasion requires. These analytic and logical questions and habits of thought can be introduced as early as preschool – as Matthew Lipman and the Philosophy for Children programmes attest – and they can be refined as children mature.[10]

Every topic in a science curriculum, from the more obvious such as evolution, genetics, cosmology, nuclear energy, photosynthesis, atomic theory, continental drift, to the less obvious and mundane, such as Newton's laws, oxidation and pendulum motion, can be the occasion for fruitful historical and philosophical investigation. With all science topics, students can be introduced to basic philosophical notions and procedures – evidence, hypothesis testing, explanation, theory dependence and so on.

Consider the Law of Inertia and its related concept of force. The law is the foundation stone of classical physics, which is taught in school to every science student. A representative textbook statement is:

> Every body continues in its state of rest or of uniform motion in a straight line except in so far as it is compelled by external impressed force to change that state.
> (Booth & Nicol 1931/1962, p. 24)

The law is so contrary to experience that it is to be expected that students find it difficult or impossible to believe; and, if they do believe it, is only for the purpose of examination and calculation. All physics teachers have this expectation confirmed on a daily basis. Students can conclude, as one German student did, that, 'physics is not about the world' (Schecker 1992, p. 75). Newton's first law might be 'demonstrated' by means of sliding a puck on an air table or on an ice sheet, or by utilising a version of Galileo's inclined plane demonstration.[11] In a purely technical science education, the law is learned by heart, and problems are worked out using its associated formulae: $F = ma$. Technical purposes might be satisfied with correct memorisation and mastery of the quantitative skills – 'a force of X newtons acts on a mass of Y kilograms: what acceleration is produced?' – but the goals of liberal education cannot be so easily satisfied. The above students deserve to be given some reasonable explanation for their being required to believe in counter-experiential statements, much less always-falsified laws. Ultimately, this means good reasons for believing in Newtonianism, and this will hinge on some modicum of HPS. Not to provide or intimate such reasons is akin to former Soviet Union students having to believe that, 'communism is the world's best social system', despite overwhelming contrary evidence.

Just a little philosophical reflection and historical investigation on this routine topic of inertia open up whole new scientific and educational vistas. The medieval natural philosophers were in the joint grip of Aristotle's physics and of common-sense beliefs resulting from their routine everyday experience; indeed, Aristotle's physics was more or less just the sophisticated articulation of common sense. A contemporary Aristotelian says that:

> Aristotle began where everyone should begin – with what he already knew in the light of his ordinary, commonplace experience. . . . Aristotle's thinking began with common sense, but it did not end there. It went much further. It added to and surrounded common sense with insights and understandings that are not common at all.
> (Adler 1978, pp. xi, xiii)

These understandings resulted in the medieval commitment to the principle of *omne quod movetur ab alio movetur* – the famous assertion of Aristotle,

Aquinas and all the scholastics that translates as: 'Whatever is moved is moved by another (the motor)'; and its inverse: if a motor ceases to act, then motion ceases. The principle grew out of daily experience, common sense and Aristotle's physics. Clagett summarised Aristotle's conviction as follows:

> For Aristotle motion is a process arising from the continuous action of a source of motion or 'motor' and a 'thing moved'. The source of motion or motor is a force – either internal as in natural motion or external as in unnatural [violent] motion – which during motion must be in contact with the thing moved.
>
> (Clagett 1959, p. 425)

Given the fact of motion in the world, then the principle led Aristotle to the postulation of a first mover. Aquinas and the scholastics took over this argument and made it an argument for the existence of a prime mover, whom they identified as God.[12]

Medieval impetus theory was an elaboration of Aristotelian physics: the mover gave something (impetus) to the moved that kept it in motion when the mover was no longer acting (the classic case of a thrown projectile). Some, such as da Marchia, thought this impetus naturally decayed, and, hence, the projectile's motion eventually ceased. Others, such as Buridan, thought that the transferred power was only diminished when it performed work, and, as pushing aside air was work, then the projectile's motion would also eventually cease. Both theories were consistent with the phenomena: when a stone is thrown from the hand, it goes only so far, then drops to the ground.[13] Galileo performed a thought experiment by thinking through Buridan's theory to the circumstance of there being no work performed, in which case the projectile, once impressed with impetus (force in modern speak), would continue moving forever. But, for Galileo, it would follow the Earth's contour. He repeated this circumstance with his experiment of a ball rolling down one incline and up another; as the second plane was gradually lowered towards horizontal, the ball moved further and further along it. He supposed that, with the smoothest plane and the most polished ball, the ball would just keep moving on the second plane when horizontal; this was the visualisation of his theory of circular inertia.[14]

Galileo had no idea of a body being able to move off the Earth in a straight line, away into an infinite void. Like everyone else, Galileo was both physically and conceptually anchored to the Earth. It was only Newton who would make this massive conceptual leap, sufficient to have a projectile leave the Earth and move in an infinite void; he moved conceptually from a 'Closed World to the Infinite Universe', to use the expressive title of Alexandre Koyré's (1892–1964) masterpiece on the subject (Koyré 1957). Newton's conceptual leap was the foundation of modern mechanics. The whole 2,000-year history of the development of the law of inertia reveals a good deal about the structure and mechanisms of the scientific enterprise, including the process of theory generation and theory choice.[15] Working through this history of argument bears fruit for arguments about worldviews and science.

For example, after the usual 'failure to convince' lesson on inertia, students typically say that, 'Newton's laws do not describe the world we experience', with some saying that, 'physics is about a special world and I do not know why we are studying it'. It was, of course, Newtonian idealisation that the Romantic reaction was directed against. For them (Keats, Coleridge, Goethe and so on), the rich world of human lived experience was not captured by the colourless point masses of Newton, or the emotion- and feeling-less mechanical world of the new science. In the twentieth century, Marcuse, Husserl, Tillich and others have repeated versions of this charge. However, what needs to be recognised, and what a competent HPS-informed teacher can point out, is what Aldous Huxley correctly observed, namely:

> The scientific picture of the world is inadequate, for the simple reason that science does not even profess to deal with experience as a whole, but only with certain aspects of it in certain contexts. All of this is quite clearly understood by the more philosophically minded men of science. . . . [Unfortunately] our time contains a large element of what may be called 'nothing but' thinking.
>
> (Huxley 1947, p. 28)

Apart from interesting and important history, basic matters of philosophy arise in any good classroom treatment of the law of inertia and the concept of force:

- *epistemology*: we never see force-free behaviour in nature, nor can it be experimentally induced, so what are the source and justification of our knowledge of bodies acting without impressed forces?
- *ontology*: we do not see or experience force apart from its manifestation, so does it have independent existence? What is mass? What is a measure of mass as distinct from weight?
- *cosmology*: does such an inertial object go on forever in an infinite void? Were bodies created with movement? If so, does it naturally decay (impetus theory), or only when work is done (Newtonian)?

These are the sorts of consideration that prompted the French polymath Henri Poincaré (1854–1912) to say: 'When we say force is the cause of motion, we are talking metaphysics' (Poincaré 1905/1952, p. 98). And, as every physics class talks of force being the cause of acceleration, then there is metaphysics present in every classroom, just waiting to be exposed by students who are encouraged to think carefully about what they are being taught, and by teachers who know something of the history and philosophy of the subject they teach. Such exposition and engagement of school classes in the fundamental ontological, epistemological and methodological matters of philosophy that are occasioned by teaching and learning the law of inertia can be seen in a number of excellent texts.[16] All of this prepares the ground for a more nuanced and informed discussion of the big issues of science and metaphysics and helps cultivate a body of 'more philosophically minded' students of science.[17]

Thinking carefully and historically about basic principles and concepts is a quite general point about the intelligent and competent mastery of any discipline, be it mathematics, history, psychology, literature, theology, economics or anything else. They all have their own, and overlapping, concepts and standards for identifying good and bad practice and judgements; consequently, there are philosophical questions (epistemological, ontological, methodological and ethical) about each discipline; there is a philosophy of each discipline. The intelligent learning of any discipline requires some appropriate interest and competence in its philosophy; that is simply what 'learning with understanding' means – an obvious educational point made by Ernst Mach (1886/1986) and, more recently, by Israel Scheffler (1970).[18] If serious scientists, such as listed earlier in this chapter, feel it important to write books on the philosophy of their subject, then assuredly science teachers and students will benefit from following their example and engaging with the same questions.

The arguments of Mach and Scheffler have, belatedly and independently, found expression in the wide international calls for students to learn about the NOS while learning science. One cannot learn about the NOS without learning philosophy of science, which was precisely Mach and Scheffler's argument.

Thought Experiments in Science

Thought experimentation is one rich and productive scientific/philosophical subject to explore with classes. Thought experiments have had an important role in the history of science – witness their use by pre-Socratics, medievals, Galileo, Leibniz, Newton, Carnot and, in the past century, by Einstein, Poincaré, Schrödinger, Eddington and Heisenberg. Neither relativity nor quantum theory could have been developed or conceptually tested without recourse to thought experiment; but whether thought experiment alone could confirm the respective theories is a much-argued point in the epistemology of science.[19] Thought experiments clearly connect science with philosophy, with some observing that the history of philosophy is just one long thought experiment. Thought experiments illuminate a significant dimension of scientific thinking and they can be utilised to great effect in classrooms. For many, they open up a wholly new scientific terrain. The following is one characterisation:

> Thought experiments are devices of the imagination used to investigate the nature of things. Thought experiments often take place when the method of variation is employed in entertaining imaginative suppositions. They are used for diverse reasons in a variety of areas, including economics, history, mathematics, philosophy, and physics. Most often thought experiments are communicated in narrative form, sometimes through media like a diagram. Thought experiments

should be distinguished from thinking about experiments, from merely imagining any experiments to be conducted outside the imagination, and from any psychological experiments with thoughts. They should also be distinguished from counterfactual reasoning in general, as they seem to require an experimental element.

(Brown & Fehige, 2011, p. 1)

Galileo

Ernst Mach (1838–1916), in his *Mechanics*, draws attention to one of the greatest thought experiments in the history of science, the thought experiment in Day One of Galileo's *Dialogues Concerning Two New Sciences* (1638/1954), which is directed at disproving the Aristotelian thesis that bodies in free fall descend with a speed that is proportional to their weight. In Galileo's text, the Aristotelian, Simplicio, stated the received view that, 'bodies of different weight move in one and the same medium with different speeds which stand to one another in the same ratio as the weights' (Galileo 1638/1954, p. 60).[20] There follows inconclusive talk about dropping cannon balls and musket balls from great heights and the claimed differences in time between when they hit the ground. The dialogue continues as follows, with Salviatti the spokesperson for Galileo.

SALV: But, even without further experiment, it is possible to prove clearly, by means of a short and conclusive argument, that a heavier body does not move more rapidly than a lighter one provided both bodies are of the same material and in short such as those mentioned by Aristotle. But tell me, Simplicio, whether you admit that each falling body acquires a definite speed fixed by nature, a velocity which cannot be increased or diminished except by the use of force [*violenza*] or resistance.

SIMP: There can be no doubt that one and the same body moving in a single medium has a fixed velocity which is determined by nature and which cannot be increased except by the addition of momentum [*impeto*] or diminished except by some resistance which retards it.

SALV: If then we take two bodies whose natural speeds are different, it is clear that on uniting the two, the more rapid one will be partly retarded by the slower, and the slower will be somewhat hastened by the swifter. Do you not agree with me in this opinion?

SIMP: You are unquestionably right.

SALV: But if this is true, and if a large stone moves with a speed of, say, eight while a smaller moves with a speed of four, then when they are united, the system will move with a speed less than eight; but the two stones when tied together make a stone larger than that which before moved with a speed of eight. Hence the heavier body moves with less speed than the lighter, an effect which is contrary to your supposition. Thus you see how, from your assumption that the heavier body moves more rapidly than the lighter one, I infer that the heavier body moves more slowly.

> SIMP: I am all at sea because it appears to me that the smaller stone when added to the larger increases its weight and by adding weight I do not see how it can fail to increase its speed or, at least, not to diminish it.

Galileo's argument is short, it is conclusive and it is elegant. Karl Popper described it as:

> One of the most important imaginary experiments in the history of natural philosophy, and one of the simplest and most ingenious arguments in the history of rational thought about our universe.
>
> (Popper 1934/1959, p. 442)

Next, consider how Galileo criticised, or dissolved, the distinction between natural and violent motions that was firmly embedded in Aristotelian physics – where natural motion occurs when bodies move towards their 'natural' places, and violent motion occurs when they move away from these places. Circular motion was natural for planets; motion towards the centre of the Earth was natural for terrestrial heavy bodies. Galileo conjectured an experiment in which a well was bored through the centre of the Earth to the other side. He asked Aristotelians to envisage what would happen when a stone was dropped down the well. Clearly, it would travel 'naturally' at increasing speed to the centre of the Earth. But what happens when it gets there? Does it stop? Does it keep going and so 'naturally' travel away from the Earth's centre? Does it somehow turn natural into violent motion? The thought experiment was used to investigate the inadequacy of the fundamental Aristotelian distinction.[21] The actual experiment, of course, could never be performed, but its power to illuminate conceptual problems in the old physics was not compromised by that fact.

This, and other such thought experiments, led Alexandre Koyré to claim for Galileo, somewhat exaggeratedly, 'the glory and the merit of having known how to dispense with experiments' (Koyré 1968, p. 75).

Newton

Isaac Newton's *Principia* (Newton 1729/1934) and *Opticks* (Newton 1730/1979) were the foundation stones of modern science; they provided the conceptual and methodological exemplars for early-modern physics and chemistry, then history, social sciences and beyond (Butts & Davis 1970, Cohen 1980). Enlightenment philosophers such as John Locke, Baruch Spinoza, David Hume, Voltaire, Jean d'Alembert and countless others wanted to extend their hypothetico-experimental methods to the study of morals, politics, religion, scripture, law and other intellectual endeavours, with the expectation that comparable knowledge and communal agreements between serious investigators might be reached in those fields (Hyland *et al.* 2003, Porter 2000). As Jan Golinski stated the matter:

> Newton provided cultural as well as theoretical and methodological resources for his followers. His natural philosophy inspired scientists to emulate it and writers and lecturers to popularize it; once popularized, it could convey messages about religion and society as well as about nature. Thus the men of the Enlightenment made of Newton a cultural symbol that answered to many of their diverse intellectual needs.
>
> (Yolton *et al.* 1991, p. 369)

As well as 'practical' experimental work, such as is seen in his pendulum and prism investigations, Newton's science and his system depended heavily on thought experiments; something recognised by Ernst Mach and less recognised by modern science teachers. Perhaps Newton's most widely known thought experiment is his 'cannon ball satellite' conjecture in his 'System of the World', which constitutes Volume 2 of his *Principia.* This is a key part of this epochal unification of celestial and terestrial mechanics: his demonstration that the laws governing the fall of bodies on Earth also govern the movement of planets and comets in the heavens. Newton wrote:

> That by means of centripetal forces the planets may be retained in certain orbits, we may easily understand, if we consider the motions of projectiles; for a stone that is projected is by the pressure of its own weight forced out of the rectilinear path, which by the initial projection alone it should have pursued, and made to describe a curved line in the air; and through that crooked way is at last brought down to the ground; and the greater the velocity is with which it is projected, the farther it goes before it falls to earth. We may therefore suppose the velocity to be so increased, that it would describe an arc of 1, 2, 5, 10, 100, 1,000 miles before it arrived at the earth, till at last, exceeding the limits of the earth, it should pass into space without touching it.
>
> (Newton 1729/1934, p. 551)

This thought experiment prepares the intellectual grounds for the unification of celestial and terrestrial mechanics, but it does not in itself do so. Importantly, the diagram (Figure 5.1) enabled folk to visualise the physics of satellites, but Newton still had to show that the cannon ball could be the Moon. This was done using the ubiquitious pendulum, 'the single most significant tool used in the first edition of *Principia mathematica*' (Meli 2006, p. 269). Newton used the pendulum to show that the distance fallen by the Moon towards the Earth in 1 second was almost exactly the distance fallen by an object to the Earth's surface in 1 second (Boulos 2006, Matthews 2000, pp. 188–193).

Perhaps the second-best-known of Newton's thought experiments, and the one that had the greatest and most enduring impact on physics and philosophy, as well as on theology, was his 'bucket experiment', where he interpreted the relative movements of water and container in a stationary bucket of water that was then set rotating to argue for absolute space and the reality of forces

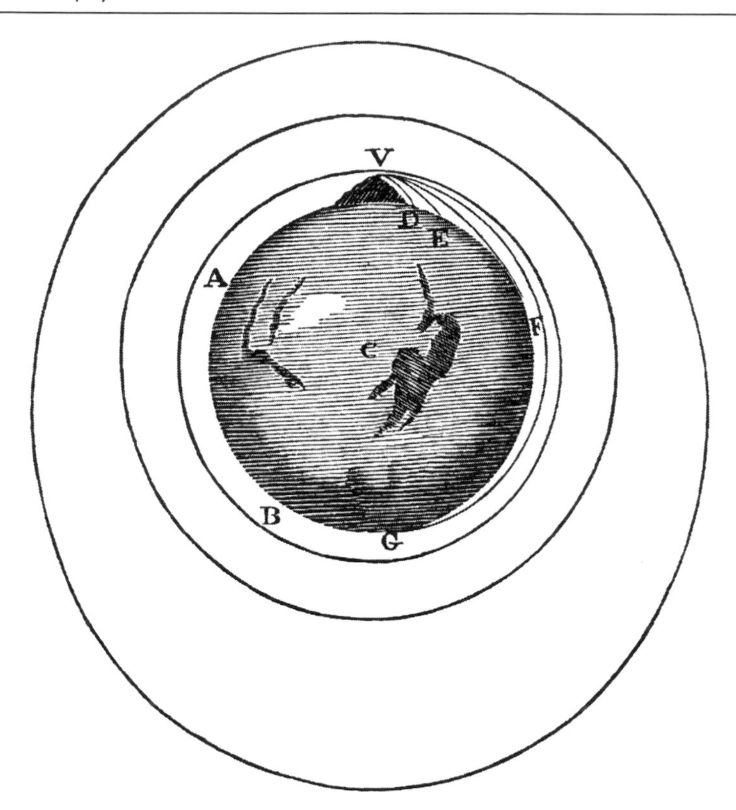

Figure 5.1 Newton's Cannon Ball Satellite
Source: Newton 1729/1934, p. 551

(Newton 1729/1934, pp. 10–11). Newton's system of natural philosophy and his 'system of the world' depended on, assumed or incorporated the idea of absolute space (and time) and, thus, real movement, not just relative movement; for Newton, space had a real existence: there was no 'empty' space between bodies. This was contrary to the cosmology of Descartes and Leibniz, for whom just bodies existed, and space was a non-existent 'relation' between them, and for whom all movement was relative to another body – which body was moving depended just on the conventional nomination of one as the 'unmoving' body. So the universe did not exist in space; space was just a consequence of the existence of the universe's bodies (planets, stars, comets, trees, clouds, etc.). We do not and cannot 'see' the universe; there is no 'outer' view or 'view from nowhere'. Thus, the disagreement seems not to be susceptible to scientific investigation; there seems no experiment that can be conducted on the universe. However, Newton engaged in a thought experiment to settle the issue in his favour. He imagined a rotating bucket of water in an empty universe (see Figure 5.2), which he believed established the existence of absolute space:

> If a vessel, hung by a long cord, is so often turned about that the cord is strongly twisted, then filled with water, and held at rest together with the water; thereupon, by the sudden action of another force, it is whirled about the contrary way, and while the cord is untwisting itself, the vessel continues for some time in this motion; the surface of the water will at first be plain, as before the vessel began to move; but after that, the vessel, by gradually communicating its motion to the water, will make it begin sensibly to revolve, and recede by little and little from the middle, and ascend to the sides of the vessel, forming itself into a concave figure (as I have experienced), and the swifter the motion becomes, the higher will the water rise, till at last, performing its revolutions in the same times with the vessel, it becomes relatively at rest in it. . . . At first, when the relative motion of the water in the vessel was greatest, it produced no endeavor to recede from the axis . . . and therefore its true circular motion had not yet begun. But afterwards, when the relative motion of the water had decreased, the ascent thereof towards the sides of the vessel proved its endeavor to recede from the axis; and this endeavor showed the real circular motion of the water continually increasing, till it had acquired its greatest quantity, when the water rested relatively in the vessel. . . . There is only one real circular motion of any one revolving body, corresponding to only one power of endeavoring to recede from its axis of motion, as its proper and adequate effect . . . the several parts of those heavens, and the planets, which are indeed relatively at rest in their heavens, do yet really move. For they change their position one to another (which never happens to bodies truely at rest), and being carried together with their heavens, partake of their motions, and as parts of revolving wholes, endeavor to recede from the axis of their motions.
>
> (Newton 1729/1934, pp. 10–11)

In both the first (water and bucket stationary) and third instances (water and bucket both moving), the water is at rest with respect to the bucket; yet the water surface is flat in the first and concave in the third. Newton maintains that only real movement of the water (a centrifugal force) can account for the difference in shape. The water must be moving with respect to absolute space. Hence, there can be, and is, real movement of bodies in the universe, and indeed movement of the universe as a whole in space. So, by thought experiment, Newton moves from observations of an actual rotating bucket of water to an unobservable, non-empirical cosmological conclusion. Newton's cosmological picture underwrites his dynamics and his view of force and acceleration: where there are real accelerations, there are real forces; where there are no real, only apparent or relative, accelerations, there are no forces. This is a presupposition of his law of inertia and the preference for inertial frames in specifying laws of nature; the laws are only true in an inertial frame.

Newton's view was rejected first by the proto-positivist Bishop George Berkeley (1685–1753) (1721/1965, p. 270), then by the thorough positivist Ernst Mach (1893/1974, pp. 277–287). The latter dismissed Newton's notions of absolute space and time as more metaphysics rather than physics; he maintained that the concavity of the surface was due to the water rotating with respect to distant celestial bodies, not with respect to any absolute space; if the bucket and water were the only bodies in the universe, the water surface would remain flat throughout the sequences.

We understand these debates about absolute space and time as bearing on the conceptual foundations of physics; but, for Newton and his contemporaries, the controversy had clear theological import, bearing upon accounts of God and his creation of the material world. If there was no empty space when God created matter, what did he create it in? And what then was the non-matter? The latter question invites the heretical, pantheist answer: God himself (Friedman 2009).

Einstein

Albert Einstein (1879–1955) eventually followed his teacher, Mach, in rejecting Newton's account of the bucket experiment. In his 1916 popular exposition of the theory of relativity, he wrote, of the Newtonian claim, that

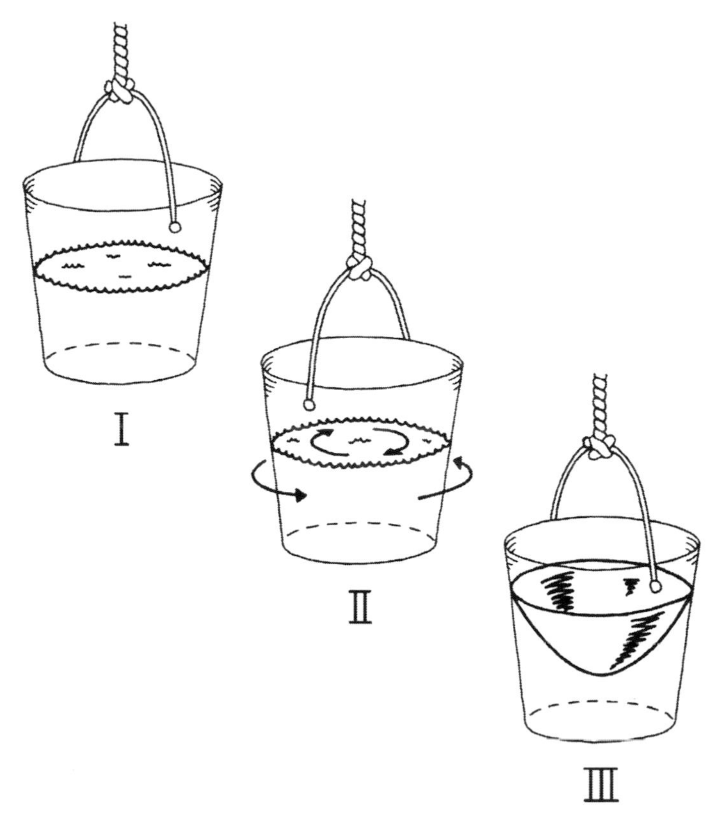

Figure 5.2 Newton's Bucket Experiment
Source: Brown 1991/2010, p. 9

physical laws are only true in inertial frames: 'no person whose mode of thought is logical can rest satisfied with this condition of things' (Einstein 1916/1961, p. 71). He went on to write:

> I seek in vain for a real something in classical mechanics (or in the special theory of relativity) to which I can attribute the different behaviour of bodies considered with respect to the reference systems K and K^1. Newton saw this objection and attempted to invalidate it, but without success. But E. Mach recognised it most clearly of all, and because of this objection he claimed that mechanics must be placed on a new basis. It can only be got rid of by means of a physics which is conformable to the general principle of relativity, since the equations of such a theory hold for every body of reference, whatever may be its state of motion.
>
> (Einstein 1916/1961, pp. 72–73)

Einstein realised that physics advances by bold conjectures and related experimenting in thought. In his 1951 'Autobiographical Essay', he states how as a teenager he felt ill at ease about the then-dominant physical interpretation of Maxwell's equations for electromagnetism. This feeling was vanquished, as was the mechanical interpretation of Maxwell's equations, by a thought

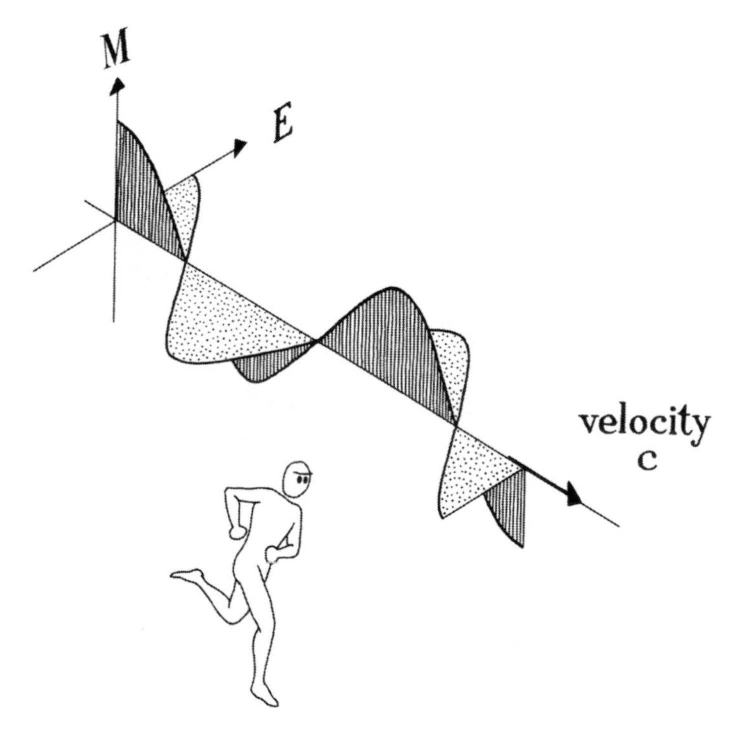

Figure 5.3 Einstein's Light Beam

Source: Brown 1991/2010, p. 17

experiment: Einstein imagined himself running along in front of a light beam and looking back at it (see Figure 5.3). He says:

> I should observe such a beam of light as a spatially oscillatory electromagnetic field at rest. However, there seems to be no such thing, whether on the basis of experience or according to Maxwell's equations.
>
> (Schilpp 1951, p. 53)

He says that, 'in this paradox the germ of the special relativity theory is already contained',[22] and goes on to say:

> The type of critical reasoning which was required for the discovery of this central point was decisively furthered, in my case, especially by the reading of . . . Ernst Mach's philosophical writings.
>
> (Schilpp 1951, p. 53)

The foregoing should suffice to establish the centrality of thought experiments in the history of science; given they have this centrality, they should have a more central position in science classrooms.

Thought Experiments in Science Teaching

The positivist Ernst Mach introduced thought experiments (*Gedankenexperimente*) to science education. He said of thought experiments that, 'The method of letting people guess the outcome of an experimental arrangement has didactic value too. . . . Experimenting in thought is important not only for the professional inquirer, but also for mental development as such'; not only the student but also 'the teacher gains immeasurably by this method' (Mach 1905/1976, pp. 142–143). Thought experiments enabled the teacher to know what grasp students had on the fundamental concepts of a discipline. And Mach meant any discipline: science, mathematics, economics, history, theology or politics.[23]

Each edition of his *Zeitschrift* carried thought experiments for his readers to perform. For instance, he asks, what is expected to happen to a beaker of water in equilibrium on a balance when a suspended mass is lowered into it? Or, in another issue, what happens when a stoppered bottle with a fly on its base is in equilibrium on a balance and then the fly takes off? These examples are of thought experiments of an anticipatory type – the actual experiment can be performed. They engage the mind and they reveal what a student believes about the relevant concepts being investigated. However, some thought experiments are not anticipatory but idealised, because the circumstances postulated cannot be produced – Newton's cannon ball experiment, Galileo's ball dropped into a well through the centre of the Earth and so on. Mach encouraged such exercises, believing that the exercise of imagination and creativity was another way of bridging the gap between humanities and the sciences: 'The planner, the builder of castles in the air, the novelist, the author of social and technological utopias is experimenting with thought' (Mach 1896/1976, p. 136).

Mach's advocacy of thought experiments did not gain many adherents among the science teachers of his day. Imagination, hypothesising and creative thought were no more characteristic of late-nineteenth-century science pedagogy than they are of contemporary science pedagogy. Einstein, who was to place thought experiments on the centre stage of modern physics, made the oft-quoted remark about his own schooling that: 'after I passed the final examination, I found the consideration of any scientific problems distasteful to me for an entire year', and 'It is, in fact, nothing short of a miracle that the modern methods of instruction have not entirely strangled the holy curiosity of inquiry' (Schilpp 1951, p. 17).

Some teachers have brought thought experiments into classrooms in the manner that Mach suggests, and these efforts have been well documented and researched. It is fruitful for teachers to understand the contribution of thought experiments to conceptual change in science and follow through with their use in promotion of conceptual change in classrooms. There has been informed research on this nexus.[24] Nancy Nersessian's work is well known. She writes:

> Thought experimenting is a principal means through which scientists change their conceptual structures. I propose that thought experimenting is a form of 'simulative model-based reasoning'. That is, thought experimenters reason by manipulating mental models of the situation depicted in the thought experiment narrative.
>
> (Nersessian, 1993, p. 292)

What holds for scientists can hold for children learning science. This is the area of which Mach said that experimenting in thought enables teachers to learn how well students understand concepts. Well before constructivism, Mach recognised that effective teaching required that teachers know the concepts as understood by students.

In Ontario, some school courses have used thought experimentation in conjunction with science fiction themes – if the Bionic Man accelerated at a certain rate would his feet melt? (Stinner 1990). These types of 'thinking physics' problem allow teachers and students to determine what they mean by fundamental concepts such as gravity, force, pressure and so on, and to think about the correct conditions for the applicability of concepts.

A textbook that exemplifies this approach to physics is L.C. Epstein's *Thinking Physics* (Epstein 1979), which contains numerous exercises such as the following:

- '*Sputnik I*, the first artificial earth satellite, fell back to earth because friction with the outer part of the earth's atmosphere slowed it down. As *Sputnik* spiralled closer and closer to the earth its speed was observed to: decrease, remain constant, increase?' (Epstein 1979, p. 157).

- 'Over a century ago, J.C. Maxwell calculated that if Saturn's rings were cut from a piece of sheet metal they would not be strong enough to withstand the tidal tension or gravitational gradient tension that Saturn would put on them and would, therefore, rip apart. But suppose the rings were cut from a piece of thick-plate, rather than thin-sheet iron. Might the thick-plate: fail as easily as the thin-sheet, fail more easily than the thin-sheet, not fail as easily as the thin-sheet?' (Epstein 1979, p. 161).

These exercises require students to think about the meaning of concepts used in describing phenomena. They are thought experiments in the tradition of Mach's *Gedankenexperimente*. They engage the mind of the learner in ways that mere calculations or 'recipe book' carrying out of an empirical experiment fail to do. They support Mach's claim that: 'Experimenting in thought is important not only for the professional inquirer, but also for mental development as such.'

A number of thought-experiment units have been taught and researched by Athanasios Velentzas and Krystallia Halkia in Greek schools. In one case, forty16-year-olds were engaged in lessons requiring them to work through Newton's cannon ball thought experiment as a way for them to understand gravitation and satellite motion (Velentzas & Halkia 2013). Part of their encouraging conclusion was:

> The process implemented in this study helped students both to overcome barriers in understanding the laws of physics due to their ideas derived from everyday experience and to modify their ideas related to the gravitational field of the Earth, as well as to retain these ideas and apply them in a new TE in the same knowledge domain two weeks after the teaching intervention.
>
> (Velentzas & Halkia 2013, p. 2637)

Computers are a boon for school thought experimentation. Computers overcome one of the standard problems with routine laboratory work and pupil experimentation: namely, that teachers have to do all the experimental design and planning (Hodson 1988). Problems of time, unfamiliar equipment, safety and so on have meant that students are often reduced to executing a teacher's preplanned experiment. Students learn manipulation skills, develop observation techniques, learn to be attentative – all of which are important – but they rarely learn the conceptual, creative skills that are the hallmark of good science.[25] Computers can remove these practical obstacles to the generation and testing of hypotheses and allow extrapolation to the idealised test situations characteristic of thought experiments. A previous note (19) has listed some of the central literature in the philosophical analysis of thought experiments. Philosophers have also elaborated and debated the epistemology of computer simulation in scientific research.[26] As with thought experiments, this philosophical debate can inform classroom use of computer simulation, virtual labs and computer-based modelling.[27] An HPS-informed teacher can elaborate and engage students in epistemological questions as they arise – for instance, what do we learn, if anything, from a 'black-box', mechanism-free model that makes confirmed predictions about events in the world? Are these acceptable in economics but not in biology?

Argumentation and Logical Reasoning in Science Classrooms

All science programmes aim to develop students' scientific thinking and reasoning, to make them more scientific, as well as having them know more science. However, scientific thinking is multifaceted. The discovery or learning side involves being attentive, observing, hypothesising, being creative, reading well, making inferences and so on. All of this is involved with obtaining and representing data, evidence or information. This discovery side also involves sharing or communicating this 'evidential' material, making suitable inductive and deductive inferences from it and so on. The justification side of scientific thinking involves arguing or establishing how the data bear upon one or more hypotheses or theories, or why some evidence is to be preferred to other evidence and so on. Scientific thinking thus connects to and involves logical thought, critical thinking and argumentation, with, hopefully, the former two preceding the third and then being developed as social engagement and argument develop. Analysis of these processes has been part of the philosophical tradition since at least Plato, and, with knowledgeable teachers, science programmes provide ample opportunity to cultivate these thinking skills, which hopefully have flow-on effects for other studies and for decision-making in life.

In recent decades, this whole constellation of scientific thinking skills has been subsumed under the umbrella title of 'scientific argumentation', as is clear from the cover text of one recent anthology on the subject, where scientific argumentation is said to involve: 'arriving at conclusions on a topic through a process of logical reasoning that includes debate and persuasion' (Khine 2012). Two researchers in the field comment that:

> Scientific arguments are hardly ever strictly formal (logical or mathematical); they are generally analogical, causal, hypothetico-deductive, probabilistic, abductive, inductive . . . One of their functions is to make a theoretical model plausible, convincingly connecting it to a growing number of phenomena.
>
> (Izquierdo-Aymerich & Adúriz-Bravo 2003, p. 38)

This more expansive and realistic account of scientific thinking can be well contributed to by philosophers. Philosophers reasonably would say that such accounts subsume normative or philosophical elements: How will logical reasoning (of a deductive or inductive kind) be separated from illogical reasoning? How will reasonable persuasion be separated from unreasonable persuasion? How is the plausible distinguished from the implausible?

What constitutes a warrant for belief as distinct from just a reason for belief? Not all persuasion is scientific, and not all exchange is debate. That the stars are in alignment might be a reason for someone's belief, but it is simply not a warrant. Here again, philosophy has a clear contribution to make to educational discussion.[28] This contribution connects to a more general one reminding folk that, in order to be educationally useful, sociology of science needs to be connected to philosophy of science; studies of changing belief need to be linked to epistemology; and studies in the psychology of reasoning need to be connected to studies of logic. Logicians have long pointed out that logic is the study of how people ought to think, not the study of how they do think; the latter is the concern of psychology, not of philosophy. Educationalists clearly need to work in both fields, but doing so without losing sight of the distinction. Not all changes of belief are educational, nor are all reasons good reasons.

Theoretical and pedagogical studies of argumentation have abounded, with one recent review article saying that there have been 'more than three hundred references in English, French and Spanish' (Adúriz-Bravo 2014, p. 1448). Some well-known anthologies[29] and review articles[30] discuss the published English research.

Logical Thinking

Thinking logically is certainly not all of scientific thinking, but it is a part of it: persistent illogical thinking advances neither humanities, politics nor religion, and certainly not science. The beginning of logical thinking is clear thinking, and this can be promoted in schools by encouraging clear writing and routinely doing 'structure of argument' exercises, where blocks of prose are given, and students are asked to identify what are premises, what are intermediate steps and what are conclusions; then to draw argument diagrams showing how each is related; and then answer questions about what claims are required to be denied in order to avoid the conclusion, what claims are sufficient to be denied, for coupled claims whether both have to be denied or just one member, what is the difference between conjoint and disjunct claims, and so on.[31]

A good deal of the evidence suggests that schools and, more particularly, science programmes need to do more to promote logical thought. A small Australian study by Gordon Cochaud (1989) is suggestive of the problem that students have with basic logic. Cochaud gave a brief, ten-item logic test to first-year science students at an Australian university. Among the items was this one, where students had to fill in the conclusion:

> If one adds chloride ions to a silver solution then a white precipitate is produced.
> Addition of chloride ions to solution K produced a white precipitate.
> Therefore . . .

In his group of sixty-five students, forty-eight concluded that solution K contained silver. Thus, nearly three-quarters of a group of high-achieving high-school graduates, who had studied science for at least 6 years, went along with fundamentally flawed reasoning. Little wonder that, as citizens, they are easily swayed by arguments such as:

> Communists support unionism.
> Fred supports unionism.
> Therefore Fred is a Communist.

The results on another item were staggering. Students were asked to complete the following syllogism:

> If an element has a low electronegativity then it is a metal.
> Element sodium is a metal.
> Therefore . . .

Fifty-nine of the sixty-five students concluded, supposedly on the basis just of the information provided, that sodium has low electronegativity. Their answer happens to be correct, but it does not follow on the basis of the information given. It would only follow if the first premise were 'If and only if an element. . . .' Thus, fully 90 per cent of the cream of high-school graduates are prone to making basic logical errors. It is little wonder that arguments of the following form are very common and persuasive:

> If people are cunning and deceitful they can obtain welfare payments.
> Fred obtains welfare payments.
> Therefore Fred is cunning and deceitful.

This is a matter of some concern for jury trials and much else in societies based on democratic decision-making.

The reasoning dimension of science competence has been recognised in curriculum documents. Ehud Jungwirth (Jungwirth 1987), in a comprehensive study of the issue, lists a number of curriculum statements that make reference to critical–logical–analytical thinking skills. Among them are:

- to enable pupils to grasp the scientific method of approach and to cultivate habits of logical and systematic thinking in them (Senior biology, Cape of Good Hope, South Africa, 1977);
- to look for and identify logical fallacies in arguments and invalid conclusions (Queensland Board of Secondary School Studies, Australia, 1983);
- the scientifically literate person has a substantial knowledge base . . . and process skills, which enable the individual . . . to think logically (National Science Teachers Association, USA, 1982).

Jungwirth studied the reasoning processes of 600 school students and 400 trainee teachers (science graduates) and university science students in three countries. He used curriculum and extracurricular (life) test items that embodied the following kinds of faulty reasoning:

1 assuming that events that follow others are caused by them;
2 drawing conclusions on the basis of an insufficient number of instances;
3 drawing conclusions on the basis of non-representative instances;
4 assuming that something that is true in specific circumstances is true in general;
5 imputing causal significance to correlations;
6 tautological reasoning.

His results were less than encouraging, given the importance of reasoning, not just to science, but to social and personal functioning more generally – voting in an election, buying a car, deciding a school-board policy, determining what went wrong with the baked cake and so on. His results can be rounded out and summarised in Table 5.1, where the percentages refer to percentages of the appropriate population who make mistakes of the above kind (1–4).

Jungwirth reports that the results on test items 5 and 6 were comparable to the above, and, aggregating all the test results, he provides the following summary of his findings. For adults, only the postgraduates performed above the 50 per cent level, with the other post-secondary groups below, or very much below, that level. On the life items, none of the adult groups averaged at more than the two-thirds level. For school students, on the curriculum items, none of the groups averaged more than slightly above the 25 per cent level; on the life items, the scores were roughly twice as high (Jungwirth 1987, p. 51). Not unreasonably, Jungwirth concludes that time should be spent in science lessons on the rudiments of correct reasoning, and that, 'teachers' pre- and in-service training should convey the message that the "covering of a large corpus of information" does not constitute the only, and not even the major component of science education' (Jungwirth 1987, p. 57).

Jungwirth draws attention to the fact that, in this reasoning domain, the blind are leading the blind. Or, as Marx said, 'who will educate the educators?' He surveys the research on teachers' thinking skills, none of which is any more

Table 5.1 Occurrence of Invalid Reasoning

Type of faulty reasoning	*School grades 9–12*		*University students*	
	Curriculum (%)	*Life (%)*	*Curriculum (%)*	*Life (%)*
1	40	50	30	25
2	30	40	30	40
3	15	50	60	60
4	35	50	30	60

encouraging than the findings above on student thinking skills. Arnold Arons has pointed out that:

> We force a large fraction of students into blind memorisation by imposing on them . . . materials requiring abstract reasoning capacities they have not yet attained – and of which many of their teachers are themselves incapable.
>
> (Arons 1974)

After administering the *Test On Logical Thinking*, Garnett and Tobin (1984) concluded that, 'many of these teachers do not possess the reasoning patterns, which activity centred science curriculum seek to develop'.

What Cochaud's and Jungwirth's studies suggest is that formal and informal logical reasoning should be taught as part of a science course. Scientists always have to make inferences and draw conclusions; doing both at school in an illogical or invalid manner advances nothing, and does not give the desired flow-on from science to other studies and to the rest of life that most hope that a good science education will achieve.

Students can be given examples of the following formal logical fallacies, which are common in everyday publications as well as in scholarly texts, and they can be trained in recognising and avoiding these fallacies.

- the fallacy of affirming the consequent:
  if *P* then *Q*
  Q
  therefore *P*

- the fallacy of denying the antecedent:
  if *P* then *Q*
  not *P*
  therefore not *Q*

- The fallacy of asserting an alternative:
  P or *Q*
  P
  therefore not *Q*

The last chapter illustrated the advantages of a historical dimension in science education. In addition, history of science provides a vehicle to introduce some basic logic to students and to show how logic is, and is not, connected to concrete scientific reasoning. Consider, for instance, debates over the photoelectric effect that have previously been discussed. The photoelectric effect and even Millikan's experiment can easily be shown in the classroom. Instead of imposing Einstein's photon theory and his equation as the scientific 'royal highway' linking the two famous experiments, something else can be made of the occasion. The historical approach shows the hesitancy with which even great scientists propose their ideas; it illustrates the variety of sensible and rational interpretations of data possible at any time; and, finally, it allows the crucial distinction between mathematical equations or models and their physical interpretation to be portrayed.

History shows the variety of relations that were postulated in the early years of this century between emitted photoelectron energy and incoming light frequency. Philosophy can raise the question of whether and how data can prove a particular theory. Many different theories, or equations, can imply the same set of data points. These points do not uniquely determine a particular curve, or equation, much less a particular physical interpretation of the equation. If students can be led to appreciate this, they are recognising what Aristotle recognised in the fourth century BC: the fallacy of affirming the consequent, as he called it. Aristotle, and of course others before him, showed that an argument of the following form is invalid:

T implies *O*	(a theory *T* implies an observation *O*)
O	(the observation *O* is made)
Therefore *T*	(the theory *T* is true)

The conclusion does not follow, because, as well as *T* implying *O*, any number of other known or unknown theories (*T*s) can also imply the same observation. The hypothesis that it rained last night implies that the road will be wet; but, equally, the hypotheses that the sanitation truck went past, that a water main broke or that a lawn hose was turned on also imply the same thing. Thus, the mere observation that the road is wet does not prove any particular hypothesis.

This simple point of logic has been a stumbling block for empiricist approaches to natural science from the time of Aristotle to the present. It is common for people to feel that confirmed predictions provide some warrant for belief in a theory; the logical point is that such confirmations cannot establish the truth of the theory, and so there is a need to reconsider the type of truth that confirmed predictions imply. In the Middle Ages, it was known as the problem of 'Saving the Appearances'.[32]

The basic point is that facts are open to a variety of interpretations, or that a scientific theory is underdetermined by its evidence, commonly referred to as the 'Duhem–Quine Thesis'. Pierre Duhem (1861–1916) highlighted this logical point in his *The Aim and Structure of Physical Theory* (Duhem 1906/1954); Karl Popper (1902–1994) elaborated some of its consequences for science in his *Logic of Scientific Discovery* (Popper 1934/1959); and Willard van Orman Quine (1908–2000) further developed it in his *From a Logical Point of View* (Quine 1953). There are two forms of the thesis. One is stated by Aquinas, in which positive outcomes of a prediction cannot be used to establish the truth of a theory; the other is stated by Duhem, in which the failure of predictions to be borne out does not allow us to conclude that the theory is false, because the prediction results, not just from the theory under consideration, but from that theory plus statements of background information. For Duhem, there were limits to the 'background information' or 'get out of jail' assumptions that could rescue a theory from falsifying empirical evidence; for Quine, there were no such limits.[33] In Quine's words:

> The totality of our so-called knowledge or beliefs, from the most casual matters of geography and history to the profoundest laws of atomic physics even of pure mathematics logic, is a man-made fabric which impinges on experience only along the edges.
>
> (Quine 1951/1953, p. 43)

Simple student experiments, 'black-box' exercises and other activities where students guess unseen connections from the behaviour of seen variables can highlight most of the logical fallacies and illustrate different interpretations of events, but these activities do not raise the important question of how science actually progresses and settles upon the best of rival theories. Historical studies provide one context in which the elements of good reasoning can be illuminated. Often, the same historical examples can also exhibit for students the 'extralogical' dimension of science: the place that commitment to metaphysics plays in the determination of theory and research programmes, the legitimate and illegitimate uses of analogy and metaphor in scientific argument and so on.

Sociological Challenges to the Rationality of Science

Since at least Aristotle, philosophers have been concerned with identifying and promoting rational thinking, while not assuming that it was the only kind of human thinking to be valued or cultivated. The long-standing view has been that, though there is no grand rational road to scientific theorising and hypothesis generation, nevertheless, science pre-eminently represents a sphere of rational appraisal of claims and competing beliefs; where there are departures from rational thinking in science, such departures have to be identified and justified. This commitment to rationalism in science typifies the Enlightenment's, or more generally the modern, view of science (Siegel 1988, 1989). However, in the past decades, this rationalist understanding has come under attack: from within the philosophy of science, from some sociologists of science and from postmodernist French philosophers and others inspired by them.

These attacks on rationality are of consequence to science educators, because they would challenge, if not undermine, one of the central justifications for the teaching of science, namely, that science teaching introduces children to a sphere of rational thought and debate that has laudable 'carry-over' effects in the rest of their studies and in life. If the adjudication of scientific dispute is truly a matter of 'mob psychology', and if scientific advances are just whatever the most economically or politically powerful group decree them to be, independent of their epistemic worth, then the rationale for the inclusion of science in the curriculum is greatly diminished.

Just as Immanuel Kant was famously awakened from his dogmatic slumber by reading David Hume, so too modern rationalist philosophers of science were awakened by Thomas Kuhn's 1962/1970 *The Structure of Scientific Revolutions*. The latter was widely interpreted as saying that scientific theory change depends as much upon mob psychology and the mortality of the aged as it does upon rational persuasion,[34] and that progress in science need not be construed as advancement towards a fixed goal of the truth about nature. Paul Feyerabend extended this thesis in his *Against Method* (1975). Many of these irrationalist charges were answered by philosophers of science,[35] but no sooner had this been done than the Edinburgh School of sociologists of science further criticised rationalism in their thoroughly externalist account of scientific change, the so-called 'strong programme' in the sociology of knowledge. The harbinger of this programme was David Bloor's *Knowledge and Social Imagery* (1976/1991); then, in a few short years followed Barry Barnes's *Interests and the Growth of Knowledge* (1977) and Bruno Latour and Stephen Woolgar's *Laboratory Life: The Social Construction of Scientific Facts* (1979/1986).[36] Seemingly, HPS had taken a radical sociological turn.

The antecedents of the strong programme included the 'weak programme' of Karl Mannheim (1936/1960) and Robert Merton (1957), but these two historical sociologists only wished to identify cultural, social and ideological factors that shaped science; they did not believe that these external circumstances brought about the specific contents of science or resulted in specific discoveries. Science is dependent on mathematics, technology, education, funding, means of communication, philosophy and so on; the weak programme's identification of all of this is admirable. This programme was announced by Karl Marx (1818–1883), who, in the opening lines of *The Eighteenth Brumaire of Louis Bonaparte*, famously wrote that:

> Men make their own history, but they do not make it just as they please; they do not make it under circumstances chosen by themselves, but under circumstances directly found, given and transmitted from the past.
>
> (Marx 1852/1969, p. 398)

Neither Marx, Mannheim nor Merton would have believed that the inverse square law of attraction would have become an inverse cube law if Newton had been raised a Hindu, or if capitalism had already triumphed in seventeenth-century England. However, this is suggested, if not implied, by the Edinburgh sociologists of scientific knowledge whose research programme set out to give an account of the external causation of the content of science.

Alongside these currents, French postmodernist philosophy, particularly that influenced by Michel Foucault, was asserting that all systems of ideas, science included, were intimately connected to the distribution of power in society, and that changes in ideas were not to be accounted for by epistemological factors, but by sociological ones. This is a continuation of the Marxist idea of social superstructure (systems of ideas and ideologies) being determined by a material base. The Marxist tradition has struggled with the problem of whether mathematics and natural science are subject to this kind of determination, with some exempting the former. Foucault, and his tradition, faces the same problem.[37]

Science teachers and teacher educators need to be aware of this multi-fronted assault on the rational assumptions of science. Any casual glance through educational theory books or science-education journals will turn up all of the above names, with various conclusions about 'the nature of science' and good science teaching being drawn from their work. There is much in the sociological work that is informative and that enlarges our understanding of how decision-making and theory change actually occur in science. The role of elites in a scientific community and their control over organs of publication, the function of rhetoric in scientific argument, the influence that economic power and interests have in the funding of research and the determination of which problems to investigate and which to avoid – these are all matters that need to be taken into account and that provide a richer and more realistic view of the scientific enterprise. Furthermore, it is correct to point out that conceptions of rationality have changed over time: The Aristotelian ideal is different from that of the British empiricists, which is different from that of modern falsificationists or Bayesian probabilists. There is a historical dimension to rationality, but there is also much that is fundamentally mistaken, and educationally deleterious, in the strong programme and postmodernist attacks on scientific rationality (Shackel 2005).

David Bloor lists four commitments that characterise the core of the strong programme's theory of science and scientific knowledge (Bloor 1976/1991, p. 7):

- *Causality*: A proper account of science would be causal; that is concerned with the conditions that bring about individual beliefs or scientific theories.
- *Impartiality*: It is impartial with respect to truth and falsity, rationality or irrationality, accepted or rejected theories.
- *Symmetry*: It would be symmetrical in its style of explanation. The same types of cause would explain true and false, rational and irrational, beliefs.
- *Reflexivity*: It would be reflexive. In principle, its patterns of explanation would have to be applicable to sociology itself.

To recognise that there are necessary conditions for the conduct of science and even for the emergence of scientific theories and beliefs is an important step, but to convert these to sufficient conditions is a mistake. Both Newtonianism and Darwinism had, respectively, connections with seventeenth- and nineteenth-century English life and culture – one might even say they grew out of their cultures – but the latter did not explain the former; the origins of beliefs do not account for either their truth or their falsity. This is the genetic fallacy. There were countless others who shared the exact life circumstance of Newton and of Darwin, but the respective theories were not forced out of them, so to speak, by their circumstances. The crucial thing overlooked by the strong-programme sociologists is the truthfulness or reasonableness of the theory as a ground for belief in it. A lot of the heat goes out of the dispute if reasons are recognised as causes. If having a good reason for the belief can be regarded as a cause of the belief, then sociologists, philosophers and educators are all in the same boat.

There is no reason to look for external sociological causes for people believing that '2 + 2 = 4'. If there were widespread belief that '2 + 2 = 5', then one would start looking for external causes: 'Is the latter what the King believes?' 'Is the latter what the ruling party demands?' 'Is 4 an unlucky number that cannot be mentioned?' 'Did the person have an especially incompetent teacher?' The more improbable, mistaken and irrational the belief, the more energetically one should seek external causes for it. However, the strong programme's impartiality thesis rules out this common-sense option. In Chapter 8, it will be argued that constructivism is one such improbable educational/philosophical theory. David Geary, a psychologist who agrees with this estimation of constructivism, proffers an externalist account of its popularity:

> In sum, constructivism is largely a reflection of current American cultural beliefs and, as such, involves the development of instructional techniques that attempt to make the acquisition of complex mathematical skills an enjoyable social enterprise that will be pursued on the basis of individual interest and choice.
>
> (Geary 1995, p. 32)

At the highest level in the sociological programme, there is confusion between ascertaining what actually happens in some specific instances and pronouncements about what should generally happen in the conduct of science. Francis Bacon, in the early seventeenth century, alerted his readers to the operation of what he called 'The Idols of the Mind'. These were the various ways in which the effort to understand the world can be thwarted: by the inadequate language available to think and write in, by the corrosive effects of self-interest, whereby people more readily believe what they want to believe, and by the direct exercise of social power wielded by dominant groups. Much contemporary sociology of science is an extension of this early Baconian investigation. However, Bacon took pains to identify the operation of 'Idols' precisely in order to overcome or compensate for their effects. That is, he distinguished between what sometimes happens and what should happen in the pursuit of knowledge. Whenever sociologists point to the operation of contemporary idols, or old ones in new dress, it is possible to ask: Are such mechanisms, procedures or influences desirable in science? This latter normative question is the one asked by philosophers of science, and it is one that students can be encouraged to ask. Then, if the effects of class, gender, race, power, religion, self-interest and so on are identified as pernicious and contrary to the scientific endeavour, it is reasonable to delineate what the ideal is against which these failings are judged.

Now, it certainly would be an embarrassment to the rational cause if no instances of scientific change could be found in which epistemological or evidential considerations were determinant, but there is no embarrassment to the rational cause if some instances of non-epistemological determination are uncovered. Indeed, a good many of these have been documented: the long history of intelligence testing and its associated theory is now seen to be almost entirely driven by class, race and gender interests that mostly are hidden, submerged or implicit,[38] and the decades of Lysenkoist genetics, or more correctly anti-genetics, in the USSR are now seen to have been in part driven by Communist Party interests, and in part by a mistaken metaphysics.[39] However, these cases can be identified, and they command our attention, because we have some sense that they are departures from proper scientific procedure.

Further, without such normative convictions, we would be in no position to complain about the above aberrations: If power is knowledge, then the white, male ruling class and the Communist Party certainly had power, and, consequently, the operation of this power must, by definition, result in knowledge. Few people, least of all minorities and those without power, would want to accept this conclusion.

Enough has been said to indicate that argument about the rationality of science is pertinent to science teaching. Harvey Siegel defends rationality and the giving of reasons as the hallmark of science education (Siegel 1989, 1993). Martin Eger addresses the question of how such a conception can allow for the role of commitment, or faith, that has been so important to the development of science (Eger 1988, 1989). Faith, or philosophical commitment, need not be irrational. Such commitments can be tested by evaluation of their scientific or experimental achievements or implications. Siegel (1993) addresses the quest for a naturalised philosophy of science and how such a quest, if successful, would impact on our understanding of the rationality of science and on the classroom teaching of science. These debates, and others, should find a place in teacher education programmes and can inform school science teaching.

Ethics, Values and Science Education

Ethical questions increasingly arise in the science classroom. The following matters are examples of those raised by students and that appear in new national science curricula: the greenhouse effect, global warming, pollution, extinction of species, genetic engineering, genetic testing of embryos for 'undesired' maladies or gender, military technology and the employment of scientists in the defence industries, the cost and direction of scientific research, nuclear energy, nuclear war and so on. These topics explicitly appear where STS and SSI orientations inform curricula and programmes, and students are expected to engage with them as part of the curricula.[40] In addition, they are rightly part of most NOS learning objectives. Development of 'ethical judgement' is an agreed aim of the 'Dublin Descriptors', widely adopted by European universities for incorporation into all university prorgrammes (Aalberts *et al.* 2012).

Teachers need to strive to make the ethical discussion as sophisticated as the classroom's scientific discussion. Again, this requires that teachers be familiar with the history and philosophy of their discipline and have some familiarity with informed ethical reasoning. Something, but not much, is served by simply rehashing or asserting popular nostrums. Teachers can benefit, and their classes can be enriched, by serious grappling with these ethical and social questions.[41] As Anna Couló comments:

> For science education, it is relatively easy to find interesting and relevant material from a socio-scientific point of view (SSI), on the role of non-cognitive values in the funding of scientific research, and on the technological consequences of scientific inquiry. It is much harder, though not impossible, to find related works framed in a more closely philosophical perspective.
>
> (Couló 2014, p. 1090)

Socio-technical–ethical questions are inevitable and should be addressed; both HPS and moral and political philosophy can contribute to their more nuanced and informed discussion; neither teachers nor students gain much if such discussions are sloganistic or merely repeat the current common prejudice.

The interconnection of science and ethics is particularly clear in contemporary human genetics programmes. The Human Genome Project has 3 per cent ($90 million) of its $3 billion budget allocated to ethical and legal ramifications. In the US, there are at least three state and national genetic education programmes that explicitly address the ethical and religious dimensions of the Genome Project.[42] The BSCS programme is outlined in a ninety-four-page document sent to all US biology teachers. In addition to the science of the Genome Project, it has students engaging in analysis and debate over the ethical and policy issues generated by genetic screening and other techniques occasioned by the Genome Project. Should employers be allowed to screen prospective employees for the Huntington's disease gene? Should those identified as genetically disposed to alcoholism be forbidden to drink? It says of these situations that:

> Individuals, institutions (schools, businesses, and other organisations), and society will have to deal with situations in which some interests are advanced and others are impaired. When the interests of everyone cannot be advanced, and when some interests are advanced at the expense of others, whose interests ought to receive

> priority? Questions about 'oughts' properly are addressed by ethics and public policy.
>
> (BSCS 1992, p. 15)

More recently, the US National Institute of Health has released a bioethics course for grades 9–12, which is explicitly constructed around ethical issues raised by and embedded in biology (NIH 2009).

In England and Wales, since 2005, there has been a senior biology course for 16–18-year-olds – Salters–Nuffield Advanced Biology – that is taught through contexts and has a strong emphasis on social aspects of biology and the ethical analysis of biological issues. Michael Reiss, one of the developers of the course, identifies four reasons for teaching ethics in the science classroom: heightening students' ethical sensitivity; increasing students' ethical knowledge; improving students' ethical judgement and making students better people (Reiss 2008).

Norman Lederman, in an influential article advocating the teaching of NOS, says that all students should learn the 'values and beliefs inherent to scientific knowledge and its development' (Lederman 2004, p. 303). However, this admonition is left hanging: the inherent values to be taught are not spelled out. Nor are they separately listed as one of the seven much-repeated characteristics of NOS (Lederman *et al.* 2002). HPS can contribute to fleshing out this crucial feature of science and, thus, enhance its teaching.

Values as External to Science

Students can perhaps recognise that economic, political, religious and philosophical values have an external role in the conduct of science. This in itself is salutary; it puts something of a brake on scientistic excess. What research areas are funded depends on values and the outcome of competing interests. Such influence is obvious in medical, agricultural, communications and weapons research. In 1961, soon to be ex-President Dwight Eisenhower famously referred to the danger presented to US society by the 'military–industrial complex'; more accurately, he could have referred to the 'military–industrial–scientific complex'. In Nazi Germany,[43] the Soviet Union[44] and the US,[45] an enormous, if not overwhelming, amount of scientific resources went into military research and into big-business and state-business research; and of course the same for the United Kingdom, China, Japan, Canada and just about all other advanced economies. In all of these countries, the military–industrial–scientific complex is dependent upon school science classrooms: if the latter are unthinking and uncritical, then the former are most likely to be as well.

As a micro example, consider how, in the late 1970s, US agri-business funded the University of California with hundreds of thousands of dollars to produce a tomato that could be mechanically harvested and so would thwart the strike efforts of the United Farm Workers, led by César Chávez. At the time, tomatoes were almost the last part of commercial agriculture that was intimately dependent on large-scale human labour and thus ripe, so to speak, for strike action. The university scientists could have worked on making tomatoes that grew better in urban backyards or on rooftops, but they were funded to work for agri-business, making more robust, thicker-skinned, longer-shelf-life and more-uniform tomatoes that hopefully were cubic, for ease of packing and commercial sandwich making. Science had been co-opted, and it was fairly clear whose interests were being served. The widespread realisation of this nexus led to the formation of such groups as Science for the People, Scientists for Nuclear Disarmament and so on. And it is partly why the International Organisation of Science and Technology Education (IOSTE) says that:

> Consistent with our mission to encourage the peaceful and ethical use of science and technology in the service of humankind, IOSTE opposes the use of science and technology by government or other organizations for military purposes against civilians.
>
> (http://ioste.nmmu.ac.za)

As well as military, political and business interests impacting on science, so also has philosophy. This is a well-known contention, aptly stated by Alexandre Koyré:

> It is, indeed, my contention that the role of this 'philosophic background' has always been of utmost importance, and that, in history, the influence of philosophy upon science has been as important as the influence – which everyone admits – of science upon philosophy.
>
> (Koyré 1954, p. 192)

It is a much-discussed issue whether such metaphysics is actually external to science; the case for metaphyics being internal to science is overwhelming and has been discussed in the opening pages of this chapter.

Religious interests are one of the other important external influences on the shape and trajectory of science. In countries where the Roman Catholic Church is politically powerful, there is little, if any, scientific research on contraception, safe abortion, simple euthanasia and so on; cloning and stem-cell research is closely controlled. Where Islam dominates, research on human origins will be negligible. As with metaphysics, there is a case for theological positions being internal to different scientific paradigms and investigations within those paradigms – witness the above-discussed case of Newtonian absolute space and his theological commitments. Newton said that he wrote the *Principia* so that 'men might believe'; here, religion provides a clear, external purpose or motivation for the work, but many say that his religion was not left outside the *Principia* but featured in its arguments. The same role was played by Joseph Priestley's belief in providence in leading to his discovery of photosynthesis. The continuing intellectual (and political) struggle has been over the possibility of science correcting the religious commitments that have guided or motivated it.

Alaistair Crombie, the noted historian of science, addressed the influence of religion on science when he surmised why the scientific revolution did not occur in the thirteenth or fourteenth centuries, despite there being no shortage of bright and scholarly natural philosophers:

> Although some of the best medieval scientific work was done on particular problems studied without any reference to theology or philosophy or even methodology, it was within a general framework of philosophy closely bearing on theology, and specifically within the system of university studies run by clerics, that the central development of medieval science took place.
>
> (Crombie 1956, p. 114)

An intellectual environment dominated by an all-powerful Roman Catholic Church that had embraced a particular philosophy (Thomism) as the 'handmaiden' of its theology simply prevented the rise of modern science. Crombie goes on to say:

> It explains much that is puzzling and seemingly downright perverse in otherwise excellent work. It helps explain, for example, the gap between the repeated insistence on . . . empirical verification and the many general assertions never tested by observation; worse, the satisfaction with imaginary experiments either incorrect or impossible.
>
> (Crombie 1956, p. 116)

This is not the well-known and direct instruction to Galileo not to teach or promote the Copernican system of the world; two centuries earlier, the influence was less immediate and more part of the intellectual fabric of the times, but the influence was no less effective. This is comparable to the situation of science in all reactionary or despotic regimes, where metaphysics external to science controls and limits scientific theorising and appraisal. This has been the situation in Roman Catholic, Islamic, Communist and Hindu states and has been a centuries-old battle in the US, where fundamentalist Christians control local school boards and thwart the teaching of evolution on religious grounds.

At the first level of analysis, these influences on, and impacts of, science are external to science; this separation is one reason for distinguishing pure from applied science. It is to everyone's benefit that external influences be identified and appraised. Such appraisals and discussions need not, and should not, be restricted to science classes; they can be undertaken in history, social science, economics, literature and religion classes. A common assumption has been that such values (political, commercial, philosophical, religious, personal) belong to the environment of science, not to the conduct of science; that science itself is free of values. Using Hume's terminology, science is concerned with what is, and values are concerned with what ought to be, and there is no connection either way between the first and second; the latter is the domain of ethics, politics, religion and custom. Some problems with this common assumption will be outlined below.

Value-free Science

The 'value-free science' position has its origins with Max Weber's (1864–1920) at-the-time progressive view that sociologists should report on the structure and functioning of society, not on how their own politics, religion or class envisages or judges society (Weber 1917/1949). Weber was a brave and committed political liberal and democrat in a very reactionary Prussia. He recognised that 'progressive' movements were advantaged by knowing how reality was; not by dreaming about it or seeing it as we would like it to be. This was the same stance taken by the equally liberal, democratic and socialist early positivists, who took their intellectual and political guidance from Ernst Mach. For these positivists, there was also a cleavage between fact and value; good science was value free. Indeed, for hardcore positivists, value statements were literally nonsensical; they thus could not be part of science. Many science teachers share, and indeed embrace, this 'value-free science' position: they believe that the pursuit of knowledge does not in itself embody values. It is the position made famous, or infamous, in Tom Lehrer's lyric:

> 'Once the rockets are up, who cares where they come down?
> That's not my department', says Wernher von Braun.

But considerations in HPS make the 'value-free' position more problematic and more interesting, and its appraisal more educationally beneficial.

External influence on the conduct of science, or more generally on understanding the world, has long and widely been recognised. Francis Bacon, in his *Novum Organum* (1620/1960), wrote about the Idols of the Mind and the need to recognise and correct for psychological, linguistic, economic and cultural influences that might distort understanding.[46] The first instalment of the debate has been about whether all the influences have been identified – feminists have argued that gender was, understandably, not listed by Bacon, but subsequently not much listed by anyone else; Marxists make the same claim about class; queer theorists make the same claim about heterosexual assumptions functioning in science; indigenous science proponents point to 'Western' assumptions that go unexamined, and so on. The second instalment has been about whether such influences can be corrected for, in particular the influence of social and personal values on the conduct of science. If the influences are local and incorrigible, then the impossibility of correction or 'making allowances for' impinges on the objectivity or universality of science.

Bacon recognised that ethical and civil values needed to underwrite the conduct of natural philosophy in his *New Atlantis*; these found expression in the functioning of the Royal Society, where honesty, civility, toleration and modesty were ideals and even rules (Sargent 2005). Three hundred years after Bacon, Robert Merton famously codified the ethical values required for the conduct of science. Merton said that, for the communal enterprise of science to be able to successfully pursue its goal of finding out about the world and effectively engaging with it, it needed to embody four values (Merton 1942/1973): *communalism* (scientific results were to be public and published for all), *universalism* (anyone could contribute to science, regardless of their religion, race, class, gender, etc.), *disinterestedness* (participants were to tell the truth, regardless of private gain or loss) and *scepticism* (all claims were to be open to criticism, debate and possible renunciation or revision). John Ziman, in a popular work subtitled *The Social Dimension of Knowledge*, added *originality* (science sought new truths, embraced new methods, utilised better explanations and theories) to this list (Ziman 1968). This enabled the convenient acronym CUDOS (communalism, universalism, disinterestedness, originality and scepticism) to be used for the Mertonian scientific norms.

Of course, science did not have to await Merton's publication to know of the presuppositions for its own success; Merton merely synthesised and condensed long-extant practices. But recognition of the norms did bring values across the moat and into the scientific castle; the idea or ideology of value-free science could not be maintained. Or the position could only be maintained by retreating further and going into the castle's keep. This was done by agreeing that science was a value-laden pursuit, characterised as Merton outlined, but the values governed the organisation and conduct of science, the construction of the castle, but not its core decision-making. When it came to decisions about scientific claims, hypotheses and theories that had been generated in a Mertonian environment, then the decision-making was free of values. In Reichenbach's terms, values certainly governed the context of discovery, but they had no place in the context of justification (Reichenbach 1938 pp. 6–7).

The value-free defenders could only take temporary rest with this position. Sixty years ago, in a book edited by the positivist Philipp Frank (Frank 1954), Barrington Moore Jr wrote:

> Few people today are likely to argue that the acceptance of scientific theories, even by scientists themselves, depends entirely upon the logical evidence adduced in support of these theories. Extraneous factors related to the philosophical climate and society in which the scientist lives always play at least some part. The interesting problem, therefore, becomes not one of ascertaining the existence of such factors but one of appraising the extent of their impact under different conditions.
>
> (Moore 1954, p. 29)

Moore's paper was focused on science in the Soviet Union and it was about the acceptance of theories, not just the direction of research, or what fields to investigate. He recognised that all large human societies require a 'set of beliefs about the purposes of life and the ways it is legitimate and not legitimate to achieve these purposes. This set of beliefs constitutes the political truths of the society' (Moore 1954, p. 35). He could, perhaps, have said they constitute the 'worldview' of the society. Although acknowledging external influence and admitting that there were similarities between the United States and the Soviet Union, Moore wanted, Bacon-like, to keep the influences apart from the science:

> Therefore as science develops its own canons for validating its propositions, there is likely to come a time when the political creed and the scientific creed conflict with each other.
>
> (Moore 1954, p. 35)

Values Internal to Science

Barrington Moore left open whether the scientific creed itself embodied values. Richard Rudner did not leave open this question in his anthology chapter titled: 'Value Judgments in the Acceptance of Theories', where he wrote:[47]

> I think that such validations do essentially involve the making of value judgments in a typically ethical issue. And I emphasize essentially to indicate by feeling that not only do scientists, as a matter of psychological fact, make value judgments in the course of such validations – since as human beings they are so constituted as to make this virtually unavoidable – but also that the making of such judgments is logically involved in the validation of scientific hypotheses; and consequently that a logical reconstruction of this process would entail the statement that a value judgment is a requisite step in the process.
>
> (Rudner 1954, p. 24)

This was a direct challenge to the long-held positivist view that values were outside the scientific kernel of theory validation; Moore brought values into the scientific castle's keep. He concluded:

> What is proposed here is that objectivity for science lies at the least in becoming precise about what value judgments are being made and might have been made in a given inquiry – and stated in the most challenging form, what value decisions ought to be made.
>
> (Rudner 1954, p. 28)

On one interpretation, the claim was straightforward: scientists who make or advise on science-related policy clearly have to make value judgements. What is the trade-off between money and health in advising on how much fluorine to add to a town's water supply to ensure effective protection against teeth cavities? When recommending a new drug for government approval, scientists need to quantify the trade-off between securing immediate, known, good effects and delaying approval because of possible, unknown, bad consequences. These are indisputably value judgements made by scientists.

Fifty years ago, philosophers made this point in response to Rudner, saying that science covers the estimated probability of some illness in some circumstance; the 'policy' advice will raise or lower that probability, depending on cost and threat to life, but the policy advice is not given qua scientist (Levi 1960). The issue has subsequently been clarified, but 'value freedom' is still debated in the philosophy and policy communities.[48]

Another way of construing the arguments, from Bacon through to the early positivists, is that they were about identifying and separating values that deformed and inhibited science from those that enhanced science. Yes, there could be values (non-empirical claims and judgements) in science, but it was important to isolate the illegitimate ones from the legitimate, science-enabling ones.

Thomas Kuhn took on this task in his 1973 Machette lecture on 'Objectivity, Value Judgment, and Theory Choice' (Kuhn 1977). He began by stating the obvious: scientists want to formulate or adopt good scientific theories, and goodness is inescapably evaluative. He listed five charactistics of good theory:

- *Accuracy*: consequences deducible from a good theory should be in demonstrated agreement with existing experiments and observations.
- *Consistency*: good theories are internally consistent and consistent with currently accepted theories.
- *Breadth*: good theories are widely applicable; they need to extend beyond local or particular cases.
- *Simplicity*: good theories should bring order to phenomena that, in their absence, would be isolated or confused.
- *Fruitfulness*: good theories should disclose new phenomena and suggest new relationships.

These were all values and they are internal to science, not external. Importantly, Kuhn is at pains to point out that they are not rules: 'the criteria of choice . . . function not as rules, which determine choice, but as values, which influence it' (Kuhn 1977, p. 331). And, as with all applications of ethical value, there is room to move: people, even experts, can disagree in how they weigh up competing values – simplicity versus breadth – or just how consistent are two competing theories.

Ernan McMullin's 1983 elaboration and critique of Kuhn's account is much anthologised. He identifies Kuhn's list as epistemic values, because they are concerned with the pursuit of knowledge, truth or the improvement of cognition that aims at understanding the natural and social worlds. He contrasts them with the non-epistemic values. The latter include the features indentified by Merton as necessary for the social pursuit of science – honesty, openness, inclusiveness – and also the enormously wide list of 'political, social, moral and religious' values that can be appealed to to close the gap between underdetermined theory and evidence in explaining a scientist's choice of theory (McMullin 1983, p. 19).

McMullin 'reworks his [Kuhn's] list just a little' and proffers the following list of epistemic or cognitive values:

- *Predictive accuracy* – but only to a degree; in the early stages, all precise theories are refuted by evidence.
- *Internal coherence* – there should be no contradictions or unexplained coincidences.
- *External consistency* – theories should be expected to cohere with current best theories and assumed ontology.
- *Unifying power* – as witnessed in Maxwell's electromagnetic theory or plate tectonics.
- *Fertility* – sustained ability to generate and incorporate new findings.
- *Simplicity* – a desirable feature, but more easily expressed than embodied.[49]

McMullin agrees with Kuhn that these values function just as values do: they are not rules that can be applied without judgement, and there are, typically, trade-offs between one and the other that reasonable scientists can disagree over (McMullin 1983, p. 16).

McMullin rightly draws attention to 'non-standard' epistemic values that are pervasive in science; foremost among these are metaphysical and religious commitments. The case of Newton has been discussed above; the Einstein–Bohr debates on the interpretation of quantum theory, the 'punctuated equilibria' debate in evolutionary biology and numerous other such debates also illustrate the operation of 'non-standard' values.

Peter Kosso introduces his discussion of 'internal and external virtues' in these terms:

> Theories are like apples; there are good ones and bad ones. . . . Apples have all sorts of features that are indicative of goodness . . . Similarly, theories have features that are indicative of their truth, and the task of justification is to identify these features and use them to guide choices as to which theories to believe.
>
> (Kosso 1992, p. 27)

Kosso provisionally separates internal from external features of good theories. By 'internal', he means features of theories that can be identified and evaluated by looking just at the theory, not at the world. So, logical consistency 'is a clear example of an internal feature' (Kosso 1992, p. 30). Other internal truth-conducive virtues that Kosso lists are:

- *Entrenchment* – a conservative value where credit is given for a theory's consistency with established theory and knowledge.

- *Explanatory cooperation* – credit accrues because a theory explains why established theories can themselves explain phenomena.
- *Testability* – any good theory must have empirically testable consequences; this does not guarantee truth, but is a necessary condition for truthful theories.
- *Generality* – assuming a simple, natural and social world of few basic mechanisms, then the more general over space and time, the more likely the theory could be true; without the assumption of simplicity, then generality is just a pragmatic or aesthetic virtue.
- *Simplicity* – if the world is simple, then simple theories are more likely to be true; this is the reason why 'lines of best fit' and their equations are preferred over lines that join all data points; it is the reason behind the use of Ockham's razor.

All of the above, and additional ones such as aesthetics, have to do with the appearance of the theory; degrees of them can be determined by looking at the theory and its intellectual milieu. In contrast, external virtues 'are relevant to the theory's relation to the world' (Kosso 1992, p. 31). For Kosso, these are explanation and confirmation.

Clearly, the identification, weighing, commitment and inculcation of epistemic values are dependent on particular visions of science. All of the foregoing virtues listed by Kuhn, McMullin and Kosso are contingent on agreement about the truth-seeking goal of science. This might be seen as a more fundamental value, one that authorises some or all of the variously listed epistemic virtues. If folk are committed to other goals for science – maintenance of traditional culture, boosting of company profits, enrichment of the state, support of a religion, 'knowing where to put the soufflé in an oven', etc. – then the associated virtues will change. Of course, some of these extra-epistemic goals will require truthfulness, and so even having a social goal as fundamental will not negate the operation of epistemic virtues. The state will not be too long enhanced if its subservient science fails to tell the truth about the world: the state is not enhanced by rockets failing to lift off (North Korea) or by massive crop failure (Soviet Union). Realism about science and rationalism about scientific decision-making assist with the easier identification and inculcation of epistemic values. And the values have to be established on empirical grounds, not on psychological or ideological grounds; their operation has to be seen as contributing to scientific success. Hence, recognising science as value-laden does not mean saying that it cannot be universal or objective.

Feminist Theory and Science Education

In the past half-century, many significant feminist critiques of science have been published; there has developed an identifiable 'feminism and science' research programme whose arguments have had a significant impact in philosophy, in philosophy of science and in science education (Noddings 2009). This feminist research programme has had three broad streams. First, a 'practical' stream, which is concerned with increasing the participation of women in school, university and industrial science and with the recovery of 'lost' women scientists in textbooks and histories of science. This stream accepts orthodox science and seeks to increase the participation of women in it by having 'girls only' science classes, kitchen-focused experiments and so on (Rosser 1986, 1993).

Second, there is a 'critical' or 'empiricist' stream that seeks to improve orthodox science by recognising and correcting mechanisms whereby masculine bias has corrupted different elements of science. These feminists accepted the standard normative picture of science as being a rational pursuit of universal and objective truths about the world; their critiques concentrated on how, in particular cases, this normative ideal was compromised by unacknowledged gender biases in the identification of research fields, the framing of research questions, the methods employed and the interpretation of results. The studies of Ruth Hubbard (1979), Carol Gilligan (1982), Donna Haraway (1989), Nancy Tuana (1989a) and Kathleen Okruhlik (1994) on bias and gender ideology in, respectively, evolutionary theory, Kolberg's moral development research, primatology, reproductive theory and biological sciences are well-known examples of this genre.

Empirical feminists believed that current science was corrigible and that feminist understandings could advance its truth-seeking goals. This work connected with, and drew some inspiration from, Marxist studies of the putative ideological deformation of economics and social science by 'hidden' class assumptions (Hartsock 1983). For Marxists, their own political economy was simply better economical science because it had a more informed 'grip' on the world, it attended to factors that routine, 'bourgeois' economics neglected. For empiricist feminists, their science was likewise better science because it recognised more about mechanisms in the world and it had better ways to study them.

These empiricist feminists were scientific realists, and, as such, the worth of their specific analyses was an empirical question about how much understanding was gained about their subject matters by feminist-informed and corrected science. Some of the particular studies were lauded; others were energetically contested. Susan Cachel, in a critical review of Haraway's very popular and widely prescribed primatology book, writes that, for Haraway:

> Primatology is not science but narrative or story-telling. The author interprets 'texts' and 'subtexts' in the way that literary critics offer readings or multiple perspectives on a text. The idea that facts should be the focus of attention is foreign to this book – the author treats 'facts', instead. One theory is as valid as another, and there is nothing to check or constrain narratives except personal taste or political agendas. . . . If one did not already possess some background, this book would give no lucid history of anthropology or primatology.
>
> (Cachel 1990, pp. 139, 141)

For Cachel, Harraway's feminism assuredly did not enable her to make any distinct contribution to our understanding of primates or to the science of primatology. As with all scientific claims, other feminist contributions need to be looked at on a case-by-case basis and appraised against nature: has the work expanded knowledge of the subject matter?

Third, there has been a more 'philosophical' tradition in feminism in which a good deal of the orthodox understanding of science and its goals is simply rejected. Early contributions to the feminist philosophical programme were made by, among others, Ruth Bleier (1984), Evelyn Fox Keller (1985, 1987), Sandra Harding (1986), Helen Longino (1989) and Jane Roland Martin (1989).[50] Sandra Harding and Merrill Hintikka promised that feminist theory would bring about 'lock, stock and barrel' changes in science, in its methodology as well as its choice of problems to investigate. In the introduction to an influential anthology, they wrote:

> A more fundamental project now confronts us. We must root out sexist distortions and perversions in epistemology, metaphysics, methodology and the philosophy of science – in the 'hard core' of abstract reasoning thought most immune to infiltration by social values.
>
> (Harding & Hintikka 1983, p. ix)

A few years later, Harding would write that this improved, 'feminised' science brings what were previously considered Baconian 'idols' into the core of the enterprise, but this is not a problem, because feminist science:

> Seeks a unity of knowledge combining moral and political with empirical understanding. And it seeks to unify knowledge of and by the heart with that which is gained by and about the brain and hand. It sees inquiry as comprising not just the mechanical observation of nature and others but the intervention of political and moral illumination 'without which the secrets of nature cannot be uncovered'.
>
> (Harding, S.G. 1986, p. 241)

It was this orientation that led Harding directly to her adoption of Marxist-influenced 'standpoint epistemology' (Harding 1991, Chapter 5). In this theory of science, bias cannot be overcome; there is no 'view from nowhere'; knowledge can only be advanced by a multiplicity of views; and the feminist view or standpoint has special epistemological merit. Harding wrote:

> The distinctive features of women's situation in a gender-stratified society are being used as resources in the new feminist research . . . to produce empirically more accurate descriptions and theoretically richer explanations than does conventional research.
>
> (Harding 1991, p. 119)

The commitments of this programme are well captured by Nancy Brickhouse, who wrote:

> Feminist epistemologies [have had] a significant impact on science education. The work of feminists such as Evelyn Fox Keller, Donna Haraway, and Sandra Harding showed the ways in which scientific knowledge, like other forms of knowledge, is culturally situated and therefore reflects the gender and racial ideologies of societies. Scientific knowledge, like other forms of knowledge, is gendered. Science cannot produce culture-free, gender-neutral knowledge because Enlightenment epistemology of science is imbued with cultural meanings of gender. This feminist critique of Enlightenment epistemology describes how the Enlightenment gave rise to dualisms (e.g., masculine/feminine, culture/nature, objectivity/subjectivity, reason/emotion, mind/body), which are related to the male/female dualism . . . in which the former (e.g., masculine) is valued over the latter (e.g., feminine). These dualisms are of particular significance to scholars writing about science because culturally defined values associated with masculinity (i.e., objectivity, reason, mind) are also those values most closely aligned with science (Keller, 1985). As such, not only was masculine culturally defined in opposition to feminine, but scientific was also defined in opposition to feminine.
>
> (Brickhouse 2001, p. 283)

In this single paragraph, Brickhouse makes at least twelve distinct and much-disputed, indeed widely rejected, historical and philosophical claims. These claims are commonplaces in feminist educational writing; merely stating them should suffice to show that their appraisal involves historical and philosophical considerations, because the claims themselves are all about HPS. It then follows that science teachers and administrators should have some familiarity with HPS in order to critically engage with and evaluate such claims.

Another feminist writing about science education and citing Sandra Harding and Evelyn Fox Keller says:

> Radical feminism is important from an historical perspective because second and third generation feminist perspectives have been influenced by it. Radical feminism argues that in the case of science, scientific ideologies and philosophies are based on androcentric foundations. This has led to a masculine way of viewing science which in most cases also means Eurocentric science, a predominately white Anglo-Saxon male perspective.
>
> (Parsons 1999, p. 991)

Understandably, there is a connection between feminism and constructivism; initially, they reinforce each other. As Nancy Brickhouse attests: ‘feminists have found constructivist views of learning to be compatible with feminist epistemologies or pedagogies’ (Brickhouse 2001, p. 284). As will be elaborated in Chapter 8, constructivism has many philosophical problems. There are only a few short epistemological steps from a multiplicity of views, to a multiplicity of knowledges, to a plurality of sciences, and then to complete relativism about science. Many take the steps in one bound, with some taking a further ontological bound to the claim that there are as many worlds as there are knowers:

> According to radical constructivism, we live forever in our own, self-constructed worlds; the world cannot ever be described apart from our frames of experience. This understanding is consistent with the view that there are as many worlds as there are knowers. . . . Our universe consists of a plenitude of descriptions rather than of an ontological world *per se*.
>
> (Roth 1999, p. 7)

It is a puzzle that such claims are made in the name of 'authentic science education'. Was there a plenitude of descriptions at the beginning of the universe? Whose descriptions were they? It might be that, in Christian cosmology, 'In the Beginning was the Word', but this does not mean 'in the beginning were descriptions'. Doubtless, it is of zero comfort to Japanese people to know that it was a wave of 'descriptions' that destroyed their towns, homes and people in the 2011 Tōhoku tsunami, much less that it was a cloud of descriptions that broke over Hiroshima and Nagasaki in August 1945. It is mysterious why science educators would talk this way; more explicit, meaningful and sensible things could be said. However, not all feminists are constructivists, with many warning against the feminist embrace of constructivism (Koertge 2000, Pati & Koertge 1994).

Taking pedagogical or curriculum initatives based on such claims, without the claims being closely evaluated, is a great disservice to all students, not just women. Many women philosophers of science have taken the lead in critical evaluation of all of the above feminist claims. Not all feminists ascribe to feminist epistemology.

Some Evaluations of Feminism in Science Education

Many women reject the claim that objectivity, rationality and analytic thinking are alien to them. Norette Koertge, a prominent philosopher of science who has also written on science education (Koertge 1969, 1998), maintains that science needs more unorthodox ideas and a greater plurality of approaches. This is a standard Popperian position, which does not in itself constitute an argument for a new epistemology of science. Against certain feminists, Koertge warns that:

> If it really could be shown that patriarchal thinking not only played a crucial role in the Scientific Revolution but is also necessary for carrying out scientific inquiry as we now know it, that would constitute the strongest argument for patriarchy that I can think of.
>
> (Koertge 1981, p. 354)

And then: 'I continue to believe that science – even white, upperclass, male-dominated science – is one of the most important allies of oppressed people' (Koertge 1981, p. 354).

Cassandra Pinnick, also a philosopher of science who has written on science education (Pinnick 2008a), echoed this observation when she wrote that the history of science provides no grounds for Sandra Harding's epistemological privileging of the feminist 'standpoint':

> Rather, the history of science, patently dominated by male achievers, amounts to a thumping good induction to the conclusion that male bias – whatever it is and to the exclusion of identifiably different kinds of bias – ought to be maximised in science.
>
> (Pinnick 2008b, p. 187)

There may be other, non-epistemological, social or political reasons for the privileging of a feminist standpoint, but these need to be separately argued for and not confused with epistemological grounds. However, such arguments are themselves going to be disputed, beginning with the basic problem of which women's standpoint will be taken as the feminist standpoint. Class, ethnicity, religion, nationality, all overlay any particular woman's standpoint or even interest. And the philosophical arguments for standpoint epistemology have also been disputed. As Pinnick concludes:

> But the idea of a gendered standpoint on science is bankrupt, beset with formal contradictions and wholly lacking an empirical track record to provide even weak inductive support. Despite the failure of the arguments associated with it to rest on anything other than unsubstantiated promises about the significant positive impact that women will have on science, the continued high profile of feminist standpoint theory risks the conclusion that hard won efforts to promote women in science – in education and in careers – amount to misallocated scarce resources.
>
> (Pinnick 2008a, p. 1062)

Conclusion

This chapter has discussed some aspects of the intimate connection of science and philosophy and has suggested that this interaction be suitably introduced to students in science classrooms and in teacher education programmes. This is part of learning about the nature of science, a topic in all science curricula. It is not just a matter of learning about philosophy, although this is important, but also the opportunity can be provided to do philosophy. Philosophy begins with the questions such as, 'What do you mean by?' and 'How do you know?'. Students can be encouraged to ask these questions at each stage of their education, in whatever subject. Such questions lead naturally into the sphere of logic and the appraisal of arguments; there is a good deal of evidence that students' naive thinking in these areas needs to be trained and informed; logical thinking is not natural. The philosophical questions listed by Robert Ennis in 1979 – concerning explanation, structure of disciplines, values, theory and observation, scientific method – are of perennial interest to science teachers and can be made interesting to students. This chapter has indicated a number of areas in contemporary curricula that have a rich philosophical dimension to which students can easily be introduced; these include: ethics and values, logical and critical thinking, science and metaphysics, thought experiments. The chapter has also introduced some debates in which teachers and curriculum writers are frequently engaged, in particular those occasioned by sociological and feminist critiques of science. Here, philosophy has an explicit role to play, and teachers need to be so informed and prepared.

Beyond these topics, there are other lively areas of theoretical debate among science educators to which philosophy of science can contribute. Argument over constructivism, particularly its epistemological claims, is an obvious area, and will be dealt with in Chapter 8. Arguments about science and worldviews, about religious belief and scientific commitment, and deliberation about appropriate multicultural science education are other areas to which philosophy can contribute, and will be so seen in Chapter 10. All of this taken together supports the core thesis of this book: that science education and HPS need to develop a more intimate relationship.

Notes

1 Some useful studies on the philosophical dimension of science are: Amsterdamski (1975), Buchdahl (1969), Burtt (1932), Cushing (1998), Dilworth (1996/2006), Gjertsen (1989), Mayr (1988), Shimony (1993), Smart (1968), Trusted (1991) and Wartofsky (1968).
2 The famous Paul Arthur Schilpp anthology of commentary on Einstein is titled *Albert Einstein: Philosopher–Scientist* (Schilpp 1951).
3 See, for instance, Bohm (1980), Bohr (1958), Boltzmann (1905/1974), Born (1968), Duhem (1906/1954), Eddington (1939), Heisenberg (1962), Jeans (1943/1981), Mach (1883/1960), Planck (1936) and von Helmholtz (1995).
4 See, for instance, Bridgman (1950), Bunge (1973, 1998a, 1998b), Campbell (1921/1952), Chandrasekhar (1987), Cushing (1998), d'Espagnat (2006), Holton (1973), Margenau (1950, 1978), Rabi (1967), Rohrlich (1987), Weinberg (2001) and Shimony (1993).
5 For instance, Bernal (1939), Birch (1990), Haldane (1928), Hull (1988), Mayr (1982), Monod (1971), Polanyi (1958) and Wilson (1998). One recent contribution to the genre is by Francis Collins, the geneticist and leader of the Human Genome Project (Collins 2007).
6 See, for instance, Susan Stebbing's classic critique of the idealist philosophical conclusions drawn by the renowned British physicists James Jeans and Arthur Eddington (Stebbing 1937/1958). See also Mario Bunge's critiques of the idealist and subjectivist conclusions drawn from quantum mechanics by David Bohm, Niels Bohr and many proponents of the Copenhagen school (Bunge 1967, 2012). The criticisms underline the point that the scientists were primarily scientists engaged with philosophy, not professional philosophers.
7 There are countless books on the worldview of modern physics: see, for example, contributions to Cushing and McMullin (1989), especially Abner Shimony's contribution 'Search for a Worldview Which Can Accommodate Our Knowledge of Microphysics' (Shimony 1989). See also the contributions to the special issue of *Science & Education* dealing with 'Quantum Theory and Philosophy' (Vol.12, Nos.5–6, 2003) and the special issue on 'Science and Worldviews' (Vol.18, Nos.6–7, 2009).
8 For informative discussion, see at least: Agassi (1964), Burtt (1932), Collingwood (1945), Holton (1988) and Wartofsky (1968).
9 It needs to be said that the recognition of the cultural dependence of science does not mean acceptance of relativity or incommensurability between sciences or theories,

although this inference is frequently drawn. This is akin to saying that, because there is a cultural input into Korean school mathematical performance, then this performance cannot be compared with other nations.

10 See Lipman (1991), Lipman and Sharp (1978), Matthews, G.B. (1982) and Sprod (2011, 2014).

11 See McDermott and Physics Education Group (1995) for such demonstrations. On Galileo's inclined plane experiments, see Palmieri (2011); on their classroom use, see Turner (2012).

12 For an elaborate and informative discussion of this argument, see Buckley (1971).

13 On medieval impetus theory, see Clagett (1959) and Moody (1975).

14 A classic discussion is Clavelin (1974).

15 See Ellis (1965) and Hanson (1965) for excellent discussions of Newton's formulation of inertia. On force, see Ellis (1976), Hesse (1961), Hunt and Suchting (1969) and Jammer (1957).

16 See especially those of Arnold Arons (Arons 1977, Chapters 14–15, 1990, Chapter 3), Gerald Holton and Stephen Brush (Holton & Brush 2001, Chapter 9), James Trefil (1978) and the Harvard Project Physics texts (Holton *et al.* 1974).

17 Ricardo Lopes Coelho, in a recent publication, discusses both the historical and pedagogical literature on this topic (Coelho 2007); Calvin Kalman (2009) elaborates and adjusts some of Coelho's arguments.

18 Mach's argument is discussed in Matthews (1989), and Scheffler's argument is discussed in Matthews (1997).

19 The classic treatments of thought experiments in the history of science are Koyré (1953/1968, 1960), Kuhn (1964) and Mach (1896/1976). More recently, their historical and philosophical function has been discussed by Bokulich (2001), Brown (1991), Brown and Fehige (2011), Gendler (2000, 2004), Norton (2004), Schlesinger (1996) and Sorensen (1992). Martin Cohen (2005) lists and discusses twenty-six thought experiements; see also contributions to the *Thought Experiments in Science and Philosophy* anthology edited by Horowitz and Massey (1991).

20 It is important to distinguish the then 'received view' from that of Aristotle, with which it is often confused. The common view is perhaps an Aristotelian one, but, as Lane points out, there is little textual evidence to attribute it to Aristotle himself. Galileo, on p. 68 of the *New Sciences*, attributes to Aristotle the claim that a 'hundred-pound iron ball falling from the height of a hundred braccia hits the ground before one of just one pound has descended a single braccio'. No one has been able to find this text in Aristotle. Lane calls it 'a sheer invention' by Galileo. The episode is discussed in Brackenridge (1989).

21 For Aristotle's conception of natural motion, see Graham (1996).

22 For Einstein's use of thought experiment, see Brown (1991/2010, pp. 15–20) and Norton (1991).

23 Mach's own account of thought experiments is in Chapter 11 of his *Knowledge and Error* (Mach 1905/1976). Expositions of his views can be found in Hiebert (1974).

24 Particularly useful articles are: Blown and Bryce (2013), Galili (2009), Helm and Gilbert (1985), Helm *et al.* (1985), Özdemir and Kösem (2014), Reiner and Burko (2003), Reiner and Gilbert (2000), Stephens and Clement (2012), Velentzas and Halkia (2011, 2013) and Winchester (1990). The whole tradition of thought-experiment research is reviewed in Asikainen and Hirvonen (2014).

25 It is depressing to read what is learned in typical school experiment classes. See at least: Carey *et al.* (1989), Hodson (1993, 1996) and Jenkins (1999).

26 See at least: Hooker (2011), Parker (2008), Wimsatt (2007) and Winsberg (2010).

27 See at least: Arriassecq *et al.* (2014), De Jong *et al.* (2013), Scalise *et al.* (2011) and Smetana and Bell (2012).

28 See, for instance, Siegel (1995).

29 Erduran and Jiménez-Aleixandre (2008) and Khine (2012).

30 See at least: Böttcher and Meisert (2011), Bricker and Bell (2008), Jiménez-Aleixandre and Erduran (2008), Kuhn (2010), Nielsen (2012, 2013) and Sampson and Clark (2008). The tradition of research is reviewed in Adúriz-Bravo (2014).
31 On these exercises, see Scriven (1976, Chapter 3). Half of David Stove's first-year logic course at University of Sydney, in 1966, consisted of just such exercises; they were invaluable for atuning teenage minds to the structure and evaluation of argument inside and outside philosophy.
32 Pierre Duhem's *To Save the Phenomena: An Essay on the Idea of Physical Theory from Plato to Galileo* (1908/1969) is an excellent source book on this tradition.
33 See Gillies (1998) on the difference between Duhem's thesis and Quine's thesis. A collection of papers on the Duhem–Quine thesis is Harding (1976).
34 As early as the 1970 Postscript to his *Structures*, Kuhn acknowledged that there were 'aspects of its initial formulation that create gratuitous difficulties and misunderstandings'. He later explicitly regreted writing these 'purple passages' (Kuhn 1991/2000), but, by then, the irrationalist horse had bolted into uncountable philosophical, sociological and educational paddocks.
35 For a sample of cogent defences of rationalism and the truth-seeking function of science see, among many: Brown (1994), Devitt (1991), Musgrave (1999), Newton-Smith (1981), Nola and Irzik (2005), Scheffler (1982), Shapere (1984), Shimony (1976), Siegel (1987) and Smith and Siegel (2004).
36 Two reviews of the first wave of sociological literature, each citing hundreds of articles, are Mulkay (1982) and Shapin (1982). For critiques of the strong programme, see, among many: Brown (1984), Bunge (1991, 1992), Nola (1991, 2000) and Slezak (1994a, 1994b).
37 Some commentators maintain that Foucault considers his 'power is knowledge' thesis applies only to the social sciences and humanities, and it was not meant to cover the natural sciences (Gutting 1989, p. 4).
38 An account of the development of intelligence testing and theory can be found in Matthews (1980, Chapter 6).
39 See Birstein (2001), Graham (1973, 1998), Joravsky (1970), Lecourt (1977) and Lewontin and Levins (1976).
40 See Cross and Price (1992), Musschenga and Gosling (1985), Ratcliffe and Grace (2003) and contributions to Zeidler (2003) and Zeidler and Sadler (2008).
41 The exchange between Eger, Hesse, Shimony and others (*Zygon* 23(3), 1988) on 'rationality in science and ethics' (reproduced in Matthews 1991) shows the benefits of striving for a modicum of philosophical sophistication in these matters. The philosophers Alberto Cordero (1992) and Michael Martin (1986/1991) provide insightful and disciplined discussion of the interplay of science, ethics and education. The research field is reviewed in Couló (2014).
42 These are 'Mapping and Sequencing the Human Genome: Science, Ethics and Public Policy' (BSCS 1992), 'Genethics – Ball State Model' (Ball State University, Muncie, IN) and 'Teacher Education in Biology' (San Francisco State University, San Francisco, CA). The programmes are discussed in Blake (1994).
43 The Nazi case is depressingly documented in Diarmuid Jeffreys' 2008 book, the subtitle of which is: *IG Farben and the Making of Hilter's War Machine* (Jeffreys 2008). See also John Cornwell's 2003 book, the subtitle of which is *Science, War and the Devil's Pact* (Cornwell 2003).
44 The Soviet Union case is equally depressingly documented in Vadim Birstein's *The Perversion of Knowledge* (Birstein 2001).
45 Among countless good books, see: Robert Bell's *Impure Science* (Bell 1992), D.S. Greenberg's *Science, Money and Politics* (Greenberg 2001) and David Noble's *America by Design* (Noble 1979).
46 Bacon did not believe that such influences needed to be denied or eradicated, just that they had to be recognised and corrected for where necessary; he was happy for the new science to serve social ends and promote welfare. For an informative discussion of

Bacon's texts and their philosophical context, see Gaukroger (2001, pp. 118–131) and Urbach (1987, Chapter 4).

47 The chapter repeats the more-discussed argument of Rudner (1953).

48 See at least: Carrier (2013), Davson-Galle (2002), Develaki (2008), Doppelt (2008), Douglas (2009), Kitcher (2001), Lacey (2005), Longino (1990, 2008), Machamer and Douglas (1999), Resnik (1998, 2007), Rooney (1992) and Ruphy (2006). See also contributions to Carrier *et al.* (2008) and Dupré *et al.* (2007).

49 Mario Bunge warns: 'The simpler hypothesis may also be the most simple-minded, and the simpler methods the least exact and exacting' (Bunge 1963, p. 86).

50 Other early contributions can be seen in anthologies such as J. Harding (1986), Harding and Hintikka (1983), Keohane *et al.* (1982), Lowe and Hubbard (1983) and Tuana (1989b).

References

Aalberts, J., Koster, E. and Boschuizen, R.: 2012, From Prejudice to Reasonable Judgement: Integrating (Moral) Value Discussions in University Courses, *Journal of Moral Education* 41(4), 437–455.

Adler, M.J.: 1978, *Aristotle for Everybody*, Macmillan, New York.

Adúriz-Bravo, A.: 2014, 'Revisiting School Scientific Argumentation from the Perspective of the History and Philosophy of Science'. In M.R. Matthews (ed.) *International Handbook of Research in History, Philosophy and Science Teaching*, Springer, Dordrecht, The Netherlands, pp. 1443–1472.

Agassi, J.: 1964, 'The Nature of Scientific Problems and Their Roots in Metaphysics'. In M. Bunge (ed.) *The Critical Approach*, Free Press, Glencoe, IL. Reprinted in J. Agassi, *Science in Flux*, Reidel, Boston, MA, 1975, pp. 208–239.

Amsterdamski, S.: 1975, *Between Experience and Metaphysics: Philosophical Problems in the Evolution of Science*, Reidel Publishing Company, Dordrercht, The Netherlands.

Arons, A.B.: 1974, 'Education Through Science', *Journal of College Science Teaching* 13, 210–220.

Arons, A.B.: 1977, *The Various Language, An Inquiry Approach to the Physical Sciences*, Oxford University Press, New York.

Arons, A.B.: 1990, *A Guide to Introductory Physics Teaching*, John Wiley, New York.

Arriassecq, I., Greca, I. and Eugenia, S.: 2014, 'Epistemological Issues Concerning Computer Simulations in Science and Their Implications for Science Education', *Science & Education* 23, 897–921.

Asikainen, M.A. and Hirvonen, P.E.: 2014, 'Thought Experiments in Science and in Science Education'. In M.R. Matthews (ed.) *International Handbook of Research in History, Philosophy and Science Teaching*, Springer, Dordrecht, The Netherlands, pp. 1235–1256.

Bacon, F.: 1620/1960, *Novum Organum and Related Writings*, F.H. Anderson (ed.), New York.

Barnes, B.: 1977, *Interests and the Growth of Knowledge*, Routledge & Kegan Paul, London.

Bell, B.F. (ed.): 1992, *I Know About LISP But How Do I Put it Into Practice: Draft Report*, Centre for Science and Mathematics Education Research, University of Waikato, Hamilton, New Zealand.

Bergmann, P.: 1949, *Basic Theories of Physics*, Prentice-Hall, New York.

Berkeley, G.: 1721/1965, *De Motu*, in D.M. Armstrong (ed.) *Berkeley's Philosophical Writings*, Macmillan, New York, pp. 251–273.

Bernal, J.D.: 1939, *The Social Function of Science*, Routledge & Kegan Paul, London.

Birch, L.C.: 1990, *On Purpose*, University of New South Wales Press, Sydney.

Birstein, V.J.: 2001, *The Perversion of Knowledge: The True Story of Soviet Science*, Westview, Cambridge, MA.

Blake, D.: 1994, 'Revolution, Revision, or Reversal: Genetics–Ethics Curriculum', *Science & Education* 3(4), 373–391.
Bleier, R.: 1984, *Science and Gender*, Pergamon Press, New York.
Bloor, D.: 1976/1991, *Knowledge and Social Imagery*, Routledge & Kegan Paul, London (2nd edition, 1991).
Blown, E.J. and Bryce, T.G.K.: 2013, 'Thought-Experiments about Gravity in the History of Science and in Research into Children's Thinking', *Science & Education* 22(3), 419–481.
Bohm, D.: 1980, *Wholeness and the Implicate Order*, Ark Paperbacks, London.
Bohr, N.: 1958, *Atomic Physics and Human Knowledge*, Wiley, New York.
Bokulich, A.: 2001, 'Rethinking Thought Experiments', *Perspectives on Science* 9(3), 285–207.
Boltzmann, L.: 1905/1974, *Theoretical Physics and Philosophical Problems*, Reidel, Dordrecht, The Netherlands.
Booth, E.H. and Nicol, P.M.: 1931/1962, *Physics: Fundamental Laws and Principles with Problems and Worked Examples*, Australasian Medical Publishing Company, Sydney (16th edn 1962).
Born, M.: 1968, *My Life & My Views*, Scribners, New York.
Böttcher, F. and Meisert, A.: 2011, 'Argumentation in Science Education: A Model-Based Framework', *Science & Education* 20(2), 103–140.
Boulos, P.J.: 2006, 'Newton's Path to Universal Gravitation: The Role of the Pendulum', *Science & Education* 15(6), 577–595.
Brackenridge, J.B.: 1989, 'Education in Science, History of Science and the "Textbook"', *Interchange* 20(2), 71–80.
Bricker, L.A. and Bell, P.: 2008, 'Conceptualizations of Argumentation From Science Studies and the Learning Sciences and Their Implications for the Practices of Science Education', *Science Education* 92(3), 473–498.
Brickhouse, N.W.: 2001, 'Embodying Science: A Feminist Perspective on Learning', *Journal of Research in Science Teaching* 38(3), 282–295.
Bridgman, P.W.: 1950, *Reflections of a Physicist*, Philosophical Library, New York.
Brown, J.R. (ed.): 1984, *Scientific Rationality: The Sociological Turn*, Reidel, Dordrecht, The Netherlands.
Brown, J.R.: 1991/2010, *The Laboratory of the Mind: Thought Experiments in the Natural Sciences*, 2nd edn, Routledge, New York.
Brown, J.R.: 1994, *Smoke and Mirrors: How Science Reflects Reality*, Routledge, New York.
Brown, R.J. and Fehige, Y.: 2011, 'Thought Experiments'. *The Stanford Encyclopedia of Philosophy*, available at: http://plato.stanford.edu/entries/thought-experiment
BSCS (Biological Science Curriculum Committee): 1992, *Mapping and Sequencing the Human Genome: Science, Ethics and Public Policy*, BSCS, Colorado Springs, CO.
Buchdahl, G.: 1969, *Metaphysics and the Philosophy of Science*, Basil Blackwell, Oxford, UK.
Buckley, M.J.: 1971, *Motion and Motion's God*, Princeton University Press, Princeton, NJ.
Bunge, M.: 1963, *The Myth of Simplicity*, Prentice-Hall, Englewood Cliffs, NJ.
Bunge, M.: 1967, 'Analogy in Quantum Mechanics: From Insight to Nonsense', *The British Journal for Philosophy of Science* 18, 265–286.
Bunge, M.: 1973, *The Philosophy of Physics*, Reidel, Dordrecht, The Netherlands.
Bunge, M.: 1991, 'A Critical Examination of the New Sociology of Science: Part 1', *Philosophy of the Social Sciences* 21(4), 524–560.
Bunge, M.: 1992, 'A Critical Examination of the New Sociology of Science: Part 2', *Philosophy of the Social Sciences* 22(1), 46–76.
Bunge, M.: 1998a, *Philosophy of Science*, Vol.1, Transaction Publishers, New Brunswick, NJ.
Bunge, M.: 1998b, *Philosophy of Science*, Vol.2, Transaction Publishers, New Brunswick, NJ.

Bunge, M.: 2012, 'Does Quantum Physics Refute Realism, Materialism and Determinism?', *Science & Education* 21(10), 1601–1610.

Burtt, E.A.: 1932, *The Metaphysical Foundations of Modern Physical Science*, 2nd edn, Routledge & Kegan Paul, London (1st edition, 1924).

Butts, R.E. and Davis, J.W. (eds): 1970, *The Methodological Heritage of Newton*, University of Toronto Press, Toronto.

Cachel, S.: 1990, 'Partisan primatology. Review of *Primate Visions: Gender, Race, and Nature in the World of Modern Science*', *American Journal of Primatology* 22(2), 139–142.

Campbell, N.R: 1921/1952, *What Is Science?* Dover, New York.

Carey, S., Evans, R., Honda, M., Jay, E. and Unger, C.: 1989, 'An Experiment Is When You Try It and See If It Works', *International Journal of Science Education* 11, 514–529.

Carrier, M., Howard, D. and Kourany, J. (eds): 2008, *The Challenge of the Social and the Pressure of Practice. Science and Values Revisited*, University of Pittsburgh Press, Pittsburgh, PA.

Carrier, M.: 2013, 'Values and Objectivity in Science: Value-Ladenness, Pluralism and the Epistemic Attitude', *Science & Education* 22(10), 2547–2568.

Chandrasekhar, S.: 1987, *Truth and Beauty: Aesthetics and Motivations in Science*, University of Chicago Press, Chicago, IL.

Clagett, M.: 1959, *The Science of Mechanics in the Middle Ages*, University of Wisconsin Press, Madison, WI.

Clavelin, M.: 1974, *The Natural Philosophy of Galileo. Essay on the Origin and Formation of Classical Mechanics*, MIT Press, Cambridge, MA.

Cochaud, G.: 1989, 'The Process Skills of Science', unpublished paper, Australian Science Teachers Association Annual Conference.

Coelho, R.L.: 2007, 'The Law of Inertia: How Understanding Its History Can Improve Physics Teaching', *Science & Education* 16(9–10), 955–974.

Cohen, I.B.: 1980, *The Newtonian Revolution*, Cambridge University Press, Cambridge, UK.

Cohen, M.: 2005, *Wittgenstein's Beetle and Other Classic Thought Experiments*, Blackwell, London.

Collingwood, R.G.: 1945, *The Idea of Nature*, Oxford University Press, Oxford, UK.

Collins, F.S.: 2007, *The Language of God: A Scientist Presents Evidence for Belief*, Free Press, New York.

Cordero, A.: 1992, 'Science, Objectivity and Moral Values', *Science & Education* 1(1), 49–70.

Cornwell, J.: 2003, *Hitler's Scientists: Science, War and the Devil's Pact*, Penguin, London.

Couló, A.C.: 2014, 'Philosophical Dimensions of Social and Ethical Issues in School Science Education: Values in Science and in Science Classrooms'. In M.R. Matthews (ed.) *International Handbook of Research in History, Philosophy and Science Teaching*, Springer, Dordrecht, The Netherlands, pp. 1087–1117.

Crombie, A.C.: 1956, *Medieval and Early Modern Science. Vol.11, Science in the Later Middle Ages and Early Modern Times: XII-XVII Centuries*, Harvard University Press, Cambridge, MA.

Cross, R.T. and Price, R.F.: 1992, *Teaching Science for Social Responsibility*, St Louis Press, Sydney.

Cushing, J.T.: 1998, *Philosophical Concepts in Physics: The Historical Relation between Philosophy and Scientific Theories*, Cambridge University Press, Cambridge, UK.

Cushing, J.T. and McMullin, E. (eds): 1989, *Philosophical Consequences of Quantum Theory*, University of Notre Dame Press, Notre Dame, IN.

d'Espagnat, B.: 2006, *On Physics and Philosophy*, Princeton University Press, Princeton, NJ.

Davson-Galle, P.: 2002, 'Science, Values and Objectivity', *Science & Education* 11(2), 191–202.

De Jong, T., Linn, M.C. and Zacharia, Z.C.: 2013, 'Physical and Virtual Laboratories in Science and Engineering Education', *Science* 340(6130), 305–308.

Develaki, M.: 2008, 'Social and Ethical Dimension of the Natural Sciences, Complex Problems of the Age, Interdisciplinarity, and the Contribution of Education', *Science & Education* 17(8–9), 873–888.

Devitt, M.: 1991, *Realism and Truth*, 2nd edn, Basil Blackwell, Oxford, UK.

Dilworth, C.: 1996/2006, *The Metaphysics of Science. An Account of Modern Science in Terms of Principles, Laws and Theories*, Kluwer Academic Publishers, Dordrecht, The Netherlands (2nd edn 2006).

Doppelt, G.: 2008, 'Values in Science'. In S. Psillos and M. Curd (eds) *The Routledge Companion to Philosophy of Science*, Routledge, New York, pp. 302–313.

Douglas, H.E.: 2009, *Science, Policy, and the Value-Free Ideal*, University of Pittsburgh Press, Pittsburgh, PA.

Duhem, P.: 1906/1954, *The Aim and Structure of Physical Theory* (trans. P.P. Wiener), Princeton University Press, Princeton, NJ.

Duhem, P.: 1908/1969, *To Save the Phenomena: An Essay on the Idea of Physical Theory from Plato to Galileo*, University of Chicago Press, Chicago, IL.

Dupré, J., Kincaid, H. and Wylie, A. (eds): 2007, *Value-Free Science? Ideals and Illusions*. Oxford University Press, New York.

Eddington, A.: 1928/1978, *The Nature of the Physical World*, University of Michigan Press, Ann Arbor, MI.

Eddington, A.: 1939, *The Philosophy of Physical Science*, Cambridge University Press, Cambridge, UK.

Eger, M.: 1988, 'A Tale of Two Controversies: Dissonance in the Theory and Practice of Rationality', *Zygon* 23(3), 291–326. Reprinted in M.R. Matthews (ed.) *History, Philosophy and Science Teaching: Select Readings*, OISE Press, Toronto, 1991.

Eger, M.: 1989, 'The "Interests" of Science and the Problems of Education', *Synthese* 80(1), 81–106.

Einstein, A.: 1916/1961, *Relativity: The Special and General Theory*, Crown Publishers, New York.

Ellis, B.D.: 1965, 'The Origin and Nature of Newton's Laws of Motion'. In R.G. Colodny (ed.) *Beyond the Edge of Certainty*, Englewood Cliffs, NJ, pp. 29–68.

Ellis, B.D.: 1976, 'The Existence of Forces', *Studies in History and Philosophy of Science* 7(2), 171–185.

Ennis, R.H.: 1979, 'Research in Philosophy of Science Bearing on Science Education'. In P.D. Asquith and H.E. Kyburg (eds) *Current Research in Philosophy of Science*, PSA, East Lansing, MI, pp. 138–170.

Epstein, L.C.: 1979, *Thinking Physics*, 2nd edn, Insight Press, San Francisco, CA.

Erduran, S. and Jiménez-Aleixandre, M.P. (eds): 2008, *Argumentation in Science Education: Perspectives From Classroom-Based Research*, Springer, Dordrecht, The Netherlands.

Feyerabend, P.K.: 1975, *Against Method*, New Left Books, London.

Frank, P. (ed.): 1954, *The Validation of Scientific Theories*, The Beacon Press, Boston, MA.

Friedman, M.: 2009, 'Newton and Kant on Absolute Space: From Theology to Transcendental Philosophy'. In M. Bitbol, P. Kersberg and J. Petitot (eds) *Constituting Objectivity: Transcendental Perspectives on Modern Physics*, Springer, Dordrecht, The Netherlands, pp. 35–50.

Galileo, G.: 1638/1954, *Dialogues Concerning Two New Sciences* (trans. H. Crew and A. de Salvio), Dover, New York (originally published 1914).

Galili, I.: 2009, 'Thought Experiments: Determining Their Meaning', *Science & Education* 18(1), 1–23.

Garnett, J.P. and Tobin, K.G.: 1984, 'Reasoning Patterns of Preservice Elementary and Middle School Science Teachers', *Science Education* 68, 621–631.

Gaukroger, S.: 2001, *Francis Bacon and the Transformation of Early-Modern Philosophy*, Cambridge University Press, Cambridge, UK.

Geary, D.C.: 1995, 'Reflections of Evolution and Culture in Children's Cognition: Implications for Mathematical Development and Instruction', *American Psychologist* 50(1), 24–37.
Gendler, T.S.: 2000, *Thought Experiment: On the Powers and Limits of Imaginary Cases*, Garland Press, London.
Gendler, T.S.: 2004, 'Thought Experiments Rethought – and Reperceived', *Philosophy of Science* 71, 1152–1163.
Gillies, D.: 1998, 'The Duhem Thesis and the Quine Thesis'. In M. Curd and J. Cover (eds) *Philosophy of Science: The Central Issues*, Norton, New York, pp. 302–319.
Gilligan, C.: 1982, *In a Different Voice: Psychological Theory and Women's Development*, Harvard University Press, Harvard, MA.
Gjertsen, D.: 1989, *Science and Philosophy: Past and Present*, Penguin, Harmondsworth, UK.
Graham, D.W.: 1996, 'The Metaphysics of Motion: Natural Motion in Physics II and Physics VIII'. In W. Wians (ed.) *Aristotle's Philosophical Development: Problems and Prospects*, Rowan & Littlefield, Lanham, MD.
Graham, L.R.: 1973, *Science and Philosophy in the Soviet Union*, Alfred A. Knopf, New York.
Graham, L.R.: 1998, *What Have We Learned About Science and Technology from the Russian Experience?*, Stanford University Press, Stanford, CA.
Greenberg, D.S.: 2001, *Science, Money, and Politics*, University of Chicago Press, Chicago, IL.
Gutting, G.: 1989, *Michel Foucault's Archelogy of Scientific Reason*, Cambridge University Press, Cambridge, UK.
Haldane, J.S.: 1928, *The Sciences and Philosophy*, Hodder & Stoughton, London.
Hanson, N.R.: 1965, 'Newton's First Law: A Philosopher's Door into Natural Philosophy'. In R.G. Colodny (ed.) *Beyond the Edge of Certainty*, Prentice Hall, Englewood-Cliffs, NJ, pp. 6–28.
Haraway, D.: 1989, *Primate Visions: Gender, Race and Nature in the World of Modern Science*, Routledge, New York.
Harding, J. (ed.): 1986, *Perspectives on Gender and Science*, Falmer Press, East Sussex, UK.
Harding, S.G.: 1986, *The Science Question in Feminism*, Cornell University Press, Ithaca, NY.
Harding, S.G.: 1991, *Whose Science? Whose Knowledge? Thinking from Women's Lives*, Cornell University Press, Ithaca, NY.
Harding, S.G: 1976, *Can Theories Be Refuted? Essays on the Duhem-Quine Thesis*, Reidel, Dordrecht, The Netherlands.
Harding, S.G. and Hintikka, M.B. (eds): 1983, *Discovering Reality: Feminist Perspectives on Epistemology, Metaphysics, Methodology, and Philosophy of Science*, Reidel, Dordrecht, The Netherlands.
Hartsock, N.: 1983, 'The Feminist Standpoint: Developing the Grounds for a Specifically Feminist Historical Materialism'. In S. Harding and M.B. Hintikka (eds) *Discovering Reality*, Reidel, Dordrecht, The Netherlands, pp. 283–310.
Heisenberg, W.: 1962, *Physics and Philosophy*, Harper & Row, New York.
Helm, H. and Gilbert, J.: 1985, 'Thought Experiments and Physics Education – Part I', *Physics Education* 20, 124–131.
Helm, H., Gilbert, J. and Watts, D.M.: 1985, 'Thought Experiments and Physics Education – Part 2', *Physics Education* 20, 211–217.
Helmholtz, H. von: 1995, *Science and Culture: Popular and Philosophical Essays* (edited with Introduction by David Cahan; original essays 1853–1892), Chicago University Press, Chicago, IL.
Hesse, M.B.: 1961, *Forces and Fields: The Concept of Action at a Distance in the History of Physics*, Thomas Nelson & Sons, London.

Hiebert, E.N.: 1974, 'Mach's Conception of Thought Experiments in the Natural Sciences'. In Y. Elkana (ed.) *The Interaction Between Science and Philosophy*, Humanities Press, Atlantic Highlands, NJ, pp. 339–348.
Hodson, D.: 1988, 'Experiments in Science and Science Teaching', *Educational Philosophy and Theory* 20(2), 53–66.
Hodson, D.: 1993, 'Re-thinking Old Ways: Towards a More Critical Approach to Practical Work in Science', *Studies in Science Education* 22, 85–142.
Hodson, D.: 1996, 'Laboratory Work as Scientific Method: Three Decades of Confusion and Distortion', *Journal of Curriculum Studies* 28, 115–135.
Holton, G.: 1973, *Thematic Origins of Scientific Thought*, Harvard University Press, Cambridge, MA.
Holton, G.: 1988, *Thematic Origins of Scientific Thought: Kepler to Einstein*, 2nd edn, Harvard University Press, Cambridge, MA.
Holton, G. and Brush, S.G.: 2001, *Physics, the Human Adventure. From Copernicus to Einstein and Beyond*, Rutgers University Press, New Brunswick, NJ.
Holton, G., Rutherford, F.J. and Watson, F.G.: 1974, *The Project Physics Course: Motion*, Horwitz Group, Sydney.
Hooker, C.A.: 2011, *Philosophy of Complex Systems*, Elsevier, Amsterdam.
Horowitz, G. and Massey, G. (eds): 1991, *Thought Experiments in Science and Philosophy*, Rowman & Littlefield, Lanham, MD.
Hubbard, R.: 1979, 'Have Only Men Evolved?'. In R. Hubbard, M.S. Henifin and B. Fried (eds) *Women Look at Biology Looking at Women*, Schenkman Press, Cambridge, MA, pp. 7–35.
Hull, D.L.: 1988, *Science as a Process: An Evolutionary Account of the Social and Conceptual Development of Science*, University of Chicago Press, Chicago, IL.
Hunt, I.E. and Suchting, W.A.: 1969, 'Force and "Natural Motion"', *Philosophy of Science* 36, 233–251.
Huxley, A.: 1947, *Science, Liberty and Peace*, Chatto & Windus, London.
Hyland, P., Gomez, O. and Greensides, F. (eds): 2003, *The Enlightenment: A Source Book and Reader*, Routledge, New York.
Izquierdo-Aymerich, M. and Adúriz-Bravo, A.: 2003, 'Epistemological Foundations of School Science', *Science & Education* 12(1), 27–43.
Jammer, M.: 1957, *Concepts of Force: A Study in the Foundations of Dynamics*, Harvard University Press, Cambridge, MA.
Jeans, J.: 1943/1981, *Physics and Philosophy*, Dover, New York.
Jeffreys, D.: 2008, *Hell's Cartel: IG Farben and the Making of Hitler's War Machine*, Bloomsbury, London.
Jenkins, E.W.: 1999, 'Practical Work in School Science: Some Questions to Be Answered'. In J. Leach and A.C. Paulsen (eds) *Practical Work in Science Education: Recent Research Studies*, Kluwer Academic Publishers, Dordrecht, The Netherlands, pp. 19–32.
Jiménez-Aleixandre, M.P. and Erduran, S.: 2008, 'Argumentation in Science Education: An Overview'. In S. Erduran and M.P. Jiménez-Aleixandre (eds) *Argumentation in Science Education: Perspectives From Classroom-Based Research*, Springer, Dordrecht, The Netherlands.
Joravsky, D.: 1970, *The Lysenko Affair*, University of Chicago Press, Chicago, IL.
Jungwirth, E.: 1987, 'Avoidance of Logical Fallacies: A Neglected Aspect of Science-Education and Science-Teacher Education', *Research in Science & Technological Education* 5(1), 43–58.
Kalman, C.S.: 2009, 'A Role for Experiment in Using the Law of Inertia to Explain the Nature of Science: A Comment on Lopes Coelho', *Science & Education* 18(1), 25–31.
Keller, E.F.: 1985, *Reflections on Gender and Science*, Yale University Press, New Haven, CT.
Keller, E.F.: 1987, 'Feminism and Science'. In S.G. Harding and J.F. O'Barr (eds) *Sex and Scientific Inquiry*, University of Chicago Press, Chicago, IL.

Keohane, N.O., Roasaldo, M.Z. and Gelpi, B.C. (eds): 1982, *Feminist Theory: A Critique of Ideology*, University of Chicago Press, Chicago, IL.

Khine, M.S. (ed.): 2012, *Perspectives on Scientific Argumentation: Theory, Practice and Research*, Springer, Dordrecht, The Netherlands.

Kitcher, P.: 2001, *Science, Truth, and Democracy*, Oxford University Press, Oxford, UK.

Koertge, N.: 1969, 'Towards an Integration of Content and Method in the Science Curriculum', *Curriculum Theory Network* 4, 26–43. Reprinted in *Science & Education* 1996, 5(4), 391–402.

Koertge, N.: 1981, 'Methodology, Ideology and Feminist Critiques of Science'. In P.D. Asquith and R.N. Giere (eds), *Proceedings of the Philosophy of Science Association 1980*, Edwards Bros, Ann Arbor, MI, pp. 346–359.

Koertge, N.: 1998, 'Postmodernisms and the Problem of Scientific Literacy'. In her *A House Built on Sand: What's Wrong with the Cultural Studies Account of Science*, Oxford University Press, New York, pp. 257–271.

Koertge, N.: 2000, '"New Age" Philosophies of Science, Constructivism, Feminism and Postmodernism', *British Journal for the Philosophy of Science* 51, 667–683.

Kosso, P.: 1992, *Reading the Book of Nature: An Introduction to the Philosophy of Science*, Cambridge University Press, New York.

Koyré, A.: 1953/1968, 'An Experiment in Measurement', *Proceedings of the American Philosophical Society* 7, 222–237. Reproduced in his *Metaphysics and Measurement*, Harvard University Press, Cambridge, MA, 1968, pp. 89–117.

Koyré, A.: 1954, 'Influence of Philosophic Trends on the Formulation of Scientific Theories'. In P.G. Frank (ed.) *The Validation of Scientific Theories*, The Beacon Press, Boston, MA, pp. 192–203.

Koyré, A.: 1957, *From the Closed World to the Infinite Universe*, The Johns Hopkins University Press, Baltimore, MD.

Koyré, A.: 1960, 'Galileo's Treatise "*De Motu Gravium*": The Use and Abuse of Imaginary Experiment', *Revue d'Histoire des Sciences* 13, 197–245. Reprinted in his *Metaphysics and Measurement*, 1968, Harvard University Press, Cambridge, MA, pp. 44–88.

Koyré, A.: 1968, *Metaphysics and Measurement*, Harvard University Press, Cambridge, MA.

Kuhn, D.: 2010, 'Teaching and Learning Science as Argument', *Science Education* 94(5), 810–824.

Kuhn, T.S.: 1962/1970, *The Structure of Scientific Revolutions*, 2nd edn, Chicago University Press, Chicago, IL (1st edition, 1962).

Kuhn, T.S.: 1964, 'A Function for Thought Experiments' in his *The Essential Tension*, University of Chicago Press, Chicago, IL, 1977, pp. 28–43.

Kuhn, T.S.: 1977, 'Objectivity, Value Judgement, and Theory Choice'. In his *The Essential Tension*, University of Chicago Press, Chicago, IL, pp. 320–339.

Kuhn, T.S.: 1991/2000, 'The Trouble with Historical Philosophy of Science', The Robert and Maurine Rothschild Lecture, Department of History of Science, Harvard University. In J. Conant and J. Haugeland (eds) *The Road Since Structure: Thomas S. Kuhn*, University of Chicago Press, Chicago, IL, pp. 105–120.

Lacey, H.: 2005, *Values and Objectivity in Science*, Lexington Books, Lantham, MD.

Latour, B. and Woolgar, S.: 1979/1986, *Laboratory Life: The Social Construction of Scientific Facts*, 2nd edn, SAGE, London.

Lecourt, D.: 1977, *Proletarian Science? The Case of Lysenko*, Manchester University Press, Manchester, UK.

Lederman, N.G.: 2004, 'Syntax of Nature of Science within Inquiry and Science Instruction'. In L.B. Flick and N.G. Lederman (eds) *Scientific Inquiry and Nature of Science*, Kluwer Academic Publishers, Dordrecht, The Netherlands, pp. 301–317.

Lederman, N.G., Abd-el-Khalick, F., Bell, R.L. and Schwartz, R.S.: 2002, 'Views of Nature of Science Questionnaire: Towards Valid and Meaningful Assessment of Learners' Conceptions of the Nature of Science', *Journal of Research in Science Teaching* 39, 497–521.

Levi, I.: 1960, 'Must the Scientist Make Value Judgments?', *The Journal of Philosophy* LVII, 345–357.
Lewontin, R. and Levins, R.: 1976, 'The Problem of Lysenkoism'. In H. Rose and S. Rose (eds) *The Radicalisation of Science*, Macmillan, London, pp. 32–64.
Lipman, M.: 1991, *Thinking in Education*, Cambridge University Press, Cambridge, UK.
Lipman, M. and Sharp, A.M. (eds): 1978, *Growing Up with Philosophy*, Temple University Press, Philadelphia, PA.
Longino, H.E.: 1989, 'Can There Be a Feminist Science?'. In N. Tuana (ed.) *Feminism & Science*, Indiana University Press, Bloomington, IN pp. 45–57.
Longino, H.E.: 1990, *Science as Social Knowledge: Values and Objectivity in Scientific Inquiry*, Princeton University Press, Princeton, NJ.
Longino, H.E.: 2008, 'Values, Heuristics, and the Politics of Knowledge'. In M. Carrier, D. Howard and J. Kourany (eds) *The Challenge of the Social and the Pressure of Practice. Science and Values Revisited*, University of Pittsburgh Press, Pittsburgh, PA, pp. 68–86.
Lowe, M. and Hubbard, R. (eds): 1983, *Women's Nature: Rationalizations of Inequality*, Pergamon Press, New York.
Mach, E.: 1883/1960, *The Science of Mechanics*, Open Court Publishing, LaSalle, IL.
Mach, E.: 1886/1986, 'On Instruction in the Classics and the Sciences'. In his *Popular Scientific Lectures*, Open Court Publishing, LaSalle, IL, pp. 338–374.
Mach, E.: 1893/1974, *The Science of Mechanics*, 6th edn, Open Court Publishing, LaSalle, IL.
Mach, E.: 1896/1976, 'On Thought Experiments'. In his *Knowledge and Error*, Reidel, Dordrecht, The Netherlands, pp. 134–147.
Mach, E.: 1905/1976, *Knowledge and Error*, Reidel, Dordrecht, The Netherlands.
Machamer, P. and Douglas, H.: 1999, 'Cognitive and Social Values', *Science & Education* 8(1), 45–54.
McDermott, L.C. and Physics Education Group: 1995, *Physics by Inquiry*, 3 volumes, John Wiley, New York.
McMullin, E.: 1983, 'Values in Science'. In P.D. Asquith and T. Nickles (eds) *PSA 1982*, Vol. 2, PSA, East Lansing, MI, pp. 3–28.
Mannheim, K.: 1936/1960, *Ideology and Utopia*, Routledge & Kegan Paul, London.
Margenau, H.: 1950, *The Nature of Physical Reality: A Philosophy of Modern Physics*, McGraw-Hill, New York.
Margenau, H.: 1978, *Physics and Philosophy: Selected Essays*, Kluwer Academic Publishers, Dordrecht, The Netherlands.
Martin, J.R.: 1989, 'Ideological Critiques and the Philosophy of Science', *Philosophy of Science* 56, 1–22.
Martin, M.: 1986/1991, 'Science Education and Moral Education', *Journal of Moral Education* 15(2), 99–108. Reprinted in M.R. Matthews (ed.) *History, Philosophy and Science Teaching: Selected Readings*, OISE Press, Toronto, 1991, pp. 102–114.
Marx, K.: 1852/1969, *The Eighteenth Brumaire of Louis Bonaparte*. In *Karl Marx and Frederick Engels: Selected Works*, Vol.1, Progress, Moscow, pp. 398–487.
Matthews, G.B.: 1982, *Philosophy and the Young Child*, Harvard University Press, Cambridge, MA.
Matthews, M.R.: 1980, *The Marxist Theory of Schooling: A Study in Epistemology and Education*, Harvester Press, Brighton, UK.
Matthews, M.R.: 1989, 'Ernst Mach and Thought Experiments in Science Education', *Research in Science Education* 18, 251–258.
Matthews, M.R. (ed.): 1991, *History, Philosophy and Science Teaching: Selected Readings*, OISE Press, Toronto.
Matthews, M.R.: 1997, 'Israel Scheffler on the Role of History and Philosophy of Science in Science Teacher Education', *Studies in Philosophy and Education* 16(1–2), 159–173.

Matthews, M.R.: 2000, *Time for Science Education: How Teaching the History and Philosophy of Pendulum Motion Can Contribute to Science Literacy*, Kluwer Academic Publishers, New York.
Mayr, E.: 1982, *The Growth of Biological Thought*, Harvard University Press, Cambridge, MA.
Mayr, E.: 1988, *Toward a New Philosophy of Biology: Observations of an Evolutionist*, Harvard University Press, Cambridge, MA.
Meli, D.B.: 2006, *Thinking with Objects: The Transformation of Mechanics in the Seventeenth Century*, Johns Hopkins University Press, Baltimore, MD.
Merton, R.K.: 1942/1973, 'The Normative Structure of Science'. In his *The Sociology of Science: Theoretical and Empirical Investigations* (N.W. Storer, ed.), University of Chicago Press, Chicago, IL, 1973, pp. 267–280.
Merton, R.K.: 1957, 'The Sociology of Knowledge'. In his *Social Theory and Social Structure*, Free Press, New York.
Monod, J.: 1971, *Chance and Necessity: An Essay on the Natural Philosophy of Modern Biology*, Knopf, New York.
Moody, E.A.: 1975, *Studies in Medieval Philosophy, Science and Logic*, University of California Press, Berkeley, CA.
Moore Jr, B.: 1954, 'Influence of Political Creeds on the Acceptance of Theories'. In P.G. Frank (ed.) *The Validation of Scientific Theories*, The Beacon Press, Boston, MA, pp. 29–36.
Mulkay, M.: 1982, 'Sociology of Science in the West', *Current Sociology* 28(3), 1–116.
Musgrave, A.: 1999, *Essays on Realism and Rationalism*, Rodopi, Amsterdam.
Musschenga, B. and Gosling, D. (eds): 1985, *Science Education and Ethical Values*, Georgetown University Press, Washington, DC.
Nersessian, N.J.: 1993, 'In the Theoretician's Laboratory: Thought Experimenting as Mental Modelling', *Philosophy of Science Association Proceedings* (2), 291–301.
Newton, I.: 1729/1934, *Mathematical Principles of Mathematical Philosophy* (trans. A. Motte, revised F. Cajori), University of California Press, Berkeley, CA.
Newton, I.: 1730/1979, *Opticks or A Treatise of the Reflections, Refractions, Inflections and Colours of Light*, Dover, New York.
Newton-Smith, W.H.: 1981, *The Rationality of Science*, Routledge & Kegan-Paul, Boston, MA.
Nielsen, J.A.: 2013, 'Dialectical Features of Students' Argumentation: A Critical Review of Argumentation Studies in Science Education', *Research in Science Education* 43(1), 371–393.
Nielsen, K.H.: 2012, 'Scientific Communication and the Nature of Science', *Science & Education* 22(9), 2067–2086.
NIH (National Institute of Health): 2009, *Bioethics Yrs 9–12 Course*, NIH, Washington, DC.
Noble, D.F.: 1979, *America by Design: Science, Technology, and the Rise of Corporate Capitalism*, Alfred A. Knopf, New York.
Noddings, N.: 2009, 'Feminist Philosophy and Education'. In H. Siegel (ed.) *The Oxford Handbook of Philosophy of Education*, Oxford University Press, Oxford, UK, pp. 508–523.
Nola, R.: 1991, 'Ordinary Human Inference as Refutation of the Strong Programme', *Social Studies of Science* 21, 107–129.
Nola, R.: 2000, 'Saving Kuhn from the Sociologists of Science', *Science & Education* 9(1–2), 77–90.
Nola, R. and Irzik, G.: 2005, *Philosophy, Science, Education and Culture*, Springer, Dordrecht, The Netherlands.
Norton, J.D.: 1991, 'Thought Experiments in Einstein's Work'. In T. Horowitz and G. Massey (eds) *Thought Experiments in Science and Philosophy*, Rowman & Littlefield, Lanham, MD, pp. 129–148.
Norton, J.D.: 2004, 'On Thought Experiments: Is There More to the Argument?', *Philosophy of Science* 71, 1139–1151.

Okruhlik, K.: 1994, 'Gender and the Biological Sciences', *Canadian Journal of Philosophy* 20, 21–42.

Özdemir, Ö.F. and Kösem, S.D.: 2014, 'The Nature and Role of Thought Experiments in Solving Conceptual Physics Problems', *Science & Education* 23(4), 865–895.

Palmieri, P.: 2011, *A History of Galileo's Inclined Plane Experiment and Its Philosophical Implications*, The Edwin Mellen Press, Lewiston, NY.

Parker, W.S.: 2008, 'Does Matter Really Matter? Computer Simulations, Experiments, and Materiality', *Synthese* 169(3), 483–496.

Parsons, S.: 1999, 'Feminisms and Science Education: One Science Educator's Exploration of Her Practice', *International Journal of Science Education* 21(9), 989–1005.

Pati, D. and Koertge, N.: 1994, *Professing Feminism: Cautionary Tales from the Strange World of Women's Studies*, Basic Books, New York.

Petersen, A.: 1985, 'The Philosophy of Niels Bohr'. In A.P. French and P.J. Kennedy (eds) *Niels Bohr: A Centenary Volume*, Harvard University Press, Cambridge, MA, pp. 299–310.

Pinnick, C.L.: 2008a, 'Science Education for Women: Situated Cognition, Feminist Standpoint Theory, and the Status of Women in Science', *Science & Education* 17(10), 1055–1063.

Pinnick, C.L.: 2008b, 'The Feminist Approach to Philosophy of Science'. In S. Psillos and M. Curd (eds) *The Routledge Companion to Philosophy of Science*, Routledge, London, pp. 182–192.

Planck, M.: 1936, *The Philosophy of Physics*, W.W. Norton, New York.

Poincaré, H.: 1905/1952, *Science and Hypothesis*, Dover, New York.

Polanyi, M.: 1958, *Personal Knowledge*, Routledge & Kegan Paul, London.

Popper, K.R.: 1934/1959, *The Logic of Scientific Discovery*, Hutchinson, London.

Porter, R.: 2000, *The Enlightenment: Britain and the Creation of the Modern World*, Penguin, London.

Quine, W.V.O.: 1951/1953, 'Two Dogmas of Empiricism', *Philosophical Review*. Reprinted in his *From a Logical Point of View*, Harper & Row, New York, 1953, pp. 20–46.

Quine, W.V.O.: 1953, *From a Logical Point of View*, Harper & Row, New York.

Rabi, I.I.: 1967, *Science the Centre of Culture*, World Publishing, New York.

Ratcliffe, M. and Grace, M.: 2003, *Science Education for Citizenship. Teaching Socio-scientific Issues*, Open University Press, Maidenhead, UK.

Reichenbach, H.: 1938, *Experience and Prediction: An Analysis of the Foundations and the Structure of Knowledge*, University of Chicago Press, Chicago, IL.

Reiner, M. and Burko, L.M.: 2003, 'On the Limitations of Thought Experiments in Physics and the Consequences for Physics Education', *Science & Education* 12, 365–385.

Reiner, M. and Gilbert, J.: 2000, 'Epistemological Resources for Thought Experimentation in Science Education', *International Journal of Science Education* 22(5), 489–506.

Reiss, M.: 2008, 'The Use of Ethical Frameworks by Students Following a New Science Course for 16–18 Year-Olds', *Science & Education* 17(8–9), 889–902.

Resnik, D.B.: 2007, *The Price of Truth*, Oxford University Press, Oxford, UK.

Resnik, D.B.: 1998, *The Ethics of Science*, Routledge, New York.

Rohrlich, F.: 1987, *From Paradox to Reality: Our Basic Concepts of the Physical World*, Cambridge University Press, Cambridge, UK.

Rooney, P.: 1992, 'On Values in Science: Is the Epistemic/Non-Epistemic Distinction Useful?'. In D. Hull, M. Forbes and K. Okruhlik (eds) *Proceedings of the 1992 Biennial Meeting of the Philosophy of Science Association*, PSA, 1, 13–22.

Rosser, S.V.: 1986, *Teaching Science and Health from a Feminist Perspective*, Pergamon Press, Exeter, UK.

Rosser, S.V.: 1993, 'Female Friendly Science: Including Women in Curricular Content and Pedagogy in Science', *Journal of General Education* 42, 191–220.

Roth, M.-W.: 1999, 'Authentic School Science: Intellectual Traditions'. In R. McCormick and C. Paechter (eds) *Learning and Knowledge*, SAGE, London, pp. 6–20.

Rudner, R.: 1953, 'The Scientist Qua Scientist Makes Value Judgments', *Philosophy of Science* 20(1), 1–6.
Rudner, R.: 1954, 'Value Judgments in the Acceptance of Theories'. In P.G. Frank (ed.) *The Validation of Scientific Theories*, Beacon Press, Boston, MA, pp. 24–28.
Ruphy, S.: 2006, '"Empiricism all the Way Down": A Defense of the Value-Neutrality of Science in Response to Helen Longino's Contextual Empiricism', *Perspectives on Science* 14, 189–214.
Sampson, V.D. and Clark, D.B.: 2008, 'Assessment of the Ways Students Generate Arguments in Science Education: Current Perspectives and Recommendations for Future Directions', *Science Education* 92(3), 447–472.
Sargent, R.-M.: 2005, 'Virtues and the Scientific Revolution'. In N. Koertge (ed.) *Scientific Values and Civic Virtues*, Oxford University Press, Oxford, UK, pp. 71–80.
Scalise, K., Timms, M., Moorjani, A., Clark, L., Holtermann, K. and Irvin, P.S.: 2011, 'Student Learning in Science Simulations: Design Features that Promote Learning Gains', *Journal of Research in Science Teaching* 48, 1050–1078.
Schecker, H.: 1992, 'The Paradigmatic Change in Mechanics: Implications of Historical Processes on Physics Education', *Science & Education* 1(1), 71–76.
Scheffler, I.: 1970, 'Philosophy and the Curriculum'. In his *Reason and Teaching*, London, Routledge, 1973, pp. 31–44. Reprinted in *Science & Education* 1(4), 385–394.
Scheffler, I.: 1982, *Science and Subjectivity*, 2nd edn, Hackett, Indianapolis, IN (1st edition, 1966).
Scheibe, E.: 2000, 'The Origin of Scientific Realism: Boltzman, Planck, Einstein'. In E. Agazzi and M. Pauri (eds) *The Reality of the Unobservable*, pp. 31–44.
Schilpp, P.A. (ed.): 1951, *Albert Einstein: Philosopher–Scientist*, 2nd edn, Tudor, New York.
Schlesinger, G.N.: 1996, 'The Power of Thought Experiments', *Foundations of Physics* 26(4), 467–482.
Scriven, M.: 1976, *Reasoning*, McGraw-Hill, New York.
Shackel, N.: 2005, 'The Vacuity of Postmodernist Methodology', *Metaphilosophy* 36(3), 295–320.
Shapere, D.: 1984, *Reason and the Search for Knowledge*, Reidel, Dordrecht, The Netherlands.
Shapin, S.: 1982, 'History of Science and Its Sociological Reconstructions', *History of Science* 22, 157–211.
Shimony, A.: 1976, 'Comments on Two Epistemological Theses of Thomas Kuhn'. In R.S. Cohen, P.K. Feyerabend and M.W. Wartofsky (eds) *Essays in Memory of Imre Lakatos*, Reidel, Dordrecht, The Netherlands, pp. 569–588.
Shimony, A.: 1983, 'Reflections on the Philosophy of Bohr, Heisenberg, and Schrödinger'. In R.S. Cohen and L. Laudan (eds) *Physics, Philosophy and Psychoanalysis*, Reidel, Dordrecht, The Netherlands, pp. 209–221.
Shimony, A.: 1989, 'Search for a Worldview Which Can Accommodate Our Knowledge of Microphysics'. In J.T. Cushing and E. McMullin (eds) *Philosophical Consequences of Quantum Physics*, University of Notre Dame Press, Notre Dame, IN, pp. 25–37.
Shimony, A.: 1993, *Search for a Naturalistic World View, Vol.I Scientific Method and Epistemology*, Cambridge University Press, Cambridge, UK.
Siegel, H.: 1987, *Relativism Refuted*, Reidel, Dordrecht, The Netherlands.
Siegel, H.: 1988, *Educating Reason: Rationality, Critical Thinking, and Education*, Routledge, London.
Siegel, H.: 1989, 'The Rationality of Science, Critical Thinking, and Science Education', *Synthese* 80(1), 9–42. Reprinted in M.R. Matthews (ed.) *History, Philosophy and Science Teaching: Selected Readings*, OISE Press, Toronto and Teachers College Press, New York, 1991.
Siegel, H.: 1993, 'Naturalized Philosophy of Science and Natural Science Education', *Science & Education* 2(1), 57–68.
Siegel, H.: 1995, 'Why Should Educators Care About Argumentation', *Informal Logic* 17(2), 159–176.

Slezak, P.: 1994a, 'Sociology of Science and Science Education: Part I', *Science & Education* 3(3), 265–294.
Slezak, P.: 1994b, 'Sociology of Science and Science Education. Part 11: Laboratory Life Under the Microscope', *Science & Education* 3(4), 329–356.
Smart, J.J.C.: 1968, *Between Science and Philosophy: An Introduction to the Philosophy of Science*, Random House, New York.
Smetana, L.K. and Bell, R.L.: 2012, 'Computer Simulations to Support Science Instruction and Learning: A Critical Review of the Literature', *International Journal of Science Education* 34(9), 1337–1370.
Smith, M.U. and Siegel, H.: 2004, 'Knowing, Believing and Understanding: What Goals for Science Education?', *Science & Education* 13, 553–582.
Sorensen, R.A.: 1992, *Thought Experiments*, Oxford University Press, Oxford, UK.
Sprod, T.: 2011, *Discussions in Science: Promoting Conceptual Understanding in the Middle School Years*, ACER Press, Melbourne.
Sprod, T.: 2014, 'Philosophical Inquiry and Critical Thinking in Primary and Secondary Science Education'. In M.R. Matthews (ed.) *International Handbook of Research in History, Philosophy and Science Teaching*, Springer, Dordrecht, The Netherlands, pp. 1531–1564.
Stebbing, L.S.: 1937/1958, *Philosophy and the Physicists*, Dover Publications, New York.
Stephens, A.L. and Clement, J.: 2012, 'Role of Thought Experiments in Science and Science Learning'. In K. Tobin, C. McRobbie and B. Fraser (eds) *Second International Handbook of Science Education*, Springer, Dordrecht, The Netherlands, pp. 157–175.
Stinner, A.: 1990, 'Philosophy, Thought Experiments and Large Context Problems in the Secondary School Physics Course', *International Journal of Science Education* 12(3), 244–257.
Trefil, J.S.: 1978, *Physics as a Liberal Art*, Pergamon Press, Oxford, UK.
Trusted, J.: 1991, *Physics and Metaphysics: Theories of Space and Time*, Routledge, London.
Tuana, N. (ed.): 1989a, *Feminism & Science*, Indiana University Press, Bloomington, IN.
Tuana, N.: 1989b, 'The Weaker Seed: The Sexist Bias in Reproductive Theory'. In N. Tuana (ed.) *Feminism & Science*, Indiana University Press, Bloomington, IN pp. 147–171.
Turner, S.C.: 2012, 'Changing Images of the Inclined Plane: A Case Study of a Revolution in American Science Education', *Science & Education* 21(2), 245–270.
Urbach, P.: 1987, *Francis Bacon's Philosophy of Science*, Open Court, LaSalle, IL.
Velentzas, A. and Halkia, K.: 2011, 'The "Heisenberg's Microscope" as an Example of Using Thought Experiments in Teaching Physics Theories to Students of the Upper Secondary School', *Research in Science Education* 41, 525–539.
Velentzas, A. and Halkia, K.: 2013, 'From Earth to Heaven: Using the "Newton's Cannon" Thought Experiment for Teaching Satellite Physics', *Science & Education* 23, 2621–2640.
Wartofsky, M.W.: 1968, *Conceptual Foundations of Scientific Thought: An Introduction to the Philosophy of Science*, Macmillan, New York.
Weber, M.: 1917/1949, 'The Meaning of "Ethical Neutrality" in Sociology and Economics'. In E.A. Shils and H.A. Finch (trans.) *The Methodology of the Social Sciences*, The Free Press, Glencoe, IL, pp. 1–47.
Weinberg, S.: 2001, *Facing Up. Science and Its Cultural Adversaries*, Harvard University Press, Cambridge, MA.
Wilson, E.O.: 1998, *Consilience: The Unity of Knowledge*, Little, Brown, London.
Wimsatt, W.C.: 2007, 'False Models as Means to Truer Theories'. In his *Re-engineering Philosophy for Limited Beings: Piecewise Approximations to Reality*, Chapter 6. Harvard University Press, Cambridge, MA.
Winchester, I.: 1990, 'Thought Experiments and Conceptual Revision in Science', *Studies in Philosophy and Education* 10(1), 73–80. Reprinted in M.R. Matthews (ed.) *History, Philosophy and Science Teaching: Selected Readings*, OISE Press, Toronto and Teachers College Press, New York, 1991.

Winsberg, E.: 2010, *Science in the Age of Computer Simulation*, University of Chicago Press, Chicago, IL.

Yolton, J.W., Porter, R., Rogers, P. and Stafford, B.M. (eds): 1991, *The Blackwell Companion to the Enlightenment*, Basil Blackwell, Oxford, UK.

Zeidler, D.L. (ed.): 2003, *The Role of Moral Reasoning on Socioscientific Issues and Discourse in Science Education*, Kluwer Academic Publishers, Dordrecht, The Netherlands.

Zeidler, D.L. and Sadler, T.D. (eds): 2008, 'Social and Ethical Issues in Science Education', Special issue of *Science & Education* 17(8–9).

Ziman, J.: 1968, *Public Knowledge: The Social Dimension of Science*, Cambridge University Press, Cambridge, UK.

Chapter 6

History and Philosophy in the Classroom

Pendulum Motion[1]

In this chapter, the teaching of a single topic, pendulum motion, will be used to illustrate the claims being made about the benefits of a liberal or contextual approach to science education. Pendulum motion is chosen in part because it has a place in nearly all science programmes, and also because it is a relatively pedestrian topic. In 'hot' topics, such as evolution, genetic engineering, nuclear energy, climate change or acid rain, historical and philosophical considerations are obviously useful. If the case for HPS can be made with a 'boring' topic, then the usefulness of HPS for science teaching is better established. Furthermore, the science of pendulum motion illustrates important general topics alluded to in this book, including:

- the interplay of mathematics, observation and experiment in the development of modern science;
- the reciprocal impacts of science on culture and society;
- the interactions of philosophy and science;
- the distinction between material objects and these objects as treated by science;
- the ambiguous role of empirical evidence in the justification or falsification of scientific claims;
- the contrast between modern scientific conceptualisations and those of common sense.

The Pendulum and the Foundation of Modern Science

The pendulum has played a major role in the development of Western society, science and culture. The pendulum was central to the studies of Galileo, Huygens, Newton, Hooke and all the leading figures of the scientific revolution. The study and manipulation of the pendulum led to many things: an accurate method of timekeeping, hence, leading to solving the longitude problem; discovering the conservation and collision laws; and ascertaining the value of the acceleration due to gravity g, showing the variation of g from equatorial to polar regions and, hence, determining the oblate shape of the earth. It provided the crucial evidence for Newton's synthesis of terrestrial and celestial mechanics, showing that fundamental laws are universal in the solar system; a dynamical proof for the rotation of the earth on its axis; the equivalence of inertial and gravitational mass; an accurate measurement of the density and, hence, mass of the earth; and much more. Pendulum motion was central to the argument between Aristotelians and Galileo over the role of experience in settling conflicting claims about the world, and it figured in Newton's major metaphysical dispute with the Cartesians, namely the dispute concerning the existence of the ether (Westfall 1980, p. 376). Domenico Bertoloni Meli observed that:

> Starting with Galileo, the pendulum was taking a prominent place in the study of motion and mechanics, both as a time-measuring device and as a tool for studying motion, force, gravity, and collision.
>
> (Meli 2006, p. 206)

With good reason, the historian Bertrand Hall attested:

> In the history of physics the pendulum plays a role of singular importance. From the early years of the seventeenth century, when Galileo announced his formulation of the laws governing pendular motion, to the early years of this century, when it was displaced by devices of superior accuracy, the pendulum was either an object of study or a means to study questions in astronomy, gravitation and mechanics.
>
> (Hall 1978, p.4 41)

The importance of the pendulum in science and philosophy was exceeded only by its importance to commerce, navigation, exploration and Western expansion. A convenient and accurate measure of the passage of time was crucial for the pressing commercial and military problem of determining longitude at sea, as well as for everyday economic and social affairs. The pendulum answered these problems. Unfortunately, the centrality and importance of the pendulum for the development of modern science are not reflected in textbooks and school curricula, where it appears as an 'exceedingly arid' subject and is mostly, even in the best classes, dismissed with well-remembered formulae ($T = 2\pi\sqrt{(l/g)}$) and some routine mathematical exercises and maybe some practical classes.

The Textbook Myth and Prehistory of the Pendulum

The standard textbook treatment of pendulum motion features the story of Galileo's discovery of the isochronous movement of the pendulum. One such account is:

> When he [Galileo] was barely seventeen years old, he made a passive observation of a chandelier swinging like a pendulum in the church at Pisa where he grew up. He noticed that it swung in the gentle breeze coming through the half-opened

> church door. Bored with the sermon, he watched the chandelier carefully, then placed his fingertips on his wrist, and felt his pulse. He noticed an amazing thing. . . . Sometimes the chandelier swings widely and sometimes it hardly swings at all . . . [yet] it made the same number of swings every sixty pulse beats.
>
> (Wolf 1981, p. 33)

This same story appears in the opening pages of the most widely used high-school physics text in the world – the PSSC's *Physics* (PSSC 1960).[2]

If the textbook account is to be believed, then a basic question is why it was that the supposed isochronism of the pendulum was only seen in the sixteenth century, when countless thousands of people of genius and with acute powers of observation had, for thousands of years, been pushing children on swings, and looking at swinging lamps and swinging weights, and using suspended bobs in tuning musical instruments, without seeing their isochronism. For centuries, people had been concerned to find a reliable measure of time, both for scientific purposes and also in everyday life, to determine the duration of activities and events, and the vital navigational matter of determining longitude at sea. As the isochronic pendulum was the answer to all these questions, the widespread failure to recognise something so apparently obvious is informative. It suggests that there is not just a problem of perception, but a deeper problem is involved, a problem of epistemology, or how things are seen.

Nicole Oresme (1320–1382), in the fourteenth century, discussed pendulum motion. In his *On the Book of the Heavens and the World of Aristotle*, he entertained the thought experiment of a body dropped into a well that had been drilled from one side of the Earth, through the centre and out the other side (a thought experiment repeated by Galileo in his 1635 *Dialogues Concerning the Two Chief World Systems*). Oresme likens this imaginary situation to that of a weight that hangs on a long cord and swings back and forth, each time nearly regaining its initial position (Clagett 1959, p. 570). Albert of Saxony, Tartaglia and Benedetti all discussed the same problem in the context of impetus theory. Leonardo da Vinci, a most acute observer, dealt on many occasions with pendulum motion and, in the late 1490s, sketched two pendulums, one on a reciprocating pump; the other on what appears to be a clock. He recognised that the descent along the arc of a circle is quicker than that along the shorter corresponding chord, an anticipation of Galileo's later Law of Chords. In 1569, Jacques Besson published a book in Lyon detailing the use of the pendulum in regulating mechanical saws, bellows, pumps and polishing machines.[3]

Thomas Kuhn, in his *Structure of Scientific Revolutions*, famously used Galileo's account of the pendulum to mark the epistemological transformation from the old to the new science. Kuhn wrote:

> Since remote antiquity most people have seen one or another heavy body swinging back and forth on a string or chain until it finally comes to rest. To the

> Aristotelians, who believed that a heavy body is moved by its own nature from a higher position to a state of natural rest at a lower one, the swinging body was simply falling with difficulty. . . . Galileo, on the other hand, looking at the swinging body, saw a pendulum, a body that almost succeeded in repeating the same motion over and over again ad infinitum. And having seen that much, Galileo observed other properties of the pendulum as well and constructed many of the most significant and original parts of his new dynamics around them.
>
> (Kuhn 1970, p. 118–119)

The question of how Galileo came to recognise and prove the laws of pendulum motion is germane to the teaching of the topic. Teachers want students to recognise and prove the properties of pendulum motion – period being independent of mass and amplitude, and varying inversely as the square root of length. How these properties were initially discovered can throw light on current attempts to teach and learn the topic; as with all other subjects in science, their history has intrinsic as well as pedagogical value.

Galileo's Account of Pendulum Motion

In a letter of 1632, 10 years before his death, Galileo surveyed his achievements in physics and recorded his debt to the pendulum for enabling him to measure the time of free fall, which, he said, 'we shall obtain from the marvellous property of the pendulum, which is that it makes all its vibrations, large or small, in equal times' (Drake 1978, p. 399). To use pendulum motion as a measure of the passage of time was a momentous enough achievement, but the pendulum is also central to Galileo's treatment of free fall, the motion of bodies through a resisting medium, the conservation of momentum, and the rate of fall of heavy and light bodies. His analysis of pendulum motion is, thus, central to his overthrow of Aristotelian physics and the development of the modern science of motion, a development of which the historian Herbert Butterfield has said:

> Of all the intellectual hurdles which the human mind has confronted and has overcome in the last fifteen hundred years, the one which seems to me to have been the most amazing in character and the most stupendous in the scope of its consequences is the one relating to the problem of motion.
>
> (Butterfield 1949/1957, p. 3)

Galileo, in his final work, *The Two New Sciences*, written during the period of house arrest after the trial that, for many, marked the beginning of the Modern Age, wrote:

> We come now to the other questions, relating to pendulums, a subject which may appear to many exceedingly arid, especially to those philosophers who are continually occupied with the more profound questions of nature. Nevertheless,

> the problem is one which I do not scorn. I am encouraged by the example of Aristotle whom I admire especially because he did not fail to discuss every subject which he thought in any degree worthy of consideration.
>
> (Galileo 1638/1954, pp. 94–95)

Galileo's comment that pendulum investigations appear 'exceedingly arid' has, unfortunately, been echoed by science students over the following 400 years.

Galileo's Early Pendulum Investigations

Galileo's new science had engineering or practical roots: he worked with and was familiar with machines, and saw that pendulums were utilised in these machines; he did not have to await the breeze blowing the chandelier in the cathedral for his first experience of pendulum motion. And it was not just observing or looking at machines, or even reading about them – inclined planes, screws, levers, pendulums – but working with machines that provided the grounds for Galileo's transformation of extant science.[4]

While the youthful Galileo was briefly a medical student at Pisa, he utilised the pendulum to make a simple diagnostic instrument for measuring pulse beats. This was the *pulsilogium*. Medical practitioners in Galileo's day realised that pulse rate was of great significance, but there was no objective, let alone accurate, measurement of pulse beat. Galileo's answer to the problem was ingenious and simple: he suspended a lead weight on a short length of string, mounted the string on a scaled board, set the pendulum in motion and then moved his finger down the board from the point of suspension (thus effectively shortening the pendulum) until the pendulum oscillated in time with the patient's pulse. As the period of oscillation depended only on the length of the string, and not on the amplitude of swing or the weight of the bob, the length of the string provided an objective and repeatable measure of pulse speed that could be communicated between doctors and patients, and kept as a record.

The *pulsilogium* provides a useful epistemological lesson (and it is easily made by students). Initially, something subjective, the pulse, was used to measure the passage of time – occurrences, especially in music, were spoken of as taking so many pulse beats. With Galileo's *pulsilogium*, this subjective measure itself becomes subject to an external, objective, public measure – the length of the *pulsilogium*'s string. This was a small step in the direction of objective and precise measurement, upon which scientific advance in the seventeenth and subsequent centuries would depend.

After his appointment to a lectureship in mathematics at the University of Pisa in 1588, Galileo quickly became immersed in the mathematics and mechanics of the 'Superhuman Archimedes', whom he never mentions 'without a feeling of awe' (Galileo 1590/1960, p. 67). Galileo's major Pisan work is his *On Motion* (1590/1960). In it, he deals with the full range of problems being discussed among natural philosophers – free-fall, motion on balances, motion on inclined planes and circular motion. In these discussions, the physical circumstances are depicted geometrically, and mathematical reasoning is used to establish various conclusions in physics: Galileo here begins the thorough mathematising of physics, which is entirely modern.[5] Galileo's genius was to see that all of the above motions could be dealt with in one geometrical construction. That is, motions that appeared so different in the world could all be depicted and dealt with mathematically in a common manner.[6]

Consider a pendulum suspended at B, moving through C, F, L, J. This is a most fruitful construction. It allowed Galileo to analyse pendulum motion as motion in a circular rim and as motion on a suspended string. By considering initial, infinitesimal motions, he was able to consider pendulum motion as a series of tangential motions down inclined planes. Two years later, he was to write an important letter to his patron, Guidobaldo del Monte, about these propositions.

Galileo made use of the diagram in Figure 6.1 to prove properties of pendulum motion. It is important to note that Galileo then qualifies this proof, saying:

> But this proof must be understood on the assumption that there is no accidental resistance (occasioned by roughness of the moving body or of the inclined plane, or by the shape of the body). We must assume that the plane is, so to speak, incorporeal or, at least, that it is very carefully smoothed and perfectly hard . . . and that the moving body must be perfectly smooth . . . and of the hardest material.

Galileo here introduced crucial idealising conditions. His new science was not going to be simply about how the world behaves, but rather how it should behave. Or, to put it another way, his science was to be about how the world would behave if various conditions were fulfilled: for the pendulum, if the string were weightless, if the bob occasioned no air resistance, if the fulcrum were frictionless and so on. In controlled experiments, some of these conditions can be fulfilled, but other conditions cannot be fulfilled, and yet they were crucial to Galileo's science.[7]

Guidobaldo del Monte: Galileo's Patron and Critic

The most significant opponent of Galileo's nascent views about the pendulum was his own academic patron, the distinguished Aristotelian engineer Guidobaldo del Monte (1545–1607). Del Monte was one of the great mathematicians and mechanics of the late sixteenth century. He was a translator of the works of Archimedes, the author of a major book on mechanics (Monte 1581/1969), a book on geometry (*Planispheriorum universalium theorica*, 1579), a book on perspective techniques, *Perspectiva* (1600), and an unpublished book on timekeeping, *De horologiis*, that discussed the theory and

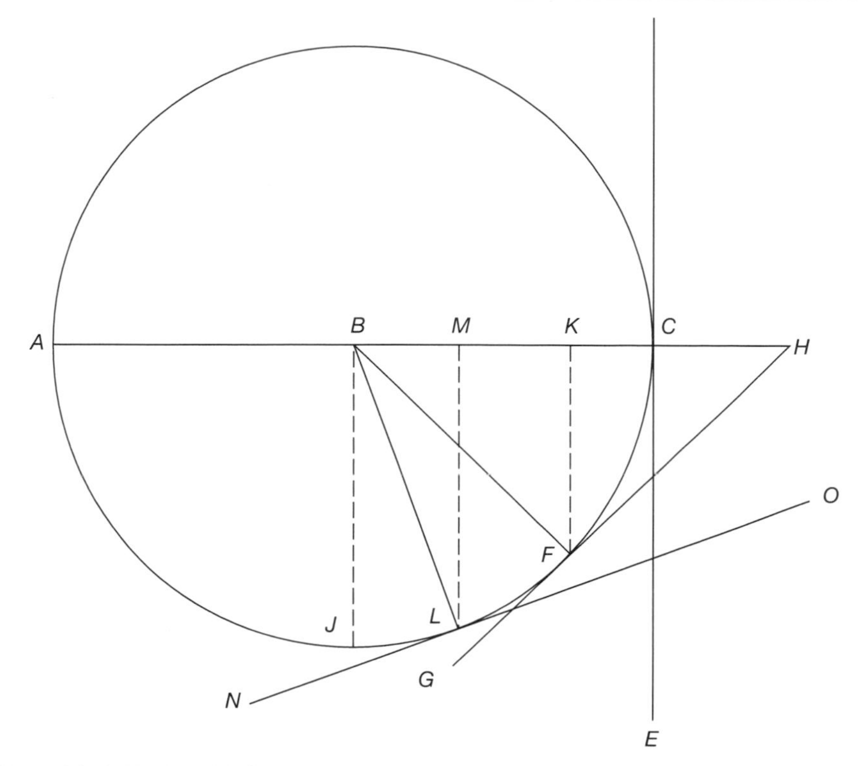

Figure 6.1 Galileo's 1600 Composite Diagram of Lever, Inclined Plane, Vertical Fall and Pendulum

Source: Galileo 1590/1960, p. 173

construction of sundials. He was a highly competent mechanical engineer and director of the Venice Arsenal, an accomplished artist, a minor noble and the brother of a prominent cardinal. He was also a patron of Galileo who secured for Galileo his first university position as a lecturer in mathematics at Pisa University (1588–1592) and his second academic position as a lecturer in mathematics at Padua University (1592–1610).

The crucial surviving document in the exchange between Galileo and his patron is a letter dated 29 November 1602, where Galileo writes of his discovery of the isochrony of the pendulum and conveys his mathematical proofs of the 'pendulum laws'.[8] As the letter is a milestone in the history of timekeeping and in the science of mechanics, and as it illustrates a number of other things about the methodology of Galileo and his scientific style, it warrants reproduction:

> You must excuse my importunity if I persist in trying to persuade you of the truth of the proposition that motions within the same quarter-circle are made in equal times. For this having always appeared to me remarkable, it now seems even more

remarkable that you have come to regard it as false. Hence I should deem it a great error and fault in myself if I should permit this to be repudiated by your theory as something false; if it does not deserve this censure, nor yet to be banished from your mind – which better than any other will be able to keep it more readily from exile by the minds of others. And since the experience by which the truth has been made clear to me is so certain, however confusedly it may have been explained in my other [letter], I shall repeat this more clearly so that you, too, by making this [experiment], may be assured of this truth.

Therefore take two slender threads of equal length, each being two or three braccia long [four to six feet]; let these be AB and EF. Hang A and E from two nails, and at the other ends tie two equal balls (though it makes no difference if they are unequal). Then moving both threads from the vertical, one of them very much as through the arc CB, and the other very little as through the arc IF, set them free at the same moment of time. One will begin to describe large arcs like BCD while the other describes small ones like FIG. Yet in this way the moveable [that is, movable body] B will not consume more time passing the whole arc BCD than that used up by the other moveable F in passing the arc FIG (see Figure 6.2). I am made quite certain of this as follows.

The moveable B passes through the large arc BCD and returns by the same DCB and then goes back toward D, and it goes 500 or 1,000 times repeating its oscillations. The other goes likewise from F to G and then returns to F, and will similarly make many oscillations; and in the time that I count, say, the first 100 large oscillations BCD, DCB and so on, another observer counts 100 of the other oscillations through FIG, very small, and he does not count even one more – a most evident sign that one of these large arcs BCD consumes as much time as each of the small ones FIG. Now, if all BCD is passed in as much time [as that]

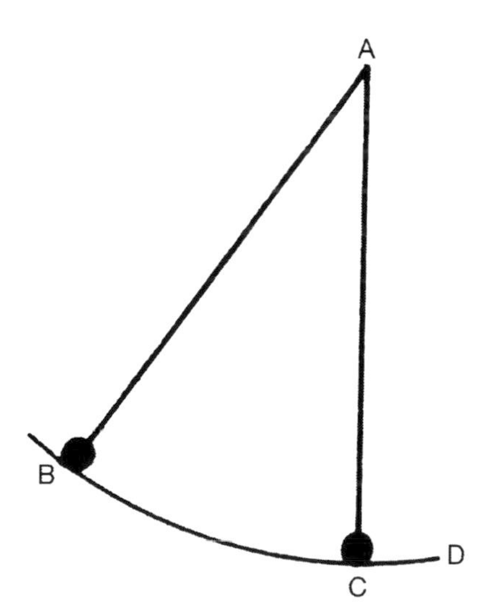

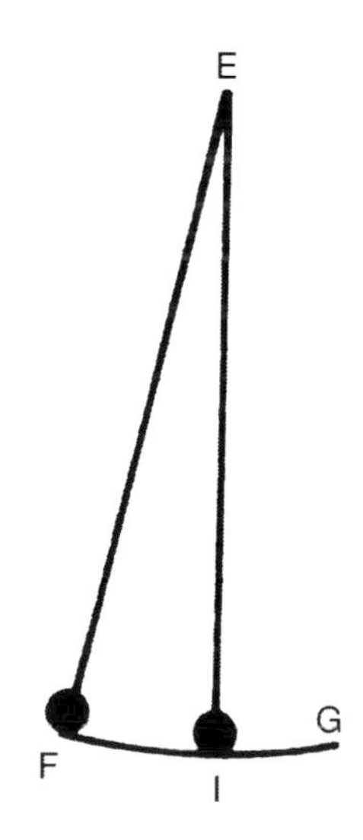

Figure 6.2 Large- and Small-Amplitude Pendulums
Source: Galileo 1602/1978, p. 69

in which FIG [is passed], though [FIG is] but one-half thereof, these being descents through unequal arcs of the same quadrant, they will be made in equal times. But even without troubling to count many, you will see that moveable F will not make its small oscillations more frequently than B makes its larger ones; they will always be together.

The experiment you tell me you made in the [rim of a vertical] sieve may be very inconclusive, perhaps by reason of the surface not being perfectly circular, and again because in a single passage one cannot well observe the precise beginning of motion. But if you will take the same concave surface and let ball B go freely from a great distance, as at point B, it will go through a large distance at the beginning of its oscillations and a small one at the end of these, yet it will not on that account make the latter more frequently than the former (see Figure 6.3).

Then as to its appearing unreasonable that given a quadrant 100 miles long, one of two equal moveables might traverse the whole and [in the same time] another but a single span, I say that it is true that this contains something of the wonderful, but our wonder will cease if we consider that there could be a plane as little tilted as that of the surface of a slowly running river, so that on this [plane] a moveable will not have moved naturally more than a span in the time that on another plane, steeply tilted (or given great impetus even on a gentle incline), it will have moved 100 miles. Perhaps the proposition has inherently no greater improbability than that triangles between the same parallels and on equal bases are always equal [in area], though one may be quite short and the other 1,000 miles long. But keeping to our subject, I believe I have demonstrated that the one conclusion is no less thinkable than the other.

Let BA be the diameter of circle BDA erect to the horizontal, and from point A out to the circumference draw any lines AF, AE, AD, and AC. I show that equal moveables fall in equal times, whether along the vertical BA or through the inclined planes along lines CA, DA, EA and FA. Thus leaving at the same moment from points B, C, D, E, and F, they arrive at the same moment at terminus A; and line FA may be as short as you wish.

And perhaps even more surprising will this, also demonstrated by me, appear: That line SA being not greater than the chord of a quadrant, and lines SI and IA being any whatever, the same moveable leaving from S will make its journey SIA

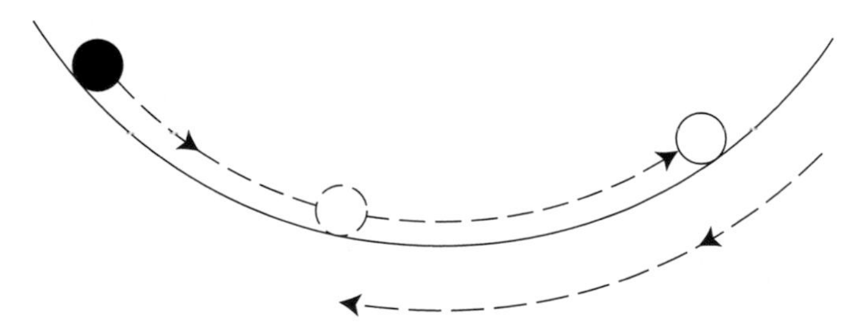

Figure 6.3 Ball in Hoop
Source: Galileo 1602/1978, p. 70

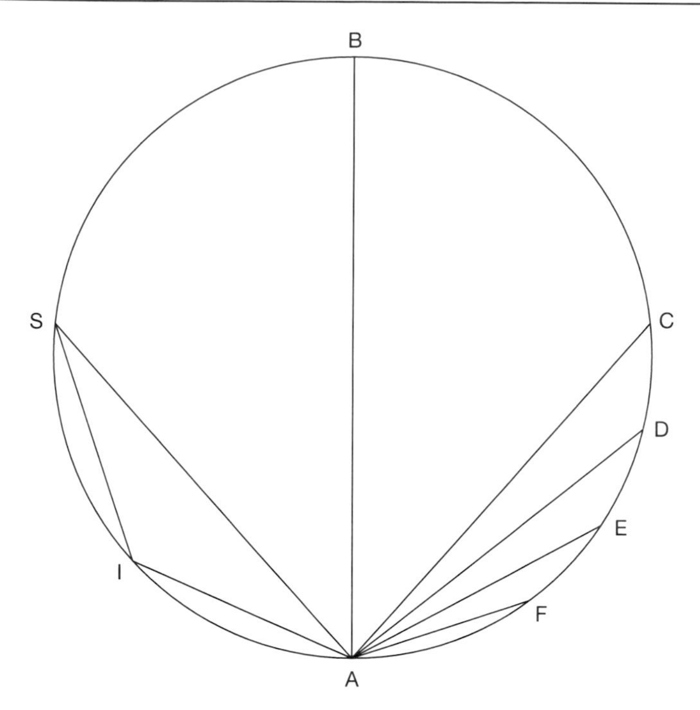

Figure 6.4 Law of Chords
Source: Galileo 1602/1978, p. 71

> more swiftly than just the trip IA, starting from I. This much has been demonstrated by me without transgressing the bounds of mechanics. But I cannot manage to demonstrate that arcs SIA and IA are passed in equal times, which is what I am seeking.
>
> Do me the favor of conveying my greetings to Sig. Francesco and tell him that when I have a little leisure I shall write to him of an experiment that has come to my mind for measuring the force of percussion. And as to his question, I think that what you say about it is well put, and that when we commence to deal with matter, then by reason of its accidental properties the propositions abstractly considered in geometry commence to be altered, from which, thus perturbed, no certain science can be assigned – though the mathematician is so absolute about them in theory. I have been too long and tedious with you; pardon me, and love me as your most devoted servitor.
>
> (Drake 1978, p. 71)

Thus, in 1602, Galileo is claiming two things about motion on chords within a circle:

1 That in a circle, the time of descent of a body free-falling, along all chords terminating at the nadir, is the same regardless of the length of the chord.

2 In the same circle, the time of descent along a chord is longer than along its composite chords, even though the former route is shorter than the latter.

This gets him tantalisingly close to a claim about motion along the arcs of the circle, the pendulum case, but not quite there. He does allude to the pendulum situation and says that two 6-foot pendulums keep in synchrony through 1,000 swings, one being displaced widely, the other barely displaced from vertical. He is not prepared to make the leap, saying, 'But I cannot manage to demonstrate that arcs SIA and IA are passed in equal times, which is what I am seeking'. Galileo 'sees' that they are passed in equal times; he has empirical proof – if one can take hypothetical behaviour in ideal situations as empirical proof, but he lacks a 'demonstration'. This is something that he believes only mathematics can provide.

Del Monte was not impressed by these proofs, claiming that Galileo was a better mathematician than a physicist. Reasonably enough, del Monte could not believe that one body would move through an arc of 10 or 20 metres in the same time as another, suspended by the same length of chord, would move through only 1 or 2 cm. Further, as a mechanic, he conducted experiments on balls rolling within iron hoops and found that Galileo's claims were indeed false: balls released from different positions in the lower quarter of the hoop reached their nadir at different times.

This was yet another case where experience and common sense were at odds with science. And, given the centrality of the pendulum in the foundation of modern science, this case goes some way to explaining why modern science was so late in appearing in human history, and why, when it did appear, it was so geographically localised; it was western European science that, within a few short decades, became universal science. Pleasingly for subsequent scientific and social history, Galileo was not moved by del Monte's or by common-sense objections.

Galileo's Mature Pendulum Claims

Galileo's physics, with his notion of circular inertia and including his pendulum analysis, was complete by 1610, when he was in his mid forties, but this physics was not widely published or disseminated; fatefully, he was introduced, in 1608, to the telescope, which side-tracked him into astronomical observations and into decades-long defence of the Copernican world system. Galileo's pendulum investigations would appear in public in his later, celebrated defences of Copernicanism: his 1633 *Dialogue* and his 1638 *Discourse*. The well-known claims about pendulums were:

- *Law of Weight Independence*: period is independent of weight.
- *Law of Amplitude Independence*: period is independent of amplitude.
- *Law of Length*: period varies directly as length; specifically the square root of length.

- *Law of Isochrony*: for any pendulum, all swings take the same time; pendulum motion is isochronous.

Although now routine and repeated in textbooks and 'replicated' in school practical classes, these claims, when made, were, with good reason, very contentious and disputed. Much about science and the nature of science can be learned from the disputes about the legitimacy of mathematisation and idealisation in science that they occasioned.

The pendulum claims can be briefly documented as follows. In the First Day of his 1638 *Dialogue*, Galileo expresses his law of weight independence as follows:

> Accordingly I took two balls, one of lead and one of cork, the former more than a hundred times heavier than the latter, and suspended them by means of two equal fine threads, each four or five cubits long. Pulling each ball aside from the perpendicular, I let them go at the same instant, and they, falling along the circumferences of circles having these equal strings for semi-diameters, passed beyond the perpendicular and returned along the same path. This free vibration repeated a hundred times showed clearly that the heavy body maintains so nearly the period of the light body that neither in a hundred swings nor even in a thousand will the former anticipate the latter by as much as a single moment, so perfectly do they keep step.
>
> (Galileo 1638/1954, p. 84)

In the Fourth Day of the 1633 *Dialogue*, Galileo states his law of amplitude independence, saying:

> [It is] truly remarkable . . . that the same pendulum makes its oscillations with the same frequency, or very little different – almost imperceptibly – whether these are made through large arcs or very small ones along a given circumference. I mean that if we remove the pendulum from the perpendicular just one, two, or three degrees, or on the other hand seventy degrees or eighty degrees, or even up to a whole quadrant, it will make its vibrations when it is set free with the same frequency in either case.
>
> (Galileo 1633/1953, p. 450)

In his final great work in mechanics, *Dialogues Concerning Two New Sciences* (1638), Galileo says that:

> It must be remarked that one pendulum passes through its arcs of 180°, 160°, etc in the same time as the other swings through its 10°, 8°, degrees. . . . If two people start to count the vibrations, the one the large, the other the small, they will discover that after counting tens and even hundreds they will not differ by a single vibration, not even by a fraction of one.
>
> (Galileo 1638/1954, p. 254)

In the First Day of his 1638 *Dialogues*, Galileo states his law of length when, in discussing the tuning of musical instruments, he says:

> As to the times of vibration of bodies suspended by threads of different lengths, they bear to each other the same proportion as the square roots of the lengths of the thread; or one might say the lengths are to each other as the squares of the times; so that if one wishes to make the vibration-time of one pendulum twice that of another, he must make its suspension four times as long. In like manner, if one pendulum has a suspension nine times as long as another, this second pendulum will execute three vibrations during each one of the first; from which it follows that the lengths of the suspending cords bear to each other the [inverse] ratio of the squares of the number of vibrations performed in the same time.
>
> (Galileo 1638/1954, p. 96)

His fourth pendulum 'law', isochronous motion, is of the greatest importance for the subsequent scientific and social utilisation of the pendulum. In the late fifteenth century, the great observer Leonardo da Vinci extensively examined, manipulated and drew pendulums, but, as one commentator remarks: 'He failed, however, to recognize the fundamental properties of the pendulum, the isochronism of its oscillation, and the rules governing its period' (Bedini 1991, p. 5).

In the Fourth Day of the 1633 *Dialogue*, Galileo approaches his law of isochrony by saying:

> Take an arc made of a very smooth and polished concave hoop bending along the curvature of the circumference *ADB* [Figure 6.5], so that a well-rounded and smooth ball can run freely in it (the rim of a sieve is well suited for this experiment). Now I say that wherever you place the ball, whether near to or far from the ultimate limit *B* . . . and let it go, it will arrive at the point *B* in equal times . . . a truly remarkable phenomenon.
>
> (Galileo 1633/1953 p. 451)

In the First Day of the 1638 *Dialogues*, Galileo writes of his law of isochrony that:

> But observe this: having pulled aside the pendulum of lead, say through an arc of fifty degrees, and set it free, it swings beyond the perpendicular almost fifty degrees, thus describing an arc of nearly one hundred degrees; on the return swing it describes a little smaller arc; and after a large number of such vibrations it finally comes to rest. Each vibration, whether of ninety, fifty, twenty, ten or four degrees occupies the same time: accordingly the speed of the moving body keeps diminishing since in equal intervals of time, it traverses arcs which grow smaller and smaller. . . . Precisely the same thing happens with the pendulum of cork.
>
> (Galileo 1638/1954, p. 84)

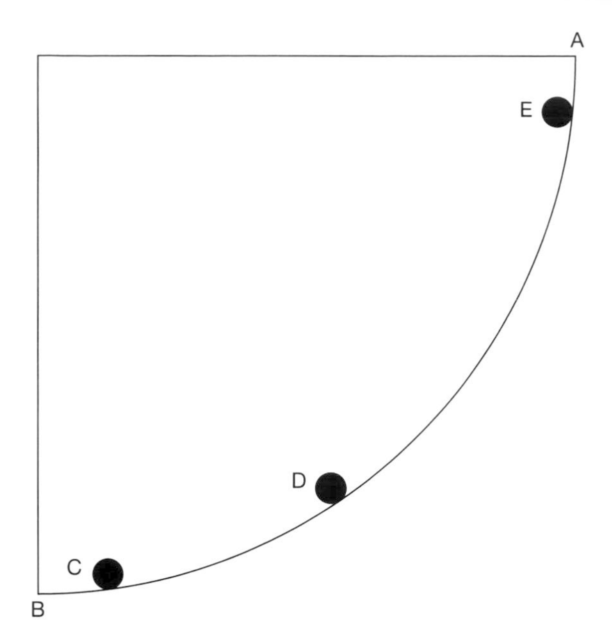

Figure 6.5 Ball in Hoop

Source: Galileo 1633/1953, p. 451

Since Galileo's time, there have been countless empirical demonstrations or tests of these mathematical proofs. This is a staple item in school physics programmes. The diagram in Figure 6.6 is from an eighteenth-century physics text (Guadagni 1764, p. 32), and the photo is of a modern reproduction exhibited in the Pavia University museum (Falomo *et al.* 2014). Both the figure and the photo derive from Galileo's geometrical proof of his chords theorem, shown in Figure 6.4. It is the crucial scientific sequence of moving from thought to reality via artisan craft and technology. In the text picture, two balls are simultaneously released from the apex A; one travels the chord AEF, one free-falls ADB, and the bells at the end of each traverse sound at the same time. If they do not so sound, then the question becomes one of either finding material 'accidents', including experimenter inadequacies, that interfere with the 'ideal' or 'world on paper' or deciding that the theorem does not apply in the world.

This Law of Chords is close to a proof for isochrony of pendulum motion. The law has shown that the time of descent down inclined planes (chords) is the same, provided the planes are inscribed in a circle and originate at the apex or terminate at the nadir. This means that amplitude does not affect the time. This is highly suggestive of a Law of Arcs, where amplitude should not affect the time of descent or time of swing. This law is proved later in Theorem XXII of the *Dialogues*, which also demonstrates the counter-intuitive

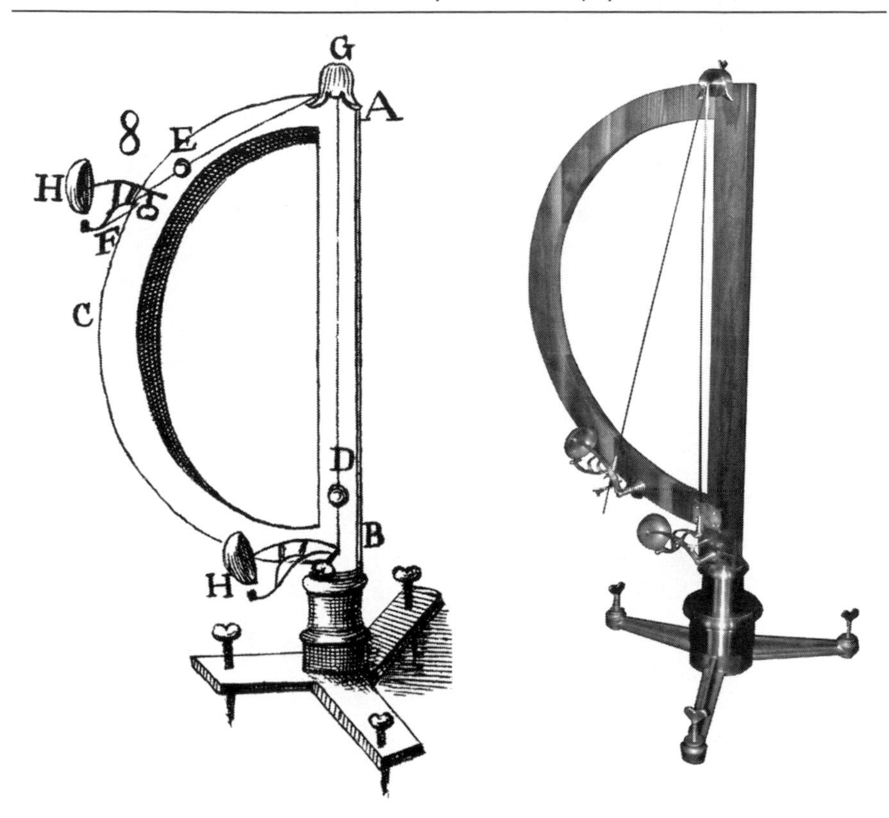

Figure 6.6 Law of Chords Demonstration; Eighteenth-Century Text and Modern Reproduction

proposition that the quickest time of descent in free fall is not along the shortest path (see Figure 6.7). He says:

> From the preceding it is possible to infer that the path of quickest descent from one point to another is not the shortest path, namely a straight line, but the arc of a circle. In the quadrant BAEC, having the side BC vertical, divide the arc AC into any number of equal parts, AD, DE, EF, FG, GC, and from C draw straight lines to the points A, D, E, F, G; draw also the straight lines AD, DE, EF, FG, GC. Evidently descent along the path ADC is quicker than along AC alone or along DC from rest at D. . . . Therefore, along the five chords, ADEFGC, descent will be more rapid than along the four, ADEFC. Consequently the nearer the inscribed polygon approaches a circle, the shorter is the time required for descent from A to C.
>
> (Galileo 1638/1954 p. 239)

Galileo realised that any truly isochronous movement could be used as a clock; one only has to calibrate the movement against the duration of a

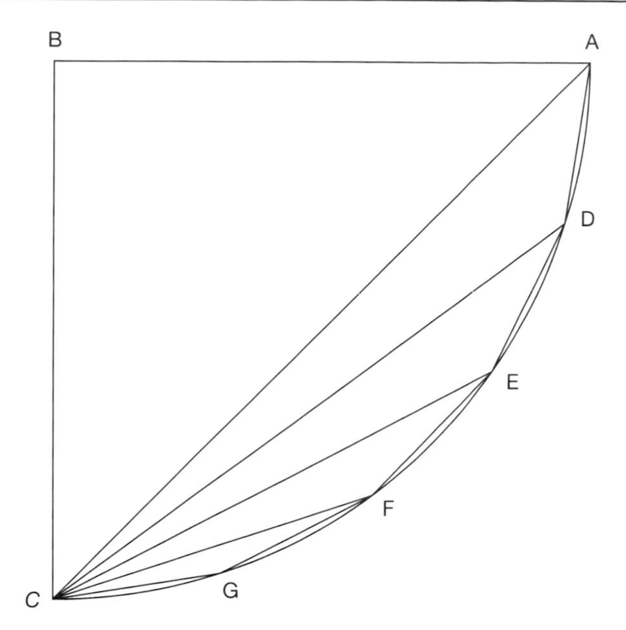

Figure 6.7 Galileo's Law of Chords Proof
Source: Galileo 1638/1954 p. 239

sidereal day and create some mechanism and technology for displaying the steady isochronous movement. He further recognised that the swinging pendulum was the ideal of such movement. Late in his life, Galileo proposed using the pendulum as a clock, and his son Vincenzio produced sketches of the proposal.[9]

Problems with Galileo's Account and the Limits of Empiricism

These marvellous proofs of Galileo did not receive universal acclaim: on the contrary, learned scholars were quick to point out considerable empirical and philosophical problems with them. The empirical problems were examples where the world did not 'correspond punctually' to the events demonstrated mathematically by Galileo. In his more candid moments, Galileo acknowledged that events do not always correspond to his theory; that the material world and his so-called 'world on paper', the theoretical world, did not correspond. Immediately after mathematically establishing his famous law of parabolic motion of projectiles, he remarks that:

> I grant that these conclusions proved in the abstract will be different when applied in the concrete and will be fallacious to this extent, that neither will the horizontal

> motion be uniform nor the natural acceleration be in the ratio assumed, nor the path of the projectile a parabola.
>
> (Galileo 1638/1954, p. 251)

One can imagine the reaction of del Monte and other hardworking Aristotelian natural philosophers and mechanicians when presented with such a qualification. It confounded the basic Aristotelian and empiricist objective of science, namely to tell us accurately about the world in which we live. The law of parabolic motion was supposedly true, but not of the world we experience: this was indeed as difficult to understand for del Monte as it is for present-day students.

As early as 1636, the notable mathematician, theologian and 'net worker' of all contemporary natural philosophers, Marin Mersenne (1588–1648), reproduced Galileo's experiments and not only agreed with del Monte, but doubted whether Galileo had ever conducted the experiments (Koyré 1968, pp. 113–117). Modern researchers have duplicated the experimental conditions described by Galileo and have found that they do not give the results that Galileo claimed (Ariotti 1968, Naylor 1974, 1980, 1989).

Del Monte and others repeatedly pointed out that pendulums do not behave as Galileo maintained; Galileo never tired of saying that ideal pendulums would obey the mathematically derived rules. Del Monte retorted that physics was to be about this world, not an imaginary, mathematical world. Opposition to the mathematising of physics was a deeply held, Aristotelian, and more generally empiricist, conviction (Lennox 1986). The British empiricist Hutchinson would later say of the geometrical constructions of Newton's *Principia* that they were just 'cobwebs of circles and lines to catch flies in' (Cantor 1991, p. 219).[10]

It is easy to appreciate the empirical reasons for opposition to Galileo's law. The overriding argument was that, if the law were true, pendulums would be perpetual motion machines, which clearly they are not. An isochronic pendulum is one in which the period of the first swing is equal to that of all subsequent swings: this implies perpetual motion. We know that any pendulum, when let swing, will very soon come to a halt: the period of the last swing will be by no means the same as the first. Furthermore, it was plain to see that cork and lead pendulums have a slightly different frequency, and that large-amplitude swings do take somewhat longer than small-amplitude swings for the same pendulum length. All of this was pointed out to Galileo, and he was reminded of Aristotle's basic methodological claim that the evidence of the senses is to be preferred over other evidence in developing an understanding of the world.

The fundamental laws of classical mechanics are not verified in experience; further, their direct verification is fundamentally impossible. Herbert Butterfield (1900–1979) conveys something of the problem that Galileo and Newton had in forging their new science:[11]

> They were discussing not real bodies as we actually observe them in the real world, but geometrical bodies moving in a world without resistance and without gravity

> – moving in that boundless emptiness of Euclidean space which Aristotle had regarded as unthinkable. In the long run, therefore, we have to recognise that here was a problem of a fundamental nature, and it could not be solved by close observation within the framework of the older system of ideas – it required a transposition in the mind.
>
> (Butterfield 1949/1957, p. 5)

An objectivist, non-empiricist account of science stresses that the transposition in the mind is really the creation of a new theoretical object or system. Even for Galileo, the pendulum seemed to stop at the top of its swing; it was only in his theory, not his perceptual mind, that it continued in smooth motion.[12]

The Pendulum and Timekeeping

It is useful to outline some of the later developments in the science of pendulum motion. They show the interaction of mathematics and experiment in scientific development, and the importance to science of the development of theoretical systems and of conceptual frameworks within which to interpret and interrogate nature. Both of these points are important for the teaching of pendulum motion.

The pendulum played more than a scientific role in the formation of the modern world. The pendulum was central to the horological revolution that was intimately part of the scientific revolution. Huygens, in 1673, following Galileo's epochal analysis of pendulum motion, utilised the pendulum in clockwork and so provided the world's first accurate measure of time (Yoder 1988). The accuracy of mechanical clocks went, in the space of a couple of decades, from plus or minus half-an-hour per day to a few seconds per day. This abrupt increase in accuracy of timing enabled hitherto unimagined degrees of precision measurement in mechanics, navigation and astronomy. It ushered in the world of precision characteristic of the scientific revolution (Wise 1995). Time could then confidently be expressed as an independent variable in the investigation of nature.

Christiaan Huygens (1629–1695) refined Galileo's pendulum laws and was the first to use these refined laws to create a pendulum clock. Huygens modified Galileo's analysis by showing, mathematically, that it was movement on the cycloid, not the circle, that was isochronous. He provides the following account of this discovery:

> We have discovered a line whose curvature is marvellously and quite rationally suited to give the required equality to the pendulum. . . . This line is the path traced out in air by a nail which is fixed to the circumference of a rotating wheel which revolves continuously. The geometers of the present age have called this line a cycloid and have carefully investigated its many other properties. Of interest to

> us is what we have called the power of this line to measure time, which we found not by expecting this but only by following in the footsteps of geometry.
>
> (Huygens 1673/1986, p. 11)

The cycloid is the curve described by a point P rigidly attached to a circle C that rolls, without sliding, on a fixed line AB. Thc full arc ADB has a length equal to 8r (r = the radius of the generating circle). A heavy point that travels along an arc of cycloid placed in a vertical position, with the concavity pointing upwards, will always take the same amount of time to reach the lowest point, independent of the point from which it is released.

Having shown mathematically that the cycloid was isochronous, Huygens then devised a simple way of making a suspended pendulum swing in a cycloidal path – he made two metal cycloidal cheeks and caused the pendulum to swing between them. Huygens' first pendulum clock (Figure 6.9) was accurate to 1 minute per day; working with the best clockmakers, he soon made clocks accurate to 1 second per day.

After showing that the period of a simple pendulum varied as the square root of its length, Huygens (Huygens 1673/1986, pp. 169–170) then derived the familiar equation for small-amplitude pendulums, that is, ones moving on the arc of a circle:

$$T = 2\pi\sqrt{(l/g)}$$

This development by Huygens of the theory of the pendulum is an example that fits objectivist accounts of scientific development. Despite Galileo's personal brilliance, he nevertheless did not appreciate or work out correctly the implications of the theoretical object that he himself created. The theory had unseen or unintended consequences that needed others to work out or discover. Huygens discovered that the cycloid (Figure 6.8), not the circle, was the vital tautochronous curve by 'following in the footsteps of geometry', a guide that Aristotelian philosophy distrusted in physical affairs. This discovery of the tautochronous curve had very little to do with sensory input. Its justification had even less to do with sense data or other putative empirical foundations for belief so extolled by positivists.

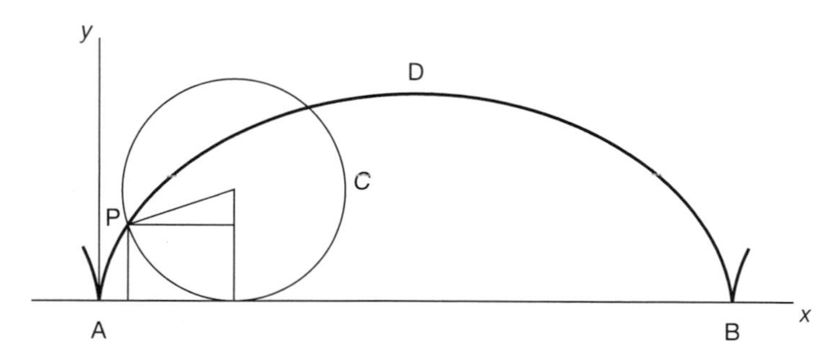

Figure 6.8 Cycloid Generated by a Moving Circle

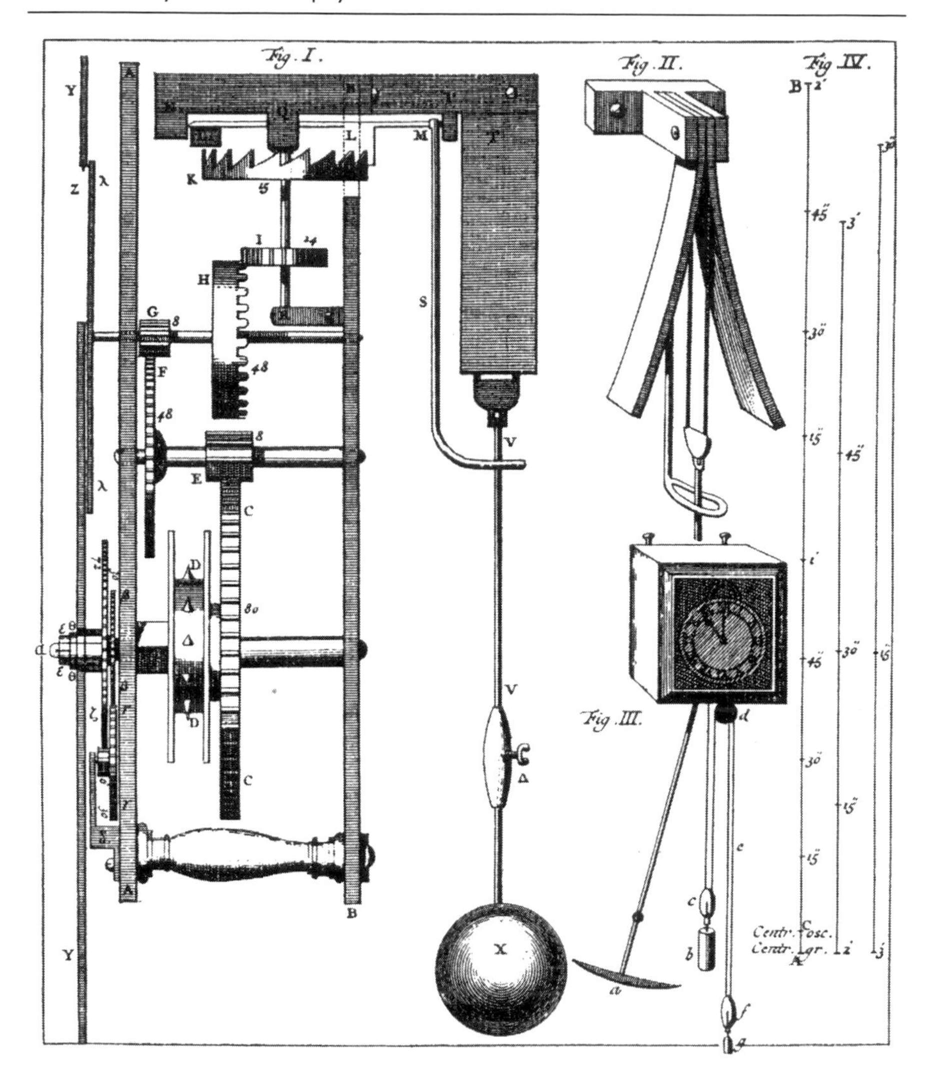

Figure 6.9 Huygens' Pendulum Clock Mechanism

Source: Huygens 1673/1986, p. 14

Huygens' Proposal of an International Standard of Length

Huygens saw that, in his pendulum equation, T = 2π√(l/g), the only variable was l, as π was a constant and, provided that the Earth was a sphere and one stayed near to sea level, g was also constant, and mass did not figure in the equation at all. So, all pendulums of a given length will have the same period, whether they be in France, England, Russia, Latin America, China or Australia. And given l and T, then g could be determined for any location. Huygens was clever enough to see that the pendulum would solve, not only the timekeeping and longitude problems, but an additional vexing problem, namely establishing an international length standard, and, in 1673, he proposed the length of a seconds pendulum (a pendulum that beats in seconds; that is, whose period is 2 seconds) to be the international unit.[13] The length of the seconds pendulum was experimentally determined by adjusting a pendulum so that it oscillated 24 × 60 × 30 times in a sideral day (each oscillation takes 2 seconds); that is, between successive transits of a fixed star across the centre of a graduated telescope lens (the sideral day being slightly longer than a solar day).

This seems like a daunting task, but it was not so overwhelming. Huygens and others knew that the length l of a pendulum varied as the square of period or T^2. So the length of a seconds pendulum is to the length of any arbitrary pendulum as $1/T^2$. However, $(1/T^2) \propto n^2$, where n is the number of oscillations of a pendulum in 1 hour. As Meli observes: 'therefore, by counting the number of oscillations and the length of his pendulum, Huygens could determine the length of the seconds pendulum' (Meli 2006, p. 205).

Having an international unit of length, or even a national unit, was a major contribution to simplifying the chaotic state of measurement existing in science and everyday life. Within France, as in other countries, the unit of length varied from purpose to purpose (timber to cloth), city to city, and even within cities. This was a significant problem for commerce, trade, construction, military hardware and technology, to say nothing of science. Many attempts had been made to simplify and unify the chaotic French system. One estimate is that, in France alone, there were 250,000 different, local measures of length, weight and volume (Alder 1995, p. 43), and each European state had comparable confusion of abundance, as did, of course, all other nations and cultures.

In the standard formulas, using standard approximations for g and π, it is easy to show that the length of a seconds pendulum will be 1 metre.

$T = 2\pi\sqrt{(l/g)}$
So $T^2 = 4\pi^2(l/g)$
So $l = T^2g/4\pi^2$
Substitute $T = 2s$ (each beat is a second), $g = 9.8\ ms^{-2}$, $\pi^2 = 9.8$
Then $l = 0.993577$ m, or very approximately 1 m

This result can reliably be demonstrated with even the crudest 1 m pendulum – a heavy nut on a piece of string suffices: ten complete swings will take 20 seconds; twenty complete swings will take 40 seconds. A great virtue of the seconds pendulum as the international length standard was that it was a fully 'natural' standard; it was something fixed by nature, unlike standards based on the length of a king's arm or foot. And, of course, an international length standard would provide a related volume standard and, hence, a mass standard when the unit volume was filled with rainwater. A kilogram is the weight of 1 litre (1,000 cc) of water. All of this can be engagingly reproduced with classes; they can be challenged to think through ways in which a mass standard can be derived from a length standard.

It is not accidental that, 200 years after Huygens, the General Conference on Weights and Measures meeting in Paris defined the standard universal metre as 'the length of path travelled by light in vacuum during a time interval of 1/299,792,458 of a second'.[14] This seemingly bizarre and arbitrary figure is within 1 millimetre of Huygens' original and entirely natural length standard, and it was so chosen precisely to replicate the length of the seconds pendulum. Unfortunately, it is the former, not the latter, that students meet in the opening pages of their science texts, which confirms their worst fears about the 'strangeness' of science. In the definition of standards, science gets off to a bad pedagogical start.

The Pendulum and Determination of the Shape of the Earth

Huygens' proposal depended on g being constant around the world (at least at sea level); it depended on the Earth being spherical. This seemed a most reasonable assumption. Indeed, to say that the Earth was not regular and spherical was tantamount to casting aspersions on the Creator: surely God the Almighty would not make a misshapen earth. But, in 1673, contrary to all expectation, this assumption was brought into question by the behaviour of the pendulum.

When Jean-Dominique Cassini (1625–1712) became director of the French *Académie Royale des Sciences* in 1669, he sent expeditions into different parts of the world to observe the longitudes of localities for the perfection of geography and navigation. The second such voyage was Jean Richer's to Cayenne in 1672–1673 (Olmsted 1942). Cayenne was in French Guiana, at a latitude of approximately 5° N. It was chosen as a site for astronomical observations because equatorial observations were minimally affected by refraction of light passing through the Earth's atmosphere – the observer, the sun and the planets were all in the same plane.

The primary purpose of Richer's voyage was to ascertain the value of solar parallax and to correct the tables of refraction used by navigators and astronomers. A secondary consideration was checking the reliability of Huygens' marine pendulum clocks, which were being carried for the purpose of establishing Cayenne's exact longitude. Richer surprisingly found that a pendulum, set to swing in seconds at Paris, had to be shortened in order to swing in seconds at Cayenne. Not shortened by much – 2.8 mm, about the thickness of a matchstick – but nevertheless shortened. Richer found that a Paris seconds clock lost two and a half minutes daily at Cayenne; the time from noon to noon was 23 hours, 57 min, 32 s. The only apparent explanation for the slowing of the pendulum at the equator was that g is less at the equator than at Paris and the poles; in other words, that the Earth is a 'flattened' sphere, an oblate, with the equator being further from the centre and so having less gravitational attraction than the poles. The methodological lessons of this episode will be fleshed out below.[15]

The Pendulum in Newton's Mechanics

The pendulum played a comparable role in Newton's work to what it had for Galileo and Huygens. Newton used the pendulum to determine the gravitational constant g, to improve timekeeping, to disprove the existence of the mechanical philosophers' ether presumption, to show the proportionality of mass to weight, to determine the coefficient of elasticity of bodies, to investigate the laws of impact and to determine the speed of sound. Richard Westfall, a distinguished Newtonian scholar, wrote that: 'the pendulum became the most important instrument of seventeenth-century science . . . Without it the seventeenth century could not have begot the world of precision' (Westfall 1990, p. 67). Concerning the pendulum's role in Newton's science, Westfall has said that, 'It is not too much to assert that without the pendulum there would have been no *Principia*' (Westfall, 1990, p. 82). No small claim to fame!

Determination of the Gravitational Constant

Marin Mersenne (1588–1648) used Galileo's circular pendulum, and its theoretical underpinnings, to ascertain the value of the gravitational constant g. Neither Galileo, nor other seventeenth-century scientists, used the idea of g in its modern sense of acceleration. Their gravitational constant was the distance that all bodies fall in the first second after their release – this is, numerically, one-half of the modern constant of acceleration.

Mersenne's earlier investigations pointed to 3 Parisian feet being the length of the seconds pendulum, and so, in a famed experiment (1647), he held a 3-horological-foot (≈ 39 inch) pendulum (a seconds pendulum) out from a wall and released it along with a free-falling mass (see Figure 6.10). He adjusted a platform under the mass until he heard just one sound (the pendulum striking the wall and the mass striking the platform simultaneously). He reasoned that this should give him the length of free fall in half-a-second (a complete one-way swing of the pendulum taking 1 second), and so, by the times-squared rule (four times the length so determined), he could calculate the length of free fall in 1 second, the gravitational constant of the period.

Huygens and Mersenne thought that the ear could not separate sounds to better than 6 inches of free fall, and so thought that inaccuracy was going to be 'built into' mechanics.[16] All realised that a seconds pendulum was the only way of getting a duration of 1 second; this was to be the key to obtaining a measurement of the gravitational constant, the length fallen in the first second of free fall.

Demonstration of Newton's Laws

Three properties made the pendulum an ideal vehicle for the demonstration of Newton's laws and the investigation of collisions: pendulums of the same length reach their nadir at the same time, irrespective of where they are

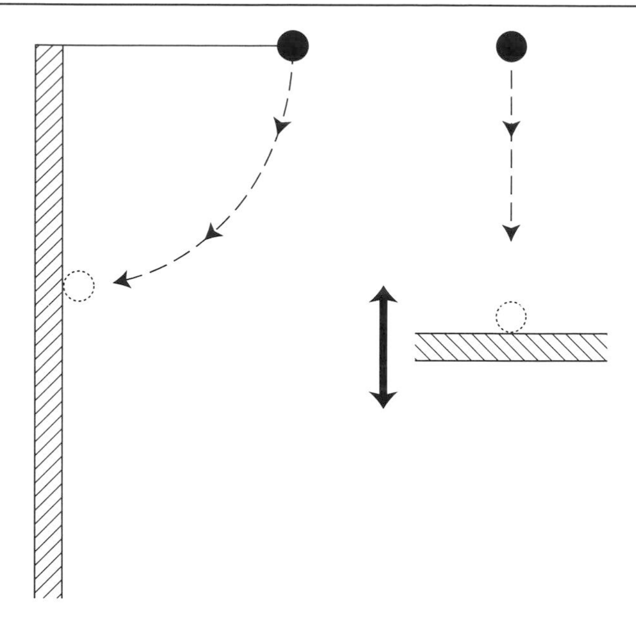

Figure 6.10 Mersenne's Gravitational-Constant Determination

released; they reach their nadir simultaneously, regardless of their mass; and their velocity at the nadir is proportional to the length of the chord joining the nadir to the point of release. Additionally, the paths of the colliding bodies could be constrained. Newton set up pendulum collision experiments to demonstrate his laws of motion and to explicate his nascent conservation law. He concluded one demonstration by writing:

> By the meeting and collision of bodies, the quantity of motion, obtained from the sum of the motions directed towards the same way, or from the difference of those that were directed towards contrary ways, was never changed.
>
> (Newton 1729/1934, pp. 22–24)

Newton does not label this the conservation of momentum. He speaks of conservation of 'motion', but it is modern momentum, *mv*, a vector quantity, that he is describing. In current terminology, his conclusion is given by the formula:

$$m_1v_1 + m_2v_1 = m_1v_2 + m_2v_2$$

The pendulum has here made a significant contribution to the foundation of classical mechanics. The law of conservation of momentum was true for both elastic (where there is no energy absorbed in the collision itself) and inelastic collisions (where energy is absorbed). For instance, if the above pendulums were made of putty, then, when they collided, they would deform and simply come to a halt; there would be no motion after the collision. In this situation, one could hardly talk of conservation of 'motion' – although Descartes, for instance, resolutely maintained that the quantity mv was the basic measure of 'motion', and mv was the basic entity that was conserved in the world. A theory-protecting strategy was to claim that, although gross motion (the putty balls) ceased, the motion of the invisible corpuscules of the mechanical worldview did not cease; they were energised.

Newton's third law, 'action is equal to reaction', was demonstrated by Newton using two long (3–4 m) pendulums and having them collide. He used a result of Galileo (that the speed of a pendulum at its lowest point is proportional to the chord of its arc) and applied it to the collision by comparing the quantities mass multiplied by chord length, before and after collision (Gauld 1998). For centuries, Newton's 'cradle' apparatus, which wonderfully manifests this conservation law, has intrigued students and citizens (Gauld 2006).

Unifying Terrestrial and Celestial Mechanics

The major question for Newton and natural philosophers was whether Newton's postulated attractive force between bodies was truly universal: that is, did it apply, not only to bodies on Earth, but also between bodies in the solar system? Aristotle, as with all ancient philosophers, made a clear distinction between the heavenly and terrestrial (sub-lunar) realms: the former being eternal, unchanging and perfect, the realm of the Gods; the latter being changeable, imperfect and corruptible, the realm of man. It was thus 'natural' that the science of both realms would be different, and, to speak anachronistically, laws applying to the terrestrial realm would not apply to the celestial realm. This cosmic divide lasted for 2,000 years.

It was the analysis of pendulum motion that rendered untenable the celestial/terrestrial distinction and enabled the move from 'the closed world to the infinite universe' (Koyré 1957). The same laws governing the pendulum were extended to the Moon, and then to the planets. The long-standing celestial/terrestrial distinction in physics was dissolved. The same laws were seen to apply in the heavens as on Earth: there was just one universe, a unitary solar system.

At 22 years of age, while ensconced in Lincolnshire to avoid London's Great Plague, Newton began to speculate that the Moon's orbit and an apple's fall might have a common cause (Herivel 1965, pp. 65–69). He was able to calculate that, in 1 second, while travelling about 1 kilometre in its orbit, the Moon deviates from a straight-line path by about one-twentieth of an inch. In the same period of time, an object projected horizontally on the Earth would fall about 16 feet. The ratio of the Moon's 'fall' to the apple's fall is then about 1:3,700. This was very close to the ratio of the square of the apple's distance from the Earth's centre (the Earth's radius), to the square of the Moon's distance from the Earth's centre, 1:3,600. Was this a cosmic coincidence? Or did the Earth's gravitational attraction apply equally to the apple and the Moon?

Following the dictates of his own method, Newton then experimentally investigated whether the derived consequences are seen in reality. He deferred to Huygens' experimental measurement, saying:

> And with this very force we actually find that bodies here upon earth do really descend; for a pendulum oscillating seconds in the latitude of Paris will be 3 Paris feet, and 8 lines ½ in length, as Mr Huygens has observed. And the space which a heavy body describes by falling in one second of time is to half the length of this pendulum as the square of the ratio of the circumference of a circle to its diameter (as Mr Huygens has also shown), and is therefore 15 Paris feet, 1 inch, 1 line $^7/_9$. And therefore the force by which the moon is retained in its orbit becomes, at the very surface of the earth, equal to the force of gravity which we observe in heavy bodies there.
>
> (Newton 1729/1934, p. 408)

Newton then draws his conclusion: 'And therefore the force by which the moon is retained in its orbit is that very same force which we commonly call gravity.' The pendulum had brought heaven down to Earth.[17]

Timekeeping as the Solution of the Longitude Problem

From the fifteenth century, when traders and explorers began journeying away from European shores, problems of navigation and position-finding, especially the determination of longitude, became more and more acute. Chinese and Polynesian seafarers had the same problem. Gemma Frisius (1508–1555) identified accurate timekeeping as the way to solve the longitude problem. In 1530, he advised:

> In our times we have seen the appearance of various small clocks, capably constructed, which, for their modest dimensions, provide no problem to those who travel. These clocks operate for 24 hours, in fact when convenient, they continue to operate with a perpetual movement. And it is with their help that the longitude can be found. . . . When one is on course for 15 or 20 miles, and wishing to know how distant one is from the point of departure, it would be preferable to wait until the clock is at an exact division of time, and at the same time, with the assistance of the astrolabe, as well as of our globe, to seek the hour of the place in which we find ourselves. . . . In this manner it is possible to find the longitude, even in a distance of a thousand miles, even without knowing where we have passed, and without knowing the distance travelled.
>
> (Pogo 1935, p. 470)

Frisius understood that the Earth makes one revolution of 360° in 24 hours, and thus, in 1 hour, it rotates through 15°, or 1° each 4 minutes. Thus, although the line of zero longitude is arbitrary, there is an objective relationship between time and longitude. However, the theoretical solution was considerably ahead of the technology available and the ability of sixteenth-century clockmakers. The pendulum was to play a pivotal role in solving this pressing problem.[18]

Nearly all the great scientists of the seventeenth century – including Galileo, Huygens, Newton and Hooke – worked intimately with clockmakers and used their analysis of pendulum motion, specifically its isochronism, in the creation of more accurate pendulum clocks and then portable watches, especially the marine chronometer, with a view to accurate longitude determination. Dava Sobel (1995) provides a very readable, bestseller account of the British efforts at solving the problem of longitude. Sadly, few of the hundreds of thousands of readers across twenty languages would relate the longitude story to their pendulum studies in school science, because the former was never mentioned alongside the latter, and Sobel leaves out completely the crucial methodological matters that lay at the heart of Galileo's discoveries. For Galileo and timekeeping, she disappointingly repeats the chandelier mythology (Sobel 1995, p. 37).

Huygens was aware of the clock's role in solving the problem of longitude, saying, 'they are especially well suited for celestial observations and for measuring the longitudes of various locations by navigators' (Huygens 1673/1986, p. 8). Captain Robert Holmes, in 1664, gave the clocks their most extensive sea tests, tests that vindicated Huygens' faith in their ability to solve the longitude problem. Ironically, Holmes was on a transatlantic voyage of pillage against Dutch possessions in Africa, New Amsterdam (New York) and South America. Huygens, in his *Horologium Oscillatorium*, repeats Holmes's account of the scientific part of this voyage.

The craftsman who first solved the technical problems of an accurate, reliable marine chronometer was John Harrison (1693–1776) – known as 'Longitude Harrison' – who was born in 1693, the son of a Yorkshire carpenter, and who died as a celebrated clockmaker in London in 1776. Harrison died 3 years after receiving the British Longitude Board's reluctant final payment of the £20,000 reward established by the Longitude Act of 1714. The prize required the then-unheard-of accuracy of 2 minutes during the 8–10 week voyage to the West Indies: Harrison's clock would keep within 30 seconds of correct time on a voyage out to the Indies and back! After this, the far reaches of the globe, and all islands and reefs in between, could be accurately mapped, and European commerce, colonisation and conquest could proceed apace. Students can be profitably engaged with all aspects of this marvellous story, thereby learning astronomy, physics, navigation and history (Bensky 2010).

Foucault's Pendulum and the Earth's Rotation

The pendulum provided the first-ever tangible and dynamic 'proof' of the rotation of the Earth. On Newton's theory, a pendulum set swinging in a particular plane should continue to swing indefinitely in that same plane, the only forces on the bob being the tension in the cord and its weight directed vertically downwards. Léon Foucault (1819–1868) – described as 'a mediocre pupil at school, [but] a natural physicist and an incomparable experimenter' (Dugas 1988, p. 380) – 'saw' that, if a pendulum were placed exactly at the North Pole and suspended in such a way that the point of suspension were free to rotate (that is, it did not constrain the pendulum's movement by applying torque), then:

> If the oscillations can continue for twenty-four hours, in this time the plane will execute a whole revolution about the vertical through the point of suspension . . . at the pole, the experiment must succeed in all its purity.
>
> (Dugas 1988, p. 380)

As the pendulum is moved from the pole to the equator, Foucault easily showed that, if T^1 is the time in which the plane of the pendulum rotates 360°, and T is the period of rotation of the earth, and β is the latitude where the experiment is being conducted, then:

$$T^1 = T/\sin\beta$$

From the formula, it can be seen that, at the poles, $T^1 = T$ (as $\sin\beta = \sin 90° = 1$); whereas, at the equator, $T^1 = \infty$ (or infinity, as $\sin 0° = 0$), and, thus, there is no rotation of the plane of oscillation at the equator.

On 2 February 1851, Foucault invited the French scientific community 'to come see the Earth turn, tomorrow, from three to five, at Meridian Hall of the Paris Observatory'. Foucault's long and massive pendulum provided an experimental 'proof' of the Copernican theory, something that had eluded Galileo, Newton and all the other mathematical and scientific luminaries who sought it (Aczel 2003, 2004, Tobin 2003).

Until Foucault's demonstration, all astronomical observations could be fitted, with suitable adjustments such as those made by Tycho Brahe, to the stationary Earth theory of the Christian tradition. The 'legitimacy' of such ad hoc adjustments in order to preserve the geocentric model of the solar system was exploited by the Catholic Church, which kept the works of Copernicus and Galileo on the *Index of Prohibited Books* up until 1835 (Fantoli 1994, p. 473). To most nineteenth-century physicists, the manifest rotation of Foucault's pendulum, shown in the successive knocking down of markers placed in a circle, was a dramatic proof of the Earth's rotation.

Some Features of Science

As mentioned in Chapter 1, and as will again be outlined in Chapter 11, the method of this book, and its recommendation for science teachers, is to make explicit different features of science, particularly philosophical ones, as curriculum content is taught. When treated historically, the pendulum provides a rich source for such lessons about science. Galileo is an outstanding example of the scientist–philosopher or philosophical scientist (to use current terminology). He made substantial philosophical contributions in a variety of areas: in ontology, with his distinction of primary and secondary qualities; in epistemology, both with his criticism of authority as an arbiter of knowledge claims and with his subordination of sensory evidence to mathematically informed reason; in methodology, with his development of the mathematical–experimental method; and in metaphysics, with his critique of the Aristotelian causal categories and rejection of teleology as an explanatory principle.[19] It is unfortunate that, despite his important contributions to philosophy, and despite his acknowledged influence on just about all philosophers of the seventeenth century and on such subsequent philosophers as Kant and Husserl, Galileo makes, at best, a cameo appearance in most histories of philosophy. And, of course, his philosophical achievements are ignored in science texts. However, historically and philosophically informed pendulum teaching can redress this lamentable oversight.

Well-informed teachers should also be aware of the pitfalls associated with drawing philosophical lessons from the life of Galileo; the pitfalls are akin to those involved in drawing theological and/or philosophical lessons from the life of Jesus (or any other of the great religious figures). Albert Schweitzer, in his monumental *The Quest of the Historical Jesus* (Schweitzer 1910/1954), keenly observed that:

> Thus each successive epoch of theology found its own thoughts in Jesus; that was, indeed, the only way in which it could make Him live. But it was not only each epoch that founds its reflection in Jesus; each individual created Him in accordance with his own character.
>
> (Schweitzer 1910/1954, p. 4)

The lesson from Schweitzer is not to avoid drawing lessons from historical figures and their work, but just be careful and considered when doing so: the historical figure cannot be treated as a Rorschach inkblot.[20] With this caveat, the following are some of the features of science that can arise in such teaching of the pendulum.

Scientific Methodology

The history of pendulum motion study shows the limitations of both inductivism and falsificationism as accounts of scientific method. Concerning inductivism, clearly Galileo did not induce his 'marvellous properties', or laws, of pendulum motion from looking at various pendulums – balls in hoops, chandeliers, swings, mechanical regulators, cork and iron bobs on strings and so on – and then generalise towards universal statements of what he saw in the particular instances. One enthusiast for inductivism has commented that:

> For him [Galileo], the facts based on them [observations] were treated as facts, and not related to some preconceived idea. . . . The facts of observation might or might not fit into an acknowledged scheme of the universe, but the important thing, in Galileo's opinion, was to accept the facts and build the theory to fit them.
>
> (H.D. Anthony in Chalmers 1976/2013, p. 2)

This was, on the contrary, the methodology of the Aristotelians, for whom the facts of experience were the starting points for science and who unsuccessfully urged Galileo to be true to the facts; it was also the methodology of Descartes, who concluded that, because the facts were so messy and erratic, no science of pendulum motion was possible. What was seen to happen in the world was important, but it did not have the importance that inductivism attributed to it. This is well recognised by I.E. Drabkin, who said of Aristotelian mechanics that it was,

> impeded not by insufficient observation and excessive speculation but by too close an adherence to the data of observation . . . an adherence to the phenomena of nature so close as to prevent the abstraction therefrom of the ideal case.
>
> (Drabkin 1938, pp. 69, 82)

Falsificationism, the view that the essence of science is to reject theories that are contradicted by the facts, which is the methodological position associated with Karl Popper, fares not much better than inductivism when dealing with the pendulum example. Inductivism and falsificationism are two sides of the same empiricist coin, and so it is to be expected that, where the first fails, the second will also fail.

The type of situation faced by Huygens – the revision of a theory in the light of contrary evidence – recurs constantly in the development of science. Richer's claim that the pendulum clock slows in equatorial regions nicely illustrates some key methodological matters about science and about theory testing. The entrenched belief since Erastosthenes in the second century BC was that the Earth was spherical (theory T), and, on the assumption that gravity alone affects the period of a constant-length pendulum, the observational implication was that the period at Paris and the period at Cayenne of Huygens' seconds pendulum would be the same (O). Thus, T implies O:

$$T \rightarrow O$$

However, Richer seemingly found that the period at Cayenne was longer (~ O). Thus, on simple, falsificationist views of theory testing such as were enunciated first by Huygens himself, and famously developed by the philosopher Karl Popper early in the twentieth century (Popper 1934/1959), we have:

$T \rightarrow O$
$\sim O$
$\therefore \sim T$

But theory testing is never so simple – a matter that was recognised by Popper and articulated by Thomas Kuhn (1970) and Imre Lakatos (1970). In the seventeenth century, many upholders of T just denied the second premise, ~O. The astronomer Jean Picard, for instance, did not accept Richer's findings. Rather than accept the message of varying gravitation, he doubted the messenger. Similarly, Huygens did not think highly of Richer as an experimentalist, especially as the sea captain had dropped and smashed one of Huygens' clocks.

Others saw that theories did not confront evidence on their own; there was always an 'other things being equal' assumption made in theory test; there were *ceteris paribus* clauses (C) that accompanied the theory into the experiment. These clauses characteristically included statements about the reliability of the instruments, the competence of the observer, the assumed empirical state of affairs, theoretical and mathematical devices used in deriving O, and so on. Thus:

$T + C \rightarrow O$
$\sim O$
$\therefore \sim T \text{ or } \sim C$

More and more evidence came in, and from other experimenters, including Sir Edmund Halley, confirming Richer's observations. Thus ~O became established as a scientific fact, to use Fleck's terminology (Fleck 1935/1979), and upholders of T, the spherical Earth hypothesis, had to adjust to it.

People who maintained belief in T reasonably said that the assumption that other things were equal was mistaken. These, in principle, were legitimate concerns. This was not easy; giving up established theories in science is never easy, especially as the alternative was to accept that the Earth was oblate in shape, an ungainly shape for the all-powerful, all-knowing Creator to have fashioned. There were a number of obvious items in C that could be pointed to as the cause of the pendulum slowing:

- C^1: The experimenter was incompetent.
- C^2: Humidity in the tropics caused the pendulum to slow because the air was denser.
- C^3: Heat in the tropics caused the pendulum to expand, hence, it beat slower.
- C^4: The tropical environment caused increased friction in the moving parts of the clock.

Each of these could account for the slowing and, hence, preserve the truth of the spherical Earth theory. However, each of them was in turn ruled out by progressively better-controlled and better-conducted experiments. Many, of course, would say that adjustment of the thickness of a match (3 mm) as a proportion of 1 metre (1,000 mm) was so minimal that it could just be attributed to experimental error, or simply ignored. And, if the theory is important, then that is an understandable tendency, but, for tougher-minded scientists, it seemed that the long-held, and religiously endorsed, theory of the spherical Earth had to be rejected on account of a persistent 3-mm discrepancy.

However, Huygens could see a more sophisticated explanation for the lessening of g at the equator, while still maintaining T, the theory of a spherical Earth. He argued:

- C^5: Objects at the equator rotated faster than at Paris, and, hence, the centrifugal force at the equator was greater; this countered the centripetal force of gravity, hence diminishing the net downwards force (gravity) at the equator and, hence, decreasing the speed of oscillation of the pendulum; that is, increasing its period.

This final explanation for the slowing of equatorial pendulums, while maintaining the spherical Earth theory, was quite legitimate and appeared to save the theory. Many would be happy to just pick up this 'get out of jail free' card and continue to believe that the Earth was spherical. Huygens did not do so. He calculated the actual centripetal force at the equator and, hence, its effect on gravitational attraction.

In modern terms, the calculation is as follows. The equatorial object follows a circular path covering an angle of 2π radians per day (86,400 seconds). So angular velocity, ω, equals $2\pi/86{,}400$ rad/sec, or 7.3×10^{-5} rad/sec. As the radius of the Earth is about 6.4×10^6 m, the centripetal acceleration of the object $a_c = \omega^2 r \approx 0.034$ m/sec^2. So this is the amount that the gravitational acceleration of equatorial objects was diminished in virtue of the spinning of the Earth.[21] This translated as a 1.5 mm shortening of the seconds pendulum, but it still left 1.5 mm unaccounted for. This is less than the thickness of a match, and yet, for such a minute discrepancy, Huygens and Newton were prepared to abandon the spherical Earth theory and claim that the true shape of the earth was an oblate. For the new, quantitative science, the 'near enough is good enough' mantra could not be maintained, something that students can be taught to appreciate.[22]

Indeed, the testing situation is even a bit more complicated. Metaphysics plays a role in theory appraisal. So the situation is as follows:

> If theory (T) and conditions (C) and background metaphysics (M) imply observation (O) T.C.M → O
> And, if O is not the case,
> ~O
> Therefore, T is false, or C is false, or M is false
> ~M v ~T v ~C

Willard van Orman Quine, and before him Pierre Duhem, elaborated the methodological point about theory appraisal that is so apparent in the shape-of-the-Earth debate:

> The totality of our so-called knowledge or beliefs, from the most casual matters of geography and history to the profoundest laws of atomic physics or even of pure mathematics and logic, is a man-made fabric which impinges on experience only along the edges. . . . But the total field is so underdetermined by its boundary conditions, experience, that there is much latitude of choice as to what statements to reevaluate in the light of any single contrary experience.
>
> (Quine 1953, p. 42)

The shape-of-the-Earth controversy is a wonderful episode in the history of science. A great pedagogical story can be made, even a drama. All the elements are there: powerful and prestigious figures, 'no name' outsiders, controversy over a big issue, mathematics and serious calculations, religion and, finally, decision-making, with ample opportunity and reason to preserve the status quo. Sadly, however, the episode is little known and hardly ever taught. If history and philosophy are valued, then there is good justification for teaching the episode, but if 'everyday, applied, immediate usefulness' is the guiding principle for constructing a science curriculum, then it is unlikely ever to be taught. As Noah Feinstein, an advocate of the 'usefulness' position, has written:

> It [usefulness theory of curriculum] seems to suggest that the curriculum should be stripped of canonical content that students are unlikely to find relevant to their daily lives – such as, for instance, the shape of the earth.
>
> (Feinstein 2011, p. 183)[23]

Real Versus Theoretical Objects

One way to conceptualise the methodological revolution of the seventeenth century is to recognise the difference between real objects and such objects when described from the standpoint of some scientific theory. Newton's theory of mechanics, for example, provides definitions of key concepts – momentum, acceleration, average speed, instantaneous speed, weight, impetus, force, point mass and so on. These concepts were hard-won theoretical constructs and are utilised in his account of pendulum motion. At the beginning of the scientific revolution, these concepts were only dimly seen, and were refined with time.

Consider the example of acceleration, which was initially defined by Galileo, and by all his predecessors in the 2,000 years of natural philosophy, as rate of increase of speed with respect to distance traversed – a natural enough definition, given that accelerating bodies increase speed over both time and distance, and that the passage of distance was both more measurable and more easily experienced by sight and feel. On a galloping horse, you see poles go past, not time ticking over. Distance was measurable and so given to geometrical depiction and analysis, in the way that time was not measurable and so not given to precise geometrical analysis. It was only in Galileo's middle age that he changed the definition to the modern one of rate of change of speed with respect to time elapsed. By 1604, Galileo had realised that distances covered by a uniformly accelerating body (free fall) increased as the square of the time elapsed. However, at this time and at least for another 6 years, he believed that speed increased in direct proportion to distance fallen. The two beliefs, also held by Descartes, are inconsistent. Sometime after 1610, he moved to the modern definition of acceleration and made his beliefs consistent.[24] Thirty years later, he refers to this in Day Three of his *New Sciences*, where he writes: 'A motion is said to be uniformly accelerated, when starting from rest, it acquires, during equal time-intervals, equal increments of speed' (Galileo 1638/1954, p. 162).

Without such a change of definition, the fundamental laws of free fall would not have been discovered. A good deal of a child's school life would pass while waiting for him or her to construct the modern definition of acceleration, and, hence, most teachers routinely commence mechanics classes by giving students the modern definition. The alternative is not considered, even by the strongest advocates of discovery learning or of minimally guided instruction. Yet the two options, and the historical story, certainly repay discussion and analysis; students can appreciate the centrality of conceptual 'discovery' in science. However, as with so much, this depends on teachers knowing the history and philosophy of what they are teaching.

Although less clearly defined than with Newton, interlocking concepts formed the conceptual structure of Galileo's physics and provided the meaning of key terms. This is Galileo's 'world on paper', as he referred to it, or his proto-theoretical system. This is the indispensable scaffolding of science that children need to be provided with; the scaffolding needs to be, in some sense, transmitted to them. For realists, there is also a world of material and other objects that exists apart from Galileo's theorising. We can see in Galileo's practice a most important intervening layer emerging between theory and the real world – the realm of theorised objects. These are natural objects as conceived and described by the relevant theoretical concepts. Planets and falling apples have colour, texture, irregular surfaces, heat, solidity and any number of other properties and relations, but, when they become the subject matter of mechanics, they are merely point masses with specified accelerations. When thus conceptualised, they are no longer natural objects, but theorised objects. In a similar way, when apples are considered by economists, they become theorised objects of a different sort – commodities, with specific exchange values. When botanists consider apples, they create yet other theoretical objects. For Galileo, a sphere of lead on the end of a length of rope, swinging in air, became, in his mechanical theory, a pendulum conceived as a point mass at the end of a weightless chord suspended from a frictionless fulcrum. The theorised objects are created or constructed by scientists;

contrary to the claims of many constructivists and sociologists of scientific knowledge, the real objects are not so created. Failure to recognise this is the cause of much philosophical confusion.

Experiment

Galileo did not just develop a system of rational mechanics in the same way as the medieval scientists who constructed mathematical models of physical systems and then proceeded no further. In contrast, Galileo's theoretical objects were the means for engaging with and working in the natural world. For him, the theoretical object provided a plan for interfering with the material world and, where need be, for making it in the image of the theoretical. When del Monte told Galileo that he had done an experiment with balls in an iron hoop and the balls did not behave as Galileo asserted, Galileo replied that the hoop must not have been smooth enough, that the balls were not spherical enough and so on. These suggestions for improving the experiment are driven by the theoretical object that Galileo had already constructed. This told Galileo the things that had to be corrected in the experiment. Without the theoretical object, he would not have known whether to correct for the colour of the ball, the material of the hoop, the diameter of the hoop, the mass of the ball, the time of day or any of a hundred other factors. It is this aspect of Galileo's work that moved Immanuel Kant to say that, with Galileo, 'a light broke upon all students of nature' because he demonstrated that:

> Reason has insight only into that which it produces after a plan of its own . . . It is thus that the study of nature has entered on the secure path of a science, after having for so many centuries been nothing but a process of merely random groping.
>
> (Kant 1787/1933, p. 20)

Galileo was a technician and an experimentalist. He put great effort into devising, making and popularising novel technical instruments. He was responsible for creating the *pulsilogium*, the *bilancetta*, the *compasso di proporzione*, the *thermoscopium* and the *telescope*, and he drew workable plans for the *pendulum clock* (Bedini 1986). He also measured and made calculations of pendulum swings. Stillman Drake has unearthed these in the Galilean manuscripts at Florence (Drake 1990, Chapter 1). However, Galileo's measurements and experimentation were directed; they were measurements of behaviour in circumstances dictated by his theoretical conceptualisations. Further, as we have seen in his debate with del Monte, the theoretical conceptualisation enabled him to identify 'accidental' departures from the ideal. Once this was done, allowances could be made, and the experiment could be refined. As has been remarked, the whole history of classical mechanics is a long attempt to force nature to match Newtonian theory.[25]

Scientific Laws

The regularity account of scientific law has been popular ever since David Hume's 1739 *Treatise on Human Nature*, where he attempted to refute the necessitarian view of law. For Hume, and those following him, scientific laws state constant relations between observables; they state what is uniformly seen to be the case. The pendulum laws present an overwhelming problem for the Humean account: the regularities do not occur. Under very refined experimental conditions – small oscillations, heavy weights, minimum air and fulcrum resistance – they almost occur, but 'almost occurring' is not what regularity accounts of law are about. Moreover, it was a commitment to the truth of the apparently disproved laws that enabled the approximate conditions for the law's applicability to be devised. That is, the truth of the law is a presupposition for identifying approximations to it, and for identifying when some behaviour is to be seen as 'almost law-like'. Russell Hanson has expressed the matter well:

> The great unifications of Newton, Clerk Maxwell, Einstein, Bohr and Schrödinger, were pre-eminently discoveries of terse formulae from which descriptions and explanations of diverse phenomena could be generated. They were not discoveries of undetected regularities.
>
> (Hanson 1959, p. 300)

Michael Scriven once wrote that, 'The most interesting thing about laws of nature is that they are virtually all known to be in error' (Scriven 1961, p. 91). Nancy Cartwright, in her *How the Laws of Physics Lie* (Cartwright 1983), makes a similar point: if the laws of physics are interpreted as empirical, or phenomenal, generalisations, then the laws lie. As Cartwright states the matter: 'My basic view is that fundamental equations do not govern objects in reality; they govern only objects in models' (Cartwright 1983, p. 129). The world does not behave as the fundamental equations dictate. This claim is not so scandalous: the gentle and random fall of an autumn leaf obeys the law of gravitational attraction, but its distance of fall is hardly described by the equation $s = ut + \frac{1}{2} at^2$. This equation refers to idealised situations. A true description, a phenomenological statement, of the falling autumn leaf would be complex beyond measure. The law of fall states an idealisation, but one that can be experimentally approached. These laws are usually stated with a host of explicit *ceteris paribus*, or 'other things being equal', conditions.[26] For the laws of pendulum motion, the *ceteris paribus* conditions would be:

- the string is weightless (so no dampening occurs);
- the bob does not experience air resistance;
- there is no friction at the fulcrum;
- all the bob's mass is concentrated at a point;
- the pendulum moves in a plane and does not experience any elliptical motion;
- gravity and tension are the only forces operating on the bob.

However, these conditions can only be approached, never realised – something that Ronald Giere has been at pains to point out (Giere 1988, pp. 76–78, 1999, Chapter 5). Giere believes, not only that scientific laws are false, they are also neither universal nor necessary (Giere 1999, p. 90). His account of the laws of pendulum motion is close to what has been advanced above in discussion of the real and theoretical objects of science. He says:

> On my alternative interpretation, the relationship between the equations and the world is indirect . . . the equations can then be used to construct a vast array of abstract mechanical systems. . . . I call such an abstract system a model. By stipulation, the equations of motion describe the behavior of the model with perfect accuracy. We can say that the equations are exemplified by the model or, if we wish, that the equations are true, even necessarily true, for the model.
>
> (Giere 1999, p. 92)

Objectivism

The seventeenth century's analysis of pendulum motion supports objectivist views of scientific theory and of epistemology.[27] These objectivist views are in opposition to all empiricist epistemologies and theories of scientific methodology. Empiricists, since Bacon, Locke and Berkeley in the seventeenth and eighteenth centuries, through to Alfred Ayer in the twentieth century, have dominated philosophical reflection on science. Both Ayer's *The Foundations of Empirical Knowledge* (Ayer 1955) and *The Problem of Knowledge* (Ayer 1956) provide quintessential empiricist accounts of epistemology. Both are preoccupied with the problem of perception, which is a telltale marker of empiricism; despite all the fuss and bother about theory dependence of perception, how things look is not of great scientific moment. Enough has been said above to indicate that a concentration on perception, to the exclusion of experimentation and theorising, misses the main thrust of the scientific revolution. The empiricist tradition maintains the following:

1 There is a distinction between basic, observational or intuited knowledge and theoretical knowledge.
2 Basic, observational or intuited knowledge does not involve theory.
3 Basic, observational or intuited knowledge is available or given to individual observers or thinkers (or knowers, or subjects).
4 Theoretical knowledge is derived from, or ultimately justified by reference to, basic, un-theoretical knowledge.

Rationalism is the other side of the empiricist coin. Both empiricist and rationalist epistemologies are foundationalist: they seek foundations for knowledge in the experience of the individual, but, whereas traditional empiricism defines experience as the 'outer' experience of the senses, rationalism extends the definition of experience to include the 'inner' experiences of the individual's mind or of reason. In both epistemologies, it is the experience of the individual knower that is primary. Once this empiricist problematic is accepted, in either its empiricist or rationalist guise, then there is only a short, and oft-taken, step to scepticism and relativism. Clearly, both 'outer' and 'inner' experience is affected by an individual's circumstances, language, culture, ideology and theory. The 'theory dependence of observation' has been much written upon and, when coupled with empiricist assumptions about knowledge, it often results in relativist or sceptical epistemologies. If experience is the foundation of knowledge, then the foundation is on shaky ground, and doubts about the soundness of the building are easy to induce.

One modern, and energetic, variant of this empiricist problematic is constructivism, which, as will be elaborated in Chapter 8, embraces the faulty epistemological trifecta of individualism, experience and, inevitably, scepticism. From Aristotle through to modern constructivism, a major mistake has been the elevation of passive experience (looking and observing) at the expense of intervention, instrumentation and measurement. The latter connects the investigator with causal processes in the world and thus puts limits on otherwise unboundable, Rorschach-like subjectivity.

Karl Popper signalled an objectivist break with the empiricist problematic in his 1934 *The Logic of Scientific Discovery*, but the account waited 40 years for its full development in *Objective Knowledge*, where he wrote:

> Traditional epistemology has studied knowledge or thought in a subjective sense – in the sense of the ordinary usage of the words 'I know' or 'I am thinking'. This, I assert, has led students of epistemology into irrelevances: while intending to study scientific knowledge, they studied in fact something which is of no relevance to scientific knowledge. For scientific knowledge simply is not knowledge in the sense of the ordinary usage of the words 'I know'. While knowledge in the sense of 'I know' belongs to what I call the 'second world', the world of subjects, scientific knowledge belongs to the third world, to the world of objective theories, objective problems, and objective arguments.
>
> (Popper 1972, p. 108)

Most objectivists in theory of knowledge share Popper's convictions that, first, the growth of scientific knowledge is the core subject matter of epistemology; and, second, knowledge is something other than beliefs or psychological states; it transcends individual consciousness. Popper delineated 'three worlds': the first world of objects, processes and events (material and otherwise); the second world of subjective, individual, mental operations (the life of the mind or private consciousness); and the third world of scientific, and other, theories, constructions and problem situations, which are a part of culture and which, although created by people, nevertheless exist independently of first- and second-world events.[28]

The objectivist tradition emphasises: first, that there is a separation of cognitive or theoretical discourse from the real world. The world is neither created by the discourse (as in idealism), nor does it somehow create the discourse (as in various reflection, or imprinting, theories from Locke to Lenin), nor does it anchor or provide foundations for the discourse (as in empiricism and positivism). Theoretical discourse and the world are each autonomous. In this sense, theory exists independently of individuals. Thus, scientific knowledge is, contrary to the claims of many constructivists, external to individuals.

Second, objectivist views distinguish within theory between:

(a) the conceptual foundations of the discourse, containing the definitions of theoretical and observational terms (there is no decisive distinction made between these kinds of term);
(b) the conceptual structure of the theory, which is the elaboration and manipulation of the basic concepts by techniques (mathematical and logical) that produce the structure of the theory (Galileo's theorems and propositions);
(c) the theorised objects of the theory, which are objects in the world as they are conceived and described by the theory – the balance treated as a uniform line with parallel weights suspended from it, the pendulum treated as a point mass on a weightless string, billiard balls treated as colourless, perfectly hard bodies and so forth.

Objectivism in Education

A source of much confusion arises from there being a second sense of 'objectivism', especially common in science-education literature. This is objectivism in the sense of 'objective truth', 'universalism' or 'certainty of belief'. This objectivism is completely orthogonal to the Popperian sense (indeed, Popper denied the possibility of absolute truth). In science education, objectivism, as well as having these absolutist overtones, is, with some confusion, frequently interpreted as a version of positivism. Two science educators write that:

> At present, most science teaching is based on an objectivist view of knowing and learning. . . . Here, objectivism subsumes all those theories of knowledge that hold that the truth value of propositions can be tested empirically in the natural world.
> (Roth & Roychoudhury 1994, p. 6)

Popper completely denies such a view. The index entry for 'objectivism' in Kenneth Tobin's *The Practice of Constructivism in Science Education* (Tobin 1993) reads 'see positivism'. Positivism is, of course, a philosophical position that Popper spent his life arguing against. This 'crossing of philosophical wires' is a major impediment to scholarly progress in science education and to its productive engagement with other disciplines: Time is needlessly spent in attacking straw men or enjoying 'feel good' euphoria, while significant literature and arguments are ignored.

Observation

Observation is an important element of science education, as teaching children to observe carefully is part of every science curriculum. For some time, there has been educational recognition of the more philosophical or problematic dimensions of observation.[29] The topic of observation has received renewed attention in recent time, with the much-repeated claim of the Lederman research group that an 'empirical base' is the first defining feature of the nature of science. As they state the matter:

> Students should be able to distinguish between observation and inference. . . . An understanding of the crucial distinction between observation and inference is a precursor to making sense of a multitude of inferential and theoretical entities and terms that inhabit the worlds of science.
>
> (Lederman *et al.* 2002, p. 500)

For this aim to be realised, teachers must first understand and appreciate the distinction. The thesis of this book is that HPS can contribute to science teaching and science teacher development by assisting with a better and deeper understanding of the many theoretical, curricular and pedagogical issues that engage science teachers. The thesis can be tested by elaborating the topic of observation.

The pendulum story is an occasion for raising one basic matter about this process of observation (including vision, perception, looking, noticing and seeing) that links personal psychology to the observer's social life: this is the distinction between 'object perception' and 'propositional perception'.

Galileo constantly saw things that those around him did not see; even 'the lynx', Leonardo da Vinci, did not see what Galileo saw. Clearly, in one sense, Leonardo did see what Galileo saw, but in another sense he did not. Ludwig Wittgenstein made the distinction in terms of seeing and seeing as (Wittgenstein 1958). Popper made the distinction in terms of perception and observation. Object perception occurs whenever healthy eyes pass over reasonably lit situations; they see what there is to be seen. Among the things seen, some are noticed or attended to. However, this kind of object perception has no epistemological or scientific import, nor even much personal import. The latter begins when, among the things noticed, people begin to verbalise what they see; this then is propositional perception, where people see 'that p', where p is some proposition or statement. This perception depends entirely on language, social embeddness and available theory. The proposition, p, has to be verbalised or written; it requires a language. Someone looks at a field and sees 'that there are trees'; they see, but do not notice, the fence. Someone looks at a room of people, but only notices the person they want to meet.

Consider the two inscriptions in Figures 6.11 and 6.12. Everyone, from child to adult, who passes their eyes over them can have an object perception of the marks, whether they are noticed or not noticed.

Once the marks are noticed, then propositional perception can begin. At the least sophisticated level, someone might see '*that* there are marks on the

Figure 6.11 Chinese Character

Figure 6.12 Japanese Character

paper'; at the next level, someone might see '*that* there are Asian characters on the paper'; at the next level, someone might see '*that* there is a Chinese character on the paper', whereas someone else might see '*that* there is a Japanese character on the paper'; at the next level, someone might see '*that* there are both Chinese and Japanese characters on the paper', but not know what they mean; with more sophistication, someone might see '*that* the word STOP in both Chinese and Japanese is on the paper'; finally, a more knowledgable person will recognise that each character needs to be qualified for a precise translation into STOP rather than PAUSE. The ascent up the levels of propositional perception is entirely dependent on what the person knows, on what is in their mind, on the language competencies of the individual. All of the latter is something that needs to be taught or enculturated; without this, there cannot be propositional perception, and it is the latter that is of relevance to most human life, including science and scientific investigation.

Two people might have the same propositional perception '*that* there is a man in the room', but one of them might see '*that* the prime minister of Australia is in the room'; this propositional perception requires being able to draw on certain knowledge or information in the head. To look at a chessboard and see '*that* the black rook is threatened by the white queen' depends on having been taught the rules of chess. No amount of looking, even careful looking, at the above inscriptions, people or chess pieces will reveal the word STOP, the prime minister of Australia or the threatened rook; to see that the latter is the case requires teaching, learning and acquisition of sophisticated language. As the English 'ordinary language' philosopher J.L. Austin (1911–1960) commented: 'we are using a sharpened awareness of words to sharpen our perception of, though not as the final arbiter, of the phenomena' (Austin 1961, p. 130).

For students, the sharpened awareness of words has to come from the outside, from engagement in a community, and it is teachers who are the major conduit for this engagement. Two moderately sophisticated students might share the propositional perception '*that* the body is moving smoothly in a circle'; they will only be able to see '*that* the body is accelerating' if they have been taught elementary Newtonian mechanics, and certainly not by 'brain storming' or 'negotiating' what they see. Likewise, students might see, as Aristotle did, '*that* the hot air is rising to the ceiling'; only with basic Newtonian mechanics will they see '*that* the falling cold air is pushing the less heavy hot air up'. The first description leads to a mechanics of *levitas*; the second leads to a mechanics of *gravitas*. So explanations depend on how things are described, on what propositional perception people have. The explanation for 'the British invaded Australia' is different from the explanation of 'the British discovered Australia'. The response to 'I see *that* Billy is exhibiting Attention Deficit Disorder' will be different from that to 'I see *that* Billy is exhibiting spoilt brat behaviour'. Clearly, propositional perception in politics, morals and aesthetics depends entirely on language, and the 'more sharpened' the language available, then the sharper can be the observations. Of course, language is necessary, but not sufficient, for propositional perception.

Propositional perception is not tightly tied to object perception. The former is always inferential. Two people looking at the same thing can see *that* completely different things are the case. In the Middle East, some people see *that* the US is defending democratic processes, whereas others see *that* the US is defending its strategic interests. And so on, ad infinitum. Effort has to go into seeing the 'facts of the matter', which will always be an item of propositional perception, as every statement of fact is a proposition, and the issue is to see how far down such agreement can go. The British empiricists, and more recently the positivists, thought that there were some basic sense-data statements that grounded all variants of propositional perception about the world.[30] For instance, 'I see a red patch' or 'I have a red patch sensation' is the common perceptual foundation for 'I see *that* there is a tomato', 'I see *that* there is a stop sign', 'I see *that* blood is on the floor' and so on. Having the red sensation is object perception; when it is described, propositional perception is occurring, and this is inferential.

Recall that, in Chapter 2, it was mentioned that the positivist Philipp Frank said it was quite wrong for textbooks to write of how, in the morning, we see 'the sun rising on the horizon'. He said: 'Actually our sense observation shows only that in the morning the distance between horizon and sun is increasing, but it does not tell us whether the sun is ascending or the horizon is descending' (Frank 1947/1949, p. 231). The statement 'that the sun is moving' is just *one* propositional perception grounded by the object perception we have when looking at the sun in the morning.

For these sorts of reason, Plato, 2,500 years ago, said that, 'we see through the eye, not with the eye', which could be taken as the beginning of the long history of philosophy of perception.[31] More recently, philosophy of perception has connected with philosophy of mind, cognitive psychology and phenomenology. In 1934, Karl Popper presciently warned against misunderstanding the place of observation in science:

> The doctrine that the empirical sciences are reducible to sense-perceptions, and thus to our experiences ... is here rejected ... there is hardly a problem in

> epistemology which has suffered more severely from the confusion of psychology with logic.
>
> (Popper 1934/1959, p. 93)

This was written against the commitment of early positivists to sense impressions, sense data and other putative perceptual bed-rocks of science. Wallis Suchting, in a paper on 'The Nature of Scientific Thought', has commented on these matters, saying that:

> Thus the *key* inadequacy of empiricism has really nothing to do with the centrality it accords to sense-experience; in particular, the controversy over whether the 'basic language' of science should be 'phenomenonalistic' or 'physicalistic' is irrelevant to the main question, a mere internal family dispute, as it were. The central deficiency of empiricism is one that it shares with a wide variety of other positions, namely, all those that see objects themselves, *however they are conceived*, as having epistemic significance *in themselves*, as inherently determining the 'form', as it were, of their own representation, rather than as determining the degree of applicability of representations of a given 'form', and hence, conversely, that the nature of what is represented can be more or less *directly* 'read off' its representation.
>
> (Suchting 1995, p. 13; original italics)

Data, Phenomena and Theory

The foregoing considerations concerning the real and theorised object distinction, object and propositional perception, idealisations and non-Humean accounts of law point to the importance of distinguishing data from phenomena, and both of these from theory. The data are observed; phenomena characteristically are not. Scientific laws and theories are about the phenomena, not about the data.[32] If this is understood, a number of things about science, and especially studies of pendulum motion, become clearer.

Real objects (processes, events, occurrences, states) are observed either in natural (Aristotle's preference) or experimental (Galileo's and Newton's preference) settings. The observation can be immediate (with eyes, microscopes, etc.) or inferred (meter readings, instrument displays, etc.). All of this occurs in the realm of the real, not the realm of discourse. The observations are then verbalised, described, written or tabulated. This has to be done in a language (including mathematics) and according to some theoretical standpoint. This is all done in the realm of discourse. These descriptions are characteristically sifted, sorted and selected – lots of readings and descriptions are simply thrown away, or ignored. The result is scientific data. These, then, are the raw representations of real objects (processes, events, occurrences, states). This step is clearly theory dependent. A range of falling red apples, or swinging weights on a string, are, in physics, represented as points on a graph, as printouts on a tickertape, as lines on a screen. These representations are not meant to mirror, or copy, the real. They are precisely meant to represent the real. And, as mentioned earlier, economists, artists and farmers have different ways of representing the same real events. Adequacy of representation simply does not mean correspondence of representation, in the sense of the representation mirroring the object.

For pendulums, even highly refined experimental apparatus will give a scatter of data points. In plotting period against length, the scatter is enormous; in plotting period against the square root of length, the scatter contracts markedly, and the phenomenon of direct relation is revealed. The laws of pendulum motion are not meant to, and cannot, explain these data points; they are too erratic. However, in science, from data come phenomena; and it is the phenomena that are the subject of scientific laws and theories. Often, a line of best fit is put through the data points, and the line is then taken to represent the phenomenon being investigated. Thereafter, it is the phenomenon that is discussed and debated, not the data. Any number of individual telescopic observations, when corrected and selected, constitute astronomical data. From these, we infer, construct, invent planetary phenomena: circular orbits, elliptical orbits, heliocentric or geocentric orbits. The latter are not seen. They are not observational. However, this is no scientific impediment. Once we settle on the phenomenon, it becomes the subject matter of our scientific theories. Newton, in Book II of his *Principia*, after laying out his Rules of Reasoning in Philosophy (our science), has a section on Phenomena. Among six phenomena that he believes his System of the World has to account for, are:

> That the fixed stars being at rest, the periodic times of the five primary planets, and (whether of the sun about the earth, or) of the earth about the sun, are as the 3/2th power of their mean distances from the sun.
>
> That the moon, by a radius drawn to the earth's centre, describes an area proportional to the time of description.
>
> (Newton 1729/1934, pp. 404–405)

These are not observational statements, and they are not data in the terms we are using. They are statements of the phenomena to be explained. As Newton acknowledged, these phenomena came from the work of the giants on whose shoulders he stood: Galileo, Kepler and Brahe.[33]

Kepler's 'elliptical planetary paths' were, in turn, phenomena separate from, and not necessarily implied by, his astronomical data and measurements. As William Whewell noted in the nineteenth century, in his critique of Mill's inductivist account of science, the concept of an elliptical path was supplied by Kepler's mind, not by his data. There is usually no univocal inference from data to phenomena. Phenomena are underdetermined by data, just as theory is underdetermined by evidence. In the above case, the data are probably consistent with periodic times of 5/4th power of mean distance.

The situation might usefully be represented as shown in Table 6.1.

- *Level 6* is constituted by objects, processes and events in the world.
- *Level 5* is constituted by perceptions, observations and psychological states occasioned by events in world.
- *Level 4* is constituted by the articulation or statement of observational experience.
- *Level 3* is constituted by the representation (graphs, tables, counts) of observations; this constitutes data.
- *Level 2* is constituted by phenomena such as 'isochronous motion' or 'elliptical paths' that can be represented by models, or empirical laws, or equations such as $T = 2\pi\sqrt{(l/g)}$.
- *Level 1* is constituted by fundamental scientific laws or high-level theory.

A chaotically moving, falling autumn leaf 'obeys' a number of fundamental causal mechanisms – gravitation, air resistance, etc. – but its path (data points) does not illustrate or confirm the appropriate laws. Contrary to Nancy Cartwright's claims, we need not believe that the fundamental laws of physics lie; they might lie about appearances (Level 3 items above), but, if we abandon the long-entrenched, Aristotelian-based conviction that the laws should be about Level 3 items, then we can maintain their truthfulness. They are true of phenomena, not of data. This was a point well made by James Brown, who concluded his analysis of this subject:

> Phenomena are to be distinguished from data, the stuff of observation and experience. They are relatively abstract, but have a strongly visual character. They are constructed out of data, but not just any construction will do. Phenomena are natural kinds that we can picture. They show up in thought experiments and

Table 6.1 Objects and Processes, Perception, Data, Phenomena and Laws

Level 1	Fundamental laws and mechanisms	Gravitational attraction Simple harmonic motion
Level 2	Phenomena, scientific models, idealisations	Four pendulum laws (mass and amplitude independence; period varies as square root of length; isochronous oscillation)
Level 3	Data	Individual period measurements for different masses, amplitudes, lengths; scattered points on a graph
Level 4	Propositional perception	Articulating different facts of the matter as seen
Level 5	Object perception	Seeing or observing swinging pendulum
Level 6	Objects, events and processes in world	Weight swinging on end of cord

> they play an indispensable role in scientific inference mediating between data and theory. So let's attend to them.
>
> (Brown 1996, p. 128)

Other Features

Other features of science and philosophical themes in the history of pendulum study could have been singled out. For example, debate over the very meaning of time. Any investigation or discussion of the pendulum leads quickly to the question of time: what is it, and how is it best measured? Such questions have been entertained for at least 2,500 years in the West, by such luminaries as Plato, Aristotle, St Augustine, Aquinas, Newton, Mach and Einstein. Other cultures have their own long traditions of engagement with time and its measurement. For all cultures, this engagement includes contributions from artists, poets, musicians, theologians, philosophers, playwrights and scientists. Discussion of time connects to calendars, religious and cultural ritual, cosmology and countless things about everyday life from personal punctuality to railways and airlines running 'on time'. Pendulum studies are an opportunity for students to appreciate and learn from these traditions of temporal engagement; they are traditions where science, philosophy, technology and culture intertwine.[34]

The themes identified above can be seen in the history of most other scientific fields: genetics, combustion, atomic structure, astronomy, evolution and so on. In all cases, discussion with students about the interplay of these influences and themes is profitable.

The Pendulum and Recent US Science Education Reforms

Attention to the place accorded to the pendulum in curricula provides a good testament to the degree to which science education has engaged with and utilised HPS. Jerrold Zacharias, the driving force behind the 1960s PSSC physics course, did not think highly of the educative value of the simple pendulum, saying that, for students, 'it is not very interesting' (Zacharias 1964, p. 69). He commends beginning instruction with the coupled pendulum, which gives 'lovely phenomena', and then students will want to know about the single pendulum, 'And this happens every time' (ibid.). It is an empirical question whether this approach works or does not work in classrooms, but, putting that aside, what students can learn about the simple pendulum will depend on teachers having some appreciation of the rich history of the subject. The evidence is that there is not much such appreciation.

It is instructive, if sobering, to look at the utilisation of the pendulum in the past three decades of intense efforts to improve US school science programmes. These efforts have involved thousands of individuals in bodies such as the AAAS, the NRC, the National Academy of Science, the National Academy of Engineering, the NSF; peak disciplinary bodies in physics, chemistry, biology, earth science; and all major national and state science-education organisations, including the NSTA and the National Association for Research in Science Teaching (NARST). Despite such massive and luminous oversight, the pendulum barely appears, and, when it does, so little of its potential for teaching scientific content, methodology and the 'nature of science' is utilised. This, in part, is a consequence of the unhealthy separation of the science-education and HPS communities, and not just in the US.

Scope, Sequence and Coordination

The large-scale and influential curriculum proposal of the US NSTA – *Scope, Sequence and Coordination* (Aldridge 1992) – highlights the pendulum to illustrate its claims for sequencing and coordination in science instruction. Yet nowhere in its discussion of the pendulum is history, philosophy or technology mentioned.

Project 2061

In 1989, the AAAS published its wonderfully comprehensive *Science for All Americans* report (AAAS 1989). It acknowledged that: 'schools do not need to be asked to teach more and more content, but rather to focus on what is essential for scientific literacy and to teach it more effectively' (AAAS 1989, p. 4). The report saw that students need to learn about 'The Nature of Science', and, hence, that was the title of its first chapter. The report recognised the importance of learning about the interrelationship of science and mathematics, saying: 'The alliance between science and mathematics has a long history dating back many centuries. . . . Mathematics is the chief language of science' (AAAS 1989, p. 34). It also acknowledged that some episodes in the history of science should be appreciated because, 'they are of surpassing significance to our cultural heritage' (AAAS 1989, p. 111).

Among the ten such episodes it picks out is Newton's demonstration that, 'the same laws apply to motion in the heavens and on earth' (AAAS 1989, p. 113). It provides a very rich elaboration of this episode and its scientific, philosophical and cultural impacts. Unfortunately, there is no mention of what enabled Newton to achieve this unification, namely the pendulum; had such mention been made, this 'big idea' could have been connected to something tangible in all students' experience, the place of mathematics in science could have again been underlined, and a wonderful case study in the nature of science could have been built upon.

The US National Standards

The under-utilisation of the pendulum can be gauged from looking at the recently adopted US *National Science Education Standards* (NRC 1996). The *Standards* adopt the same liberal or expansive view of scientific literacy as the NCEE did in 1983, saying that it 'includes understanding the nature of science, the scientific enterprise, and the role of science in society and personal life' (NRC 1996, p. 21). The *Standards* devote two pages to the pendulum (pp. 146–147). However, there is no mention of the history, philosophy or cultural impact of pendulum motion studies; no mention of the pendulum's connection with timekeeping; no mention of the longitude problem; and no mention of Foucault's pendulum.

Astonishingly, in the suggested assessment exercise, the obvious opportunity to connect standards of length (the metre) with standards of time (the second) and with standards of weight is not taken. Rather, students are asked to construct a pendulum that makes six swings in 15 seconds. This is a largely pointless exercise, especially when they could have been asked to make one that beats in seconds and then measure its length and enquire about the coincidence between their seconds pendulum and the metre (Matthews 1998).

Depressingly, the *Standards* document was reviewed in draft form by tens of thousands of teachers and educators. It is clear that, if even a few of the readers had a little historical and philosophical knowledge about the pendulum, this could have transformed the treatment of the subject in the *Standards* and would have encouraged teachers to realise the liberal goals of the document through their treatment of the pendulum. This would have resulted in a much richer and more meaningful science education for US students. That this historical and philosophical knowledge is not manifest in the *Standards* indicates the amount of work that needs to be done in having science educators become more familiar with the history and philosophy of the subject they teach, and of having science-education communities more engaged with the communities of historians and philosophers of science.

America's Lab Report

The US NRC commissioned a large study on practical work in US schools that was published as *America's Lab Report: Investigations in High School Science* (NRC 2006). The book has 236 pages, seven chapters and hundreds of references. The pendulum has three entries in the index. On its first appearance, it is said to be regrettable that teachers simplify pendulum experiments and ignore the 'host of variables that may affect its operation' (NRC 2006, p. 117). Teachers are advised to recognise these 'impediments', such as friction and air resistance, but the writers go on to say that this 'can quickly become overwhelming to the student and the instructor' (NRC 2006, p. 118). This is not very helpful. It could have been an occasion to say something about the fundamental importance of idealisation and abstraction to the very enterprise of science, of not letting the trees get in the way of seeing the forest. This was the problem identified by Thomas Kuhn in his discussion of the pendulum and faced by da Vinci; it is the heart of the debate between Galileo and his patron Guidabaldo del Monte, but the *Lab Report* says nothing about this fundamental scientific procedure, much less provide some historical background to its resolution. The pendulum allows students very tangibly to begin seeing the effect of 'impediments' and 'accidents' (Koertge

1977), or 'errors' in contemporary language, on the manifestation or 'visibility' of core natural processes.

On the pendulum's second appearance in the index, the 'typical pendulum experiment' is criticised, because it is 'cleaned up' and used just to teach science content – that the 'period of a pendulum depends on the length of the string and the force of gravity' – and not scientific process skills (NRC 2006, p. 126). In contrast to these 'bad' pendulum practical classes, on the pendulum's third appearance, a 'good' class is described occupying two pages in a highlighted box. In this class, teachers are first advised to demonstrate swinging pendulums, then, in a very guided fashion, to have students graph the relationships between period and mass, period and amplitude, and period and length, and finally it is suggested that the teacher discuss the importance of obtaining an adequate amount of data over a range of the independent variable (NRC 2006, pp. 128–129).

This does no harm and can do some good, but there is nothing noteworthy here. Everything about the rich history of the pendulum has been stripped out: no mention of Galileo, Huygens, Newton, Hooke, universal gravitation, timekeeping, clocks, length standards, longitude, shape of the Earth or conservation laws. No connection intimated between science, technology and society; no sense of participation in a scientific tradition. Nothing. Teachers are not even told to talk about these great scientists and their pendulum-based discoveries.

And in this set piece, nationally distributed, 'model pendulum lesson' teachers and students are told to plot period against length on a graph. This is a task with only minimally useful outcomes: such a graph provides a scatter of points that merely establishes a trend. After having done this, 17–18-year-old students could have been so easily asked to plot period against the square root of length. When this is done, a straight line is obtained from the data, not a scatter of points. As discussed above, the physical phenomenon is revealed by mathematical manipulation. Period is seen, as it was by Galileo and Huygens, not just to vary with length, but to vary directly with the square root of length; the conclusion from the data moves from inconclusive $T \propto L$ to conclusive $T = k\sqrt{L}$. The model lesson tells teachers to 'avoid introducing the formal pendulum equation, because the laboratory activity is not designed to verify this known relationship' (NRC 2006, p. 129). Final-year students in Japan, Korea, Singapore and a good deal of the rest of the world have no such problem, and US students deserve better than a dumbed-down, HPS-free curriculum.

The graph of period against square root of length shows, in a manageable way, the dramatic impact of mathematics on physics; without the mathematical notion of square root, we see qualitative trends; utilising the square root, we see a precise, quantitative relationship. Further, this precise relationship will allow the pendulum to be connected with free fall, where distance of fall varies as the square of time. All of this is missed in the *Lab Report*, and also missed is the opportunity for richer pendulum-informed teaching of physics. What appears to have happened is what the NRC recognises in another publication: 'As educators, we are underestimating what young children are capable of as students of science – the bar is almost always set too low' (NRC 2007, p. vii). A pity they did not follow their own advice.

The Next Generation Science Standards

The NRC gives three reasons for producing updated NGSS in the US, one of which is that there is a 'growing body of research on learning and teaching in science' that can be utilised. The history of supposed science curriculum 'reforms' suggests that caution should be exercised about such claims. In the 1950s and 1960s, the 'growing body of research on learning and teaching' gave us behaviourism and behavioural objectives, which have disappeared without educational trace; in the 1970s, the 'growing body of research' gave us discovery learning and 'scientist for a day' teaching, both with minimal if not deleterious effects; in the 1980s and 1990s, the 'growing body of research' gave us constructivism, which swept all before it in university schools of education, but, in more sober light, its substantial philosophical and pedagogical failings have been recognised, as will be shown in Chapter 8. Good understanding of teaching and learning is certainly needed, but the improvement of curricula does not flow just from knowledge about how to better teach and learn material, but rather it flows from knowledge of what material to teach and learn, and where to place the topics and concepts in state and national standards. This is where a richer understanding of the history and philosophy of pendulum studies and utilisation (and, of course, of all other topics) can well contribute to science education. It can make for better curricula and for better connections between disciplinary strands in curricula.

The pendulum 'ticks all of the NGSS boxes', so to speak. Very young children, as shown in Japan, Korea and numerous other countries, can profitably and enjoyably engage with pendulum activities (Kwon *et al.* 2006, Sumida 2004).[35] It is not accidental that Jean Piaget used the pendulum for his investigation of the progressive development of children's scientific reasoning ability, especially their identification and control of variables (Bond 2004). The sophistication of pendulum activities and their relation with other areas and topics in science can be enhanced with progression through school; obvious connections with music, mathematics, technology and engineering can be made, and even connections with chemistry (De Berg 2006). The full range of process skills (data collection and representation, hypothesis generation), methodological skills (generating hypotheses, evaluating these against evidence, theory testing and so on) and model construction can all be cultivated using pendulum classes.[36]

It remains to be seen how the pendulum will feature in the final NGSS document, but the signs are not good. In the current (2012) draft, the pendulum is mentioned four times, and each time it is in connection with the transformation of energy from potential to kinetic forms. This is a level of abstraction way beyond what is needed or called for, it is beyond the life experience of the students and it reifies the role played by the pendulum in the history of physics and in its social utilisation. The draft document mentions Newton's laws, his theory of gravitation and the conservation of momentum, but no mention of the pendulum, which could so easily be used to make manifest and experiential each of these learning goals.

Conclusion

The pendulum case has been introduced in this chapter as an example where the HPS can contribute to even routine science education. It provides an opportunity to learn about science at the same time as one is learning the subject matter of science. With good HPS-informed teaching the pendulum-motion case enables students to appreciate the transition from common-sense and empirical descriptions characteristic of Aristotelian science, to the abstract, idealised and mathematical descriptions characteristic of the scientific revolution. The pendulum provides a manageable, understandable and straightforward way into scientific thinking and away from everyday and empirical thinking; it shows, at the same time, how scientific, idealised thinking nevertheless is connected with the world through controlled experiment.

The 'contextual' teaching of science, as suggested here, is not a retreat from serious, or hard, science, but the reverse. To understand what happened in the history of science takes effort. Further, it is appealing to students. A frequent refrain from intelligent students who do not go on with study in the sciences is that, 'science is too boring; we only work out problems'.[37] The history of human efforts to understand pendulum motion is far from boring: it is peopled by great minds, their debates are engaging, and the history provides a storyline on which to hang the complex theoretical development of science. As well as gaining an improved understanding of science, students taught in a contextual way can better understand the nature of science, and have something to remember long after the equation for the period of a pendulum is forgotten.

The following diagram (Figure 6.13), where the columns represent curriculum subjects and the circles topics within subjects, displays the integrative curricular function of history and philosophy.[38]

The content of the school day, or at least year, can be more of a tapestry, rather than a curtain of unconnected curricular beads. The latter is a well-documented problem with US science education, with its fabled 'mile wide and one inch thick' curricula (Kesidou & Roseman 2002). However, pendulum motion, if taught from a historical and philosophical perspective, allows connections to be made with topics in religion, history, mathematics, philosophy, music and literature, as well as other topics in the science programme. And such teaching promotes greater understanding of science, its methodology and its contribution to society and culture. However, such connections first need to be recognised by curriculum writers and by teachers charged with implementing curricula or achieving standards; this raises the

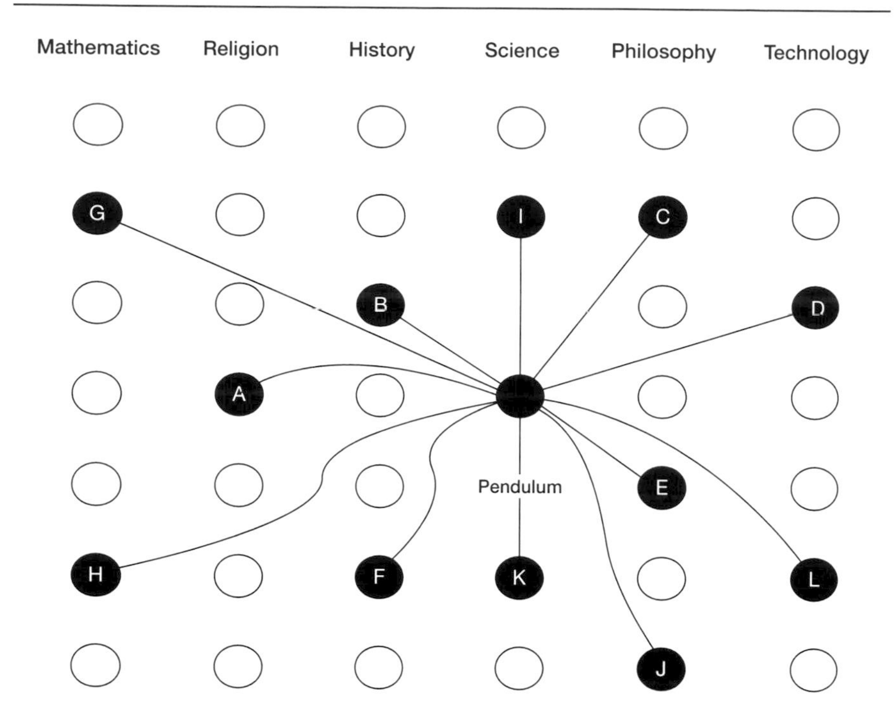

Figure 6.13 HPS-Informed Curricular Linkage. A: design argument; B: European voyages of discovery; C: Aristotelian physics and methodology; D: pendulum clock; E: idealisation and theory testing; F: timekeeping and social regulation; G: geometry of the circle; H: applied mathematics; I: measurement and standards; J: time; K: energy; L: geodesy

whole question of HPS in pre-service or in-service teacher education, but that is a different subject, for a different chapter.

Notes

1 This chapter has benefited from research published in Matthews (2000, 2001, 2014).
2 See the account of the philosophy and international impact of PSSC physics given in Chapter 3.
3 There is no adequate history of the medieval pendulum, but see Büttner (2008) and Hall (1978).
4 See Büttner (2008), Lefèvre (2001), Machamer (1998) and contributions to Renn (2001).
5 Galileo surely had precursors. There were medieval natural philosophers – John Buridan, Nicole Oresme, Thomas Bradwardine and his Merton College colleagues, and others – who utilised mathematics, but it was not fully engaged with their physics (natural philosophy) as mathematics was with Galileo. See Clagett (1959) and Moody (1975).
6 This will be a recurrent theme in the history of pendulum-related science, where it is seen that many different mechanical, biological and chemical processes will manifest the mathematical formulae for simple harmonic motion.
7 This vital idealising feature of Galileo's physics is discussed in Koertge (1977) and McMullin (1985).

8 Ronald Naylor (1980, pp. 367–371) and W.C. Humphreys (Humphreys 1967, pp. 232–234) discuss the letter in the context of Galileo's work on the law of fall.
9 Discussion and references can be found in White (1966, p. 109).
10 A reaction shared by many to the mathematisation of psychology and the 'boxes and arrows' of modern cognitive science.
11 This disjunct between 'lived experience' and scientific conceptualisation bears specifically on teaching the law of inertia, something that has been discussed in Chapter 5.
12 The foregoing problems of medieval Aristotelians are the same for contemporary 'discovery', 'enquiry' or 'minimal guidance' teaching; without guidance or instruction, there is no learning.
13 This episode is elaborated in Matthews (2001).
14 Accounts of the development of the standard metre can be found in Alder (1995, 2002) and Kula (1986, Chapters 21–23).
15 On the history of debate about the shape of the Earth, see Chapin (1994).
16 See Martins (1993).
17 On the pendulum's role in this unification, see especially Boulos (2006).
18 The history is dealt with in Matthews (2000, Chapters 6, 7).
19 Maurice Finocchiaro (1980, p. 149) has provided a table of fourteen philosophical topics to be found in Galileo's *Dialogue on the Two Chief World Systems* (1633/1953).
20 Jaroslav Pelikan provides a richly documented historical study of the same quest (Pelikan 1985).
21 For the physics and mathematics of these calculations, see Holton and Brush (2001, pp. 128–129).
22 This is a wonderful episode in the history of science. A great story can be made, even a drama. All the elements are there: powerful and prestigious figures, 'no name' outsiders, controversy over a big issue, mathematics and serious calculations, religion, final decision-making with ample opportunity to preserve the status quo. Sadly, however, the episode is little known and hardly ever taught.
23 Feinstein elaborates his personal position by writing:

> I would argue that knowing the shape of the earth is part of being 'well-read' in science – an excellent thing indeed, and one capable of giving us joy and satisfaction, but not one that all of our students will find useful.
>
> (Feinstein 2011, p. 183)

24 On this important conceptual advance, see Moody (1975, p. 403).
25 In recent decades, philosophers have returned to the Kantian high estimation of experiment in science and have written illuminating work on the topic. See at least: Hacking (1988) and contributions to Radder (2003) and Gooding *et al.* (1989).
26 An extensive study of the logic of idealisation in science is Nowak (1980). See also Laymon (1985) and Portides (2007).
27 Such objectivist epistemology is articulated in Popper (1972, Chapters 3, 4), Althusser and Balibar (1970), Baltas (1988, 1990), Chalmers (1976/2013), Mittelstrass (1972), Sneed (1979) and Suchting (1986).
28 See Irzik (1995) and Musgrave (1974).
29 See especially Hodson (1986) and Norris (1985).
30 On this tradition, see Yolton (2000).
31 Norwood Russell Hanson echoed Plato when he wrote: 'there is more to seeing than meets the eyeball' (Hanson 1958). For a wide range of twentieth-century readings in philosophy of perception, see Swartz (1965).
32 The distinction between scientific data and scientific phenomena has been developed at length by James Bogen and James Woodward (1988), James Woodward (1989), Ronald Laymon (1982, 1984) and James Brown (1996). For an interpretation of Galileo's work in terms of the data/phenomena distinction, see Hemmendinger (1984). The conceptualisation can considerably illuminate persistent issues in educational research, but it is rarely so utilised; see discussion in Brian Haig (2014, Chapter 2) and Matthews (2004).

33 For an analysis of the meaning of 'phenomena' in Newton's work, see Achinstein (1990).

34 See at least: Barnett (1998), Landes (1983), Turetzky (1998) and van Rossum (1996).

35 The front cover of an excellent pendulum booklet produced for Japanese elementary students is fully adorned with images of Galileo and Huygens – a nice comment on the universality of science and its ability to be embraced by cultures beyond its original European home.

36 On classroom utilisation of pendulum investigations as a way of teaching physics and cultivating and assessing scientific reasoning skills, see at least: Kanari and Millar (2004), Kwon *et al.* (2006), Stafford (2004), Zachos (2004) and the hundred-plus references in Gauld (2004).

37 This was the refrain repeatedly made some years ago when an Australian deans of science study reported on the top 10 per cent of school science achievers, most of whom did not go into tertiary science-related programmes.

38 The idea for this visual representation of the argument comes from an AAAS lecture of Gerald Holton, subsequently published as Holton (1995).

References

AAAS (American Association for the Advancement of Science): 1989, *Project 2061: Science for All Americans*, AAAS, Washington, DC. Also published by Oxford University Press, 1990.

Achinstein, P.: 1990, 'Newton's Corpuscular Query and Experimental Philosophy'. In P. Bricker and R.I.G. Hughes (eds) *Philosophical Perspectives on Newtonian Science*, MIT Press, Cambridge, MA, pp. 135–173.

Aczel, A.D.: 2003, *Pendulum: Léon Foucault and the Triumph of Science*, Atria Books, New York.

Aczel, A.D.: 2004, 'Leon Foucault: His Life, Times and Achievements', *Science & Education* 13(7–8), 675–687.

Alder, K.: 1995, 'A Revolution to Measure: The Political Economy of the Metric System in France'. In M.N. Wise (ed.) *The Values of Precision*, Princeton University Press, Princeton, NJ, pp. 39–71.

Alder, K.: 2002, *The Measure of All Things: The Seven-Year Odyssey that Transformed the World*, Little Brown, London.

Aldridge, B.G.: 1992, 'Project on Scope, Sequence, and Coordination: A New Synthesis for Improving Science Education', *Journal of Science Education and Technology* 1(1), 13–21.

Althusser, L. and Balibar, E.: 1970, *Reading Capital*, New Left Books, London.

Ariotti, P.E.: 1968, 'Galileo on the Isochrony of the Pendulum', *Isis* 59, 414–426.

Austin, J.L.: 1961, *Philosophical Papers*, Oxford University Press, Oxford, UK.

Ayer, A.J.: 1955, *The Foundations of Empirical Knowledge*, Macmillan, London.

Ayer, A.J.: 1956, *The Problem of Knowledge*, Penguin, Harmondsworth, UK.

Baltas, A.: 1988, 'On the Structure of Physics as a Science'. In D. Batens and J.P. van Bendegens (eds) *Theory and Experiment*, Reidel, Dordrecht, The Netherlands, pp. 207–225.

Baltas, A.: 1990, 'Once Again on the Meaning of Physical Concepts'. In P. Nicolacopoulos (ed.) *Greek Studies in the Philosophy and History of Science*, Kluwer Academic Publishers, Dordrecht, The Netherlands, pp. 293–313.

Barnett, J.E.: 1998, *Time's Pendulum: From Sundials to Atomic Clocks, the Fascinating History of Timekeeping and How Our Discoveries Changed the World*, Harcourt Brace, New York.

Bedini, S.A.: 1986, 'Galileo and Scientific Instrumentation'. In W.A. Wallace (ed.) *Reinterpreting Galileo*, *Catholic University of America Press*, Washington, DC, pp. 127–154.

Bedini, S.A.: 1991, *The Pulse of Time: Galileo Galilei, the Determination of Longitude, and the Pendulum Clock*, Olschki, Florence, Italy.
Bensky, T.J.: 2010, 'The Longitude Problem From the 1700s to Today: An International and General Education Physics Course', *American Journal of Physics* 78(1), 40–46.
Bogen, J. and Woodward, J.: 1988, 'Saving the Phenomena', *The Philosophical Review* XCVII(3), 303–350.
Bond, T.G.: 2004, 'Piaget and the Pendulum', *Science & Education* 13(4–5), 389–399.
Boulos, P.J.: 2006, 'Newton's Path to Universal Gravitation: The Role of the Pendulum', *Science & Education* 15(6), 577–595.
Brown, J.R.: 1996, 'Phenomena'. In R.S. Cohen, R. Hilpinen and Q. Renzong (eds) *Realism and Anti-Realism in the Philosophy of Science*, Kluwer Academic Publishers, Dordrecht, The Netherlands, pp. 117–129.
Butterfield, H.: 1949/1957, *The Origins of Modern Science 1300–1800*, G. Bell, London.
Büttner, J.: 2008, 'The Pendulum as a Challenging Object in Early-Modern Mechanics'. In W.R. Laird and S. Roux (eds) *Mechanics and Natural Philosophy before the Scientific Revolution*, Springer, Dordrecht, The Netherlands, pp. 223–237.
Cantor, G.: 1991, *Michael Faraday: Sandemanian and Scientist*, St. Martin's Press, New York.
Cartwright, N.: 1983, *How the Laws of Physics Lie*, Clarendon Press, Oxford, UK.
Chalmers, A.F.: 1976/2013, *What Is This Thing Called Science?* 4th edn, University of Queensland Press, St Lucia.
Chapin, S.L.: 1994, 'Geodesy'. In I. Grattan-Guinness (ed.) *Companion Encyclopedia of the History and Philosophy of the Mathematical Sciences*, Routledge, London, pp. 1089–1100.
Clagett, M.: 1959, *The Science of Mechanics in the Middle Ages*, University of Wisconsin Press, Madison, WI.
De Berg, K.C.: 2006, 'Chemistry and the Pendulum: What Have They To Do With Each Other?', *Science & Education* 15(6), 619–641.
Drabkin, I.E.: 1938, 'Notes on the Laws of Motion in Aristotle', *American Journal of Philology* 59, 60–84.
Drake, S.: 1978, *Galileo at Work*, University of Chicago Press, Chicago, IL. Reprinted Dover Publications, New York, 1996.
Drake, S.: 1990, 'The Laws of Pendulum and Fall'. In his *Galileo: Pioneer Scientist*, University of Toronto Press, Toronto, pp. 9–31.
Dugas, R.: 1988, *A History of Mechanics*, Dover, New York (originally published1955).
Falomo, L., Albanesi, G. and Bevilacqua, F.: 2014, 'Museum Heroes All: The Pavia Approach to School-Science Museum Interactions', *Science & Education* 23(4), 761–780.
Fantoli, A.: 1994, *Galileo: For Copernicanism and for the Church* (trans. G.V. Coyne), Vatican Observatory Publications, Vatican City (distributed by University of Notre Dame Press).
Feinstein, N.: 2011, 'Salvaging Science Literacy', *Science Education* 95, 168–185.
Finocchiaro, M.A.: 1980, *Galileo and the Art of Reasoning*, Reidel, Dordrecht, The Netherlands.
Fleck, L.: 1935/1979, *Genesis and Development of a Scientific Fact*, T.J. Trenn and R.K. Merton (eds), University of Chicago Press, Chicago, IL.
Frank, P.: 1947/1949, 'The Place of Philosophy of Science in the Curriculum of the Physics Student', *American Journal of Physics* 15 (3), 202–218. Reprinted in his *Modern Science and Philosophy*, Harvard University Press, Harvard, MA, pp. 228–259.
Galileo, G.: 1590/1960, '*De Motu*'. In I.E. Drabkin and S. Drake (eds) *Galileo Galilei On Motion and On Mechanics*, University of Wisconsin Press, Madison, WI, pp. 13–114.
Galileo, G.: 1602/1978, 'Letter to Guidobaldo del Monte'. In S. Drake *Galileo at Work: His Scientific Biography*, Dover Publications, Mineola, NY, pp. 69–71.
Galileo, G.: 1633/1953, *Dialogue Concerning the Two Chief World Systems* (trans. S. Drake), University of California Press, Berkeley, CA (2nd revised edition, 1967).

Galileo, G.: 1638/1954, *Dialogues Concerning Two New Sciences* (trans. H. Crew and A.de Salvio), Dover Publications, New York (originally published 1914).
Gauld, C.F.: 1998, 'Solutions to the Problem of Impact in the 17th and 18th Centuries and Teaching Newton's Third Law Today', *Science & Education* 7(1), 49–67.
Gauld, C.F.: 2004, 'Pendulums in Physics Education Literature: A Bibliography', *Science & Education* 13(7–8), 811–832.
Gauld, C.F.: 2006, 'Newton's Cradle in Physics Education', *Science & Education* 15(6), 597–617.
Giere, R.N.: 1988, *Explaining Science: A Cognitive Approach*, University of Chicago Press, Chicago, IL.
Giere, R.N.: 1999, *Science Without Laws*, University of Chicago Press, Chicago, IL.
Gooding, D., Pinch, T. and Schaffer, S. (eds): 1989, *The Uses of Experiment*, Cambridge University Press, Cambridge, UK.
Guadagni, C.A.: 1764, *Specimen Experimentorum Naturalium*, Carotti, Pisa, Italy.
Hacking, I.: 1988, 'Philosophers of Experiment'. In A. Fine and J. Leplin (eds) *PSA* 2, 147–156.
Haig, B.D.: 2014, *Investigating the Psychological World: Scientific Method in the Behavioral Sciences*, MIT Press, Cambridge, MA.
Hall, B.S.: 1978, 'The Scholastic Pendulum', *Annals of Science* 35, 441–462.
Hanson, N.R.: 1958, *Patterns of Discovery*, Cambridge University Press, Cambridge, UK.
Hanson, N.R.: 1959, 'Broad and the Laws of Dynamics'. In P.A. Schilpp (ed.) *The Philosophy of C.D. Broad*, Tudor Publishing Company, New York, pp. 281–312.
Hemmendinger, D.: 1984, 'Galileo and the Phenomena: On Making the Evidence Visible'. In R.S. Cohen and M.W. Wartofsky (eds) *Physical Sciences and the History of Physics*, Reidel, Dordrecht, The Netherlands, pp. 115–143.
Herivel, J: 1965, *The Background to Newton's 'Principia'*, Clarendon Press, Oxford, UK.
Hodson, D.: 1986, 'Rethinking the Role and Status of Observation in Science Education', *Journal of Curriculum Studies* 18(4), 381–396.
Holton, G.: 1995, 'How Can Science Courses Use the History of Science?' In his *Einstein, History and Other Passions*, American Institute of Physics, Woodbury, NY, pp. 257–264.
Holton, G. and Brush, S.G.: 2001, *Physics, the Human Adventure. From Copernicus to Einstein and Beyond*, Rutgers University Press, New Brunswick, NJ.
Hume, D.: 1739/1888, *A Treatise of Human Nature: Being an Attempt to Introduce the Experimental Method of Reasoning into Moral Subjects*, Clarendon Press, Oxford, UK.
Humphreys, W.C.: 1967, 'Galileo, Falling Bodies and Inclined Planes: An Attempt at Reconstructing Galileo's Discovery of the Law of Squares', *British Journal for the History of Science* 3(11), 225–244.
Huygens, C.: 1673/1986, *Horologium Oscillatorium. The Pendulum Clock or Geometrical Demonstrations Concerning the Motion of Pendula as Applied to Clocks* (trans. R.J. Blackwell), Iowa State University Press, Ames, IA.
Irzik, G.: 1995, 'Popper's Epistemology and World Three'. In I. Kuçuradi and R.S. Cohen (eds) *The Concept of Knowledge: The Ankara Seminar*, Kluwer Academic Publishers, Dordrecht, The Netherlands, pp. 83–95.
Kanari, Z. and Millar, R.: 2004, 'Reasoning From Data: How Students Collect and Interpret Data in Science Investigations', *Journal of Research in Science Teaching* 41(7), 748–769.
Kant, I.: 1787/1933, *Critique of Pure Reason*, 2nd edn (trans. N.K. Smith), Macmillan, London (1st edition, 1781).
Kesidou, S. and Roseman, J.E.: 2002, 'How Well Do Middle School Science Programs Measure Up? Findings from Project 2061's Curriculum Review', *Journal of Research in Science Teaching* 39(6), 522–549.
Koertge, N.: 1977, 'Galileo and the Problem of Accidents', *Journal of the History of Ideas* 38, 389–409.
Koyré, A.: 1957, *From the Closed World to the Infinite Universe*, The Johns Hopkins University Press, Baltimore, MD.

Koyré, A.: 1968, *Metaphysics and Measurement*, Harvard University Press, Cambridge, MA.

Kuhn, T.S.: 1970, *The Structure of Scientific Revolutions*, 2nd edn, Chicago University Press, Chicago, IL (1st edition, 1962).

Kula, W.: 1986, *Measures and Man*, Princeton University Press, Princeton, NJ.

Kwon, Y.-J., Jeong, J.-S. and Park, Y.-B.: 2006, 'Roles of Abductive Reasoning and Prior Belief in Children's Generation of Hypotheses about Pendulum Motion', *Science & Education* 15(6), 643–656.

Lakatos, I.: 1970, 'Falsification and the Methodology of Scientific Research Programmes'. In I. Lakatos and A. Musgrave (eds) *Criticism and the Growth of Knowledge*, Cambridge University Press, Cambridge, UK, pp. 91–196.

Landes, D.S.: 1983, *Revolution in Time. Clocks and the Making of the Modern World*, Harvard University Press, Cambridge, MA.

Laymon, R.: 1982, 'Scientific Realism and the Hierarchical Counterfactual Path from Data to Theory'. In P.D Asquith and T. Nickles (eds) *PSA*, pp. 107–121.

Laymon, R.: 1984, 'The Path from Data to Theory'. In J. Leplin (ed.) *Scientific Realism*, University of California Press, Berkeley, CA, pp.108–123.

Laymon, R.: 1985, 'Idealizations and the Testing of Theories by Experimentation'. In P. Achinstein and O. Hannaway (eds) *Observation, Experiment, and Hypothesis in Modern Physical Science*, MIT Press, Cambridge, MA, pp. 147–173.

Lederman, N., Abd-el-Khalick, F., Bell, R.L. and Schwartz, R.S.: 2002, 'Views of Nature of Science Questionnaire: Towards Valid and Meaningful Assessment of Learners' Conceptions of the Nature of Science', *Journal of Research in Science Teaching* 39, 497–521.

Lefèvre, W.: 2001, 'Galileo Engineer: Art and Modern Science'. In J. Renn (ed.) *Galileo in Context*, Cambridge University Press, Cambridge, UK, pp. 11–24.

Lennox, J.G.: 1986, 'Aristotle, Galileo, and the "Mixed Sciences"'. In W.A. Wallace (ed.) *Reinterpreting Galileo*, Catholic Univerity of America Press, Washington, DC, pp. 29–51.

Machamer, P.: 1998, 'Galileo's Machines, His Mathematics, and His Experiments'. In P. Machamer (ed.) *The Cambridge Companion to Galileo*, Cambridge University Press, Cambridge, UK, pp. 53–79.

Martins, R. de A.: 1993, 'Huygens's Reaction to Newton's Gravitational Theory'. In J.V. Field and F.A.J.L. James (eds) *Renaissance and Revolution: Humanists, Scholars, Craftsmen and Natural Philosophers in Early Modern Europe*, Cambridge University Press, Cambridge, UK, pp. 203–214.

Matthews, M.R.: 1998, 'Opportunities Lost: The Pendulum in the USA National Science Education Standards', *Journal of Science Education and Technology* 7(3), 203–214.

Matthews, M.R.: 2000, *Time for Science Education: How Teaching the History and Philosophy of Pendulum Motion can Contribute to Science Literacy*, Kluwer Academic Publishers, New York.

Matthews, M.R.: 2001, 'Methodology and Politics in Science: The Case of Huygens' 1673 Proposal of the Seconds Pendulum as an International Standard of Length and Some Educational Suggestions', *Science & Education* 10(1–2), 119–135.

Matthews, M.R.: 2004, 'Data, Phenomena and Theory: How Clarifying the Distinction can Illuminate the Nature of Science'. In K. Alston (ed.) *Philosophy of Education 2003*, US Philosophy of Education Society, Champaign, IL, pp. 283–292.

Matthews, M.R.: 2014, 'Pendulum Motion: A Case Study in How History and Philosophy can Contribute to Science Education'. In M.R. Matthews (ed.) *International Handbook of Research in History, Philosophy and Science Teaching*, Springer, Dordrecht, The Netherlands, pp. 19–56.

McMullin, E.: 1985, 'Galilean Idealization', *Studies in the History and Philosophy of Science* 16, 347–373.

Meli, D.B.: 2006, *Thinking with Objects: The Transformation of Mechanics in the Seventeenth Century*, Johns Hopkins University Press, Baltimore, MD.

Mittelstrass, J.: 1972, 'The Galilean Revolution: The Historical Fate of a Methodological Insight', *Studies in the History and Philosophy of Science* 2, 297–328.

Monte, G. del: 1581/1969, 'Mechaniche'. In S. Drake and I.E. Drabkin (eds) *Mechanics in Sixteenth-Century Italy*, University of Wisconsin Press, Madison, WI, pp. 241–329.
Moody, E.A.: 1975, *Studies in Medieval Philosophy, Science and Logic*, University of California Press, Berkeley, CA.
Musgrave, A.: 1974, 'The Objectivism of Popper's Epistemology'. In P.A. Schilpp (ed.) *The Philosophy of Karl Popper*, Open Court Publishing, LaSalle, IL, pp. 560–596.
Naylor, R.H.: 1974, 'Galileo's Simple Pendulum', *Physis* 16, 23–46.
Naylor, R.H.: 1980, 'The Role of Experiment in Galileo's Early Work on the Law of Fall', *Annals of Science* 37, 363–378.
Naylor, R.H.: 1989, 'Galileo's Experimental Discourse', in D. Gooding, T. Pinch and S. Schaffer (eds) *The Uses of Experiment*, Cambridge University Press, Cambridge, UK, pp. 117–134.
Newton, I.: 1729/1934, *Mathematical Principles of Mathematical Philosophy* (trans. A. Motte, revised F. Cajori), University of California Press, Berkeley, CA.
Norris, S.P.: 1985, 'The Philosophical Basis of Observation in Science and Science Education', *Journal of Research in Science Teaching* 22(9), 817–833.
Nowak, L.: 1980, *The Structure of Idealization*, Reidel, Dordrecht, The Netherlands.
NRC (National Research Council): 1996, *National Science Education Standards*, National Academies Press, Washington, DC.
NRC (National Research Council): 2006, *America's Lab Report: Investigations in High School Science*, National Academies Press, Washington, DC.
NRC (National Research Council): 2007, *Taking Science to School. Learning and Teaching Science in Grades K-8*, National Academies Press, Washington, DC.
Olmsted, J.W.: 1942, 'The Scientific Expedition of Jean Richer to Cayenne (1672–1673)', *Isis* 34, 117–128.
Pelikan, J.: 1985, *Jesus Through the Centuries: His Place in the History of Culture*, Yale University Press, New Haven, CT.
Pogo, A.: 1935, 'Gemma Frisius, His Method of Determining Differences of Longitude by Transporting Time-Pieces (1530) and His Treatise on Triangulation (1533)', *Isis* 22(64), 469–485.
Popper, K.R.: 1934/1959, *The Logic of Scientific Discovery*, Hutchinson, London.
Popper, K.R.: 1972, *Objective Knowledge*, Clarendon Press, Oxford, UK.
Portides, D.: 2007, 'The Relation Between Idealisation and Approximation in Scientific Model Construction', *Science & Education* 16(7–8), 699–724.
PSSC (Physical Science Study Committee): 1960, *Physics*, D.C. Heath, Boston, MA.
Quine, W.V.O.: 1953, *From a Logical Point of View*, Harper & Row, New York.
Radder, H. (ed.): 2003, *The Philosophy of Scientific Experimentation*, University of Pittsburgh Press, Pittsburgh, PA.
Renn, J. (ed.): 2001, *Galileo in Context*, Cambridge University Press, Cambridge, UK.
Roth, M.-W. and Roychoudhury, A.: 1994, 'Physics Students' Epistemologies and Views about Knowing and Learning', *Journal of Research in Science Teaching* 31(1), 5–30.
Schweitzer, A.: 1910/1954, *The Quest of the Historical Jesus: A Critical Study of its Progress from Reimarus to Wrede*, 3rd edn, Adam and Charles Black, London.
Scriven, M.: 1961, 'The Key Property of Physical Laws – Inaccuracy'. In H. Feigl and G. Maxwell (eds) *Current Issues in the Philosophy of Science*, Holt, Rinehart & Winston, New York, pp. 91–101.
Sneed, J.D.: 1979, *The Logical Structure of Mathematical Physics*, 2nd edn, Reidel, Dordrecht, The Netherlands.
Sobel, D.: 1995, *Longitude: The True Story of a Lone Genius Who Solved the Greatest Scientific Problem of His Time*, Walker Publishing Company, New York.
Stafford, E.: 2004, 'What the Pendulum can Tell Educators About Children's Scientific Reasoning', *Science & Education* 13(7–8), 757–790.
Suchting, W.A.: 1986, 'Marx and "the Problem of Knowledge"'. In his *Marx and Philosophy*, Macmillan, London, pp. 1–52.

Suchting, W.A.: 1995, 'The Nature of Scientific Thought', *Science & Education* 4(1), 1–22.
Sumida, M.: 2004, 'The Reproduction of Scientific Understanding About Pendulum Motion in the Public', *Science & Education* 13(4–5), 473–492.
Swartz, R.J. (ed.): 1965, *Perceiving, Sensing and Knowing*, Doubleday, New York.
Tobin, K. (ed.): 1993, *The Practice of Constructivism in Science and Mathematics Education*, AAAS Press, Washington, DC.
Tobin, W.: 2003, *The Life and Science of Léon Foucault: The Man Who Proved the Earth Rotates*, Cambridge University Press, Cambridge, UK.
Turetzky, P.: 1998, *Time*, Routledge, London.
van Rossum, G.D.: 1996, *History of the Hour: Clocks and Modern Temporal Orders*, Chicago University Press, Chicago, IL.
Westfall, R.S.: 1980, *Never at Rest: A Biography of Isaac Newton*, Cambridge University Press, Cambridge, UK.
Westfall, R.S.: 1990, 'Making a World of Precision: Newton and the Construction of a Quantitative Physics'. In F. Durham and R.D Purrington (eds) *Some Truer Method. Reflections on the Heritage of Newton*, Columbia University Press, New York, pp. 59–87.
White, L.: 1966, 'Pumps and Pendula: Galileo and Technology'. In C.L. Golino (ed.) *Galileo Reappraised*, University of California Press, Berkeley, CA, pp. 96–110.
Wise, M.N. (ed.): 1995, *The Values of Precision*, Princeton University Press, Princeton, NJ.
Wittgenstein, L: 1958, *Philosophical Investigations*, Basil Blackwell, Oxford, UK.
Wolf, F.A.: 1981, *Taking the Quantum Leap*, Harper & Row, New York.
Woodward, J.: 1989, 'Data and Phenomena', *Synthese* 79, 393–472.
Yoder, J.G.: 1988, *Unrolling Time: Christiaan Huygens and the Mathematization of Nature*, Cambridge University Press, Cambridge, UK.
Yolton, J.W.: 2000, *Realism and Appearances: An Essay in Ontology*, Cambridge, Cambridge University Press, UK.
Zacharias, J.R.: 1964, 'Curriculum Reform in the USA'. In S.C. Brown, N. Clarke and J. Tiomno (eds) *Why Teach Physics: International Conference on Physics in General Education*, MIT Press, Cambridge, MA, pp. 66–70.
Zachos, P.: 2004, 'Pendulum Phenomena and the Assessment of Scientific Inquiry Capabilities', *Science & Education* 13(7–8), 743–756.

Chapter 7

History and Philosophy in the Classroom

Joseph Priestley and the Discovery of Photosynthesis[1]

There are three important social and educational considerations that justify dealing with Joseph Priestley in school science programmes:

- First, schools are asked to address pressing environmental problems and especially the 'goodness of air' (to use Priestley's phrase) and, thus, they need to teach about the process of photosynthesis, something on which Priestley shed so much early understanding.
- Second, NOS goals are included in numerous international curricula, and Priestley's writings and practice well illustrate many of the essential features of NOS.
- Third, there is a widespread concern in education and in culture with reappraising and re-examining the tenets of the European Enlightenment tradition, and in particular its universalist, naturalist and secular commitments. Michael Peters, quoted earlier, well captures this widespread concern (Peters 1995, pp. 327–328). Priestley's life and achievements provide a good case for evaluating what is dead and what is living from the original Enlightenment claims and achievements.

Priestley made significant contributions to all three areas, and, pleasingly, he wrote simply and engagingly, with 'the public' in mind; he wrote in such a way that readers could themselves experiment and observe as he was doing; he might be considered the first advocate of 'science for all'. So, infusing historical and philosophical dimensions into this standard curriculum topic allows, not just the content of science to be learned, but also important things about science, about important contributors to its history, and, consequently, the nature of science can be better appreciated.

Photosynthesis is a fundamental process for life on earth, with many biologists rating it the most important natural process; as such, it has long been a core part of the school biology curriculum. However, it is well known that children at all levels have great difficulty comprehending and understanding photosynthesis: students' conceptual understanding of the process routinely lags behind what might be anticipated from their grade level and from the curriculum they have been taught.[2] Given the current, well-publicised environmental problems concerning the state of the atmosphere, carbon trading, CO_2 emissions, greenhouse gases, forest preservation and so on, then correcting the inadequate student and general public knowledge of such a fundamental natural process becomes more pressing.

The science part is tied up with two basic, complementary processes: photosynthesis[3] and respiration:

> *Photosynthesis*: carbon dioxide + water + energy (light) → organic compounds (starch) + oxygen
> *Respiration*: organic compounds + oxygen → carbon dioxide + water + energy

The first process represents both 'carbon capture' and the restoration of air; when buried coal seams are dug up, or forests are cut down and burned, there is massive carbon release via the second process. The two processes are major components of the Earth's 'carbon cycle'. Clearly, it is important for students to learn about these processes and their wider social, economic, cultural and ethical dimensions and impacts. This is part of responsible citizenship. However, additionally, there is great value in learning about how these fundamental processes came to be discovered and understood; such learning provides appreciation and understanding of the nature of science and the scientific enterprise.

As was mentioned in Chapter 2, Priestley was one of the foremost scientists (natural philosophers) of the eighteenth century, he was a lifelong devout Christian minister and he was an energetic exponent of Enlightenment principles. In particular, Priestley advocated: the application of the methodology of the new Newtonian science to all fields of enquiry – historical, theological, educational, ethical; the separation of Church and state; freedom of speech; freedom of religion; decriminalisation of religious belief and practice; and the freedom of science (including historical studies of religious scripture) from political and religious control. As with Locke, Kant, Rousseau and all Enlightenment figures, he was a ceaseless advocate of education and, specifically, of what would now be called 'science education'.[4] Thus, the teaching of photosynthesis, with attention to the historical and philosophical dimensions of Priestley's life and work, allows each of the above current educational concerns to be productively addressed.

Some Appraisals of Priestley

Modern appreciation of Priestley has been blighted by the harsh and unfair judgement of Thomas Kuhn in his best-selling *Structure of Scientific Revolutions* (Kuhn 1970). In a famous passage, Kuhn writes of the irrationality of paradigm change and of old paradigms just dying out until, 'at last only a few elderly hold-outs remain'. He then singularly names Priestley as an example 'of the man who continues to resist after his whole profession has been converted' and adds that such a man 'has *ipso facto* ceased to be a scientist' (Kuhn 1970, p. 159). This outrageous charge 'blackened' Priestley's reputation in the academic world; Kuhn's has become the widely accepted obituary for Priestley – the stubborn old man who held on to belief in a peculiar phlogiston substance and who resisted the dawning bright light of Lavoisierian chemistry. Pleasingly, some historians and philosophers have provided extensive studies that refute Kuhn's caricature of Priestley, but, unfortunately, their work is not translated into more than twenty languages, nor set as class reading in countless thousands of courses, nor read by millions.

A more generous and accurate assessment of Priestley was given by Frederic Harrison, in his Introduction to a nineteenth-century edition of Priestley's *Scientific Correspondence*, as follows:

> If we choose one man as a type of the intellectual energy of the eighteenth century, we could hardly find a better than Joseph Priestley, though his was not the greatest mind of the century. His versatility, eagerness, activity, and humanity; the immense range of his curiosity in all things, physical, moral, or social; his place in science, in theology, in philosophy, and in politics; his peculiar relation to the Revolution, and the pathetic story of his unmerited sufferings, may make him the hero of the eighteenth century.
>
> (Bolton 1892, Introduction)

Priestley's Life

There has been a good deal written about Priestley's life and accomplishments that teachers can draw on for elaborating contemporary lessons.[5] Priestley was born in Yorkshire in 1733 and died on 6 February 1804 in the United States, in the small, isolated backwoods town of Northumberland in the state of Pennsylvania. Although the bicentenary of his death was rightly marked in historical and chemical circles, it unfortunately went unnoticed in education circles. This is a pity, as Priestley was a dedicated teacher, educationalist and, perhaps, the first modern science teacher. As well as teaching, preaching and researching, Priestley wrote a number of influential works on the theory and practice of education.

Priestley had a severe and disturbing Calvinistic upbringing.[6] In his late teenage years, being a religious dissenter and, hence, barred from Oxford and Cambridge universities, he attended Daventry Academy, where, as a teenager, he read Locke, Newton, Hartley and many of the major philosophical, scientific and religious works of the time. It was an institution where the 'serious pursuit of truth' was the preoccupation (Priestley 1806/1970, p. 75).[7] This was in marked contrast to the scholarly climate in the established universities.[8] He acquired fluency in Greek, Latin, Syriac and a number of European languages, including, later, High Dutch.

At 22 years of age, he was ordained a Dissenting minister, the duties of which vocation were the central preoccupation of his adult life. He ministered in a number of small rural towns, where he also established schools, being perhaps the first ever teacher of science to engage students in laboratory work (Schofield 1997, p. 79). In his late twenties, he taught language, history, logic and literature at the Warrington Academy, where he also began reading contemporary works in chemistry and electricity that supplemented his earlier readings of Newton's optics and astronomy.

At age 34 years, he was called as minister to the Presbyterian Chapel in Leeds, which was a centre for Yorkshire's newly born and thriving commercial and industrial life. He left Leeds and worked for 5 years as a secretary and children's tutor for Lord Shelburne, a prominent liberal English politician who negotiated the Treaty of Paris that ended the American Revolution. During this employment, he travelled in Europe, famously meeting, in October 1774 in Paris, with Antoine Lavoisier, with whose name Priestley's has ever since been entwined, because of controversy over the discovery of oxygen and Priestley's dogged refusal to accept the latter's 'new' chemistry.[9] Priestley lent support to the American and French Revolutions, seeing both of them as the victory of liberty and freedom over the stultifying, autocratic power of the established church (be it Roman Catholic in the *ancien régime* of France, Anglican in the United Kingdom or Lutheran in Germany and Scandanavia) and state.[10] He publicly rejected Trinitarian belief, notwithstanding that such denial was a capital offence at the time, and founded the Unitarian sect. In detailed publications, he argued that the Church's triune doctrine was the product of Hellenistic philosophical corruption of the early Christian church (Priestley 1786). Newton, of course, held the same position, but never advertised the matter; Priestley's liberalism, and ultimately his epistemological convictions, led him to very public affirmation of Unitarianism: truth emerges from dispute and defence of positions.

England was never a comfortable place for supporters of the 1789 French Revolution, and it was distinctly less so after the initial 'middle-class' revolution gave way to the Paris Commune in 1792, and after England joined the counter-revolutionary, reactionary coalition in war against France in 1793. So, at age 61 years, after his laboratory and library were destroyed by an enraged 'King and Church' reactionary mob, and after various close dissenter friends and political liberals were transported as convicts to Botany Bay, Priestley fled England in 1794 and travelled to Northumberland in a remote rural corner of Pennsylvania, where he spent the last decade of his life writing, ministering and productively engaging with prominent politicians, especially Thomas Jefferson.[11] He died in 1804, in his own, still-standing home.[12]

Priestley's Publications

Christian ministry was the most important thing in Priestley's life, and he kept affirming this, from his ordination in 1755 at age 22 to his death at age 70.[13] However, along with his active clerical life, Priestley published an enormous number of substantial and authoritative works across a wide range of fields: these included over 200 books, pamphlets and articles on history of science (specifically of electricity and optics), political theory, theology, biblical criticism, church history, theory of language, philosophy of education, rhetoric, as well as chemistry, for which he is now best known. Included in this corpus are about twenty substantial, multivolume works, many of which went into second, third and fourth editions. Priestley's *Collected Works* (twenty-five volumes), which do not include all his scientific publications, are in Rutt (1817–1832/1972). Some of Priestley's scientific correspondence is in Robert Schofield's edited anthology (1966). The most accessible source for a range of his major writings is still the 350-page anthology edited by John Passmore (Passmore 1965).[14]

Priestley and the Enlightenment

Priestley's life spanned the core years of the European Enlightenment, which was inspired by the achievements and writings of the new science of Bacon, Galileo, Huygens and, above all, Isaac Newton.[15] Indeed, the whole Enlightenment began with the conviction of Newton, Locke, Hume and others that the methods of the new science should be applied in the moral and political sciences. Newton expressed the matter in his *Opticks* as follows: 'If natural philosophy in all its Parts, by pursuing this Method, shall at length be perfected, the Bounds of Moral Philosophy will be also enlarged' (Newton 1730/1979, p. 405).[16] David Hume echoed this expectation with the subtitle of his famous *Treatise on Human Nature*, which reads, *Being an Attempt to Introduce the Experimental Method of Reasoning into Moral Subjects*. In the Preface, he says he is following the philosophers of England who have 'began to put the science of man on a new footing' (Hume 1739/1888, p. xxi).

At Daventry Academy, Priestley read Newton and Locke and their major expositors and there began developing his particular variant of the Enlightenment worldview. He was a devout Christian, not a Deist believing in an impersonal Creator or 'Intelligent Designer', who set the world going then left it alone. He believed in a personal God and concurrently developed a rationalist and materialist worldview that was consistently brought to bear upon his scientific and other investigations and was in turn reinforced by these investigations.

Priestley was a polymath and, in current terminology, a 'public intellectual'; he had staggeringly wide interests, but, more than this, he explicitly sought for coherence and intellectual unity in his scholarly, personal, religious and political activity. Newton had established that the single law of attraction applied on Earth and in the heavens. Priestley thought the same simplicity of law would apply through the social and mental (psychological) realms as well; this, in part, because there was only a single substance, matter, throughout all realms. He was a forceful advocate of the materialist tradition in the Enlightenment. He was an ontological monist, rejecting all dualisms in natural philosophy, psychology and religion.[17] He did not believe there were a multiplicity of kinds of substance in the world: recourse to 'imponderable fluids', including Lavoisier's caloric, to explain magnetic, electric, optical or heat phenomena, was both unnecessary (as they explained nothing, and the phenomena could be explained by suitable movement of particles, as was maintained by the atomists) and fanciful, as no such entities (non-material 'fluids') existed. As he said of supposed electrical fluid: 'there is no electric fluid at all, and that electrification is only some [new] modification of the matter of which any body consisted before that operation' (Schofield 1966, p. 58).[18]

For Priestley, his epistemology (empiricism) related to his ontology (materialist monism), and both related to his theology (Unitarianism) and to his psychology (associationism). All the foregoing bore upon his political and social theory (liberalism). He was a consciously synoptic or systematic thinker: knowledge and life were a whole and had to relate consistently. Whether Priestley achieved the coherence he sought has been a matter of considerable debate. From the very outset, many have disputed the coherence of Priestley's claimed conjunction of ontological materialism and Christian belief. Aiming for coherence and arriving at it are, of course, two different things.

Priestley was committed to a Christian worldview that was informed by natural philosophy. The worldview is developed throughout his work; one partial expression is in the Preface to his *History of Electricity*:

> A philosopher [scientist] ought to be something greater, and better than another man. The contemplation of the works of God should give sublimity to his virtue, should expand his benevolence, extinguish every-thing mean, base, and selfish in his nature, give dignity to all his sentiments, and teach him to aspire to the moral perfections of the great author of all things. . . . A life spent in the contemplation of the productions of divine power, wisdom, and goodness, would be a life of devotion.
>
> (Priestley 1767/1775, Preface)

With such a worldview, the pursuit of what we now call 'scientific' knowledge was a religious virtue, indeed almost a religious obligation: to ignore the world was to ignore God's handiwork; to find out about the world was to give respect and honour to God. Certainly, there was a religious obligation not to thwart the advance of knowledge, not to stand in the way of, or suppress, truth. And, as authoritarianism and absolutism were antithetical to the pursuit of truth, both had to be opposed, in churches and states. For Priestley, religion and epistemology were combined or codependent. Contrary to many popular present-day views, for Priestley, religious knowledge was not a different kind of knowledge with a different epistemology; he was not an eighteenth-century exponent of 'non-overlapping magisteria' (NOMA).

Priestley's First Steps Towards the Discovery of Photosynthesis

Priestley did not begin serious chemical studies until his early 30s, during his ministry at the Leeds Presbyterian Chapel (1767–1773). In quick succession, by utilising a new method of collecting 'airs' over water and mercury, and by utilising a new and massive burning lens as a source of heat,[19] Priestley created, isolated and listed properties of a dozen or more of the major 'airs'.

The 1772 Royal Society Talks

The experiments and investigations of airs, conducted in Leeds by Priestley, were announced to the scholarly world in a series of talks he delivered to the Royal Society in London, in March 1772. The talks subsequently were published as his famous 118-page paper in the Society's *Philosophical Transactions* of the same year – 'Observations on Different Kinds of Air' (Priestley 1772a).[20] This paper was translated into many European languages; it was widely read, including by Lavoisier; and, in 1773, it was awarded the coveted Copley Medal of the Royal Society – the eighteenth-century equivalent of the Nobel Prize.[21] The paper was elaborated, with further experiments, in his three-volume *Experiments and Observations on Different Kinds of Air* (Priestley 1772b). These publications established Priestley as the undisputed 'father of pneumatic chemistry'.

On 21 February 1770, he wrote to his lifelong intimate friend and fellow Unitarian Theophilus Lindsey (1723–1808) that, 'he was now taking up some of Dr. Hale's inquiries concerning air' (Schofield 1997, p. 237). As he wrote in his Royal Society address:

> The quantity of air which even a small flame requires to keep it burning is prodigious. It is generally said, that an ordinary candle consumes, as it is called, about a gallon in a minute. Considering this amazing consumption of air, by fires of all kinds, volcanoes, etc. it becomes a great object of philosophical inquiry, to ascertain what change is made in the constitution of the air by flame, and to discover what provision there is in nature for remedying the injury which the atmosphere receives by this means. Some of the following experiments will, perhaps, be thought to throw a little light upon the subject.
>
> (Priestley 1772a, p. 162)

Priestley's Christian worldview motivated this quest: with centuries of animal and human respiration, plus volcanoes and natural fires, the atmosphere should be progressively rendered unfit for human life, but there were theological reasons why this could not happen. A beneficent, all-powerful Creator would not design such a world; God must have made some provision for the natural restoration of air.

Priestley's first thought, or hypothesis, was the common-sensical one: as air is necessary both for animal and vegetable life, then both animals and plants must process air in the same manner. However, experiment led him to reject this idea. As he wrote:

> One might have imagined that, since common air is necessary to vegetable, as well as to animal life, both plants and animals had affected it in the same manner, and I own that I had that expectation when I first put a sprig of mint into a glass jar standing inverted in a vessel of water; but when it had continued growing there for some months, I found that the air would neither extinguish a candle, nor was it at all inconvenient to a mouse, which I put into it.
>
> (Priestley 1772a, p. 162)

Priestley's investigations bore fruit and, on 23 August 1771, he wrote again to Lindsey saying: 'I have discovered what I have long been in quest of, viz., that process in nature by which air, rendered noxious by breathing, is restored to its former salubrious condition' (Schofield 1966, p.133). In his Royal Society address, Priestley said:

> This observation led me to conclude, that plants, instead of affecting the air in the same manner with animal respiration, reverse the effects of breathing, and tend to keep the atmosphere sweet and wholesome, which it had become noxious, in consequence of animals living and breathing, or dying and putrefying in it.
>
> (Priestley 1772a, p. 166)

Priestley did suggest a mechanism for the beneficent effect: 'this restoration of vitiated air is affected by plants imbibing the phlogistic matter with which it is overloaded by the burning of inflammable bodies' (Priestley 1775–1777, Vol.1, p. 49), but, in keeping with his strict epistemological principle of only giving cautious or provisional status to conjectured, unseen mechanisms, he added, 'whether there be any foundation for this conjecture or not, the fact is, I think indisputable' (ibid.). His distinction between observational facts, upon which there could and should be agreement, and unseen, putative mechanisms was a fundamental one for Priestley. The distinction appears many times in his writings. For example, in a 1779 letter to Giovanni Fabroni concerning plants thriving in inflammable air, he says: 'The facts appear to me to be rather extraordinary. You must help me to explain them, for I am a very bad theorist' (Schofield 1966, p. 171). His insistence on the distinction between fact and interpretation was such that he has sometimes been called a 'proto-positivist'. The philosophical issue has long been whether the 'facts of the matter' can be articulated or described without recourse to theory.

The 1772 paper is a tour de force and justly known as a landmark in the history of science. It describes Priestley's manufacture of soda water (Pyrmont water); his creation, but not recognition, of oxygen by heating saltpetre (potassium nitrate); his nitric oxide test for the 'goodness of air'; and, last but not least, his identification of the mechanisms for the restoration of 'the goodness of air'. Any one of these achievements singularly would probably have earned him the Copley medal.

Testing the Goodness of Air

An important step towards understanding the process of restoration of air by plants (photosynthesis) was having some quantitative test for the 'goodness of air'. Without such a test, it was akin to saying that some treatment made something heavier, or longer, without scales or tapes to indicate just how much heavier or how much longer. Priestley's novel nitrous air test provided such a quantitative instrument.

Priestley took a given volume of insoluble, colourless nitrous air, mixed this with double its volume of insoluble, colourless common air, waited for the reaction that formed soluble, brown nitrous vapour to take place, then shook the resultant air over water and measured how much was dissolved by noting the rise in water level in the collecting jar. In modern terms, the nitrous air (NO) was combining with oxygen (O_2) in the air to form the red, turbid and soluble nitrogen dioxide (NO_2). The more the water rose, the more oxygen had been consumed, and, hence, the better the 'goodness' of the air in the sample. This was his famous, and much used, Nitrous Air (nitric oxide, NO) Test for the goodness of air (Boantza 2007, pp. 513–516).

Priestley's Final Steps Towards Photosynthesis

When Priestley left Leeds in 1773 to begin work as a librarian, companion and child's tutor for Lord Shelburne, he had put into place a good many pieces of the 'restoration of air' puzzle. In the 1770s and 1780s, he would return to the puzzle and put other pieces into place. Priestley's 'experiments and observations' in 1778 caused him to refine his 1772 accounts of the restoration of air. He wrote:

> In general, the experiments of this year were unfavourable to my former hypothesis. For whether I made the experiments with air injured by respiration, the burning of candles, or any other phlogistic process, it [the air] did not grow better but worse.
>
> In most of the cases in which the plants failed to meliorate the air they were either manifestly sickly, or at least did not grow and thrive, as they did most remarkably in my first experiments in Leeds; the reason for which I cannot discover.
>
> (Priestley 1779–1786, Vol.I, p. 298, in Nash 1948, p. 359)

One of his problems was lack of sufficient light: experiments conducted outdoors gave better results than those conducted indoors, especially indoors away from windows. He scaled down the degree of conviction that he placed in the restoring power of vegetation, saying:

> Upon the whole, I still think it probable that the vegetation of healthy plants, growing in situations natural to them, has a salutary effect on the air in which they grow. For one clear instance of the melioration of air in these circumstances should weigh against a hundred cases in which the air is made worse by it, both on account of the many disadvantages under which all plants labour, in the circumstances in which these experiments must be made, as well as the great attention, and many precautions, that are requisite in conducting such a process. I know no experiments that require so much care.
>
> (Priestley 1779–1786, Vol.I, p. 299, in Nash 1948, p. 360)

Within a few months of penning the above letter, and while working on a range of philosophical, theological and political tasks, he was prepared to think further about the source of the pure air he saw released by green matter and plants in his phials. Initially, he thought it came from the green matter or leaves, but he was able to devise a nice experimental test of this hypothesis. In September 1779, he wrote to his good friend Benjamin Franklin (1706–1790), relating that:

> Though you are so much engaged in affairs of more consequence [drafting the Declaration of American Independence], I know it will give you some pleasure to be informed that I have been exceedingly successful in the prosecution of my experiments since the publication of my last volume [his *Experiments and Observations on Different Kinds of Airs*].
>
> I have confirmed, explained, and extended my former observations on the purification of the atmosphere by means of vegetation; having first discovered that the green matter I treat of in my last volume is a vegetable substance, and then that other plants that grow wholly in water have the same property, all of them without exception imbibing impure air, and emitting it, as excrementitious to them, in a dephlogisticated state.
>
> (Schofield 1966, pp. 178–179)

Other experiments confirmed that the green matter, along with aquatic green leaves, only produced pure air in the presence of sunlight; heat was no substitute for light. Thus, the vegetable hypothesis was restored and, indeed, extended: not only did vegetation restore atmospheric air depleted by fires and animal respiration, it also restored water that had dissolved unhealthy air and whose dissolved air was being rendered noxious by respiration of fish. Priestley's research on the restoration of air basically finished at this point. Most of the outlines of what would, in the late nineteenth century, come to be called 'photosynthesis' were in place.[22]

Priestley was aware of the need for control groups and for the identification and control of variables in experiments, and of the need to be cautious in deriving definite conclusions from experiments in natural philosophy. All of this is inconsistent with the 'narrow-minded dogmatist' image of Priestley that was so casually, and without foundation, broadcast by Thomas Kuhn – an inconsistency that should not be lost on students and that might encourage them to be careful about what they read in history, philosophy and education texts and what they are told in lectures.

Features of Science

As with the pendulum example, the suggestion here is that philosophical (methodological, ontological, metaphysical and epistemological) themes and issues be identified and discussed at whatever level is appropriate to students and their educational circumstance, when they are learning photosynthesis. As will be argued in Chapter 11, teachers should be relaxed about just what might constitute 'the nature of science' and think more about dealing with 'features of science' in their classes, and doing this as they teach scientific topics. Some of the features that clearly arise in a Priestley-informed, historical approach to teaching photosynthesis are at least the following.

Metaphysics and Science

To the end of the sixteenth century, Aristotle was the most significant figure in the history of Western thought. Along with everything else he accomplished, he was the founder of biological science. He was an amazingly acute observer of the natural world and wrote five books on animals and one on plants.[23] However, for 2,000 years, two Aristotelian scientific/philosophical positions – first, his account of plant nutrition; second, his account of the elemental nature of air – thwarted the discovery and formulation of a correct account of the 'restoration of air' (photosynthesis). Both Aristotelian positions grew out of his more fundamental embrace of observation-based common sense as the foundation of all natural philosophy. One contemporary Aristotelian writes: 'In an effort to understand nature, society and man, Aristotle began where everyone should begin – with what he already knew in the light of his ordinary, commonplace experience' (Adler 1978, p. xi).

In his *On Plants*, Aristotle sees the parallel between plants and animals, saying: 'the absorption of food is in accordance with a natural principle, and is common to both animals and plants . . . and animals and plants have to be provided with food similar in kind to themselves' (Barnes 1984, Vol.2, p. 1253). In his treatise *On the Soul*, he mentions how plants are fed: 'if we are to distinguish and identify organs according to their functions, the roots of plants are analogous to the head in animals' (Barnes 1984, Vol.1, p. 662). Plant roots and animal mouths both have the function of absorbing food. Plants have their head, so to speak, in the ground; this accords with people's naive and immediate understanding, and with their agricultural practice: plants build themselves up from seeds by taking food and water from the soil through their roots. This was the basis of the medieval 'analogist' understanding of plants. Very slowly, this common-sense view of plants began to be unravelled by seventeenth-century 'experimentalist' investigators who took as their model, not Aristotelian observation of plants, but Baconian and Galilean-like experiments on plants (Delaporte 1982).

The second conceptual barrier to an understanding of photosynthesis was the Aristotelian conception of air. Until Priestley's time, the understanding of air as a single, fundamental, non-divisible element held sway in science (natural philosophy) and, of course, in everyday life. In the Aristotelian worldview, or scheme of things, water was another such singular element (along with earth and fire). It was recognised that not all air was the same: just as water could be made dirty and fouled, so too could air be contaminated by smoke, dust, putrefaction and so on. Such was the bad air of mines, swamps, prisons, etc. However, the bad airs were not thought of as a composite, they were regarded, in modern terms, as a mixture; as with dirty water, the impurities were just added to, and carried in, the air and could be filtered out. The properties or physics of air – in particular, air pressure and its dependence on altitude, and the compressibility of air – had been investigated by Torricelli, Boyle, Pascal, von Guericke and others, but not the composition of air. As has been mentioned in Chapter 4, the Aristotelian 'elemental' category acted as an 'epistemological obstacle' to such investigations. Priestley well expressed this understanding in his justly famous *Experiments and Observations on Different Kinds of Air*:

> There are, I believe, very few maxims in philosophy that have laid firmer hold upon the mind, than that air, meaning atmospherical air (free from various foreign matters, which were always supposed to be dissolved, and intermixed with it) is a simple elementary substance, indestructible, and unalterable, at least as much as water is supposed to be.
>
> (Priestley 1775–1777, Vol.II, p. 30)

As will be further elaborated in Chapter 10, this Aristotelian and common-sense picture was beginning to break down in Priestley's time. The mechanical worldview of Galileo, Boyle, Newton and the new science was seen to render pointless the whole Aristotelian metaphysical picture and its corresponding scientific programme of explanations in terms of natures, forms and essences transforming matter in accordance with inner teleological potentials. On the Aristotelian account, an acorn seed contained the potential of the tree, and this potential directed the development of the seed into an acorn tree, not a banana tree or a rose bush. With the demise of Aristotelian metaphysics, the idea of air and water as fundamental, homogenous elements was made contingent or contestable: it was something that could be investigated by empirical procedure, and this was done.

Experiment and Science

Along with the philosophical critiques of Aristotelian metaphysics, the new science legitimated the experimental investigation of nature; it was no longer constrained by the Aristotelian strictures on interfering with nature. For Aristotle, natural philosophy was to study 'natural motions', not 'violent motions'; experiment, which constrained nature, resulted in unnatural motions and, thus, shed no light on natural processes.[24]

Johann Baptista van Helmont (1577–1644), the Flemish physician and cross-over figure between alchemy and chemistry, had published his famous willow-tree experiment, showing that, over a 5-year period, a willow-tree seedling planted in a pot gained around 164 lb of 'tree material', seemingly just from the addition of water (Helmont 1648). Thus, water was apparently being turned into wood, an earthy material; water was 'transmuted' into earth, as the alchemists expressed it.

Robert Boyle (1627–1691), the well-known English natural philosopher and less well-known alchemist, utilised van Helmont's experiments in his detailed criticism of the Aristotelian metaphysical system published in his 1661 *The Sceptical Chymist*. He bypassed Helmont's potted earth by growing plants just in water and found the same effect: the plant grew (an increase in earthy material) just by addition of water. This result strengthened the alchemist's claim that water could be transmuted into earth, thus refuting the Aristotelian view that they were separate elements. This was yet another case of scientific practice forcing an adjustment in metaphysics.

Stephen Hales (1677–1761), the English clergyman, in his 1727 *Vegetable Staticks*, recognised experimentally that air literally entered into plants when they grew, and was in turn given off by growing plants. The worldview that motivated and informed his quantitative and experimental investigation of nature was the then standard Christian one:

> Since . . . the all-wise Creator [had] observed the most exact proportions, of number, weight and measure, in the make of all things; the most likely way, therefore, to get any insight into the nature of those parts of the creation, which come within our observation, must in all reason be to number, weigh and measure.
>
> (Hales 1727, p. 1, in Scott 1970, p. 44)

Joseph Black (1728–1799), a Scottish chemist, had, by heating marble (calcium carbonate), in 1756, isolated and identified carbon dioxide or 'fixed air' as he called it.[25] He recognised that it was a thoroughly different kind of air from atmospheric air – it turned limewater milky and did not support combustion. A colleague wrote of Black's discovery that:

> He had discovered that a cubic inch of marble consisted of about half its weight of pure lime and as much air as would fill a vessel holding six wine gallons. . . . What could be more singular than to find so subtile a substance as air existing in the form of a hard stone, and its presence accompanied by such a change in the properties of the stone.
>
> (Leicester 1956/1971, p. 134).

Thus began the idea that there were a number of separate airs; the term 'gas', coined by van Helmont, was not widely used. However, despite these advances, the idea that common atmospheric air was a composite of airs (gases) was not at all widespread. As late as 1771, the French chemist Turgot was writing of air as a 'ponderable substance which constantly enters into the state of vapour or expansive fluid according to the degree of heat contained' (Brock 1992, p. 102).

Worldviews and Science

The educational importance of connecting science and worldviews was recognised by the AAAS in its *Project 2061*: 'Becoming aware of the impact of scientific and technological developments on human beliefs and feelings should be part of everyone's science education' (AAAS 1989, p. 173). A historical approach to teaching photosynthesis allows context for discussion and elaboration of these abstract themes.

Pringle's much-cited Copley Medal address well conveys the overarching sense of cosmic design, teleology and anthropocentric purpose that constituted Priestley's worldview and that of most natural philosophy of the period. The ideas of design and providence were famously articulated by another of Priestley's contemporaries, William Paley (1743–1805), whose *Natural Theology* (Paley 1802/2006) was a compulsory text for all students in Cambridge and Oxford. It was the deep-seated idea of providence that, for most, flowed naturally from belief in a beneficent Creator. God was absolutely pervasive in medieval and early-modern natural philosophy; for all natural philosophers, nature was in the foreground of their investigations, but God was the background; they simply assumed that they were studying God's handiwork, in much the same way as a person today studying a clock is aware that they are studying something that someone made, and that what they are seeing reflects good or bad design and craft skills.[26]

Providence was variously held to be operative at three levels: it governed the natural world (the occurrence of earthquakes, storms, etc.), it controlled human history (the outcome of wars, etc.), and finally it was operative in individual human lives (recovery from illness, avoidance of accidents, etc.). For Muslims, 'God willing . . .' still prefaces all claims about future events. Priestley shared this ubiquitous worldview. In his *First Principles of Government*, he wrote:

> Such is my belief in the doctrine of an over-ruling providence, that I have no doubt, but that every thing in the whole system of nature, how noxious soever it may be in some respects, has real, though unknown uses; and also that every thing, even the grossest abuses in the civil or ecclesiastical constitutions of particular states, is subservient to the wise and gracious designs of him, who, notwithstanding these appearances, still rules in the kingdoms of men.
>
> (Miller 1993, p. 6)

Priestley's investigation of the restoration of air cemented his worldview. For him, nature is shown thus to be so wonderfully formed that 'good never fails to arise out of all evils to which, in consequence of general laws, most beneficial to the whole, it is necessarily subject' (Priestley 1774–1786, Vol. II, p. 63). In his *Memoirs*, he wrote that the greatest virtue of scientific studies was their tendency 'in an eminent degree, to promote a spirit of piety, by exciting our admiration of the wonderful order of the Divine Works and Divine Providence' (Priestley 1806/1970, p. 200). And further, with the philosophical unbelievers of the Enlightenment directly in mind (just as Isaac Newton had in writing the *Principia*), Priestley adds that those of a 'speculative turn' could not avoid the perception and admiration of 'this most wonderful and excellent provision' (ibid.).

The significant issue in discussion of science and worldviews is just how, and if, any particular worldview is supported, or contradicted, by science.

Abductive Reasoning

Priestley's argument for a providential worldview is best understood as neither a *deductive* argument (which would clearly be invalid) nor as an *inductive* inference (the argument is not from a sample to a whole), but as an *abductive* argument, to use the term introduced by Charles Sanders Peirce in the late nineteenth century (Aliseda 2006). More recently, this kind of argument has been called 'inference to the best explanation' (Lipton 1991, Psillos 2004)). Its structure is:

> There is some well-documented observation O about the world.
> If some theory or supposition T were true, then O would be expected to be the case.
> Therefore O provides grounds for believing in the truth of T.

Priestley's final step from the understanding of natural processes to knowledge of divine (supernatural) properties, from knowledge of the world to knowledge of God, was the standard inference in all natural theology (theological speculation that was independent of revelation). In the eighteenth century, it was a step taken by nearly all natural philosophers.

In his *History of Electricity*, Priestley had written:

> The investigation of the powers of nature, like the study of Natural History, is perpetually suggesting to us views of the divine perfections and providence, which are both pleasing to the imagination, and improving to the heart.
>
> (Priestley 1767/1775, p. iv)

Priestley realised that this step was not logically compelling, it was not demonstrative. It was 'suggestive', but psychologically it 'could not be avoided'. In the eighteenth century, both things were true, given background knowledge and culture; but then, and now, the step was not logically compelling; it was not demonstrative. Priestley well knew that the step from observation of nature to unseen natural mechanisms did not result in indubitable knowledge. Aristotle and the medievals recognised the same limitation for this argument form; they knew that an argument of the following form is invalid:

> T (theory) implies O (observation)
> O (observation occurs)
> Therefore T (is true)

The argument commits the Fallacy of Affirming the Consequent (also known as *modus ponens*): many other Ts could also imply O, and, thus, any particular

T was not proved by occurrence of O. Consequently, the step from observation to *supernatural* mechanisms was even less compelling.

'Argument to the best explanation' can be illustrated using one of Peirce's examples: if we find fish fossils inland, far from the current ocean shore (O), then we can abduce the theory or hypothesis that the ocean once covered the area (T). This theory provides the best current explanation of O. Thus, O provides support for belief in T. This typical piece of scientific reasoning is neither deductive nor inductive; Peirce labelled it 'abductive' and, of course, he recognised that it was not demonstrative (Peirce 1931–1935, Vol.2, p. 629). The observation O provides support for belief in T; it does not prove the truth of T. The process begins with the scientific assumption that there has to be *some* explanation of O; in science, there is also the assumption that there is a *natural* explanation of O; *super*-natural best explanations are not allowed, as will be elaborated in Chapter 10; they are ruled out by commitment to *methodological naturalism*.

For Priestley, O was the restoration of air, and T was his Christian world-view. The latter provided the best explanation of the observed phenomena. He knew that inferences to the best explanation were still tentative and not demonstrative. However, the inference had another support: Priestley was an ardent believer in the Christian Revelation; his entire life was spent as a Christian clergyman and serious scholar of scripture. His view was that the two classes of premise – observations of nature plus revelation correctly interpreted – jointly justified a compelling inference to divine or supernatural agency as the best explanation of the observations and effects his science disclosed.

However, within half a century of Priestley's death, the work of Darwin would severely challenge, and for many completely undermine, the theistic picture of a providential natural world. This was, at least, the case for the *natural* level of providential operation, Darwinism having nothing to say about either *historical* or *personal* levels. Priestley's observations about natural processes, such as the role of vegetation in the restoration of air, would hold, but less and less was there agreement with his theological explanations of the observations. After Darwin, the recognition of adaptation without design was a commonplace; there was a competing 'best explanation' for the existence of O that was also a natural explanation; it did not require recourse to supernatural agency. Many gave up Judaeo-Christian (and Islamic) belief; many retained the belief *sans* providence; many, such as contemporary 'Intelligent Design' proponents, reinterpreted providence to align it with a seemingly 'independently functioning' world.[27]

Materialism

Priestley provides ample opportunity for the elaboration and appraisal of a long-standing ontological position in philosophy, namely materialism. Priestley believed in a single God above and a single matter below. He was a materialist:

his ontology did not allow spirits, souls or minds of a kind that were ontologically distinct from matter.[28] He was an ontological monist, writing:

> What peculiar excellence is there in those particles of matter which compose my body, more than those which compose the table on which I write. . . . If I knew that they were instantly, and without any painful sensation to myself, to change places, I do not think it would give me any concern.
>
> (Gibbs 1967, p. 99)

In the Introduction to his edition of *Hartley's Theory of the Human Mind*, Priestley writes:

> I rather think that the whole man is of some uniform composition, and that the property of perception, as well as the other powers that are termed mental, is the result (whether necessary or not) of such an organical structure as that of the brain.
>
> (Priestley 1775, p. xx)

Although seemingly a contradiction, his was a Christian materialism, as clearly stated when he reflected upon the impact of his writings:

> The consequence (which I now enjoy) is a great increase of materialists; not of atheistical ones, as some will still represent it, but of the most serious, the most rational and consistent christians.
>
> (Passmore 1965, p. 169)

Priestley rejected any belief in the soul as existing apart from the body; this was a 'false philosophy from the East' and was entirely without scriptural warrant. He regarded belief in an independent and individual soul, and its required ontology that separated matter and spirit, as entirely un-Hebraic, un-Christian and the root of most aberrations, fantasies and corruptions in the Christian churches. In 1778, he wrote to the Revd C. Rotherham that:

> I was an Arian till I went to Leeds, and my Materialism is but of late standing, though you see that I now consider the doctrine of the soul to have been imported into Christianity, and to be the foundation of the capital corruptions of our religion.
>
> (Priestley 1806/1970, p. 40)

For Priestley, not only was belief in a soul un-Christian, it was just philosophical folly. One had to embrace some version of either Platonism or Aristotelianism for the soul to make philosophical sense, and he rejected both.

Ethics and Animal Experimentation

Until Priestley's new nitrous air test, the only means available for ascertaining the 'goodness of air' were variants of the miner's canary test; most commonly, a laboratory mouse was placed in a sealed container of the air to be tested, and its 'goodness' was measured by how long the mouse lived. The mouse test stoked the embryonic Romantic and humanistic reactions against the new science, which were so dramatically captured by Joseph Wright of Derby in his evocative 1768 painting of *An Experiment on a Bird in the Air Pump*. There, a bird lies dead from suffocation in an evacuated jar, with a pensive audience looking on, and one woman looking away, inviting a 'science is not for women' response in viewers.

Priestley encountered this reaction: Anna Laetitia Aikin, the daughter of a close friend and an admirer of Priestley, published a book of verse in 1773 that contains a poem entitled 'The Mouse's Petition' – it concerns a mouse that she found in a cage in Priestley's study. She knew the nature of his experiments and also of his championing of human freedom, and so she wrote the piece, leaving it alongside the mouse cage to provoke him.[29]

> O hear a pensive prisoner's prayer,
> For liberty that sighs;
> And never let thine heart be shut
> Against the wretch's cries!
>
> For here forlorn and sad I sit,
> Within the wiry grate;
> And tremble at the approaching morn,
> Which brings impending fate.
>
> If e'er thy breast with freedom glowed,
> And spurned a tyrant's chain,
> Let not thou strong oppressive force
> A free-born mouse detain!

Pleasingly for countless generations of mice, for the repute of science and for precision in pneumatic measurement, Priestley was so moved by the poem that he found a new chemical test for the goodness of air.

As has been discussed in Chapter 5, questions of values, ethics, morals and science cannot be avoided in science programmes. All major universities now have ethics committees that regulate research in science and social science that impinges directly and indirectly on humans, animals and, more broadly, social welfare, and researchers need to justify their methods and aims to such committees. The once straightforward and unreflective use of animals for scientific experiments and laboratory dissections is now strictly controlled (Rollin 2009). More than this, however, concern for animals is explicitly cultivated and even demanded – an aim of the New Zealand science syllabus is 'the care of animals' and recognition of their rights. Hitherto, partly under the influence of belief in value-free science, these questions have largely been ignored in science education – rats, mice and frogs were routinely killed in furtherance of curricular objectives. The Priestley example allows this philosophical and educational subject to be placed in historic context and provides an occasion for students to analyse and debate the issues.

Commercialisation of Science

In 1767, Priestley became first person to create and bottle soda water, or 'Pyrmont water' as he called it.[30] Pyrmont was a famous medicinal spa in Hanover. Priestley saw that the Pyrmont bubbles were carbon dioxide (fixed air), which he had captured as a by-product of a Leeds brewery and which he was able to independently produce by mixing chalk and acid and capturing the emitted air in a bladder, thus putting it under pressure. There was, at the time, great interest in ascertaining the efficacious component of English and Continental mineral waters, but no one was thinking of manufacturing them. Priestley, who recognised the enormous wealth that could be made, nevertheless turned down the opportunity of commercial bottling of his Pyrmont water, saying that natural philosophers should 'search for truth, not money'. The commercial opportunities were not lost on Johann J. Schweppe (1740–1821), who, from 1793, began manufacturing and selling high-pressure soda water from his factory off Cavendish Square. As they say, the rest is history.[31]

Priestley's innocent, almost saint-like, turning away from the commercial windfall that would have followed his selling his revolutionary soda-water process can be an occasion for taking up the whole issue of science's engagement, if not entanglement, with business and commerce. This is not as simple as Romanticists might wish: Priestley was a champion of applied science and the commercial utilisation of scientific knowledge; for him, pursuing good science was both a religious obligation and the way of furthering the Enlightenment project of 'improving man's estate'; hence, his support of the manufacturing and business-focused Lunar Society (Schofield 1963, Uglow 2002). The commercialisation of science and its implications for 'the nature of science' have been much written on over the past few decades.[32] Has science 'sold its soul' or 'gained a body'? If NOS curricular objectives are extended beyond the usual methodological and epistemological core, then commercialisation warrants the attention of teachers and students.

Priestley in the Classroom

William Brock, in his massive *The Fontana History of Chemistry*, describes Priestley as 'one of the most engaging figures in the history of science' (Brock 1992, p. 99). Enough has been indicated here to give credence to Brock's claim. Enough has also been said to support the claim that the practice of science is interwoven with philosophy and worldviews; all three mutually affect each other. As mentioned at the outset of the chapter, there are three strong considerations that make easy the utilisation of Priestley's life and work in classrooms: the acknowledged environmental problem of the 'goodness of air', the inclusion of 'nature of science' in many curricula, and the widespread concern with reappraising aspects of the European Enlightenment tradition. With informed teachers, Priestley can be used to further student understanding of each of these matters.

With the lessons from the past few decades of incorporating historical and philosophical studies into science programmes,[33] there are some obvious ways in which Priestley's work might be incorporated.

Historical Vignettes

Suitable for the curriculum at any level is the presentation by students or teachers of brief historical vignettes concerning Priestley. At a minimal level, this is designed to put a human face on chemistry and biology lessons and to indicate something of the history of the subject. Such vignettes can be tailored to the interests, sophistication and grade level of the class. Topics might include Priestley's religion, his politics, his educational theory and practice, his marriage and family life, his support of the American and French Revolutions, his dealings with Lavoisier, his creation of soda water, his opposition to the new oxygen theory of combustion, his opposition to colonisation and the slave trade, his influence on the Founding Fathers of the US and so on.

Additionally, vignettes might be presented on the wider scientific, political, social, religious and intellectual circumstances of Priestley's time: the practice of religious discrimination; the intertwining of religion and state in Europe and England; the role of science in the French and English Enlightenments and its role in European Imperialism; the state of parliamentary government; European colonisation; the impact of the French Revolution; the social effects of embryonic capitalist production in England; the role of science in the furtherance of navigation, commerce and industry. Vignettes can take the form of individual or group essays that might be presented to the class as talks or powerpoint presentations. They can contribute to better understanding of scientific content; to better appreciation of the scientific tradition and, hopefully, a sense of being indebted to that tradition; to increased interest in science; and to more general educational goals concerning students' sense of place, culture and identity. One not inconsiderable advantage of vignettes is that they allow controversial matters to be dealt with in classrooms with the safety of historic remove. Whereas, for example, critical discussion of the state, of censorship and of religious entanglement in the state might be dangerous or forbidden in many contemporary Western and Islamic societies, or in Communist China, it can be relatively safe and objective to discuss these matters in the context of Priestley's life, times and arguments.[34]

Historical–Investigative Teaching

A more rigorous way of bringing Priestley to the classroom is to try to wed laboratory classes to historical stories; that is, to follow along the path of experimental science; to follow in the footsteps of the masters, as one might say. While doing this, it is possible to reproduce something of the intellectual puzzles and scientific debates that originally prompted the experiments. Participation in this sort of journey can give students a much richer appreciation of the achievements, techniques and intellectual structure of science, while developing their own scientific knowledge and competence.

Ernst Mach (1838–1916) recognised this at the end of the nineteenth century, when he wrote:

> Every young student could come into living contact with and pursue to their ultimate logical consequences merely a few mathematical or scientific discoveries. Such selections would be mainly and naturally associated with selections from the great scientific classics. A few powerful and lucid ideas could thus be made to take root in the mind and receive thorough elaboration.
>
> (Mach 1886/1986, p. 368)

With the exception of Westaway, Holmyard, Bradley and a few others in England, and Conant and some others in the US, Mach's suggestions were ignored by science teachers. Mach's approach was famously taken in Conant's *Harvard Case Studies in Experimental Science* (Conant 1948). Chapter 2 is titled, 'The Overthrow of the Phlogiston Theory: The Chemical Revolution of 1775–1789' (Conant 1948), and Chapter 5 is titled 'Plants and the Atmosphere' (Nash 1948). The chapters provide historical texts, glossaries, details of experimental apparatus and so on, all of which can be utilised in classroom discussion of Priestley.

In recent times Nahum Kipnis has promoted this historical–investigative approach (Kipnis 1996). He has, for example, based a course on optics around retracing the classic, and usually very simple, experiments and demonstrations in the history of the subject (Kipnis 1992). Students read original literature, they re-enact historical experiments and themselves elaborate and debate interpretations of what they see in the laboratory. Readings and experiments on photosynthesis could suitably be substituted for the optics material. In such courses, students do not just read history, they do practical work and carry out investigations, but, instead of the practical activities being isolated, they are connected with a tradition of scientific development.[35]

Another current example of this historical–investigative approach is at the University of Chester, where John Cartwright has taught an elective history of science course that has a 4–6-week component on *The Discovery of Oxygen*. The course aims are:

1 to promote an understanding of the historical origins of science and the distinctive nature of scientific enquiry;
2 to develop an awareness of the interaction between scientific thinking and the wider culture;
3 to foster an empathetic understanding of ideas from previous cultures;
4 to develop an awareness of the nature of historical enquiry;

5 to enable students to appreciate the force and impact of scientific thinking and ideas.

The course and its *Student Guide* (Cartwright 2004) are a nice example of a wider, contextual approach to the teaching and learning of chemistry. It could provide a template for a comparable course on *The Discovery of Photosynthesis*.

Priestley's studies on the restoration of air are well suited to this historical–investigative approach. A good many of his discoveries are relatively simple to reproduce: making soda water, producing oxygen by heating metal oxides, darkness versus light conditions for effectiveness of green plants in restoring air, the nitrous air test, the observation of 'green matter' and the conditions under which it creates pure air and so on.

What makes the use of Priestley attractive is that he wrote very complete and readable accounts of his work. Priestley meant for his experiments to be reproduced by readers. His writings were a means for the education of the populous and, thus, to realising Priestley's conception of the true goal of the Enlightenment – the development of an informed citizenry who respected reason, were distrustful of authority, prized autonomy and recognised an open society and public debate as the preconditions of knowledge growth in all fields of endeavour, but especially for scientific and religious understanding. Priestley was an advocate of 'science for all', some two centuries before it became an educational slogan.

Interdisciplinary Teaching

Priestley's intellectual engagements were wide-ranging – science, theology, education, politics, history, philosophy – it is unrealistic to think that all this can be covered in a science course. However, it is not unrealistic to hope that some coordination between subject areas can be achieved in a school, or college, and, thus, for teachers in related fields to work together on the 'big picture' presented by Priestley's work.

Such coordination is, of course, almost unheard of in school systems. History, science, mathematics, music, social studies, literature, religion and philosophy – all go their own way, with barely a passing curricular nod to each other. From the students' point of view, and even from the teachers', knowledge is truly fragmented. However, well-chosen themes, such as 'the restoration of air', that are heuristically rich, can organise a curriculum to maximise the degree to which the interdependence of knowledge becomes more transparent. It may be, minimally, a matter of looking at existing, independently generated curricula and simply pulling the related parts together and arranging for some coordination and cross-referencing, but it can be more than this.

A praiseworthy example, and potential model, of coordination between disciplines occurs at the Oberstufen Kolleg of the University of Bielefeld,

Germany. The college utilises a 'historical–genetical approach to science teaching'. At the Oberstufen Kolleg:

> There is attention given to the historical, social and philosophical dimension of science. Frequently, historical examples are presented in a rather anecdotal fashion in courses of science, in order to motivate students for the 'real thing', the scientific content. History and philosophy are merely instrumentalised and serve to 'sell the product'. Our intention differs: we consider the historical and philosophical dimension to be an essential part of science and of instruction in science, that aims to present science in a social and historical context.
>
> (Misgeld *et al.* 2000)

Further examples of such cross-disciplinary and coordinated teaching are the various science, technology, engineering, arts and mathematics (STEAM) curricula in the US, Korea and some other countries,[36] and some such efforts towards coordination are part of the US *Next Generation Science Standards* statements, where 'cross-cutting' concepts and curricula are identified and encouraged.

Reading, understanding in their context and appreciating selections from the range of Priestley's work on restoration of air, on methodology, on philosophy and theology, along with some historical–investigative redoing of his simple experiments, would make excellent material for such cooperative, interdisciplinary curricula. The school curriculum might then look something like the following table (Figure 7.1) with subjects or disciplines in the columns and topics in the rows.

Conclusion

Priestley is an under-utilised figure in science education. Although his contribution to the discovery of oxygen is recognised, this is usually glossed by comment about him being an obscurantist concerning Lavoisier's new chemistry and a dogmatist concerning his own adherence to the phlogiston account of combustion and respiration. Unfortunately, Priestley's contribution to the modern understanding of photosynthesis is seldom mentioned in school curricula. This is a pity, as his role was pivotal, and students can very easily be led through many of the same steps that he took. There is the opportunity for students to 'walk in the footsteps' of a great scientist and, thereby, not only learn scientific content, method and methodology, but also get a sense of participation in a tradition of thought and analysis that is at the core of the modern world.

Such Priestley-guided participation allows students to appreciate and understand key elements of the scientific tradition: hard work, experimentation, independence of mind, a respect for evidence, a preparedness to bring scientific modes of thought to the analysis and understanding of more general social and cultural problems, a deep suspicion of authoritarianism and

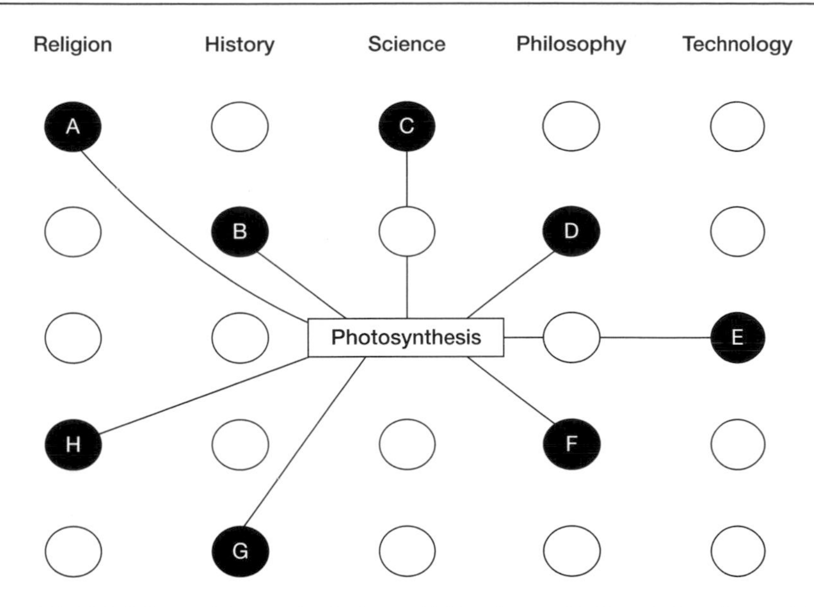

Figure 7.1 HPS-Informed Curricular Linkage. (a) Revelation, (b) French Revolution, (c) composition of air, (d) experiment, (e) soda water, (f) epistemology, (g) the Enlightenment, (h) providence

dogmatism, and the concern for promotion of an open society as the condition for the advance of knowledge.

Bringing Priestley into education allows light to be shed upon the mutual interaction of worldviews and science; it allows the scientific sources of the European Enlightenment to be investigated; and it allows the evaluation of the special Enlightenment niche occupied by Priestley, namely the theistic, albeit dissenting, strand of the Enlightenment. Understanding and appreciating this connection between science and the Enlightenment and having the opportunity to examine what is dead and what is living in that tradition can be a major contribution of science classes to the general education of students in the modern world.

Notes

1 This chapter is dependent on research published in Matthews (2009).

2 Among countless studies documenting children's inadequate (given age and grade level) understanding of photosynthesis see: Cañal (1999), Eisen and Stavy (1988), Wandersee (1985) and references therein.

3 The term 'photosynthesis' was coined in 1898 by the Englishman Charles Barnes (1858–1910) to denote the complex biological–chemical process of the 'synthesis of complex carbon compounds out of carbonic acid, in the presence of chlorophyll, under the influence of light' (Gest 2002, p. 7).

The equation for the process is:

$$6CO_2 + 6H_2O + \text{solar energy} \rightarrow C_6H_{12}O_6 + 6O_2$$

4 For details, see Chapter 2 and the references therein.
5 The definitive and exhaustive biographical study of Priestley is Robert Schofield's two-volume work (Schofield 1997, 2004). See also the studies of William Brock (Brock 2008) and John McEvoy (McEvoy 1978–79, 1990, McEvoy & McGuire 1975) and contributions to Rivers and Wykes (2008) and Anderson and Lawrence (1987).
6 In his *Memoirs*, Priestley writes: 'I felt occasionally such distress of mind as it is not in my power to describe, and which I still look back upon with horror' (Priestley 1806/1970, p. 71).
7 For accounts of the dissenting academies, see Wykes (1996).
8 Richard Westfall, in his biography of Newton, says that Cambridge at the time was 'fast approaching the status of an intellectual wasteland' (Westfall 1980, p. 190).
9 A good popular account of the intellectual entanglement of Priestley and Lavoisier is Jackson's *A World on Fire* (Jackson 2005).
10 For a selection of Priestley's political writings, see Miller (1993).
11 For studies of Priestley in America, see Graham (2008).
12 Priestley's house in the town long functioned as the Priestley Museum and Research Centre. Sadly, state budget cuts have now closed it.
13 There are numerous studies of Priestley's theological and religious life; see especially Brooke (1990) and Wykes (2008).
14 A full bibliographic listing of Priestley's books, pamphlets and articles is contained in Schofield (2004, pp. 407–422).
15 See references in Chapter 2.
16 At the time, 'moral philosophy' covered a broad field; it meant more or less all studies other than 'natural philosophy' (or science, in our terms).
17 For Priestley's materialism, see Priestley (1778). For critical exposition and discussion of his position, see Schofield (1970, pp. 261ff.) and Dybikowski (2008).
18 Modern teachers trying to dissuade students of their 'fluid' view of electricity are following in Priestley's footsteps.
19 This was a 12-inch (30-cm) magnifying glass, with a 20-inch (50-cm) focal distance, that gave more heat than any other means available. Continental chemists were purportedly using it to melt diamonds.
20 All Royal Society *Transactions* papers are now available on the Society's web page.
21 For further details of the paper and the Copley Medal, see Guerlac (1957) and McKie (1961).
22 After the passage of 60 years, a still excellent treatment of the historical development of early photosynthesis studies is the essay of Leonard Nash in James Conant's *Harvard Case Studies in Experimental Science* (Nash 1948, pp. 369–434). For more recent work, see Magiels (2010).
23 See the excellent, two-volume Jonathan Barnes edition of Aristotle's *Collected Works* (Barnes 1984).
24 For qualification of this standard interpretation of Aristotle, see Newman (2004, Chapter 5).
25 Called 'fixed' because he thought it was, as a whole air, trapped or 'fixed' in calcium and other metal carbonates; the heating released the air; when dissolved in limewater and precipitated, it again became fixed.
26 For informed discussion of providence and science, see at least: Funkenstein (1986) and contributions to Lindberg and Numbers (1986).
27 These are all philosophically and theologically complex options. Clearly, Christian and Islamic belief among philosophically and scientifically sophisticated people survived Darwin, with many retaining some conception of providence.
28 On Priestley's materialism, see Schwartz (1990) and Yolton (1983, Chapter 6).
29 The poem is in O'Brien (1989, p. 62). This book also contains a wealth of material on Warrington Academy and Priestley's teaching career there.
30 See Priestley (1772a). An informative discussion of the soda-water episode, with diagrams of apparatus, can be found in Gibbs (1967, pp. 57–58, 69–70). See also Coley (1984) and Golinski (1999, pp. 112–117).

31 This explains why '1793' is stamped on the tops of Schweppe's drink bottles.
32 See at least: Kitcher (2001), Resnik (2007) and contributions to Irzik (2013) and to Radder (2010).
33 See contributions to the journal *Science & Education* from its first volume in 1992 to the present.
34 See Wandersee and Roach (1998) for examples of types and effectiveness of such vignettes.
35 For an extensive overview and appraisal of this 'historical–investigative' tradition, see Heering and Höttecke (2014).
36 These are coordinated and integrated coordinated science, technology, engineering, art and mathematics programmes. See Tang (2012).

References

AAAS (American Association for the Advancement of Science): 1989, *Project 2061: Science for All Americans*, AAAS, Washington, DC. Also published by Oxford University Press, 1990.

Adler, M.J.: 1978, *Aristotle for Everybody*, Macmillan, New York.

Aliseda, A.: 2006, *Abductive Reasoning: Logical Investigations Into Discovery and Explanation*, Springer, Dordrecht, The Netherlands.

Anderson, R.G.W. and Lawrence, C. (eds): 1987, *Science, Medicine and Dissent: Joseph Priestley (1733–1804)*, Wellcome Trust and Science Museum, London.

Barnes, J. (ed.): 1984, *The Complete Works of Aristotle*, 2 volumes, Princeton University Press, Princeton, NJ.

Boantza, V.D.: 2007, 'Collecting Airs and Ideas: Priestley's Style of Experimental Reasoning', *Studies in History and Philosophy of Science* 38(3), 506–522.

Bolton, H.C. (ed.): 1892, *Scientific Correspondence of Joseph Priestley*, New York.

Brock, W.H.: 1992, *The Fontana History of Chemistry*, Harper Collins, London.

Brock, W.H.: 2008, 'Joseph Priestley, Enlightened Experimentalist'. In I. Rivers and D.L. Wykes (eds) *Joseph Priestley: Scientist, Philosopher, and Theologian*, Oxford University Press, Oxford, UK, pp. 49–79.

Brooke, J.H.: 1990, '"A Sower Went Forth": Joseph Priestley and the Ministry of Reform'. In A.T. Schwartz and J.G. McEvoy (eds) *Motion Towards Perfection: The Achievement of Joseph Priestley*, Skinner House Books, Boston, MA, pp. 21–56.

Cañal, P.: 1999, 'Photosynthesis and "Inverse Respiration" in Plants: An Inevitable Misconception?' *International Journal of Science Education* 21(4), 363–372.

Cartwright, J.: 2004, *The Discovery of Oxygen: Student Guide*, Department of Chemistry, University of Chester, UK.

Coley, N.G.: 1984, 'The Preparation and Uses of Artificial Mineral Waters (ca. 1680–1825)', *Ambix* 21, 32–48.

Conant, J.B.: 1948, 'The Overthrow of the Phlogiston Theory: The Chemical Revolution of 1775–1789'. In J.B. Conant (ed.) *Harvard Case Histories in Experimental Science*, Harvard University Press, Cambridge, MA, pp. 67–115.

Delaporte, F.: 1982, *Nature's Second Kingdom: Explorations of Vegetality in the Eighteenth Century*, MIT Press, Cambridge, MA.

Dybikowski, J.: 2008, 'Joseph Priestley, Metaphysician and Philosopher of Religion'. In I. Rivers and D.L. Wykes (eds) *Joseph Priestley: Scientist, Philosopher, and Theologian*, Oxford University Press, Oxford, UK, pp. 80–112.

Eisen, Y. and Stavy, R.: 1988, 'Students' Understanding of Photosynthesis', *The American Biology Teacher* 50(4), 208–212.

Funkenstein, A.: 1986, *Theology and the Scientific Imagination: From the Middle Ages to the Seventeenth Century*, Princeton University Press, Princeton, NJ.

Gest, H.: 2002, 'History of the Word Photosynthesis and the Evolution of its Definition', *Photosynthesis Research* 73, 7–10.

Gibbs, F.W.: 1967, *Joseph Priestley: Revolutions of the Eighteenth Century*, Doubleday, New York (first published as *Joseph Priestley: Adventurer in Science and Champion of Truth*, Thomas Nelson, London, 1965).

Golinski, J.: 1999, *Science as Public Culture: Chemistry and Enlightenment in Britain, 1760–1820*, Cambridge University Press, Cambridge, UK.

Graham, J.: 2008, 'Joseph Priestley in America'. In I. Rivers and D.L. Wykes (eds) *Joseph Priestley: Scientist, Philosopher, and Theologian*, Oxford University Press, Oxford, UK, pp. 203–230.

Guerlac, H.: 1957, 'Joseph Priestley's First Papers on Gases and Their Reception in France', *Journal of the History of Medicine* 12, 1–12.

Heering, P. and Höttecke, D.: 2014, 'Historical–Investigative Approaches in Science Teaching'. In M.R. Matthews (ed.) *International Handbook of Research in History, Philosophy and Science Teaching*, Springer, Dordrecht, The Netherlands, pp. 1473–1502.

Helmont, van J.B.: 1648, *Ortus Medicinae*, Leyden. English translation by J. Chandler, Oriatrike, London 1662.

Hume, D.: 1739/1888, *A Treatise of Human Nature: Being an Attempt to Introduce the Experimental Method of Reasoning Into Moral Subjects*, Clarendon Press, Oxford, UK.

Irzik, G. (ed.): 2013, 'Commercialisation and Commodification of Science: Educational Responses', *Science & Education* 22(10).

Jackson, J.: 2005, *A World on Fire: A Heretic, an Aristocrat, and the Race to Discover Oxygen*, Penguin, New York.

Kipnis, N.: 1992, *Rediscovering Optics*, BENA Press, Minneapolis, MN.

Kipnis, N.: 1996, 'The "Historical–Investigative" Approach to Teaching Science', *Science & Education* 5(3), 277–292.

Kitcher, P.: 2001, *Science, Truth, and Democracy*, Oxford University Press, Oxford, UK.

Kuhn, T.S.: 1970, *The Structure of Scientific Revolutions*, 2nd edn, Chicago University Press, Chicago, IL (1st edition, 1962).

Leicester, H.M.: 1956/1971, *The Historical Background of Chemistry*, Dover, New York.

Lindberg, D.C. and Numbers, R.L. (eds): 1986, *God and Nature: Historical Essays on the Encounter between Christianity and Science*, University of California Press, Berkeley, CA.

Lipton, P.: 1991, *Inference to the Best Explanation*, Routledge, London.

McEvoy, J.G.: 1978–1979, 'Joseph Priestley, "Aerial Philosopher": Metaphysics and Methodology in Priestley's Chemical Thought from 1762–1781', Pt.I *Ambix* 25 1–255, Pt.II 93–116, Pt.III 1 53–175; Pt.IV 26 16–38.

McEvoy, J.G.: 1990, 'Joseph Priestley and the Chemical Revolution: A Thematic Overview'. In A.T. Schwartz and J.G. McEvoy (eds) *Motion Toward Perfection: The Achievement of Joseph Priestley*, Skinner House Books, Boston, MA, pp. 129–160.

McEvoy, J.G. and McGuire, J.E.: 1975, 'God and Nature: Priestley's Way of Rational Dissent', *Historical Studies in the Physical Sciences* 6, 325–404.

Mach, E.: 1886/1986, 'On Instruction in the Classics and the Sciences'. In his *Popular Scientific Lectures*, Open Court Publishing Company, LaSalle, IL, pp. 338–374.

McKie, D.: 1961, 'Joseph Priestley and the Copley Medal', *Ambix* 9(1), 1–22.

Magiels, G.: 2010, *From Sunlight to Insight: Jan IngenHousz, the Discovery of Photosynthesis and Science in the Light of Ecology*, Brussels University Press, Brussels.

Matthews, M.R.: 2009, 'Science and Worldviews in the Classroom: Joseph Priestley and Photosynthesis', *Science & Education* 18(6–7), 929–960.

Miller, P. (ed.): 1993, *Priestley: Political Writings*, Cambridge University Press, Cambridge, UK.

Misgeld, W., Ohly, K.P. and Strobl, G.: 2000, 'The Historical–Genetical Approach to Science Teaching at the Oberstufen-Kolleg', *Science & Education* 9(4), 333–341.

Nash, L.K.: 1948, 'Plants and the Atmosphere'. In J.B. Conant (ed.) *Harvard Case Histories in Experimental Science*, 2 volumes, Harvard University Press, Cambridge, MA, pp. 325–436.

Newman, W.R.: 2004, *Promethean Ambitions: Alchemy and the Quest to Perfect Nature*, University of Chicago Press, Chicago, IL.
Newton, I.: 1730/1979, *Opticks or A Treatise of the Reflections, Refractions, Inflections & Colours of Light*, Dover Publications, New York.
O'Brien, P.: 1989, *Warrington Academy 1757–86*, Owl Books, Wigan, UK.
Paley, W.: 1802/2006, *Natural Theology; or Evidence for the Existence and Attributes of the Deity Collected From the Appearances of Nature*, Oxford University Press, Oxford, UK.
Passmore, J.A. (ed.): 1965, *Priestley's Writings on Philosophy, Science and Politics*, Collier Macmillan, London.
Peirce, C.S.: 1931–1935, *Collected Papers of Charles Sanders Peirce*, Charles Hartshorne and Paul Weiss (eds) Harvard University Press, Cambridge, MA.
Peters, M.: 1995, 'Philosophy and Education "After" Wittgenstein'. In P. Smeyers and J.D. Marshall (eds) *Philosophy and Education: Accepting Wittgenstein's Challenge*, Kluwer Academic Publishers, Dordrecht, The Netherlands, pp. 189–328.
Priestley, J. (ed.): 1775, *Hartley's Theory of the Human Mind, on the Principle of the Association of Ideas; with Essays Relating to the Subject of it*, J.Johnson, London.
Priestley, J.: 1767/1775, *The History and Present State of Electricity, with Original Experiments*, 2nd edn, J. Dodsley, J. Johnson and T. Cadell, London; 3rd edition, 1775, reprinted Johnson Reprint Corporation, New York, 1966, with Introduction by Robert E. Schofield.
Priestley, J.: 1772a, 'Observations on Different Kinds of Air', *Philosophical Transactions* 60, 147–264.
Priestley, J.: 1772b, *Directions for Impregnating Water with Fixed Air, In Order to Communicate to it the Peculiar Spirit and Virtue of Pyrmont Water, and Other Mineral Waters of a Similar Nature*, J. Johnson, London. Reprinted in his *Experiments and Observations on Air*, Vol.2, 1775. The pamphlet was reprinted by the American Bottlers of Carbonated Beverages, Washington, DC, 1945.
Priestley, J.: 1774–1786, *Experiments and Observations on Different Kinds of Air*, 6 volumes, Vol.1, 1774, J.J. Johnson, London.
Priestley, J.: 1775–1777, *Experiments and Observations on Different Kinds of Air*, 2nd edn, 3 Vols, J. Johnson, London. Sections of the work have been published by the Alembic Club with the title *The Discovery of Oxygen*, Edinburgh, 1961.
Priestley, J.: 1778, *A Free Discussion of the Doctrines of Materialism and Philosophical Necessity, In a Correspondence between Dr. Price, and Dr. Priestley*, J. Johnson & T. Cadell, London.
Priestley, J.: 1786, *History of Early Opinions Concerning Jesus Christ, Compiled From Original Writers; Proving that the Christian Church Was at First Unitarian*, 4 volumes, Birmingham.
Priestley, J.: 1806/1970, *Memoirs of Dr. J. Priestley to the Year 1795 Written by Himself, with a Continuation by his Son, J. Priestley*, 2 volumes, J. Lindsay (ed.), Philadelphia, PA. Reprinted Adams & Dart, Bath, 1970.
Psillos, S.: 2004. 'Inference to the Best Explanation and Bayesianism'. In F. Stadler (ed.) *Induction and Deduction in the Sciences*, Kluwer, Dordrecht, The Netherlands, pp. 83–91.
Radder, H. (ed.): 2010, *The Commodification of Academic Research*, University of Pittsburgh Press, Pittsburgh, PA.
Resnik, D.B.: 2007, *The Price of Truth*, Oxford University Press, Oxford, UK.
Rivers, I. and Wykes, D.L. (eds): 2008, *Joseph Priestley: Scientist, Philosopher, and Theologian*, Oxford University Press, Oxford, UK.
Rollin, B.E.: 2009, 'The Moral Status of Animals and Their Use as Experimental Subjects'. In H. Kuhse and P. Singer (eds) *A Companion to Bioethics*, Wiley-Blackwell, Singapore.
Rutt, J.T. (ed.): 1817–1832/1972, *The Theological and Miscellaneous Works of Joseph Priestley*, 25 volumes, London. Reprinted by Kraus, New York, 1972.

Schofield, R.E.: 1963, *The Lunar Society of Birmingham*, Oxford University Press, Oxford, UK.
Schofield, R.E. (ed.): 1966, *A Scientific Autobiography of Joseph Priestley (1733–1804): Selected Scientific Correspondence*, MIT Press, Cambridge, MA.
Schofield, R.E.: 1970, *Mechanism and Materialism: British Natural Philosophy in an Age of Reason*, Princeton University Press, Princeton, NJ.
Schofield, R.E.: 1997, *The Enlightenment of Joseph Priestley: A Study of His Life and Work from 1733 to 1773*, Penn State Press, University Park, PA.
Schofield, R.E.: 2004, *The Enlightened Joseph Priestley: A Study of His Life and Work from 1773 to 1804*, Penn State Press, University Park, PA.
Schwartz, A.T.: 1990, 'Priestley's Materialism: The Consistent Connection'. In A.T. Schwartz and J.G. McEvoy (eds) *Motion Toward Perfection: The Achievement of Joseph Priestley*, Skinner House Books, Boston, MA, pp. 109–127.
Scott, E.L.: 1970, 'The McBridean Doctrine of Air. An Eighteenth-Century Explanation of Some Biochemical Processes including Photosynthesis', *Ambix* 27, 43–57.
Tang, W.T.: 2012, 'Building Potemkin Schools: Science Curriculum Reform in a STEM school', *Journal of Curriculum Studies* 44(5), 659–678.
Uglow, J.: 2002, *The Lunar Men: Five Friends Whose Curiosity Changed the World*, Faber & Faber, London.
Wandersee, J.H.: 1985, 'Can the History of Science Help Science Educators Anticipate Students' Misconceptions?', *Journal of Research in Science Teaching* 23(7), 581–597.
Wandersee, J.H. and Roach, L.M.: 1998, 'Interactive Historical Vignettes'. In J.J. Mintzes, J.H. Wandersee and J.D.Novak (eds) *Teaching Science for Understanding. A Human Constructivist View*, Academic Press, San Diego, CA, pp. 281–306.
Westfall, R.S.: 1980, *Never at Rest: A Biography of Isaac Newton*, Cambridge University Press, Cambridge, UK.
Wykes, D.L.: 1996, 'The Contribution of the Dissenting Academy to the Emergence of Rational Dissent'. In K. Haakonssen (ed.) *Enlightenment and Religion: Rational Dissent in Eighteenth-Century Britain*, Cambridge University Press, Cambridge, UK, pp. 99–139.
Wykes, D.L.: 2008, 'Joseph Priestley, Minister and Teacher'. In I. Rivers and D.L. Wykes (eds) *Joseph Priestley: Scientist, Philosopher, and Theologian*, Oxford University Press, Oxford, UK, pp. 20–48.
Yolton, J.W.: 1983, *Thinking Matter: Materialism in Eighteenth-Century Britain*, University of Minnesota Press, Minneapolis.

Chapter 8

Constructivism and Science Education

The thesis of this book is that HPS can usefully be brought to bear on theoretical, curricular and pedagogical issues in science teaching, and consequently that some learning of HPS should be a natural part of science-teacher education programmes. The contribution of HPS to an evaluation of the strengths and weaknesses of educational constructivism well illustrates the thesis.

Constructivism, as a theory of knowledge and learning, has been the major theoretical influence in contemporary science and mathematics education and, in its postmodernist and deconstructionist form, it is a significant influence in contemporary mathematics, literary, artistic, social studies and religious education. Its impact is evident in theoretical debates, curriculum construction and pedagogical practice in all of these subjects. Constructivism as a psychological, educational and philosophical orientation fuels the learner-centred, teacher-as-facilitator, localist, 'progressive' side of the educational maths wars, phonics debates and discovery-learning disputes. It was appealed to in the hugely popular 'Problem-Based Learning' (PBL) curricula and pedagogy that began at the McMaster University Medical School in 1969 and that, for 20 years, swept through US, UK, European and Australasian medical training institutions (Colliver 2000, 2002, Neville 2009).[1] PBL was adopted by the World Federation of Medical Education in advance of any empirical evidence that PBL-trained learners would become better doctors (Neville 2009, p. 1). This was the same pattern as was evident in the formal adoption of constructivism in science and mathematics education. As one editorial in the *Journal of Teacher Education* declares:

> Constructivism is the new rallying theme in education. Its popularity derives from its origins in a variety of disciplines, notably philosophy of science, psychology, and sociology. The implications of a constructivist perspective for education differ depending on its disciplinary foundation, but professional education groups as diverse as the National Association for the Education of Young Children and the National Council of Teachers of Mathematics have based revisions of their standards for practice on the constructivist assumption that learners do not passively absorb knowledge but rather construct it from their experiences.
>
> (Ashton 1992, p. 322)

The Rise and Fall of Constructivism

During the 1980s and 1990s, there were countless professional development workshops, conference presentations, journal articles and books, all articulating constructivist theory and developing its pedagogical implications. This is not the place for a detailed history of constructivism in education,[2] but, to justify the claim about its domination of educational theory in the last decades of the twentieth century, it suffices to mention a few well-regarded review articles.

Peter Fensham claimed that, 'The most conspicuous psychological influence on curriculum thinking in science since 1980 has been the constructivist view of learning' (Fensham 1992, p. 801). A former president of the US NARST said that: 'A unification of thinking, research, curriculum development, and teacher education appears to now be occurring under the theme of constructivism . . . there is a lack of polarised debate' (Yeany 1991, p. 1). Another past president of the same organisation wrote that:

> There is a paradigm war waging in education. Evidence of conflict is seen in nearly every facet of educational practice . . . [but] there is evidence of widespread acceptance of alternatives to objectivism, one of which is constructivism.
>
> (Tobin 1993, p. ix)

In 2000, two researchers quantified the impact of constructivism in educational research and practice and reported that there were over 1,000 items in the Education Resource Information Center (ERIC) database and:

> As for the quantities of materials intended for or developed by practicing teachers, a sense of their proliferation might be gleaned from the internet, where the hits for 'constructivism + education' number in the tens or hundreds of thousands, depending on the search engine used.
>
> (Davis & Sumara 2003, p. 409)

The most recent version of the authoritative 'constructivism and research' bibliography, prepared by Reinders Duit and colleagues at the University of Kiel, is available online and contains more than 1,000 entries (Duit 2009). Duit says of the bibliography that:

> This research has been carried out within what is called the constructivist view including individual constructivist and social constructivist perspectives. The bibliography may therefore now be viewed as an attempt to document constructivist research in science education.

Constructivist theory has had an impact on education way beyond the confines of research journals and scholarly conferences that can be documented in ERIC searches; it is adopted as the 'official' pedagogical theory in a number of countries, states and provinces. These include at least the following: Ontario Province of Canada, Thailand, Greece, Turkey, New Zealand, India, Taiwan, Spain, the Australian state of Western Australia and some states and school districts of the US; it is ubiquitous in North American teacher education programmes.

Although seeming to sweep all before it, constructivism has had its critics: a good many psychologists, philosophers, educators, teachers and parents have raised their concerns over diverse parts of the constructivist programme. From the outset, there have been philosophical criticisms[3] and, more recently, detailed criticism of its claims to being a guide for successful pedagogy.[4] Pleasingly, there are signs that at least 'serious' or 'radical' constructivist influence is waning, and that more realistic, limited and better-grounded philosophical and instructional claims are being advanced in the name of constructivism. Indeed, one prominent advocate has published an article titled, 'Constructivism in Education: Moving On' (Tobin 2000). Similarly, in medical schools, PBL is moving on and out. One researcher concluded there was:

> No convincing evidence that PBL improves the knowledge base and clinical performance, at least not of the magnitude that would be expected given the resources required for a PBL curriculum.

So, when enthusiasm for an educational programme, much less a 'world-view', goes from 'winning a paradigm war' to 'moving on', all within two decades, it behoves researchers to take stock of why this has happened and draw some lessons about the discipline of science education. What needs to be overcome is the field's propensity for importing philosophical, psychological and political theories, while only minimally understanding them, and for having enthusiasm substitute for conceptual coherence and empirical evidence.

Versions of Constructivism

Constructivism is a heterogeneous movement. One review has identified at least the following varieties: contextual, dialectical, empirical, information-processing, methodological, moderate, Piagetian, post-epistemological, pragmatic, radical, realist, social and socio-historical (Good *et al.* 1993). To this list could be added humanistic constructivism (Cheung & Taylor 1991) and didactic constructivism (van den Brink 1991). From its origins in developmental psychology, constructivism has spread to encompass, often naively, many domains of educational enquiry. The range of constructivist concerns can be seen in the subheadings of one science-education article: 'A Constructivist View of Learning', 'A Constructivist View of Teaching', 'A View of Science', 'Aims of Science Education', 'A Constructivist View of Curriculum' and 'A Constructivist View of Curriculum Development' (Bell 1991).

There are basically two major traditions of constructivism. The first is psychological constructivism, originating with Jean Piaget's account of children's learning as a process of personal, individual, intellectual construction arising from their activity in the world. This tradition bifurcates into, on the one hand, the more personal, subjective tradition of Piaget that can be seen in von Glasersfeld's work, and, on the other hand, into the social constructivism of the Russian Vygotsky and his followers, who stress the importance of language communities for the cognitive constructions of individuals, as can be seen in the work of Duckworth (1996), Gergen (1994, 1999) and Lave (1988).

The second major tradition is sociological constructivism, originating with Emile Durkheim and augmented by sociologists of culture such as Peter Berger and, more recently, by sociologists of science in the Edinburgh School, such as Barry Barnes, David Bloor, Harry Collins and Bruno Latour. This sociological tradition maintains that scientific knowledge is socially constructed and vindicated, and it investigates the circumstances and dynamics of science's construction. In contrast to Piaget and Vygotsky, it ignores the individual psychological mechanisms of belief construction and focuses upon the extra-individual social circumstances that, it claims, determine the beliefs of individuals; the individual becomes a sort of 'black box' for the theory. Extreme forms of sociological constructivism claim that science is nothing but a form of human cognitive construction, comparable to artistic or literary construction, and having no particular claim to truth; dominant theories are the theories of dominant scientists.[5]

For many, constructivism has ceased being just a learning theory, or even an educational theory, but rather it constitutes a worldview, or *Weltanschauung*,[6] as suggested in remarks such as:

> To become a constructivist is to use constructivism as a referent for thoughts and actions. That is to say when thinking or acting, beliefs associated with constructivism assume a higher value than other beliefs. For a variety of reasons the process is not easy.
>
> (Tobin 1991, p. 1)

The Constructivist Foundations website identifies the common features of all constructivist positions as being as follows:[7]

- Constructivist approaches question the Cartesian separation between objective world and subjective experience.
- Consequently, they demand the inclusion of the observer in scientific explanations.
- Representationalism is rejected; knowledge is a system-related cognitive process rather than a mapping of an objective world on to subjective cognitive structures.
- According to constructivist approaches, it is futile to claim that knowledge approaches reality; reality is brought forth by the subject rather than passively received.

- Constructivist approaches entertain an agnostic relationship with reality, which is considered beyond our cognitive horizon; any reference to it should be refrained from.
- Therefore, the focus of research moves from the world that consists of matter to the world that consists of what matters.
- Constructivist approaches focus on self-referential and organisationally closed systems; such systems strive for control over their inputs rather than their outputs.
- With regard to scientific explanations, constructivist approaches favour a process-oriented approach rather than a substance-based perspective, e.g. living systems are defined by processes whereby they constitute and maintain their own organisation.
- Constructivist approaches emphasise the 'individual as personal scientist' approach; sociality is defined as accommodating within the framework of social interaction.
- Finally, constructivist approaches ask for an open and less dogmatic approach to science, in order to generate the flexibility that is needed to cope with today's scientific frontier.

It should be clear that the appraisal of each of these foundations of constructivism requires some competence in HPS; it is the latter that the claims are about. Philosophers would regard most of the above claims as being unintelligible, false or at best highly contentious. However, many educators 'take them on board' in an entirely naive and ill-informed manner, and they build educational edifices, curricula and teaching practices upon them.

Constructivism as Psychology and Philosophy

Constructivism is standardly presented as both as a theory of learning (a psychological theory) and a theory of knowledge (a philosophical, and specifically epistemological, theory). It is self-consciously a composite theory. A typical account of the theory is given by Catherine Fosnot in a much-cited constructivist anthology:

> Constructivism is a theory about knowledge and learning; it describes both what 'knowing' is and how one 'comes to know'. Based on work in psychology, philosophy, science and biology, the theory describes knowledge not as truths to be transmitted or discovered, but as emergent, developmental, non-objective, viable constructed explanations by humans engaged in meaning-making in cultural and social communities of discourse. Learning from this perspective is viewed as a self-regulatory process of struggling with the conflict between existing personal models of the world and discrepant new insights, constructing new representations and models of reality as a human meaning-making venture with culturally developed tools and symbols, and further negotiating such meaning through cooperative social activity, discourse, and debate in communities of practice.

> Although constructivism is not a theory of teaching, it suggests taking a radically different approach to instruction from that used in most schools.
>
> (Fosnot 2005, p. ix)

This characterisation of constructivism as being a composite of learning theory (psychology) with theory of knowledge (philosophy) is apparent in the writing of the founders of educational constructivism – Piaget, Vygotsky and Bruner. Piaget called his own theory 'Genetic Epistemology', and this philosophical concern is reflected in the title of one of his books – *Psychology and Epistemology* (Piaget 1972). Jerome Bruner, speaking of his famous *The Process of Education* book (Bruner 1960), which presented a constructivist alternative to didactic, transmissionist, behaviourist-informed 'banking' pedagogy, wrote that:

> Its ideas sprang from epistemology and the sciences of knowing . . . all of us were, I think, responding to the same 'epistemic' malaise, the doubts about the nature of knowing that had come first out of the revolution in physics and then been formalized and amplified by philosophy.
>
> (Bruner 1983, p. 186)

Clearly, claims, of whatever kind, about the impact on epistemology of the revolution in physics are matters for historians and philosophers of science to appraise. And they have done so. Educators, of necessity, are thrust into this field and need to be cognisant of at least its contours.

It is important to recognise a persistent ambiguity in the constructivist linking of learning theory to epistemology. The founders of constructivism regarded epistemology seriously, as a philosophical endeavour; they offered accounts of what constitutes human knowledge and how knowledge claims were compared and tested. Subsequently, however, many in the constructivist tradition simply collapse epistemology into psychology and, although they talk about studying the acquisition of knowledge, they really mean studying the acquisition of beliefs. One such example occurs in a recent book of Andreas Quale, *Radical Constructivism* (Quale 2008). Quale says learning is the process through which we gain knowledge, and knowledge is the product of the learning process (p. 45). He of course recognises, with Plato and the bulk of the philosophical and common-sensical tradition, that 'it is possible to learn things that are not true' (p. 45), but this is not a bother to him, because 'such an association of knowledge with truth is not made in constructivism' (p. 45). However, merely saying that there is no problem in identifying learning with knowledge does not mean that there is none. In children's books, a threat might go away because a person closes their eyes, but this comforting event seldom happens in the real world, or in the world of philosophical argument.

On the basis of 'armchair' psychology, one can assert that the neurological and psychological processes whereby knowledge is acquired are the same as those processes whereby ignorance is acquired. Learning is truth-neutral: one can equally learn good and bad habits, reasonable and unreasonable opinions, true and false beliefs. Over time, and perhaps even in the present, the bulk of what humans have learned about the world, their societies and perhaps even themselves has been false (think how many people believed, and still believe, the sun rotates around the earth; how many US citizens believe in Special Creation; how many US citizens believe the Iraq invasion was about spreading democracy, and so on), and any decent learning theory will account for this learning.

For a psychologist of learning, the learning of Islam in a Pakistani mosque, the learning of Judaism in a Yeshiva, the learning of Roman Catholicism in a seminary, the learning of Marxism–Leninism in a Politburo training school will all be considered cases of learning a subject matter. The neurological, psychological and behavioural processes of learning the particular content will be the same in all cases: learning Islam will be the same as learning Christianity. The psychological processes involved in becoming a Republican are going to be pretty much the same as becoming a Democrat (or whatever the local political options might be). The processes of learning Creation science and of learning Darwinian science will be the same. The psychologist, as learning theorist, has no special interest in appraising the truthfulness or adequacy of what is learned. A clinical psychologist might have such an interest, and hopefully they would, but the learning theorist need not have such interest in the epistemological status of what is learned.

Likewise, theories of teaching will be subject matter blind: there can be good teachers of falsehoods as well as of truths, of ideology as well as of science; the characteristics of both will be the same. Goebbels could not be faulted for being a bad teacher: what he taught was bad, but his teaching was good.[8]

However, for science education, psychology needs to be linked to and informed by philosophy. In science classes, it is putative truths about the world that are being taught, and it is rational learning that is sought.[9] Understanding what constitutes truth and rationality is a philosophical endeavour, and HPS has an important contribution to the educator's deliberations.

An Evidential Dilemma

There is a clear 'evidential dilemma' for constructivists who try to support their theory by empirical research. On the one hand, they wish to appeal to the nature of cognitive realities (learning processes) and epistemological realities (especially the history of science) to support their pedagogical, curricular and epistemological proposals. Yet, on the other hand, they say that such realities cannot be known, or are forever inaccessible to us. For many constructivists, reality collapses into the completely subjective 'my experience of reality'.[10]

Thus, one researcher, who champions 'socio-transformative constructivism' and who supports the position with a study of eighteen students in a secondary science-methods class, is impelled to remark that:

> Note that by using the term empirical evidence, I am not taking a realist or empiricist stance, nor any other Western orientation. I use the term 'empirical evidence' with the understanding that knowledge is socially constructed and always partial. By 'empirical evidence' I mean that information was systematically gathered and exposed to a variety of methodology checks. Hence in this study I do not pretend to capture the real world of the research participants (realism), nor do I pretend to capture their experiential world (empiricism). What I do attempt is to provide spaces where the participants' voices and subjectivities are represented along with my own voice and subjectivities.
>
> (Rodriguez 1998, p. 618)

One can get the general drift of what is being asserted here, namely that there is no 'uncontested' evidence, but, instead of trying to get more adequate evidence, the author proposes that research should: 'provide spaces where the participants' voices and subjectivities are represented along with my own voice and subjectivities'. As a guide to educational research, this is completely opaque and positively invites a massive Hawthorne effect into every piece of research; indeed, if the effect is not there, then the research has been poorly (positivistically) conducted. One can understand the reluctance of funding bodies to finance the creation of such spaces when real problems abound.[11]

Unfortunately, such mystification has become the coin of the constructivist research realm and, more broadly, of the educational research realm; the latter being firmly established with the widely embraced research handbooks of Yvonna Lincoln and Ergon Guba (Guba and Lincoln 1989, Lincoln and Guba 1985). The authors write of the second that:

> The constructivist paradigm is espoused by the authors and shown to offer multiple advantages, including empowerment and enfranchisement of stake-holders, as well as an action orientation that defines a course to be followed. Not merely a treatise on evaluation theory, Guba and Lincoln also comprehensively describe the differences between the positivist and constructivist paradigms of research, and provide a practical plan of the steps and processes in conducting a fourth generation evaluation.
>
> (Guber & Lincoln 1989, back cover)

Constructivist Epistemology and Its Problems

Constructivism emphasises that science is a creative human endeavour that is historically and culturally conditioned, and that its knowledge claims are not absolute. This is certainly worth saying, but it is a truism shared by most philosophers and historians of science. Beyond this truism, constructivism is committed to certain epistemological positions that are widely disputed and rejected; given the educational influence of the doctrine, these warrant close scrutiny. At their core, both personal and social constructivism have a subjectivist and empiricist understanding of human knowledge and, consequently, of scientific knowledge. As one of the most influential constructivists in science and mathematics education has put it:

> Knowledge is the result of an individual subject's constructive activity, not a commodity that somehow resides outside the knower and can be conveyed or instilled by diligent perception or linguistic communication.
>
> (von Glasersfeld 1990a, p. 37)

Since Plato, no one has thought that knowledge comes just from looking, even diligent looking. As Plato remarked, 'We see through the eye, not with the eye'. But, equally clearly, knowledge can and must be conveyed by linguistic communication.[12] Each time one asks directions to a café or a restroom in a new town, knowledge is conveyed linguistically from the local who has it to the out-of-towner who does not have it: 'take the second on the right and go along 50 metres'. Knowing *how* might be difficult to transmit linguistically, but knowing *that* can only be transmitted linguistically.[13] When obvious realities are denied, it is a signal that the person is captured by an ideology.

Some extracts from various sources can give a sense of the epistemological and ontological positions adopted by constructivists in science education; they are all variants of a subject-centred, empiricist theory of knowledge:

> Although we may assume the existence of an external world we do not have direct access to it; science as public knowledge is not so much a discovery as a carefully checked construction.
>
> (Driver & Oldham 1986, p. 109)

> Put into simple terms, constructivism can be described as essentially a theory about the limits of human knowledge, a belief that all knowledge is necessarily a product of our own cognitive acts. We can have no direct or unmediated knowledge of any external or objective reality. We construct our understanding through our experiences, and the character of our experience is influenced profoundly by our cognitive lens.
>
> (Confrey 1990, p. 108)

Steven Lerman (1989), following Kilpatrick (1987) and, earlier, von Glasersfeld, suggests that the core epistemological theses of constructivism are as follows:

1 Knowledge is actively constructed by the cognising subject, not passively received from the environment.
2 Coming to know is an adaptive process that organises one's experiential world; it does not discover an independent, preexisting world outside the mind of the knower.

Relativism

All constructivists are epistemological relativists: To deny that one account or theory can be better than another (with 'better' inevitably placed in scare quotes by constructivists) and, likewise, to deny that one account might be more true than another simply go with the constructivist territory.[14] Such relativism has its philosophical problems.[15] Clearly, lots of different things can make sense to people, and people can disagree about whether a particular proposition makes sense to them or does not make sense. The ways in which a proposition can make sense is independent of the reference of the proposition; matters about the truth of a proposition are not so liberal: they depend upon how the world is and what claims we make about it. Consequently, 'making sense' is a very unstable plank with which to prop up curriculum proposals and adjudicate debates about curriculum content.

Furthermore, most scientific advances have entailed commitment to propositions that literally defied sense – Copernicus's rotating Earth, Galileo's point masses and colourless bodies; Newton's inertial systems that, in principle, cannot be experienced and also his ideas of action at a distance; Darwin's gradualist evolutionary assumptions, so at odds with the fossil record; Einstein's mass–energy equivalence and so forth. Indeed, the topic of pendulum motion, as we have seen, exhibits the problems with using 'making sense' as a goal and arbiter in science education. In the theoretical object of classical mechanics, the bob at its highest point is both at rest and accelerating with the acceleration of gravity; at its lowest point, it is moving with maximum speed in a tangential direction, and yet its acceleration is vertically upwards. Neither of these propositions makes immediate sense, and yet they are consequences of the physical theory that allows construction of the pendulum clock and successful predictions to be made about the behaviour of the real, material objects that constitute pendulums. Within the Newtonian theory of circular motion, the propositions 'make sense'. But the theory does not emerge from sensations, and, not only is it not traceable to experience, but it contradicts immediate experience and is only roughly in accord with refined, experimental experience. This is why Wolpert, among others, comments that, 'if something fits in with common sense it almost certainly isn't science ... the way in which the universe works is not the way in which common sense works' (Wolpert 1992, p. 11).

Flowing directly from constructivism's individualistic empiricism is the neglect of the inherently social aspect of scientific development. It is not just that individuals are dependent upon others for their language and conceptual furniture, but, as far as science is concerned, the growth of scientific understanding goes hand in hand with initiation into a scientific tradition, a tradition within which point masses and instantaneous accelerations make sense. A valuable tradition is passed on, not reinvented by each generation. There are serious educational questions posed by the business of selecting those aspects of a tradition worthy of transmission, and the processes whereby they are passed on. However, these questions only arise and can be addressed if this apprenticeship dimension of education is recognised. Subjective, or psychological, constructivism only dimly recognises this. Social constructivism sees it more clearly, but then needs to address the epistemological or normative elements in the social construction of knowledge.

Children's thoughts are private, but their concepts are public. Whether or not particular thoughts are going to constitute knowledge is not a matter for the individual to determine; or rather, if they do so determine, then it is against a public standard. These, and other considerations, led D.W. Hamlyn to say:

> Any view which in effect construes the child as a solitary inquirer attempting to discover the truth about the world must be rejected. (What after all could be meant by 'truth' in these circumstances?)
>
> (Hamlyn 1973, p. 184)

It is usually teachers who mediate between students and this public standard. Without such public criteria, the word 'knowledge' is reducible to 'belief'. What constitutes knowledge and what makes a claim knowledgeable are issues of great epistemological and political importance. In the facile personal constructivist view of knowledge, these questions evaporate. For social constructivism, they also evaporate, but just more slowly: What will be the social group whose agreement will make some proposition an item of knowledge?

Scepticism

Relativism is one problem, and serious enough for science teachers, but of orders more serious is when constructivism segues into complete scepticism, the view that we cannot have any knowledge of nature, its structure or properties. This is not scepticism about any particular claim (that a gremlin ate the student's essay), but global scepticism about all claims concerning the world. Constructivists constantly assert that we have no direct access to reality, that reality remains forever hidden. Antonio Bettencourt, for example, puts the matter this way: 'constructivism, like idealism, maintains that we are cognitively isolated from the nature of reality. . . . Our knowledge is, at best, a mapping of transformations allowed by that reality' (Bettencourt, 1993, p. 46). Leaving aside the problem of understanding what is meant by the second half of the claim, the first half – 'cognitive isolation' – resonates throughout constructivist writing.

Cognitive isolation from the world is a fundamental tenet of Ernst von Glasersfeld's radical constructivism. It is affirmed in just about all of his publications, with one clear statement being:

> To claim true knowledge of the world, you would have to be certain that the picture you compose on the basis of your perceptions and conceptions is in every respect a true representation of the world as it really is. But in order to be certain that it is a good match, you should be able to compare the representation to what

> it is supposed to represent. This, however you cannot do, because you cannot step out of your human ways of perceiving and conceiving.
>
> (von Glasersfeld 1995, p. 26)

Philip Kitcher calls this assertion the *The Inaccessibility of Reality Argument*, or IRA for short. He says of it that: 'the IRA is a terrorist weapon which anti-realists employ with enormous confidence' (Kitcher 2001, p. 156). It has, of course, been utilised by many in the philosophic tradition: the British empiricists, continental idealists, logical positivists and, more recently, Nelson Goodman, Hilary Putnam and Richard Rorty. So the constructivist recourse to IRA is not without honourable pedigree.

However, there have also been equally honourable opponents of IRA in philosophy. The opposing, 'common-sense realism' view was nicely stated by Moritz Schlick, in 1935. Against Carnap and Neurath, his fellow positivists, he stated:

> I have been accused of maintaining that statements can be compared with facts. I plead guilty. I have maintained this. But I protest against my punishment: I refuse to sit in the seat of the metaphysicians. I have often compared propositions to facts; so I had no reason to suppose that it couldn't be done. I found, for instance, in my Baedeker the statement: 'this cathedral has two spires'. I was able to compare it with 'reality' by looking at the cathedral, and this comparison convinced me that Baedeker's assertion was true.
>
> (Schlick 1935, pp. 65–66, in Nola 2003, p. 146)

Schlick's 'tourist' argument, of course, applies at the next level down. Viruses, bacteria, molecules and a host of microscopic entities were once only postulated and were indeed inaccessible to scientists and everyone else, but, with refined technology, they become as visible to students in laboratories as were Schlick's cathedral spires to the tourist walking through town, or, now, the details of the Moon's surface. The IRA thesis is not as 'uncontestable' as constructivists make it out to be; it has been contested and found severely wanting.[16]

Constructivist Ontology and Its Problems

Constructivists often embrace an idealist ontology, or idealist theory about the existential status of scientific and everyday objects; that is, they variously maintain that the world is created by, and dependent upon, human thought. Various Kuhn-inspired sociologists of science repeatedly state that different observers 'live in different worlds' and that they create those worlds. These astounding claims pass over the major ambiguity: on the one hand, the complete truism that different observers and different groups have different experiences; on the other, that the world in which they live varies from observer to observer and group to group. The latter is not a truism and requires some argument, as does the more advanced claim that these various worlds are created by the observer. Kenneth Gergen, an influential social constructivist, expresses this position, saying there is: 'a multiplicity of ways in which "the world" is, and can be, constructed' (Gergen 1994, p. 82). Karin Knorr-Cetina's formulation is:

> It is the thrust of the constructivist conception to conceive of scientific reality as progressively emerging out of indeterminacy and (self-referential) constructive operations, without assuming it to match any pre-existing order of the real.
>
> (Knorr-Cetina 1983, p. 135)

Educational Idealism

Ernst von Glasersfeld's radical constructivism is the best-known idealist variant in educational circles. He says:

> The realist believes his constructs to be a replica or reflection of independently existing structures, while the constructivist remains aware of the experiencer's role as originator of all structures . . . for the constructivist there are no structures other than those which the knower constitutes by his very own activity of coordination of experiential particles.
>
> (von Glasersfeld 1987, p. 104)

As will be detailed in Chapter 9, realists need not make any such claims about 'replication' and 'reflection'; they indeed make claims about the world, but recognise that 'there is more to seeing than meets the eyeball', and the claims are the outcome of social, personal and cultural circumstance.

Elsewhere, von Glasersfeld writes:

> I can no more walk through the desk in front of me than I can argue that black is white at one and the same time. What constrains me, however, is not quite the same thing in the two cases. That the desk constitutes an obstacle to my physical movement is due to the particular distinctions my sensor system enables me to make and to the particular way in which I have come to coordinate them. Indeed, if I now could walk through the desk, it would no longer fit the abstraction I have made in prior experience.
>
> (von Glasersfeld 1990b, p. 24)

This argument is flawed, and obviously so. For the realist, the inability of our body to 'walk through' another body has nothing to do with our sensory powers, but everything to do with the composition and structures of the bodies. Changing our sensory powers will no more allow us to walk through a hitherto impenetrable table than changing our shirt would allow us to do so. Upon dying, we lose all sensory powers, but this does not mean our body can then penetrate a table. Our having or not having sensory powers makes no difference to the penetrability of the table; to think that it does is just philosophical idealism.

John Staver, a deservedly prominent science educator, stated the ontological idealist position as follows:

> For constructivists, observations, objects, events, data, laws, and theory do not exist independently of observers. The lawful and certain nature of natural phenomena are properties of us, those who describe, not of nature, that is described.
>
> (Staver 1998, p. 503)

Again, this is a flawed position. Observations and theory clearly depend upon us, but not the objects observed, nor their structures. Philosophical alarm bells should ring when an author runs together 'observations' with 'events' and 'objects'. For a realist, and for any serious scientist, there are categorical differences between these classes. Only a philosophical idealist can run them together without alarm bells ringing, and, when they ring, the idealist case has to be argued, not just assumed.

Rosalind Driver, a rightly famous and influential science educator, frequently affirmed the idealist position. For instance she wrote:

> Science as public knowledge is not so much a 'discovery' as a carefully checked 'construction' . . . and that scientists construct theoretical entities (magnetic fields, genes, electron orbitals . . .) which in turn take on a 'reality'.
>
> (Driver 1988, p. 137)

Here, it is being said that the Earth does not have a structure until geophysicists impose it; there is not an evolutionary structure in the animal world until biologists impose such structure; atoms have no structure until such is imposed by physicists; and so on. One might ask: if gravity waves are our creation, why spend so much time and money looking for them?

Despite Driver's basic argument form being fallacious, it is nevertheless widespread. The argument has the form:

> Premise: Some concept is a human construction.
> Conclusion: Therefore, the referent of the concept does not exist.

One only has to state this argument to see that it is an invalid inference, and its validity depends upon making explicit a suppressed premise of the form:

> Suppressed premise: All concepts that are human constructions can have no existential reference.

But this suppressed premise is simply dogma for which no evidence is provided. Not only are 'electron orbitals' and 'magnetic fields' human constructions, but so also are 'my house', 'mountain', 'table' and all the other observational terms we use. If the foregoing widespread constructivist argument, utilised by

Rosalind Driver, were valid, then not only would electron orbitals not exist, neither would our house, nor the tables in it, nor mountains that we might live near. Indeed, given that the personal pronoun 'I' is a human construction, individual cognising subjects might not exist. However, such considerations are frequently dismissed as 'philosophical quibbles'.

Sociological Idealism

The ontological idealism here embraced by educational constructivists mirrors and is encouraged by a comparable idealism common among new-style, post-Mertonian sociologists of science, particularly those associated with the Edinburgh School.[17] The influential sociologist Emile Durkheim had written, in 1955, that:

> If thought is to be freed, it must become the creator of its own object; and the only way to attain this goal is to accord it a reality that it has to make or construct itself. Therefore, thought has as its aim not the reproduction of a given reality, but the construction of a future reality. It follows that the value of ideas can no longer be assessed by reference to objects but must be determined by the degree of their utility, their more or less 'advantageous character'.
>
> (Durkheim 1972, p. 251)

This idealism has been carried through by the Edinburgh School. Latour and Woolgar at one point say that, '"out-there-ness" is the consequence of scientific work rather than its cause' (Latour & Woolgar 1986, p. 182). They go on to say that reality is the consequence rather than the cause of scientific construction. And they also assert that, 'there is little to be gained by maintaining the distinction between the "politics" of science and its "truth"' (Latour & Woolgar 1986, p. 237). Other contributors to the Edinburgh programme say such things as, the planets are 'cultural objects' (Lynch *et al.* 1983). Harry Collins says that, 'the natural world has a small or non-existent role in the construction of scientific knowledge' (Collins 1981, p. 3). Woolgar embraces idealism, saying that his research programme,

> is consistent with the position of the idealist wing of ethnomethodology that there is no reality independent of the words (texts, signs, documents, and so on) used to apprehend it. In other words, reality is constituted in and through discourse.
>
> (Woolgar 1986, p. 312)

One can see here a confusion between ideas of real and theoretical objects, and between physical and intellectual activity. All realists acknowledge that reality does not just imprint itself on the mind of scientists or observers; few have been so ignorant as to hold the 'reflection' or 'imprinting' theory of knowledge so frequently ascribed to them in education texts. Science does not deal with real objects per se, but with real objects as they are depicted by the theoretical apparatus of science – falling coloured balls become point masses with specified accelerations, fields of peas become phenotypes of particular descriptions, bubbling solutions become chemical equations and so on. An enormous amount of intellectual effort on the part of the tradition of scientists, and of individual scientists, goes into creating these theoretical objects, with their concepts of forces, masses, genes, cells, species, equilibrium conditions and so on. The fact that the theoretical apparatus is humanly constructed, and that natural objects are only considered in theoretical dress, does not imply that the real objects are human creations, or that the real objects have no part in the appraisal of the scientific worth of the conceptual structures brought to bear upon them.

The common constructivist move is from premises stating that knowledge is a human creation, that it is historically and culturally bound, and that it is not absolute, to the conclusion that knowledge claims are either unfounded or relativist. Usain Bolt's sprinting ability is undoubtedly genetically and culturally bound, but from this we need not draw the conclusion that he is not the world's fastest timed sprinter. Nor does the recognition that this is not an absolute claim (who knows what unrecognised and untimed sprinters might be around) mean that we cannot have confidence in the claim that Bolt is faster than any known alternative. In athletics, sensible fallibilism lies between absolutism and relativism; so too in science and in most other judgemental matters.

More than 20 years ago, Wallis Suchting provided a detailed, philosophically informed, line-by-line critique of von Glasersfeld's hugely popular version of constructivism, concluding that:

> First, much of the doctrine known as 'constructivism' . . . is simply unintelligible. Second, to the extent that it is intelligible . . . it is simply confused. Third, there is a complete absence of any argument for whatever positions can be made out. . . . In general, far from being what it is claimed to be, namely, the New Age in philosophy of science, an even slightly perceptive ear can detect the familiar voice of a really quite primitive, traditional subjectivistic empiricism with some overtones of diverse provenance like Piaget and Kuhn.
>
> (Suchting, 1992, p. 247)

The critique was ignored, and the constructivist caravan moved on.

Constructivist Pedagogy and Its Problems

Most constructivists see a connection between constructivist theories of learning and knowledge on the one hand and pedagogical direction for teachers on the other. Constructivist learning theory has transposed into mathematics, literacy and science pedagogy. This is why constructivism has become so widely adopted in teacher education programmes around the world. One response to criticism of constructivist theory is to say that, although the theory might be poorly articulated and might even be psychologically and philosophically problematic, nevertheless, constructivist pedagogy is valuable and should be supported (Grandy 1997). This position is understandable, but it rests on a moot point: how efficacious is constructivist pedagogy in teaching any subject, but especially science? A good deal of research says that it is not very effective at all.

Characteristics of Constructivist Teaching

There are many constructivist-inspired, student-centred teaching methods that include project learning, discovery learning and enquiry teaching. One of the most prominent constructivists in mathematics education informs teachers that:

> In constructivism, a zone of potential construction of a specific mathematical concept is determined by the modifications of the concept children might make in, or as a result of, interactive communication in the mathematical learning environment.
>
> (Steffe 1992, p. 261)

This thirty-four-word sentence is typical of much constructivist writing: clearly, more needs be added before this advice can be 'operationalised' or even made intelligible. Two leading constructivists in science education provide one expansion of the pedagogical advice. Driver and Oldham (1986) describe constructivist teaching as embodying a number of stages or steps:

1. *Orientation*, where pupils are given the opportunity to develop a sense of purpose and motivation for learning the topic.
2. *Elicitation*, during which pupils make their current ideas on the topic of the lesson clear. This can be achieved by a variety of activities, such as group discussion, designing posters or writing.
3. *Restructuring of ideas*: this is the heart of the constructivist lesson sequence. It consists of a number of stages, including:
    - *clarification and exchange of ideas*, during which pupils' meanings and language may be sharpened up by contrast with other, and possibly conflicting, points of view held by other students or contributed by the teacher;
    - *construction of new ideas* in the light of the above discussions and demonstrations; students here can see that there are a variety of ways of interpretating phenomena or evidence;
    - *evaluation of the new ideas*, either experimentally or by thinking through their implications; students should try to figure out the best ways of testing the alternative ideas; students may at this stage feel dissatisfied with their existing conceptions.
4. *Application of ideas*, where pupils are given the opportunity to use their developed ideas in a variety of situations, both familiar and novel.

5 *Review* is the final stage, in which students are invited to reflect back on how their ideas have changed by drawing comparisons between their thinking at the start of the lesson sequence and their thinking at the end.

Driver and Oldham liken the final, review stage to the learning-about-learning emphasis that Joseph Novak and Bob Gowin (Novak & Gowin 1984) claim should be a part of all teaching. That is, as they learn material, students should, at the same time, be learning something about the process of effective learning; this has also been referred to as 'metacognition' (White & Gunstone 1989). It is important to recognise that such 'metacognition' should be both psychological (how is something best learned?) and epistemological (what makes the learned material knowledge?). Often, the second dimension is overlooked or just assumed. Attention to it can be seen in the educational research tradition of 'personal epistemology'.[18]

Clearly, such research needs be informed by the tradition of philosophical epistemology, otherwise mistaken claims (leaving aside the weasel 'may be' qualifiers) such as the following can too glibly be made:

> The constructivist mode of learning may be associated with teachers having sophisticated epistemologies, and an orientation to the traditional/transmissive conception may be reflective of teachers holding naive epistemologies associated with omniscient authority and certain knowledge.
>
> (Chan & Elliot 2004, p. 819)

Elsewhere, 'epistemological development' scales are constructed to measure the efficacy of particular pedagogical strategies or interventions, and the constructivist end is labelled 'mature', and the realist end is labelled 'immature' (Guba & Lincoln 1989)! Some researchers, who seemingly mistake indoctrination for education, give 'realist' students explicit 'write-this-down' lessons on constructivism when their conversion has not occurred in the normal course of constructivist facilitation:

> If the epistemological development is partly a factor of age, then we could simply wait for the students to become constructivists, the most mature epistemological commitment. . . . However simply exposing students to an environment in which constructivist epistemology is implicit may not be sufficient.
>
> (Roth & Roychoudhury 1994, p. 28)[19]

Curriculum Planning

Driver and Oldham go on to say that constructivist curriculum planners cannot adopt the standard model of a passive student, an active teacher and the curriculum as something the latter transmits to the former. Two changes required are that the curriculum is not seen as a body of knowledge or skills, but the programme of activities from which such knowledge or skills can possibly be acquired or constructed; and also that there is to be a shift in the status of the curriculum from that which is determined prior to teaching (though negotiable between adults), to something with a problematic status.

These comments illustrate a problem with constructivism: it frequently overreaches itself. It uses claims about learning processes and developmental psychology (the original heart of constructivism) to establish wider educational and social positions. The curriculum, for instance, does not flow from learning theory alone. Learning theory may indicate how something should be taught, but what and how much should be taught and to whom follow from different or additional considerations. Among these are judgements of social needs, personal needs, the relevant merits of different domains of knowledge and experience and, finally, due political decision-making. Constructivists frequently ignore, or implicitly assume, such considerations in extrapolating from learning theory to curriculum matters, and to educational theory more generally.

The Driver and Oldham claims, for instance, do not cast much light upon the difficult matter of curriculum development. Their move from rejecting the curriculum as a body of knowledge or skills, to saying it is a programme of activities from which such knowledge and skills might be acquired does not do away with the need to specify such knowledge or skills; it merely 'kicks the task down the road', as politicians might say. It is ambiguous to say that the curriculum has a problematic status. It may be problematic whether particular components are in the curriculum, but this is another truism, as there is always debate about the contents of the curriculum – will the Depression be included in a history curriculum? However, it does not follow from this truism that specific contents are problematic. It may be problematic whether geometry is included in high-school mathematics, but it does not follow from this alone that geometry is problematic.

Other constructivists endorse the work of educational critical theorists, such as Michael Apple, Henry Giroux, Peter McLaren and others. Jane Gilbert, for instance, says: 'There are many parallels between the literature on the development of critical pedagogy [and] the literature on constructivist learning' (Gilbert 1993, p. 35). This is, in part, because critical theorists, 'question the value of such concepts as individualism, efficiency, rationality and objectivity, and the forms of curriculum and pedagogy that have developed from these concepts (Gilbert 1993, p. 20).

Just why non-conformity, inefficiency, irrationality and subjectivity should be valued in science education we are not told; but such endorsement of critical theory is frequently done without engagement with the serious criticism levelled against the theory in broader education circles. Francis Schrag, a former president of the US Philosophy of Education Society (PES), criticised Giroux's work saying: 'The article [Giroux's] shows respect neither for logic nor the English language, nor for the cause it avows, democracy' (Schrag 1988, p. 143). Schrag cites, as an example of Giroux's style, one sentence that warrants full reproduction:

> In this case, the notion of voice is developed around a politics of difference and community that is not rooted in simply a celebration of plurality, but rather in a particular form of human community that encourages and dignifies plurality as part of an ongoing effort to develop social relations in which all voices in their differences become unified in their efforts to identify and recall moments of human suffering and the need to overcome the conditions that perpetuate such suffering.

As will be shown in Chapter 12, this sentence is unfortunately representative of much constructivist, postmodernist, supposedly critical, educational writing – or educobabble, as it is disparagingly referred to. It is an intellectual illness that, once spread, becomes the norm. A linguistic form of Gresham's law operates, whereby outrages such as Sokal's hoax become possible. A contribution that philosophy can make to educational debate is to encourage clear communication; this at least allows errors and faulty reasoning to be identified, and not just persist in an obscurantist fog. The ability to write clearly, and to express an opinion in an intelligible manner, should be one of the basic outcomes of education. If that is not achieved, then just about all else is lost.

Teaching the Content of Science

Many science educators are interested in finding out how, on constructivist principles, somebody teaches a body of scientific knowledge that is in large part abstract (depending on notions such as velocity, acceleration, force, genes, vectors), that is removed from experience (propositions about atomic structure, cellular processes, astronomic events), that has no connection with prior conceptions (ideas of viruses, antibodies, molten core, evolution, electromagnetic radiation) and that is alien to common sense and in conflict with everyday experience, expectations and concepts. Joan Solomon well articulated the problem:

> Constructivism has always skirted round the actual learning of an established body of knowledge . . . students will find that words are used in new and standardised ways: problems which were never even seen as being problems, are solved in a sense which needs to be learned and rehearsed. For a time all pupils may feel that they are on foreign land and no amount of recollection of their own remembered territory with shut eyes will help them to acclimatise.
>
> (Solomon 1994, p. 16)

Teaching a body of knowledge involves, not just teaching the concepts, but also the method, and something of the methodology or theory of method. How all of this is to be taught, without teachers actually conveying something to pupils, is a moot point. It is impossible for a person to learn to play chess without the rules of chess in some way being conveyed to him or her; they cannot be made up by the individual. A student who asks if the rook can move diagonally needs to be told 'no'; this piece of knowledge can be, and has to be, transferred from someone who knows to someone who does not know.

The Efficacy of Constructivist Methods

The supposed efficacy of constructivist, or minimally guided, pedagogy has long been challenged by educational researchers. Controversy about literacy teaching, the 'Reading Wars', was the public face of this research debate. The efficacy of constructivist-inspired 'whole language' and 'reading recovery' literacy programmes has been contrasted with phonic instruction in countless studies. After three decades of studies, the overwhelming conclusion is that constructivist teaching does little to develop reading competence and is markedly less successful than its non-constructivist, phonics alternative. Although learning to speak is natural, learning to read is unnatural and needs to be explicitly taught.

Twenty years ago, it was well documented that New Zealand's own, path-breaking, constructivist-driven reading recovery (or 'whole language') programmes did not work, and where they did it was on account of the influence of 'extra-classroom' influences (Matthews 1995, Chapter 2). These results did not prevent reading recovery sweeping across the international elementary-school world and its founder, Marie Clay, being knighted. The unfortunate outcome was that it wasted millions of teacher-hours of time and thwarted the literacy of hundreds of thousands, if not millions, of children, and is still doing so.[20] The most recent New Zealand national report has looked at studies in the intervening two decades and reaches the original conclusion:

> Three interrelated factors were identified as contributing to the failure of New Zealand's literacy strategy: a rigidly constructivist orientation toward literacy education, the failure to respond adequately to differences in literate cultural capital at school entry, and restrictive policies regarding the first year of literacy teaching.
>
> (Tunmer *et al.* 2013, p. 34)

The same research arguments were thrashed out in the equally public 'Maths Wars', with the same outcome: constructivist-inspired maths teaching leaves students innumerate (Geary 1995, Klein 2007). One psychologist, introducing his appraisal of this research, says it:

> begins with an overview and critique of basic philosophical themes that currently guide educational practice in the United States, in particular the constructivist view of mathematics education. . . . At the same time, many of these educational researchers have ignored or dismissed a large body of relevant psychological research and theory.
>
> (Geary 1995, p. 31)

Richard Mayer, a past-president of the Division of Educational Psychology of the American Psychological Association, a former editor of the *Educational Psychologist* and a former co-editor of *Instructional Science*, in a landmark study, reviewed an extensive body of research on constructivist pedagogy and concluded that it did not work, and, where it did work, it was by virtue of departing from constructivist principles (Mayer 2004). His analysis was confirmed by Kirschner, Sweller and Clark, who, in a review article, argued that:

> The past half century of research on this issue has provided overwhelming and unambiguous evidence that unguided or minimally guided learning is significantly less effective and efficient than guidance that is specifically designed to support the cognitive processing necessary for learning. Not only is minimally-guided learning ineffective for most learners, it may even be harmful for some. . . . The best evidence developed over the past half century supports the view that minimally-guided learning does not enhance student achievement any more than throwing a non-swimmer out of a boat in the middle of a deep lake supports learning to swim.
>
> (Kirschner *et al.* 2006, p. 75)

Such conclusions seem obvious, and dictated by the very nature of the discipline of science. Someone learning to play chess has to be told the rules by someone who knows the rules; learners cannot make up the rules, they cannot negotiate the rules, and even if they brainstorm to the conclusion that rooks can move diagonally, this does not mean that rooks can so move. Knowledge of what is allowed and not allowed in chess has to be transmitted; further competence in chess depends, not just on knowing the rules, but on guidance and worked examples, on seeing how better players have responded to similar situations; so it is in learning science.

E.D. Hirsch Jr, in his *The Schools We Need*, documents the impact of constructivism in the USA and concludes:

> In short, the term 'constructivism' has become a kind of magical incantation used to defend discovery learning, which is no more sanctioned by psychological theory than any other form of constructed learning. To pretend that it is so sanctioned illustrates what I mean by the 'selective use of research'.
>
> (Hirsch 1996, p. 135)

Cultural Consequences of Constructivism

Constructivism is fraught with grave educational and cultural implications that are seldom recognised, much less engaged with. All cultures build up traditions and understandings that they pass on in formal and informal settings. Having such traditions is the hallmark of a healthy culture. Each new generation does not have to start completely anew the task of making meaning. Radical constructivism, with its in-principle aversion to transmission of knowledge, makes tradition nugatory; indeed, if it is seriously adopted, it destroys traditional culture. The core of traditional, indeed any healthy, culture is the transmission of the culture's beliefs and mores; it is plainly ridiculous, and culture destroying, for constructivists to maintain that putative knowledge cannot be so transmitted.[21]

On the other hand, it is notorious that people have, for centuries, thought that the grossest injustices, and the greatest evils, have all made sense. The subjection of women to men has, and still does, make perfectly good sense to millions of people and to scores of societies; explaining illness in terms of possession by evil spirits makes perfectly good sense to countless millions; the intellectual inferiority of particular races is perfectly sensible to millions of people, including some of the most advanced thinkers; to very sophisticated Nazi Germans, it made sense to regard Jewish people as subhumans and to institute extermination programmes for them; apartheid made sense to South Africans, just as racial discrimination did to US citizens until very recently. The list of atrocities and stupidities that have made perfect sense at some time or other, or in some place or other, is endless. It seems clear that the appeal to sense is not going to be sufficient to refute such views. However, the appeal to truth, or right, which is independent of human desires or power, may be able to overturn such opinions and practices. Certainly, the interests of the less powerful and marginalised are not advanced by championing the view that power is truth; minority rights have always been better advanced by holding on to the view that truth is power. Michael Devitt, recognising these and other problems, commented that:

> I have a candidate for the most dangerous contemporary intellectual tendency, it is . . . constructivism. Constructivism is a combination of two Kantian ideas with twentieth-century relativism. The two Kantian ideas are, first, that we make the known world by imposing concepts, and, second, that the independent world is (at most) a mere 'thing-in-itself' forever beyond our ken. . . . [Considering] its role in France, in the social sciences, in literature departments, and in some largely well-meaning, but confused, political movements [it] has led to a veritable epidemic of 'world-making'. Constructivism attacks the immune system that saves us from silliness.
>
> (Devitt 1991, p. ix)

The relativism, and subjectivism, of constructivism is particularly ill suited to dealing with the complex, trans-social problems facing the contemporary world. There is a need for the sustained application of Enlightenment reason and the rejection of self-interest in the attempt to deal with pressing environmental, political and social questions – think of the political situation in many parts of Africa, the Middle East, the Indian subcontinent, the Balkans and elsewhere.[22] Karl Popper recognised this socially corrosive aspect of constructivism, when he said:

> The belief of a liberal – the belief in the possibility of a rule of law, of equal justice, of fundamental rights, and a free society – can easily survive the recognition that

> judges are not omniscient and may make mistakes about facts. . . . But the belief in the possibility of a rule of law, of justice, and of freedom, can hardly survive the acceptance of an epistemology which teaches that there are no objective facts; not merely in this particular case, but in any other case.
>
> (Popper 1963, p. 5)

Conclusion

This chapter has given an indication of the enormous impact of constructivism on the theory and practice of science education, and has appraised the explicit epistemological and ontological claims made by constructivists. The general conclusion reached has been that, in as much as there are arguments advanced for the epistemological and ontological positions, they are weak arguments. Constructivism amounts to a restatement of standard empiricist theory of science and suffers all the well-known faults of that theory.

However, the interactive, anti-dogmatic teaching practices supported by constructivism need not be abandoned. Von Glasersfeld acknowledges that: 'Good teachers . . . have practised much of what is suggested here, without the benefit of an explicit theory of knowledge . . . their approach was intuitive and successful' (von Glasersfeld 1989, p. 138). Other epistemologies and other educational theories can equally suggest and demand humane, engaged, interactive, antidogmatic and intellectual teaching, aimed at the development of critical capacities and well-formed understandings. It is clear that the best of constructivist pedagogy can be had without constructivist epistemology – Socrates, Montaigne, Locke, Priestley, Mill and Russell are just some who have conjoined engaging, constructivist-like pedagogy with non-constructivist epistemology. Since Socrates, this has characterised the best of liberal education, but, as with Socrates, realists and non-sceptics have been prepared to challenge students' firmly held beliefs, which may well be reinforced by an overwhelming amount of common-sense experience and deeply held cultural values. In brief, what is good in constructivism has long been known in philosophy and in the liberal tradition of education, and that what is novel in constructivism is misguided and dangerous to both education and society.

Notes

1 An extensive PhD thesis on the pedagogical and theoretical deficiencies of PBL is being written by Gary Niven, in the School of Medicine at the University of Queensland.
2 There is yet to appear a comprehensive historical account of the ups and downs, and the external and internal dynamics, of the constructivist wave that moved through education from the 1970s to the early years of this century. Limited historical reviews can be found in Osborne (1996), Phillips (1997b, 2000) and Solomon (1994).
3 See especially: Bowers (2007), Grandy (1997), Kragh (1998), Matthews (1993, 2000), McCarty and Schwandt (2000), Nola (1997, 2003), Phillips (1997a, 1997b, 2000), Scerri (2003), Slezak (2000, 2010, 2014), Small (2003) and Suchting (1992).
4 See especially, Kirschner *et al.* (2006), Mayer (2004) and contributions to Tobias and Duffy (2009).

5 The claims of sociological constructivism and its contentious and revolutionary implications for science education are examined and mostly refuted in Slezak (1994a, 1994b).
6 The German expression for 'world outlook' is more directly connected to feelings, ethics and personal and political action than the more passive, spectator-like Anglo term 'worldview'.
7 From the Constructivist Foundations website: www.univie.ac.at/constructivism/journal
8 On the distinction between teaching and education, see Hirst (1971); on the concept of indoctrination, see contributions to Snook (1972).
9 On the notion of 'rational learning', see Hamlyn (1973).
10 On the relationship between constructivism and classic empiricism, see Matthews (1993) and Suchting (1992).
11 See contributions to NRC (2002).
12 See, for instance, Hamlyn (1978).
13 The distinction owes its modern form to Ryle (1949).
14 See Bickhard (1997) and Niiniluoto (1991).
15 See, for instance, Siegel (1987), Norris (1997), and contributions to Nola (1988).
16 See, especially, Nola (2003) and Papayannakos (2008).
17 The idealism of these sociologists has been well surveyed by Bunge (1991, 1992).
18 Some central studies are: Burr and Hofer (2002), Chinn and Malhotra (2002), Hofer and Pintrich (1997), Kuhn *et al.* (2000) and Schommer (1994).
19 This case is elaborated and discussed in Matthews (1998).
20 There is an enormous literature on this subject, with governments around the world commissioning their own reports. With qualifications, the overall conclusion that children need to be directly taught reading, phonic-by-phonic and then combinations of phonics, still holds. This does not, of course, guarantee that they will read; the latter depends on having engaging material to read.
21 This is the thrust of Chet Bowers' critique of constructivism as being a vehicle for Western imperialism (Bowers 2007). On this issue, see Taber (2009, pp. 148–160).
22 Harvey Siegel has written on these themes, with one paper appropriately titled: 'Radical Pedagogy Requires "Conservative" Epistemology' (Siegel 1995).

References

Ashton, P.T.: 1992, 'Editorial', *Journal of Teacher Education* 43(5), 322.

Bell, B.F.: 1991, 'A Constructivist View of Learning and the Draft Forms 1–5 Science Syllabus', *SAME Papers 1991*, 154–180.

Bettencourt, A.: 1993, 'The Construction of Knowledge: A Radical Constructivist View'. In K. Tobin (ed.) *The Practice of Constructivism in Science Education*, AAAS Press, Washington, DC, pp. 39–50.

Bickhard, M.H.: 1997, 'Constructivism and Relativisms: A Shopper's Guide', *Science & Education* 6(1–2), 29–42. Reprinted in M.R. Matthews (ed.) *Constructivism in Science Education: A Philosophical Examination*, Kluwer Academic Publishers, Dordrecht, The Netherlands, 1998, pp. 99–112.

Bowers, C.A.: 2007, *The False Promises of Constructivist Theories of Learning: A Global and Ecological Critique*, Peter Lang, New York.

Bruner, J.S.: 1960, *The Process of Education*, Random House, New York.

Bruner, J.S.: 1983, *In Search of Mind: Essays in Autobiography*, Harper & Row, New York.

Bunge, M.: 1991, 'A Critical Examination of the New Sociology of Science: Part 1', *Philosophy of the Social Sciences* 21(4), 524–560.

Bunge, M.: 1992, 'A Critical Examination of the New Sociology of Science: Part 2', *Philosophy of the Social Sciences* 22(1), 46–76.

Burr, J.E. and Hofer, B.K.: 2002, 'Personal Epistemology and Theory of Mind: Deciphering Young Children's Beliefs About Knowledge and Knowing', *New Ideas in Psychology* 20, 199–224.

Chan, K.W. and Elliott, R.G.: 2004, 'Relational Analysis of Personal Epistemology and Conceptions About Teaching and Learning', *Teaching and Teacher Education* 20(8), 817–831.
Cheung, K.C. and Taylor, R.: 1991, 'Towards a Humanistic Constructivist Model of Science Learning: Changing Perspectives and Research Implications', *Journal of Curriculum Studies* 23(1), 21–40.
Chinn, C.A. and Malhotra, B.A.: 2002, 'Epistemologically Authentic Reasoning in Schools: A Theoretical Framework for Evaluating Inquiry Tasks', *Science Education* 86, 175–218.
Collins, H.M.: 1981, 'Stages in the Empirical Programmes of Relativism', *Social Studies of Science* 11, 3–10.
Colliver, J.A.: 2000, 'Effectiveness of Problem-Based Learning Curricula: Research and Theory', *Academic Medicine* 75, 259–266.
Colliver, J.A.: 2002, 'Constructivism: The View of Knowledge That Ended Philosophy or a Theory of Learning and Instruction?' *Teaching and Learning in Medicine* 14(1), 49–51.
Confrey, J.: 1990, 'What Constructivism Implies for Teaching'. In R. Davis, C. Maher and N. Noddings (eds) *Constructivist Views on the Teaching and Learning of Mathematics*, National Council of Teachers of Mathematics, Reston, VA, pp. 107–124.
Davis, B. and Sumara, D.: 2003, 'Constructivist Discourses and the Field of Education: Problems and Possibilities', *Eduational Theory* 52(4), 409–428.
Devitt, M.: 1991, *Realism and Truth*, 2nd edn, Basil Blackwell, Oxford, UK.
Driver, R.: 1988, 'A Constructivist Approach to Curriculum Development'. In P. Fensham (ed.) *Development and Dilemmas in Science Education*, Falmer Press, New York, pp. 133–149.
Driver, R. and Oldham, V.: 1986, 'A Consructivist Approach to Curriculum Development in Science', *Studies in Science Education* 13, 105–122.
Duckworth, E.: 1996, *The Having of Wonderful Ideas*, 2nd edn, Teachers College Press, Columbia University, New York (1st edition, 1987).
Duit, R.: 2009, *Bibliography – STCSE*, www.ipn.uni-kiel.de/aktuell/stcse/stcse.html
Durkheim, E.: 1972, *Selected Writings* (ed. and trans. A. Giddens), Cambridge University Press, Cambridge, UK.
Fensham, P.J.: 1992, 'Science and Technology'. In P.W. Jackson (ed.) *Handbook of Research on Curriculum*, Macmillan, New York, pp. 789–829.
Fosnot, C.T. (ed.): 2005, *Constructivism: Theory, Perspectives, and Practice*, 2nd edn, Teachers College Press, New York.
Geary, D.C.: 1995, 'Reflections of Evolution and Culture in Children's Cognition: Implications for Mathematical Development and Instruction', *American Psychologist* 50(1), 24–37.
Gergen, K.: 1994, *Realities and Relations: Soundings in Social Construction*, Harvard University Press, Cambridge, MA.
Gergen, K.: 1999, *An Invitation to Social Construction*, SAGE, London.
Gilbert, J.: 1993, 'Constructivism and Critical Theory'. In B. Bell (ed.) *I Know About LISP But How Do I Put It into Practice: Final Report of the Learning in Science Project (Teacher Development)*, Centre for Science and Mathematics Education Research, University of Waikato, Hamilton, New Zealand.
Good, R., Wandersee, J. and St Julien, J.: 1993, 'Cautionary Notes on the Appeal of the New "Ism" (Constructivism) in Science Education'. In K. Tobin (ed.) *Constructivism in Science and Mathematics Education*, AAAS, Washington, DC, pp. 71–90.
Grandy R.E.: 1997, 'Constructivism and Objectivity: Disentangling Metaphysics from Pedagogy', *Science & Education* 6(1–2), 43–53. Reprinted in M.R. Matthews (ed.) *Constructivism in Science Education: A Philosophical Examination*, Kluwer Academic Publishers, Dordrecht, The Netherlands, pp. 113–123.
Guba, E.G. and Lincoln, Y.S.: 1989, *Fourth Generation Evaluation*, SAGE, Newbury Park, CA.

Hamlyn, D.W.: 1973, 'Human Learning'. In R.S. Peters (ed.) *The Philosophy of Education*, Oxford University Press, Oxford, UK, pp. 178–194.
Hamlyn, D.W.: 1978, *Experience and the Growth of Understanding*, Routledge & Kegan Paul, London.
Hirsch, E.D.: 1996, *The Schools We Need and Why We Don't Have Them*, Doubleday, New York.
Hirst, P.H.: 1971, 'What is Teaching?', *Journal of Curriculum Studies* 3(1), 5–18. Reprinted in R.S. Peters (ed.) *The Philosophy of Education*, Oxford University Press, Oxford, UK, 1973, pp. 163–177.
Hofer, B.K. and Pintrich, P.R.: 1997, 'The Development of Epistemological Theories: Beliefs About Knowledge and Knowing and Their Relation to Learning', *Review of Educational Research* 67(1), 88–140.
Kilpatrick, J.: 1987, 'What Constructivism Might Be in Mathematics Education'. In J.C. Bergeron, N. Herscovics and C. Keiran (eds) *Psychology of Mathematics Education*, Proceedings of the Eleventh International Conference, Montreal, pp. 3–27.
Kirschner, P., Sweller, J. and Clark, R.E.: 2006, 'Why Minimally Guided Learning Does Not Work: An Analysis of the Failure of Discovery Learning, Problem-Based Learning, Experiential Learning and Inquiry-Based Learning', *Educational Psychologist* 41(2), 75–96.
Kitcher, P.: 2001, 'Real Realism: The Galilean Strategy', *The Philosophical Review* 110(2), 151–197.
Klein, D.: 2007, 'A Quarter Century of US "Math Wars" and Political Partianship', *Journal of the British Society for the History of Mathematics* 22(1), 22–33.
Knorr-Cetina, K.: 1983, 'The Ethnographic Study of Scientific Work: Towards a Constructivist Interpretation of Science'. In K. Knorr-Cetina and M. Mulkay (eds) *Science Observed: Perspectives on the Social Study of Science*, SAGE, London, pp. 115–140.
Kragh, H.: 1998, 'Social Constructivism, the Gospel of Science and the Teaching of Physics', *Science & Education* 7(3), 231–243. Reprinted in M.R. Matthews (ed.) *Constructivism in Science Education: A Philosophical Examination*, Kluwer, Dordrecht, The Netherlands, pp. 125–137.
Kuhn, D., Cheney, R. and Weinstock, M.: 2000, 'The Development of Epistemological Understanding', *Cognitive Development* 15, 309–328.
Latour, B. and Woolgar, S.: 1986, *Laboratory Life: The Social Construction of Scientific Facts*, 2nd edition, SAGE, London (1st edition, 1979).
Lave, J.: 1988, *Cognition in Practice: Mind, Mathematics and Culture in Everyday Life*, Cambridge University Press, New York.
Lerman, S.: 1989, 'Constructivism, Mathematics, and Mathematics Education', *Educational Studies in Mathematics* 20, 211–223.
Lincoln, Y.S. and Guba, E.G.: 1985, *Naturalistic Inquiry*, SAGE, Newbury Park, CA.
Lynch, M., Livingstone, E. and Garfinkel, H.: 1983, 'Temporal Order in Laboratory Work'. In K.D. Knorr-Cetina and M. Mulkay (eds) *Science Observed*, SAGE, London, pp. 205–238.
Matthews, M.R.: 1993, 'A Problem With Constructivist Epistemology'. In H.A. Alexander (ed.) *Philosophy of Education 1992*, US Philosophy of Education Society, Urbana, IL, pp. 303–311.
Matthews, M.R.: 1995, *Challenging New Zealand Science Education*, Dunmore Press, Palmerston North, New Zealand.
Matthews, M.R.: 1998, 'In Defence of Modest Goals for Teaching About the Nature of Science', *Journal of Research in Science Teaching* 35(2), 161–174.
Matthews, M.R.: 2000, 'Constructivism in Science and Mathematics Education'. In D.C. Phillips (ed.) *National Society for the Study of Education 99th Yearbook*, National Society for the Study of Education, Chicago, IL, pp. 161–192.
Mayer, R.E.: 2004, 'Should There Be a Three-Strikes Rule Against Pure Discovery Learning? The Case for Guided Methods of Instruction', *American Psychologist* 59(1), 14–19.

McCarty, L.P. and Schwandt, T.A.: 2000, 'Seductive Illusions: von Glasersfeld and Gergen on Epistemology and Education'. In D.C. Phillips (ed.) *Constructivism in Education: 99th Yearbook of the National Society for the Study of Education*, NSSE, Chicago, IL, pp. 41–85.

Neville, A.J.: 2009, 'Problem-Based Learning and Medical Education Forty Years On: A Review of Its Effects on Knowledge and Clinical Performance', *Medical Principles and Practice* 18, 1–9.

Niiniluoto, I.: 1991, 'Realism, Relativism and Constructivism', *Synthese* 89(1), 135–162.

Nola, R. (ed.): 1988, *Relativism and Realism in Science*, Reidel, Dordrecht, The Netherlands.

Nola, R.: 1997, 'Constructivism in Science and in Science Education: A Philosophical Critique', *Science & Education* 6(1–2), 55–83. Reproduced in M.R. Matthews (ed.) *Constructivism in Science Education: A Philosophical Debate*, Kluwer Academic Publishers, Dordrecht, The Netherlands, 1998, pp. 31–59.

Nola, R.: 2003, '"Naked Before Reality; Skinless Before the Absolute": A Critique of the Inaccessibility of Reality Argument in Constructivism', *Science & Education* 12(2), 131–166.

Norris, C.: 1997, *Against Relativism: Philosophy of Science, Deconstruction and Critical Theory*, Blackwell, Oxford, UK.

Novak, J.D. and Gowin, D.B.: 1984, *Learning How to Learn*, Cambridge University Press, New York.

NRC (National Research Council): 2002, *Scientific Research in Education*, R.J. Shavelson and L. Towne (eds), National Academy Press, Washington, DC.

Osborne, J.: 1996, 'Beyond Constructivism', *Science Education* 80(1), 53–82.

Papayannakos, D.P.: 2008, 'Philosophical Skepticism not Relativism is the Problem with the Strong Programme in Science Studies and with Educational Constructivism', *Science & Education* 17(6), 573–611.

Phillips, D.C.: 1997a, 'Coming to Terms With Radical Social Constructivisms', *Science & Education* 6(1–2), 85–104. Reprinted in M.R. Matthews (ed.) *Constructivism in Science Education: A Philosophical Examination*, Kluwer Academic Publishers, Dordrecht, The Netherlands, pp. 139–158.

Phillips, D.C.: 1997b, 'How, Why, What, When, and Where: Perpsectives on Constructivism in Psychology and Education', *Issues In Education* 3(2), 151–194.

Phillips, D.C.: 2000, 'An Opinionated Account of the Constructivist Landscape'. In D.C. Phillips (ed.) *Constructivism in Education*, National Society for the Study of Education, Chicago, IL, pp. 1–16.

Piaget, J.: 1972, *Psychology and Epistemology: Towards a Theory of Knowledge*, Penguin, Harmondsworth, UK.

Popper, K.R.: 1963, *Conjectures and Refutations: The Growth of Scientific Knowledge*, Routledge & Kegan Paul, London.

Quale, A.: 2008, *Radical Constructivism. A Relativist Epistemic Approach to Science Education*, Sense Publishers, Rotterdam, The Netherlands.

Rodriguez, A.J.: 1998, 'Strategies for Counterresistance: Toward Sociotransformative Constructivism and Learning to Teach Science for Diversity and for Understanding', *Journal of Research in Science Teaching* 35(6), 589–622.

Roth, M.-W. and Roychoudhury, A.: 1994, 'Physics Students' Epistemologies and Views About Knowing and Learning', *Journal of Research in Science Teaching* 31(1), 5–30.

Ryle, G.: 1949, *The Concept of Mind*, Hutchinson, London.

Scerri, E.R.: 2003, 'Philosophical Confusion in Chemical Education Research', *Journal of Chemical Education* 80(20), 468–474.

Schlick, M.: 1935, 'Facts and Propositions', *Analysis* 2(5), 65–70.

Schommer, M.: 1994, 'Synthesizing Epistemological Beliefs Research: Tentative Understandings and Provocative Confusions', *Educational Psychology Review* 6(4), 293–319.

Schrag, F.: 1988, 'Response to Giroux', *Educational Theory* 38(1), 143–144.

Siegel, H.: 1987, *Relativism Refuted*, Reidel, Dordrecht, The Netherlands.

Siegel, H.: 1995, 'Radical Pedagogy Requires "Conservative" Epistemology', *Journal of Philosophy of Education* 29(1), 33–46. Reproduced in his *Rationality Redeemed? Further Dialogues on an Educational Ideal*, Routledge, New York, 1997

Slezak, P.: 1994a, 'Sociology of Science and Science Education: Part I', *Science & Education* 3(3), 265–294.

Slezak, P.: 1994b, 'Sociology of Science and Science Education. Part 11: Laboratory Life Under the Microscope', *Science & Education* 3(4), 329–356.

Slezak, P.: 2000, 'A Critique of Radical Social Constructivism'. In D.C. Phillips (ed.) *Constructivism in Education: 99th Yearbook of the National Society for the Study of Education*, NSSE, Chicago, IL, pp. 91–126.

Slezak, P.: 2010, 'Radical Constructivism, Epistemology and Dynamite', *Constructivist Foundations* 6(1), 102–111.

Slezak, P.: 2014, 'Constructivism in Science Education'. In M.R. Matthews (ed.) *International Handbook of Research in History, Philosophy and Science Teaching*, Springer, Dordrecht, The Netherlands, pp. 1023–1055.

Small, R.: 2003, 'A Fallacy in Constructivist Epistemology', *Journal of Philosophy of Education* 37(3), 483–502.

Snook, I.A. (ed.): 1972, *Concepts of Indoctrination*, Routledge & Kegan Paul, London.

Solomon, J.: 1994, 'The Rise and Fall of Constructivism', *Studies in Science Education* 23, 1–19.

Staver, J.: 1998, 'Constructivism: Sound Theory for Explicating the Practice of Science and Science Teaching', *Journal of Research in Science Teaching* 35(5), 501–520.

Steffe, L.P.: 1992, 'Schemes of action and operation involving composite units', *Learning and Individual Differences* 4, 259–309.

Suchting, W.A.: 1992, 'Constructivism Deconstructed', *Science & Education* 1(3), 223–254. Reprinted in M.R. Matthews (ed.) *Constructivism in Science Education: A Philosophical Examination*, Kluwer Academic Publishers, Dordrecht, The Netherlands, 1998, pp. 61–92.

Taber, K.S.: 2009, *Progressing Science Education: Constructing the Scientific Research Programme Into the Contingent Nature of Learning Science*, Springer, Dordrecht, The Netherlands.

Tobias, S. and Duffy, T. (eds): 2009, *Constructivism Theory Applied to Instruction: Success or Failure?* Lawrence Erlbaum, Hillsdale, NJ.

Tobin, K.: 1991, 'Constructivist Perspectives on Research in Science Education.' Paper presented at the annual meeting of the National Association for Research in Science Teaching, Lake Geneva, WI.

Tobin, K. (ed.): 1993, *The Practice of Constructivism in Science and Mathematics Education*, AAAS Press, Washington, DC.

Tobin, K.: 2000, 'Constructivism in Science Education: Moving On'. In D.C. Phillips (ed.) *Constructivism in Education*, National Society for the Study of Education, Chicago, IL, pp. 227–253.

Tunmer, W.E., Chapman, J.W., Greaney, K.T., Prochnow, J.E. and Arrow, A.W.: 2013, *Why the New Zealand National Literacy Strategy has Failed and What can be Done About It*, Massey University Institute of Education, Massey, New Zealand.

van den Brink, J.: 1991, 'Didactic Constructivism', In E.von Glasersfeld (ed.) *Radical Constructivism in Mathematics Education*, Kluwer, Dordrecht, The Netherlands, pp. 195–227.

von Glasersfeld, E.: 1987, *Construction of Knowledge*, Intersystems Publications, Salinas, CA.

von Glasersfeld, E.: 1989, 'Cognition, Construction of Knowledge and Teaching', *Synthese* 80(1), 121–140.

von Glasersfeld, E.: 1990a, 'Environment and Communication'. In L.P. Steffe and T. Wood (eds) *Transforming Children's Mathematics Education: International Perspectives*, Lawerence Erlbaum, Hillsdale, NJ, pp. 30–38.

von Glasersfeld, E.: 1990b, 'An Exposition of Constructivism: Why Some Like It Hot'. In R. Davis, C. Maher and N. Noddings (eds) *Constructivist Views on the Teaching and Learning of Mathematics*, National Council of Teachers of Mathematics, Reston, VA, pp. 19–30.

von Glasersfeld, E.: 1995, *Radical Constructivism. A Way of Knowing and Learning*, The Falmer Press, London.

White, R.T. and Gunstone, R.F.: 1989, 'Metalearning and Conceptual Change', *International Journal of Science Education* 11, 577–586.

Wolpert, L.: 1992, *The Unnatural Nature of Science*, Faber & Faber, London.

Woolgar, S.: 1986, 'On the Alleged Distinction between Discourse and Praxis', *Social Studies of Science* 16, 309–317.

Yeany, R.H.: 1991, 'A Unifying Theme in Science Education?', *NARST News* 33(2), 1–3.

Chapter 9

A Central Issue in Philosophy of Science and Science Education

Realism and Anti-Realism

There are many fundamental philosophical issues raised by science that can also be raised in science classrooms. Some of these philosophical features of science have been discussed in Chapter 4, when dealing with 'Metaphysics and Air Pressure', in Chapter 5, when dealing with 'Philosophy in the Classroom', in Chapter 6, when elaborating 'Pendulum Motion', and in Chapter 7, when elaborating 'Priestley and Photosynthesis', and others will be discussed in Chapter 11, when dealing with the 'Nature of Science'. Pleasingly, philosophy does not have to be brought into science classrooms, as it is already there; it just needs to be identified and discussed in a way whereby students can themselves begin to appreciate the philosophical dimension of science and to take beginning steps in thinking philosophically; philosophy is not an added burden for teachers: it is part of the subject they teach. Any philosophy-of-science textbook, anthology or encyclopedia will, for example, have chapters on: theory change, experimentation, idealisation, scientific revolutions, laws, reduction, metaphor, analogy, models, causation, explanation, values, methodology, observation, truth, approximate truth and so on. These philosophical features can be identified and elaborated in the classroom when teaching routine topics such as evolution, genetics, oxidation, mechanics, relativity, electricity, paleontology, photosynthesis and so on. The features or aspects can appropriately be pointed to when students make enquiries, conduct experiments, collect data, propose and appraise hypotheses and so on.

Likewise, philosophy is present in most of the theoretical issues that engage teachers, curriculum writers and administrators: religion, multiculturalism, discipline structure and so on. One such theoretical issue, constructivism, was outlined in the previous chapter and leads naturally into discussion of a central philosophical issue that has echoed through the history of science, and that bears significantly upon what is taught about the nature of science: namely, the debate between realists and anti-realists over the aims of science and the reality and knowability of theoretical entities postulated in scientific theories to explain events and phenomena. This fundamental debate has echoes in many of the other disputes in philosophy of science and is frequently played out in science education, with strong-voiced researchers and teachers found on both sides of the realist/non-realist fence. The discussion can be appreciated, and joined, by teachers and students alike, in a way that displays the philosophical dimension of science.

The Realist/Anti-Realist Divide

The basic realist conviction is that the world and our knowledge of it are two different things; how we learn about something and the thing itself are identical. Man is not the measure of all things, as Protagoras might not have said. Bas van Fraassen provides a succinct statement of realism. It is the following view:

> Science aims to give us, in its theories, a literally true story of what the world is like; and acceptance of a scientific theory involves the belief that it is true.
>
> (van Fraassen 1980, p. 8)

It needs to be noted that the arguments developed here will be about *scientific* realism and anti-realism; they will not be about 'global' anti-realism of the kind found in philosophical idealism and scepticism. In idealism, it is contended that nothing is real beyond the cognising subject; the external world is a mirage. In scepticism, it is contended that there is 'an objective reality' (as constructivists are wont to say), but that we can have no access to it, even the sensible world (tables, chairs, trees) is forever beyond our knowledge. No scientists, and few philosophers of science, hold these global anti-realist positions, and they are irrelevant to science teachers. However, some scientists and many reputable philosophers of science defend scientific anti-realism, and it is this view that will be appraised in this chapter. It should also be noted that the central, and perhaps only, debate is about what scientific *theories* tell us or do not tell us about the world. The arguments are about explanatory, unobservable, theoretical constructs and entities such as magnetic field, electron, gravitational attraction and the like. The philosophical arguments are usually not about the observational claims of science, at least not directly about data and meter readings; they are about their explanations.

It is useful to recognise that there is a 'Wittgensteinian' family of realist positions.[1] They all share:

- an *ontological* commitment to the reality and independence of the world; external things and events, including unobservables, exist independently of the cognising subject;
- a *semantic* commitment to the linkage of scientific claims to external things and events; science makes claims about the world;
- an *epistemological* commitment; namely, that science has made some truthful, or approximately truthful, claims about entities and processes in both the observed and unobserved world, the former being the everyday world revealed by ordinary vision (billiard balls, fish, clouds, etc.), the latter the world indicated by instruments and inference (molecules, atoms, magnetic fields, proteins, etc.);

- an *axiological* commitment that the aim and purpose of science is to produce statements and theories about the world that are true; other purposes, such as utility or economic gain, are secondary, or just by-products of truthfulness.[2]

Likewise, there is a family of anti-realist positions that are united by their rejection, sometimes for different reasons, of one or all of the realist's ontological, semantic, epistemological and axiological claims. The anti-realist family includes positivism, empiricism, instrumentalism, constructivism, constructive empiricism, idealism and, of course, the whole gamut of post-modernisms. Anti-realists believe that scientific knowledge is confined to the world of experience or sensory phenomena, and that any postulated theoretical entities that go beyond such experience have to be treated only as aids, tools, models or heuristic devices for coordinating sensory or observable phenomena, but they do not have any existence. Further, the aim of science is to produce theories that predict phenomena and connect economically – usually mathematically – items of experience.

Elaboration, refinement and defence of each of the realist and anti-realist positions can be read in the references footnoted below. Consistent with the overall methodology of this book, here the debate will be elaborated using historical examples that are also recognisable to science teachers, examples that are ubiquitous in school textbooks and curricula.[3] From the ancient astronomical debate about crystalline spheres, through the instrumentalist position urged upon Galileo by Cardinal Bellarmine, the bitter debates between Newtonians and Cartesians over the reality of gravitational attraction, the equally heated eighteenth-century arguments over the existence of 'imponderable fluids' such as caloric, and the debates about the reality of atoms that engaged Ernst Mach and others in the nineteenth century, to the controversy between the realist Einstein and the instrumentalist Bohr over the Copenhagen interpretation of quantum mechanics – the issue of a realist versus anti-realist interpretation of scientific theory has been at the centre of philosophical debate about science.

Astronomy: How the Heavens Work

All human societies since their beginnings have sought to understand the fabric of the heavens – stars, planets, Sun, Moon, comets and so on. This understanding has been a variable mix of religion, metaphysics, cosmology, mythology, astrology and astronomy – if one might loosely use the modern term. In most societies, most of these elements are present together, with, at different times and in different places, one or other having more or less prominence. The history of astronomy, which is a common school subject, at least as far as the transformation from heliocentric to geocentric understanding of the solar system is concerned, is a nice thread on which to lay out the basic contrast between realist and anti-realist understandings of science (and additionally recurring themes, such as the roles of mathematics and technology in science, the 'interpretation' of human observations, conflicting sources of authority and truth in science, and much more). The AAAS, in introducing its astronomy section in *Science for All Americans*, says:

> To observers on the earth, it appears that the earth stands still and everything else moves around it. Thus, in trying to imagine how the universe works, it made good sense to people in ancient times to start with those apparent truths. The ancient Greek thinkers, particularly Aristotle, set a pattern that was to last for about 2,000 years: a large, stationary earth at the center of the universe, and – positioned around the earth – the sun, the moon, and tiny stars arrayed in a perfect sphere, with all these bodies orbiting along perfect circles at constant speeds.
>
> (AAAS 1989, p. 112)

This is a nice and satisfying phenomenal picture of the heavens, but ancient natural philosophy (or nascent science) conjectured about causes and mechanisms. This is where realists and anti-realists began to separate.

Platonic Empiricism and Aristotelian Realism

The beginnings of the enduring realist versus anti-realist understanding of science can be seen in debates in the ancient world about planetary dynamics or the mechanism of their motion. First Anaximander (*c.*488–428 BCE), then Eudoxus (*c.*409–356 BCE), Callippus (370–300 BCE) and Aristotle (384–322 BCE) proposed that the regularly moving planets were embedded in rotating crystalline spheres that kept them moving steadily and at fixed distances from each other, from the Earth and from the stars, which themselves were embedded in an outmost sphere defining the limits of the world. Retrograde and 'speeding' and 'slowing' motions were accounted for by increasing the number of spheres, with twenty-six being postulated by Eudoxus, thirty-three by Callippus and forty-seven by Aristotle, in order to accommodate ever more astronomical observations.[4] These increasingly complicated mechanisms were required in order to 'save the appearances' of the heavenly bodies as seen from Earth.

There is scholarly dispute over just how realistically the pre-Aristotelians interpreted their postulated spheres, but, with Aristotle's *Metaphysics* and *On the Heavens*, they were clearly given a physical existence, being composed of the unchanging fifth element, the ether. These were the 'crystalline' spheres of later natural philosophy. The spheres, of course, needed their own mover, and, for Aristotle, this was the Prime Mover, later identified as God in the Hellenistic–Christian tradition.[5]

However, astronomical reaslism was not without its anti-realist challenges. Plato legitimised non-realistic interpretations of natural philosophy. This is known from the commentaries of Simplicius (490–560 CE), the neo-Platonist, who says of Plato that he,

> lays down the principle that the heavenly bodies' motion is circular, uniform, and constantly regular. Thereupon he sets the mathematicians the following problem: What circular motions, uniform and perfectly regular, are to be admitted as hypotheses so that it might be possible to save the appearances presented by the planets?
>
> (Duhem 1908/1969, p. 5)

Astronomers and mathematicians were asked by Plato for a model of the movement of heavenly bodies that would conform to, and enable predictions about, astronomical phenomena – daily motion of the Sun, monthly motion of the Moon, times of rising and setting of the planets, changes of season, planetary regression, periods of the planets, time of the equinoxes and other matters. The model was not meant to conform to reality; it needed only to conform to the metaphysical principle that required heavenly motions to be circular and to be consistent with, and predict, astronomical events; the mathematical model was just a calculating device.

Plato's pupil, Eudoxus, took up this challenge. Thomas Heath writes:

> It does not appear that Eudoxus speculated upon the causes of these rotational motions or the way in which they were transmitted from one sphere to another; nor did he inquire about the material of which they were made. . . . It would appear that he did not give his spheres any substance or mechanical connection; the whole system was a purely geometrical hypothesis, or a set of theoretical constructions calculated to represent the apparent paths of the planets and enable them to be computed.
>
> (Heath 1913/1981, p. 196)

Later, Claudius Ptolemy (90–168), in his famous *Almagest*, rejected the spheres in favour of a mathematical and instrumentalist astronomy. His planets moved in circular cycles, epicycles and deferents, which were so configured as to 'save the appearances' as catalogued by astronomers. He retains Plato's metaphysical privileging of circular motion, but the Aristotelian robust realism of extant crystalline spheres was abandoned in favour of mere geometric construction and, at best, just formal acknowledgement of mechanisms such as planetary souls or intelligences. After acknowledging Aristotle's division of philosophy into theoretical and practical branches, Ptolemy notes that, 'Aristotle quite properly divides also the theoretical into three intermediate genera: the physical, the mathematical, and the theological' (Ptolemy 1952, p. 5). Ptolemy embraces mathematics and shuns the first and the last, saying:

> the other two genera of the theoretical would be expounded in terms of conjecture rather than in terms of scientific understanding: the theological because it is in no way phenomenal and attainable, but the physical because its matter is unstable and obscure, so that for this reason philosophers could never hope to agree on

> them . . . only the mathematical, if approached enquiringly, would give its practitioners certain and trustworthy knowledge.
>
> (Ptolemy 1952, pp. 5–6)

Arthur Koestler (1905–1983) describes the situation as follows:

> Astronomy after Aristotle becomes an abstract sky-geometry, divorced from physical reality. . . . It serves a practical purpose as a method for computing tables of the motions of the sun, moon, and planets; but as to the real nature of the universe, it has nothing to say.
>
> (Koestler 1964, p. 77)

Some in this instrumentalist tradition were *epistemologically* anti-realist about planetary mechanisms, saying that such mechanisms might exist, but we just do not have access to them. Others were *ontologically* anti-realist about mechanisms, saying that there are no grounds for postulating their existence. All, of course, were realists about planets and heavenly bodies; it would be another 2,500 years before it became fashionable to think that bodies came in and out of existence, depending on who was thinking about, or postulating, them.

However, anti-realism did not carry the day. Because Aristotle provided an overarching philosophical and 'scientific' system within which his planetary dynamics neatly fitted, his realist astronomy, with its real concentric spheres, lived on for the subsequent 1,500 years.[6]

Copernican and Galilean Realism Against Osiander's Instrumentalism

Nicolaus Copernicus (1473–1543), in his *Six Books Concerning the Revolutions of the Heavenly Spheres* (Copernicus 1543/1952), resurrected the ancient realist programme in astronomy. Copernicus believed that both astronomy and physics should propose hypotheses that both answered Plato's demand for predictability and for saving the phenomena, but also were in accord with how the world was. To this end, he revived the ancient but overlooked heliocentric, moving-Earth model of Aristarchus of Samos. Copernicus, in the Dedication of his book to Pope Paul II, says of the Ptolemaic tradition that:

> Those, on the other hand who have devised systems of eccentric circles, although they seem in great part to have solved the apparent movements by calculations which by these eccentrics are made to fit, have nevertheless introduced many things which seem to contradict the first principles of the uniformity of motion.[7]

As Copernicus lay dying, Andreas Osiander (1498–1552), the Lutheran scholar charged with arranging publication of the *Revolutions*, inserted a preface that is the embodiment of empiricist and instrumentalist understanding of scientific theory. It says in part:

> For the astronomer's job consists of the following: To gather together the history of the celestial movements by means of painstakingly and skilfully made observations, and then – since he cannot by any line of reasoning reach the true causes of these movements – to think up or construct whatever hypotheses he pleases such that, on their assumption, the self-same movements, past and future both, can be calculated by means of the principles of geometry. . . . It is not necessary that these hypotheses be true. They need not even be likely. This one thing suffices, that the calculation to which they lead agree with the result of observation.
>
> (Duhem 1908/1969, p. 66)

This instrumentalist preface is remarkably like the statement of Ernst von Glasersfeld quoted in the last chapter: 'our knowledge is useful, relevant, viable . . . if it stands up to experience and enables us to make predictions . . . [it] gives us no clue as to how the "objective" world might be'. This similarity is not surprising: constructivism is an instrumentalist epistemology, as George Bodner, an American constructivist, so frankly admits:

> The constructivist model is an instrumentalist view of knowledge. Knowledge is good if and when it works, if and when it allows us to achieve our goals. . . . A similar view was taken by Osiander, who suggested in the preface of Copernicus' *De Revolutionibus* [that] 'There is no need for these hypotheses to be true, or even to be at all like the truth; rather, one thing is sufficient for them – that they yield calculations which agree with the observations'.
>
> (Bodner, 1986, p. 874)

Galileo (1564–1642) adopted the Copernican hypothesis sometime around 1600. The hypothesis was, of course, scientifically and theologically controversial. In 1615, the illustrious Cardinal Robert Bellarmine (1542–1621) proposed to Galileo the sort of empiricism and instrumentalism that Osiander had deviously thrust upon Copernicus. The Cardinal said:

> It seems to me that [you] are proceeding prudently by limiting yourselves to speaking suppositionally and not absolutely, as I have always believed Copernicus spoke. For there is no danger in saying that, by assuming the earth moves and the sun stands still, one saves all the appearances better than by postulating eccentrics and epicycles; and that it is sufficient for the mathematician. However it is different to want to affirm that in reality the sun is in the center of the world . . . this is a very dangerous thing.
>
> (Finocchiaro 1989, p. 67)

However, Galileo did not embrace the instrumentalist olive branch offered by Bellarmine: he maintained a resolute realism about the Copernican hypothesis. In his *Two Chief World Systems* (1633), he repeats Copernicus's claim against Ptolemy that:

> However well the astronomer might be satisfied merely as a calculator, there was no satisfaction and peace for the astronomer as a scientist . . . although the celestial appearances might be saved by means of assumptions essentially false in nature, it would be very much better if he could derive them from true suppositions.
>
> (Galileo 1633/1953, p. 341)

Galileo's realism was underwritten and reinforced by his telescopic observations. The story is fairly well known.[8] As Galileo writes, in 1610:

> About 10 months ago a rumor came to our ears that a spyglass had been made by a certain Dutchman by means of which visible objects, though far removed from the eye of the observer, were distinctly perceived as if nearby.
>
> (Galileo 1610/1989, p. 36)

In the same year, he built his own telescope and made his monumental Moon and Jupiter observations. Contemporary empiricists and instrumentalists simply denied the reality of what Galileo communicated; they maintained that what he saw was entirely a product of his own eyes and the dubious technology of the new instruments.

Additionally, Galileo's claims violated the 2,000-year-old metaphysical commitment to an ontological separation of the heavenly and the terrestrial realms. The former was unchanging, perfect and incorruptible; the latter was changing, imperfect and corruptible. The realms had nothing in common. Galileo's Moon drawings could not be interpreted realistically, as they showed the Moon to be like the Earth and, likewise, showed that Jupiter itself had moons: science and metaphysics were at odds. Galileo's seventeenth-century opponents used versions of what 400 years later would become commonplace in science-education constructivist writing: 'we have no independent access to reality; we do not know what is really there'.[9] Galileo's response to philosophical scepticism was to train his telescope on nearby churches, towers and ships outside the harbour and ask if what was seen corresponds, or nearly so, to what is seen with the naked eye, with any shortfall being made good by improvement of the instrument. Thus, the truth of his metaphysics-defying astronomical claims was established by empirical demonstrations. This is a recurring motif in the history of science.

Classical Physics: Newton's Realism and Berkeley's Empiricism

Isaac Newton (1642–1727) was a realist in the tradition of Aristotle and Galileo. He proposed a mechanism (gravitational attraction) that moved the planets and that underwrote the celestial laws of planetary motion uncovered by Kepler, and the terrestrial laws discovered by Galileo. His realism underlies his insistence on the reality of absolute space and time, in contradiction to those who maintain that only relative space and time exist, the space and time of our experience. In his *Scholium* on 'Space and Time', Newton says:

> But because the parts of space cannot be seen, or distinguished from one another by our senses, therefore in their stead we use sensible measures of them. . . . And so, instead of absolute places and motions, we use relative ones; and that without any inconvenience in common affairs; but in philosophical disquisitions, we ought to abstract from our senses, and consider things themselves, distinct from what are only sensible measures of them.
>
> (Newton 1729/1934, p. 8)

Newton was also a realist about forces: when a body accelerated, including moving steadily in an orbit, there was a real force acting upon it: something was making the body accelerate. Forces were not just mathematical conveniences or conventions useful in linking together successive locations of a moving body. Force was responsible for the body moving; it had the same ontological status as the body moved. Although, in free fall and planetary motion for instance, only the accelerating body could be seen, Newton believed that a real, unseen force was responsible for the acceleration. Famously, the mechanical philosophers Huygens and Leibniz rejected such forces: for them, forces only arose from contact, from the collision of bodies. In contrast, Newton remained realistic about these forces; for him, force was a theoretical construct postulated to explain observational occurrences;[10] it was not, to use a methodological concept common in psychology, an intervening variable that merely linked variables in a mathematical manner (Meehl & MacCorquodale 1948); nor was it, as Mach would later claim, a mere convenience for the economy of thought.

Bishop George Berkeley (1685–1753), in his 1721 *De Motu*, continued this empiricist attack on the reality of gravitational attraction, but in addition he argued against the reality of forces more generally. Berkeley said:

> Force, gravity, attraction and similar terms are convenient for purposes of reasoning and for computations of motion and of moving bodies, but not for the understanding of the nature of motion itself.
>
> (Berkeley 1721/1901, p. 506)

This is an extension of his earlier, 1710 *Principles of Human Knowledge* idealist argument for the non-reality of extrasensory existence. There he had said:

> All the choir of heaven and furniture of the earth, in a word all those bodies which compose the mighty frame of the world, have not any subsistence without a mind – that their being is to be perceived or known . . . let anyone consider those arguments which are thought manifestlyto prove that colours and tastes exist only

> in the mind, and he shall find they may with equal force be brought to prove the same thing of extension, figure, and motion.
>
> (Berkeley 1710/1962, pp. 67–71)[11]

Atomism: Realist and Non-Realist Interpretations

Atomic theory is a wonderful exemplar of the 2,500-year dispute between realist and instrumentalist/empiricist/constructivist interpretations of the aims of science and of the interpretation of scientific theory. Modern atomic theory straddles many scientific domains and disciplines; it inexorably connects science with philosophy; it underlies much modern domestic, communications, medical, industrial and military technology; it appears in senior (and some junior) science curricula around the world – thus it is appropriate for the thesis of this book to examine how historical and philosophical considerations can contribute to its better understanding and, subsequently, to students' enriched understanding of the nature of science.

The US *Next Generation Science Standards* mention atoms in their 'cross-cutting' concepts, saying:

> For example, the stability of the book lying on the table depends on the fact that minute distortions of the table caused by the book's downward push on the table in turn cause changes in the positions of the table's atoms. These changes then alter the forces between those atoms, which lead to changes in the upward force on the book exerted by the table.
>
> (NRC 2013, p. 100)

This is a deceptively simple realist picture that can be considerably enhanced and made more interesting and challenging by historical and philosophical input. When science teaching is informed by HPS input, the 'blind faith' component of science learning, so obvious in the preceding quotation, can be diminished, the sense of connection to an important tradition can be enhanced, and students' own epistemology or embryonic theory of knowledge can be cultivated. Some points in the history of atomism, and its associated philosophical debate, will be mentioned here in order to give a sense of the contribution of HPS to classroom teaching. Pleasingly, there is an embarrassment of riches when it comes to studies on the history and philosophy of atomism.[12]

Origins of Atomism

Atomism had its origin in pre-Socratic philosophy, specifically in the materialism of Leucippus and Democritus, who maintained that the basic constituent of all matter was little unseen material atoms and nothing apart from atoms and the void existed. Democritus is rightly praised for attempting 'natural', rather than mythical, animistic, religious, anthropomorphic or teleological explanations for natural events and processes, and for recognising that there is more to the world than meets the eye.

Alan Chalmers (2009), in a recent comprehensive study of the history of atomism, sees two basic problems with ancient atomism: first, such atomic explanations always follow the event: they do not predict experience or events, rather they just provide 'explanations' after the fact; second, the offered explanations are not determinate: there could be any number of permutations of atomic shapes and sizes that could equally be claimed to 'explain' the event or property. For Chalmers, this early atomism was philosophical not scientific atomism; it amounted to 'hand-waving' that did not guide any fruitful practice: 'Rather than being a source of and inspiration towards a viable scientific atomism, philosophical atomism constituted a barrier to it that needed to be transcended' (Chalmers 2009, p. 265).

As is well known, the atomist's disordered, chaotic, purposeless, mechanical, 'bumping together' worldview was supplanted by Aristotle's ordered, teleological, purposeful, 'organismic' conception of nature. The latter seemed so much better to fit people's own experience of intentionality, striving and orderly growth of all things around them. Oak seeds, no matter how much they were shaken around, turned into oak trees, not tomato plants or mice; there seemed to be something internal to the seed apart from atoms, something that 'governed' its growth and interaction with its environment. Aristotle developed a systematic philosophical edifice in which Matter, Form, Act and Potential were the basic categories, and these were to guide Aristotelian natural philosophy (science) for the following 2,500 years.

Seventeenth-Century Atomism

The seventeenth-century natural philosophers associated with the scientific revolution did resurrect versions of ancient atomism in their struggles with dominant Aristotelianism. Galileo was among the first to again embrace atomistic ontology. This was first and most famously stated in his *The Assayer* (Galileo 1623/1957), where he advances invisible 'atomic' motions as the cause of heat. He says:

> But first I must consider what it is that we call heat, as I suspect that people in general have a concept of this which is very remote from the truth. For they believe that heat is a real phenomenon, or property, or quality, which actually resides in the material by which we feel ourselves warmed.
>
> (Galileo 1623/1957, p. 274)

Galileo believed that it was the shape, size, motion and collisions of minute, unseen 'atoms' or corpuscules that determined all outward and perceivable states, processes and phenomena. There was no place here for unfolding Aristotelian Form or Potential. As will be seen in Chapter 10, the Roman Catholic Church immediately recognised the threat that such resurrected ontology posed to its own ensemble of philosophical/theological teaching.

Robert Boyle (1627–1691) famously promoted the Corpuscularian or Mechanical philosophy, but was careful to insulate theology from its reach.

So the eternal soul, the mind, the bestowing of Grace, the activity of angels and the rest of Christendom's rich constellation of existing things were beyond the reach of Boyle's new (or old) system, where everything was either atoms or void. Newton, the greatest of all seventeenth-century scientists, was a champion of the New Philosophy or the recovered Mechanical Worldview. Beginning in his student days, Newton embraced Galileo's mathematical methods, his Copernicanism, his experimentalism, his rejection of Aristotle's physics, his rejection of scholastic philosophy and his embryonic atomism.[13]

Ernst Mach's Instrumentalism

Ernst Mach (1838–1916) carried the empiricist and instrumentalist standard in the great philosophical war at the turn of the twentieth century: the reality or non-reality of atoms and molecules, the meaningfulness or meaninglessness of the atomic hypothesis. Mach's first public refutation of atomism was in his 1872 *History and Root of the Principle of the Conservation of Energy*. There, he held the hypothesis to be useless, saying that it contributed nothing to what phenomenal knowledge already told us. Mach argued:

> But let us suppose for a moment that all physical events can be reduced to spatial motions of material particles (molecules). What can we do with that supposition? Thereby we suppose that things which can never be seen or touched and only exist in our imagination and understanding can have the properties and relations only of things which can be touched. We impose on the creations of thought the limitations of the visible and tangible. . . . In a complete theory, to all details of the phenomenon details of the hypothesis must correspond, and all rules for these hypothetical things must also be directly transferable to the phenomenon. But then molecules are merely a valueless image.
>
> (Mach 1872/1911, p. 49)

Mach did not have an aversion just to atoms; he was quite catholic in his refusal to reify any theoretical construct. He said of Newtonian gravitational attraction, for instance, that it was not just unknowable, but there was no such thing: gravitation was merely a human construct useful for the economy of thought and for the mathematisation of particular experimental relationships. Mach recognised Berkeley and Hume as like-minded philosophers (although his ideas were developed in advance of his reading them).

Max Planck's Realism

Mach's greatest adversary was Max Planck (1858–1947), who, for a time, as with nearly all the physicists of his generation, shared Mach's empiricist viewpoint (Heilbron 1986, pp. 44–46). Richard Miller notes about these constructivist efforts to retain the successful theories but not be committed to their referents, that:

> Most characteristic methods and outlooks of modern philosophy of science can be traced to the last quarter of the nineteenth century, when philosophically minded physicists and chemists, especially in Austria and Germany, tried to show how the fruits of classical mechanics, Maxwellian electrodynamics and atomistic chemistry could be retained without commitment to the distinctive entities of each theory.
>
> (Miller 1987, p. 351)

Planck, during his Machian phase, opposed Boltzmann's atomic interpretation of the Second Law of Thermodynamics. After his own 1900 work on black-body radiation, he converted to the realist and atomist camp. As with many converts, he became more of a realist than most realists – he maintained that atoms were as real as planets, and probably looked the same except scaled down (Toulmin 1970, p. 24).

Planck's first public rejection of Mach's ideas occurred in his 1908 lecture 'The Unity of the Physical World-Picture', which he gave in Leyden at the invitation of Lorentz (reproduced in Toulmin 1970). Planck says, against Mach's fundamental claim that science has to be anchored in our psychological elements or experiences, that: 'The whole development of theoretical physics until now has been marked by a unification achieved by emancipating the system from its anthropomorphous elements, in particular from specific sense impressions' (Toulmin 1970, p. 6).

He also criticised the thesis of Mach's acclaimed history of physics, *The Science of Mechanics*, saying:

> When the great masters of the exact sciences introduced their ideas into science: when Nicolaus Copernicus removed the earth from the center of the universe . . . when Isaac Newton discovered the laws of gravitation . . . when Michael Faraday created the foundations of electrodynamics . . . 'economical' points of view were certainly the last to fortify these men in their battle against traditional attitudes and overriding authorities. No: it was their unshaken faith, whether based on artistic or religious foundations, in the reality of their [atomic] world-picture.
>
> (Toulmin 1970, p. 26)

Planck finished his lecture with the claim that Machian empiricism was antithetical to the progress of science:

> If the Machian principle of economy were ever to become central to the theory of knowledge, the thought processes of such leading intellects would be disturbed, the flights of their imagination would be paralyzed, and the progress of science might thus be fatally impeded.
>
> (Toulmin 1970, p. 26)

These, of course, were provocative words, and, notwithstanding his 72 years, Mach responded. His reply, titled 'The Guiding Principles of My

Scientific Theory of Knowledge', was published in 1910 (reproduced in Toulmin 1970). Mach, in polemical style, says of Planck and his supporters that they,

> are on the way to founding a church. . . . To this I answer simply: If belief in the reality of atoms is so important to you, I cut myself off from the physicists' mode of thinking, I do not wish to be a true physicist, I renounce all scientific respect – in short: I decline with thanks the communion of the faithful. I prefer freedom of thought.
>
> (Toulmin 1970, p. 37)

Some Philosophical Considerations

This sketch of the history of debate between realist and anti-realist accounts of astronomical mechanisms and of atomism shows that some basic distinctions are important in order to discuss the issue; approaching the subject with too limited, black and white contrasts will not allow the nuances of the history to be appreciated. The sketch also illustrates one of the central themes of this book: the close relationship between science and philosophy, and the importance of understanding both fields in order to understand the history of either, and indeed to properly understand the scientific theories discussed or taught.

Empiricist Arguments Against Realism

The three most powerful arguments that empiricists urge against realists are: first, the 'idleness' argument; second, the 'graveyard' argument; and, third, the 'underdetermination' argument. Arguably Larry Laudan (1984) and Bas van Fraassen (1980) have provided the most wide-ranging critiques of realism, and van Fraassen has given the most sophisticated restatement of empiricism and instrumentalism as a viable philosophy of science. Van Fraassen says that:

> To be an empiricist is to withhold belief in anything that goes beyond the actual, observable phenomena, and to recognize no objective modality in nature . . . [it] involves throughout a resolute rejection of the demand for an explanation of the regularities in the observable course of nature, by means of truths concerning a reality beyond what is actual and observable.
>
> (van Fraassen 1980, p. 202)

The first, 'idleness', argument was stated by Mach and was succinctly expressed by Carl Hempel (1905–1997) in his famous paper 'The Theoretician's Dilemma' (Hempel 1958/1965). Hempel states that empiricists (Braithwaite, Carnap, Feigl and others) regarded all scientific terms as belonging to either of two realms: the observable or the theoretical. The function of theories was to deductively explain or inductively enjoin observations, so that, given one set of observations, a second set could be predicted.

Realists believed that these explanations worked because of connections in the world between the theoretical entities and processes they postulated and visible events. The postulated theoretical entities behaved in a law-like manner (hence, excluding angels and spirits from the class of scientific theoretical entities). After quoting the behaviourists Hull and Skinner on the subject, he then poses the theoretician's dilemma as follows:

> If the terms and principles of a theory serve their purpose they are unnecessary . . . if they do not serve their purpose they are surely unnecessary. But given any theory, its terms and principles either serve their purpose or they do not. Hence the terms and principles of any theory are unnecessary.
>
> (Hempel 1958/1965, p. 186)

The theoretical terms – 'force', 'field', 'caloric', 'intelligence', 'class', 'gene' and so on – occupy scientific space but pay no rent. After due elaboration, Hempel criticises and rejects this argument, saying that the supposed observational bed-rock is 'a fiction' (Hempel 1963, p. 701), and that, as well as deductive explanations of phenomena, theories have to provide for inductive expansion of claims, and this cannot be done using just observational terms (Hempel 1963, p. 700).

The second, 'graveyard', argument against realism has traditionally been most convincing. It was given sharp formulation by Larry Laudan (Laudan 1984), who points out that the history of science is littered with discarded theoretical entities that earlier were firmly ensconced in the best and most successful science of their time – crystalline spheres, caloric, phlogiston, humours, the ether and so on – all these theoretical terms were assumed to be referential. It turns out they were not. And Laudan adds that there is no reason to think that our best current candidates for truth will have a different fate. This is his 'pessismistic meta-induction' (PMI) argument against realism (Laudan 1984).

The third, 'underdetermination', argument appeals to the fact that theoretical terms are always underdetermined by the evidence available, and, consequently, the same evidence will also support other extant or potential theoretical entities. So, the evidence provides no special basis for any particular referential claim.

Defending Realism

Realists have offered defences against the empiricist arguments outlined above, and the thrust of these can be readily understood and utilised in classroom discussion and elaborations.[14] Hilary Putnam's 'no miracles' argument is widely endorsed:

> The positive argument for realism is that it is the only philosophy that doesn't make the success of science a miracle . . . [realism] is part of any adequate scientific description of science and its relations to its objects.
>
> (Putnam 1975, p. 73)

Other realists (Psillos 1999, 2011) have outlined how Laudan's PMI argument can be accommodated.

First, by making burial conditions more stringent and, hence, reducing the number of tombs in the graveyard; not just any old discarded theory is allowed burial, but only well-confirmed theories whose confirmation came from prediction of novel facts. Empirical adequacy is not just passive agreement with facts or phenomena – astrology and natural theology are both capable of that – but to be buried in the scientific section of the cemetery requires that the theory has made novel, confirmed predictions that result from its recourse to its postulated theoretical entities. This step thins out the number of graves on which to base the pessimistic induction.

Second, by checking whether the buried theories are indeed dead. The realist assuredly needs to acknowledge theory change, even for substantial and successful theories, but it is always an open question as to what degree the new theory retains elements or entities from the old theory it replaced. To the degree in which there is continuity in theory change, then to that degree the grounds for PMI are further diminished.[15] Fresnel's theory of light warranted burial in the scientific cemetery (it was widely endorsed, successful and made confirmed predictions), but, although buried, parts of the theory did live on and inform subsequent nineteenth-century optics, so it could be exhumed and not take up so much cemetery space. Laudan has an argument, but it is not the lay down *misère* that it is oft taken to be by anti-realists.

Ian Hacking (1983) has cautioned that, if science is conceived as simply a representation of the world, then the empiricist arguments are so strong that realism has no satisfactory reply; realism has to look to new forms of justification. He finds these in the success of scientific intervention and experimentation. On the reality of electrons, Hacking endorsed a scientist's observation: 'So far as I'm concerned, if you can spray them, then they are real' (Hacking 1983, p. 23). Hacking provides philosophical support for this practitioner's intuition, as does Allan Franklin, who argues that the ongoing success of experimental practice confirms modest realism:

> Supporting a realist position does not, however, mean that I believe in either the absolute truth of the laws or in the 'real' existence of the entities. It means only that I think we have good reasons for believing in the truth of the laws and in the existence of the entities.
>
> (Franklin 1999, p. 160)

It is useful to delineate some of the forms that realism can take in order to clarify what is being defended in the name of realism, and what is not being defended. Leplin (1984, p. 1) provides a comprehensive list of theses that span the range of realist philosophical positions. He points out that realists can be said to be affirming one of a number of slightly different theses, these being:

1 The best current scientific theories are at least approximately true.
2 The central terms of the best current theories are genuinely referential.

3 The approximate truth of a scientific theory is sufficient explanation of its predictive success.
4 The approximate truth of a scientific theory is the only possible explanation of its predictive success.
5 A scientific theory may be approximately true even if referentially unsuccessful.
6 The history of at least the mature sciences shows progressive approximation to a true account of the physical world.
7 The theoretical claims of scientific theories are to be read literally, and so read they are definitively true or false.
8 Scientific theories make genuine, existential claims.
9 The predictive success of a theory is evidence for the referential success of its central terms.
10 Science aims at a literally true account of the physical world, and its success is to be reckoned by its progress towards achieving this aim.

A strong form of realism might hold the combination of theses 1, 2, 4, 7 and 10. A modest form of realism might, for instance, hold the combination of theses 6, 8 and 9. Modest realism is, in effect, saying that science aims to provide a true account of a world that is beyond and independent of our own mental states, and that well-proven scientific theories are approximately true, their postulated explanatory entities do exist – successful scientific theories 'latch on to the world'.[16]

Howard Stein argues for the importance of attending to fine detail in the realism/anti-realism debate:

> When the positions are assessed against the background of the actual history of science, (a) each of the contrary doctrines, interpreted with excessive simplicity, is inadequate as a theory of the dialectic of scientific development; (b) each, so interpreted, has contributed in important instances to actual damage to investigations by great scientists (Huygens, Kelvin, Poincaré); whereas (c) in both the theoretical statements and the actual practice of . . . the most sophisticated philosophers/scientists, important aspects of realism and instrumentalism are present together in such a way that the alleged contradiction between them vanishes.
>
> (Stein 1989, p. 47)

With recognition of this caution, modest realism can be supported. It maintains the following:

- Theoretical terms in a science attempt to refer to some reality.
- Scientific theories are, to whatever degree, successful in their attempts at reference.
- Scientific progress, in at least mature sciences, is due to their being increasingly true.

- The natural world that science investigates is independent of our thoughts and our minds.

Conclusion

The debate between realists and anti-realists over the status of scientific theory – whether theoretical statements about unobservables are meant to refer to real, existing entities, and to what degree do they so refer – has been canvassed for a number of reasons. First, it has so dominated the history of philosophical reflection on the nature of science that it ought to feature in school discussions of this subject. Second, school textbooks frequently endorse one or other of the views, but with very little understanding of the historical or philosophical issues involved. Familiarity with the debate allows teachers to be more critical of the texts and widen students' appreciation of this core matter. Third, constructivism is decidedly empiricist and instrumentalist in its view of science and the goals of scientific enquiry; indeed, in many cases, where it is asserted that people can only know about their experiences, it is outrightly positivist, despite this word being a term of abuse among educators. These empiricist commitments flow over to classroom practice, where it is commonly held, indeed, it is almost the default position among educators, that, 'constructivism is the most mature epistemological theory', and that successful science teaching can be judged by how many students adopt the position (Roth & Roychoudhury 1994, p. 28; Tsai 1999, p. 1219). Realism is by no means as discredited and without support as is commonly believed, and it is salutary for educators to recognise this. Such recognition would lessen the amount of constructivist indoctrination that occurs in teacher education and school classrooms.

Notes

1 There are countless books and articles discussing realism and anti-realism. The online *Stanford Encyclopedia of Philosophy* is a good starting point for both history and literature. Two classic books are, in support of realism, Psillos (1999) and, for anti-realism, van Fraassen (1980). Four anthologies in which both sides of the argument can be found are: Churchland and Hooker (1985), Cohen *et al.* (1996), Leplin (1984) and Nola (1988).

2 Although these commitments are listed separately, they are connected. A realist ontological claim about an entity needs to be linked with an epistemological claim about its properties; the former cannot be held without the latter. To do so would be equivalent to saying goblins exist, but nothing is known or can be said about them.

3 In Chapter 12, it will be shown that this is the most efficacious way of bringing HPS into teacher education programmes.

4 There are many sources for ancient astronomy, but see at least: Clagett (1957, Chapter 7), Heath (1913/1981) and Sambursky (1956, Chapters 3, 4).

5 For the Christian tradition's arguments from motion to motion's God, see Buckley (1971).

6 Recall that Copernicus's 1543 revolutionary treatise was titled *On the Revolutions of the Heavenly Spheres.*

7 The Dedication is much anthologised. See Matthews (1989, pp. 40–44).

8 See Drake (1970, Chapter 7) and van Helden (1985).
9 As two constructivists write: 'Generally, constructivists recognise a reality that exists independently of cognising beings, but hold that direct access to this reality is forever elusive' (Roth & Roychoudhury 1994, p. 6). This particular constructivist claim is elaborated, and convincingly criticised, in Kitcher (2001) and Nola (2003).
10 On this topic, see Cohen (2002).
11 It is noteworthy that Ernst von Glasersfeld nominates 1710 as one of the greatest years in the history of philosophy, on account of its being the year of publication of Berkeley's book, and also noteworthy that he mentions *The Principles* as the first philosophy book he read while a refugee in Ireland during the Second World War. Not surprisingly, the echo of Bishop Berkeley can be heard in much contemporary constructivist discussion.
12 Among excellent recent works are: Chalmers (2009), Pullman (1998), Pyle (1997) and Siegfried (2002). An older and rewarding work by a Thomist philosopher is van Melsen (1952).
13 For Newton's early scientific and philosophical formation, see Herivel (1965) and Westfall (1980, Chapters 3–5).
14 See, for instance, Boyd (1984), Hooker (1985, 1987), McMullin (1984), Musgrave (1996), Psillos (1999, 2011), Schlagel (1986) and Snyder (2005).
15 Alberto Cordero has well argued this case (Cordero 2013).
16 The idea of approximate truth, or 'verisimilitude' as Popper called it, has its problems, but it can be defended (Devitt 1991, Oddie 1986).

References

AAAS (American Association for the Advancement of Science): 1989, *Project 2061: Science for All Americans*, AAAS, Washington, DC. Also published by Oxford University Press, 1990.

Berkeley, G.: 1710/1962, *The Principles of Human Knowledge*, G.J. Warnock (ed.), Collins, London.

Berkeley, G.: 1721/1901, *De Motu*, in A.C. Fraser (ed.) *The Works of George Berkeley*, Oxford University Press, Oxford, UK (extracts in D.M. Armstrong (ed.) *Berkeley's Philosophical Writings*, New York, 1965).

Bodner, G.M.: 1986, 'Constructivism: A Theory of Knowledge', *Journal of Chemical Education* 63(10), 873–878.

Boyd, R.N.: 1984, 'The Current Status of Scientific Realism'. In J. Leplin (ed.) *Scientific Realism*, University of California Press, Berkeley, CA, pp. 41–82.

Buckley, M.J.: 1971, *Motion and Motion's God*, Princeton University Press, Princeton, NJ.

Chalmers, A.F.: 2009, *The Scientist's Atom and the Philosopher's Stone: How Science Succeeded and Philosophy Failed to Gain Knowledge of Atoms*, Springer, Dordrecht, The Netherlands.

Churchland, P.M. and Hooker, C.A. (eds): 1985, *Images of Science*, University of Chicago Press, Chicago, IL.

Clagett, M.: 1957, *Greek Science in Antiquity*, Abelard-Schuman, London.

Cohen, I.B.: 2002, 'Newton's Concepts of Force and Mass, with Notes on the Laws of Motion'. In I.B. Cohen and G.E. Smith (eds) *The Cambridge Companion to Newton*, Cambridge University Press, Cambridge, UK, pp. 57–84.

Cohen, R.S., Hilpinen, R. and Renzong, Q. (eds): 1996, *Realism and Anti-Realism in the Philosophy of Science: Beijing International Conference, 1992*, Kluwer Academic Publishers, Dordrecht, The Netherlands.

Copernicus, N.: 1543/1952, *On the Revolutions of the Heavenly Spheres* (trans. C.G. Wallis), Encyclopædia Britannica, Chicago, IL.

Cordero, A.: 2013, 'Conversations Across Meaning Variance', *Science & Education* 22(6), 1305–1313.

Devitt, M.: 1991, *Realism and Truth*, 2nd edn, Basil Blackwell, Oxford, UK.

Drake, S.: 1970, *Galileo Studies*, University of Michigan Press, Ann Arbor, MI.
Duhem, P.: 1908/1969, *To Save the Phenomena: An Essay on the Idea of Physical Theory from Plato to Galileo*, University of Chicago Press, Chicago, IL.
Finocchiaro, M.A.: 1989, *The Galileo Affair:A Documentary History*, University of California Press, Berkeley, CA.
Franklin, A.: 1999, *Can that be Right? Essays on Experiment, Evidence, and Science*, Kluwer Academic Publishers, Dordrecht, The Netherlands.
Galileo, G.: 1610/1989, *Sidereus Nuncius (The Sidereal Messenger)* (trans. and ed. Albert van Helden), The University of Chicago Press, Chicago, IL.
Galileo, G.: 1623/1957, *The Assayer*. In S. Drake (ed.) *Discoveries and Opinions of Galileo*, Doubleday, New York, pp. 229–280.
Galileo, G.: 1633/1953, *Dialogue Concerning the Two Chief World Systems* (trans. S. Drake), University of California Press, Berkeley, CA (2nd revised edition, 1967).
Hacking, I.: 1983, *Representing and Intervening*, Cambridge University Press, Cambridge, UK.
Heath, T.: 1913/1981, *Aristarchus of Samos: The Ancient Copernicus*, Clarendon Press, Oxford, UK (reprinted, Dover 1981).
Heilbron, J.L.: 1986, *The Dilemmas of an Upright Man: Max Planck as Spokesman for German Science*, University of California Press, Berkeley, CA.
Hempel, C.G.: 1958/1965, 'The Theoretician's Dilemma', *Minnesota Studies in the Philosophy of Science*, Vol.II. Reprinted in his *Aspects of Scientific Explanation*, Macmillan, New York, 1965, pp. 173–226.
Hempel, C.G.: 1963, ' Implications of Carnap's Work for the Philosophy of Science'. In P.A. Schilpp (ed.) *The Philosophy of Rudolf Carnap*, Open Court Publishers, LaSalle, IL, pp. 685–709.
Herivel, J: 1965, *The Background to Newton's 'Principia'*, Clarendon Press, Oxford, UK.
Hooker, C.A.: 1985, 'Surface Dazzle, Ghostly Depths: An Exposition and Critical Evaluation of van Fraassen's Vindication of Empiricism against Realism', in P.M. Churchland and C.A. Hooker (eds) *Images of Science*, University of Chicago Press, Chicago, IL, pp. 153–196.
Hooker, C.A.: 1987, *A Realistic Theory of Science*, State University of New York Press, Albany.
Kitcher, P.: 2001, 'Real Realism: The Galilean Strategy', *The Philosophical Review* 110(2), 151–197.
Koestler, A.: 1964, *The Sleepwalkers*, Penguin Books, Harmondsworth, UK.
Laudan, L.: 1984, 'A Confutation of Convergent Realism'. In J. Leplin (ed.) *Scientific Realism*, University of California Press, Berkeley, CA, pp. 218–249.
Leplin, J. (ed.): 1984, *Scientific Realism*, University of California Press, Berkeley, CA.
Mach, E.: 1872/1911, *The History and Root of the Principle of the Conservation of Energy*, Open Court Publishing Company, Chicago, IL.
McMullin, E.: 1984, 'A Case for Scientific Realism'. In J. Leplin (ed.) *Scientific Realism*, University of California Press, Berkeley, CA, pp. 8–40.
Matthews, M.R. (ed.): 1989, *The Scientific Background to Modern Philosophy*, Hackett Publishing Company, Indianapolis, IN.
Meehl, P. and MacCorquodale, K.: 1948, 'On a Distinction Between Hypothetical Constructs and Intervening Variables', *Psychological Review* 55, 95–107.
Miller, R.W.: 1987, *Fact and Method: Explanation, Confirmation and Reality in the Natural and Social Sciences*, Princeton University Press, Princeton, NJ.
Musgrave, A.: 1996, 'Realism, Truth and Objectivity'. In R.S. Cohen, R. Hilpinen and Q. Renzong (eds) *Realism and Anti-Realism in the Philosophy of Science*, Kluwer Academic Publishers, Dordrecht, The Netherlands, pp. 19–44.
NRC (National Research Council): 2013, *Next Generation Science Standards*, National Academies Press, Washington, DC.
Newton, I.: 1729/1934, *Mathematical Principles of Mathematical Philosophy* (trans. A. Motte, revised F. Cajori), University of California Press, Berkeley, CA.

Nola, R. (ed.): 1988, *Relativism and Realism in Science*, Reidel Academic Publishers, Dordrecht, The Netherlands.
Nola, R.: 2003, '"Naked Before Reality; Skinless Before the Absolute": A Critique of the Inaccessibility of Reality Argument in Constructivism', *Science & Education* 12(2), 131–166.
Oddie, G.: 1986, *Likeness to Truth*, Reidel, Dordrecht, The Netherlands.
Psillos, S.: 1999, *Scientific Realism: How Science Tracks Truth*, Routledge, London.
Psillos, S.: 2011, 'Is the History of Science the Wasteland of False Theories?' In P.V. Kokkotas, K.S. Malamitsa and A.A. Rizaki (eds) *Adapting Historical Knowledge Production to the Classroom*, Sense Publishers, Rotterdam, The Netherlands, pp. 17–36.
Ptolemy, C.: 1952, *The Almagest*, University of Chicago Press, Chicago, IL (Britannica *Great Books*, Vol.16, trans. R. Catesby Taliaferro).
Pullman, B.: 1998, *The Atom in the History of Human Thought*, Oxford University Press, Oxford, UK.
Putnam, H.: 1975, *Mathematics, Matter and Method: Philosophical Papers Volume I*, Cambridge University Press, Cambridge, UK.
Pyle, A.: 1997, *Atomism and Its Critics: From Democritus to Newton*, Thoemmes Press, Bristol, UK.
Roth, M-W. and Roychoudhury, A.: 1994, 'Physics Students' Epistemologies and Views About Knowing and Learning', *Journal of Research in Science Teaching* 31(1), 5–30.
Sambursky, S.: 1956, *The Physical World of the Greeks*, Routledge & Kegan Paul, London.
Schlagel, R.: 1986, *Contextual Realism: A Meta-physical Framework for Modern Science*, Paragon House, New York.
Siegfried, R.: 2002, *From Elements to Atoms: A History of Chemical Composition*, American Philosophical Society, Philadelphia, PA.
Snyder, L.J.: 2005, 'Confirmation for a Modest Realism', *Philosophy of Science* 72, 839–849.
Stein, H.: 1989, 'Yes, But . . . Some Skeptical Remarks on Realism and Anti-Realism', *Dialectica* 43(1–2), 47–65.
Toulmin, S.E. (ed.): 1970, *Physical Reality: Philosophical Essays on Twentieth-Century Physics*, Harper & Row, New York.
Tsai, C.-C.: 1999, 'The Progression Toward Constructivist Epistemological Views of Science: A Case Study of the STS Instruction of Taiwanese High School Female Students', *International Journal of Science Education* 21(11), 1201–1222.
van Fraassen, B.C.: 1980, *The Scientific Image*, Clarendon Press, Oxford, UK.
van Helden, A.: 1985, *The Invention of the Telescope*, Universityof Chicago Press, Chicago, IL.
van Melsen, A.G.: 1952, *From Atomos to Atom*, Duquesne University Press, Pittsburgh, PA.
Westfall, R.S.: 1980, *Never at Rest: A Biography of Isaac Newton*, Cambridge University Press, Cambridge, UK.

Chapter 10

Science, Worldviews and Education[1]

Science, formerly 'natural philosophy', has always been a dynamic part of culture; it is affected by culture and has effects on culture; thus, science and worldviews (or *Weltanschauungen*) are interrelated, and a decent science education should give students some appreciation of this interrelationship.[2]

Hugh Gauch, an agricultural scientist, wrote the lead essay in a thematic issue of *Science & Education* (2009) dedicated to *Science, Worldviews and Education* (Matthews 2009b), where he averred that questions about science's relation to worldviews, either theistic or atheistic ones, are among the most significant of contemporary issues for scientists, science teachers and culture more generally (Gauch 2009). Many people are vitally interested in questions such as whether God exists, whether the world has purpose, whether there are spiritual entities that have causal influence on the world, whether humans have spiritual souls that distinguish them from the animal world, whether the world is such that prayers can be answered and natural causal processes interrupted and so on. It is surely important for students and teachers to know if science can give answers, one way or the other, to these questions, or whether science is necessarily mute on the matters. Presumably, knowledge of the nature of science should shed some light on whether science can or cannot answer such questions. Gauch surveys opinions of scientists, philosophers and educators and, predictably, finds disagreement within each group on the question of the legitimate purview of science.

Importantly, Gauch carefully reports what position papers of the AAAS and the US NRC say about the defining characteristics of science and, thus, what they say about worldviews and science. He identifies seven 'pillars' of the scientific enterprise that the AAAS and the NRC endorse. These are:

- Pillar P1: *Realism*. The physical world, which science seeks to understand, is real.
- Pillar P2: *Presuppositions*. Science presupposes that the world is orderly and comprehensible.
- Pillar P3: *Evidence*. Science demands evidence for its conclusions.
- Pillar P4: *Logic*. Scientific thinking uses standard and settled logic.
- Pillar P5: *Limits*. Science has limits in its understanding of the world.

- Pillar P6: *Universality*. Science is public, welcoming persons from all cultures.
- Pillar P7: *Worldview*. Science, hopefully, contributes to a meaningful worldview.

Gauch sees these seven pillars as, in part, amounting to the popular view that investigation of the supernatural lies outside the domain of science; this is the widely held NOMA position put forward by the late Stephen J. Gould (Gould 1999). However, Gauch also finds an inconsistency with the AAAS position, because, at the same time, the AAAS asserts that, 'we live in a directional, although not teleological, universe'. For Gauch, this is a denial of the fundamental worldview of the Judaic–Christian–Islamic traditions for which the world is neither purposeless nor ultimately unguided, and it is thus a statement that, contra NOMA, science is not worldview-independent. He advances and defends the related thesis that:

> Science is worldview independent as regards its presuppositions and methods, but scientific evidence, or empirical evidence in general, can have worldview import. Methodological considerations reveal this possibility and historical review demonstrates its actuality.
>
> (Gauch 2009, p. 679)

The following fundamental questions arise for science teachers and curriculum writers, and have been addressed by educators and by historians and philosophers of science:

- What constitutes a worldview?
- How do worldviews impinge upon, and in turn be modified by, ontological, epistemological, ethical and religious commitments?
- What worldview commitments, if any, are presupposed in the practice of science?
- What is the overlap between learning about the NOS and learning about worldviews associated with science?
- What is the legitimate domain of the scientific method? Should scientific method be applied to historical questions, especially to historical questions concerning scriptures and sacred texts?
- To what extent should learning about the scientific worldview be a part of science instruction?
- Should science instruction inform student worldviews or leave them untouched? Should students be just 'border crossers', moving from their own culture with its particular worldviews to the science classroom in order to 'pick up' instrumental or technical knowledge and then back to their 'native' culture, without being affected by the worldviews and outlooks of science?
- What judgement do we make of science-education programmes where the scientific view of the world is not affirmed or internalised, but only learned

for instrumental or examination purposes; where learning science is akin to an anthropological study where students are not expected to believe or adopt what they are learning, but merely be able to manipulate formulae and give correct answers on exams?

As with all topics covered in this book, it is clear that the answers to these questions do not come from learning theory, classroom management skills or most of the standard content in teacher education programmes, but rather they require historical and philosophical competence.

Science, Philosophy and Worldviews: Some Historical Developments

The celebrations in 2009 of the 150th anniversary of the publication of Darwin's *The Origin of Species* generated wide recognition of the interplay of science, culture and worldviews. Internationally – by dint of popular journals, academic symposia, newspaper articles, museum displays, books and television documentaries – the general public came to see what scholars have long recognised, namely that the *Origin* not only provided a novel account of the origin of species by natural selection, but it also initiated a transformation of modern worldviews and a new understanding of the place of human beings in the natural world.[3] Versions of Darwin's evolutionary naturalism, reinforced and strengthened by modern genetics,[4] have entered into most modern worldviews, excepting those of Christian fundamentalists, Muslims and many indigenous cultures.[5]

Earlier, in 2005, with the celebration of the centenary of Einstein's *annus mirabilis*, the public also saw and appreciated the contribution of science to worldviews. People knew, perhaps less clearly and dramatically than with Darwin, that something important began to happen in 1905 with the publication of Einstein's three papers; there was some appreciation of what physicist–philosopher Fritz Rohlich wrote:

> The development of quantum mechanics led to the greatest conceptual revolution of our century and probably to the greatest that mankind had ever experienced. It most likely exceeded the great revolutions in our thinking brought about by the Copernican revolution, the Darwinian revolution, and the special as well as the general theory of relativity. Quantum mechanics forced us to reconsider our deepest convictions about the reality of nature.
>
> (Rohrlich 1987, p. 136)

The Ancient World

Although Darwin and Einstein are the most recent and most widely known cases of science impacting on philosophy and culture, these impacts go right back to the very cradle of Western natural philosophy; there has been a continuous interaction between science, philosophy, metaphysics and, ultimately, worldviews. The 'science' (natural philosophy) of the classical and Hellenic materialists and atomists – Thales, Anaximander, Leucippus, Democritus, Epicurus, Anaxagoros and others – was in constant struggle with the dualist, finalist, teleological purposeful worldviews of Platonists and Aristotelians. Karl Popper (Popper 1963, Chapter 5) drew attention to this 'struggle' between the early naturalist and materialist scientific tradition among the pre-Socratics and its dualist, teleological, philosophical opponents, chiefly Plato and Aristotle. The latter pair won, and the former group were, for nearly 2,000 years, relegated to being just 'pre-Socratics', or the philosophical 'warm-up' or targets for the main Athenian adventure. However, to a small extent, their reputation has been recovered, with one representative historian of Greek philosophy writing of the atomists Leucippus and Democritus that:

> In their atomism, their theory of motion, their distinction between primary and secondary qualities, and most of all, in their insistence that explanation of natural processes shall be mechanical, the atomists anticipated much in the world view of modern science.
>
> (Allen 1966, p. 15)

Anaximander's explanation of thunder as noise created, not by heavenly gods or spirits, but by the rubbing together of wind particles, well represents the division between materialist or naturalist explanatory systems and 'pre-scientific' ones.[6] However, in saying this, it needs to be remembered that the materialist programme of Leucippus, Democritus, Epicurus and other ancient atomists was hard to reconcile with the daily observation of animal life and of human experience, for both of which intention, purpose and teleology seemed pervasive. Whatever the strengths of the ancient (and modern) materialist programme, it just did not seem to easily encompass animal and human life; the system and categories of Aristotle seemed much more appropriate and satisfying. As one contemporary geneticist observed:

> The demolition of Aristotelism as a scientific system started in the Renaissance, and was successful in the physical disciplines. In biological sciences, however, Aristotelism lasted until the Darwinian revolution, which is still in progress in our days. . . . In fact, the Aristotelian viewpoint in biology still has a good number of supporters among modern biologists.
>
> (Montalenti 1974, p. 4)

For Popper, and many others, the scientific revolution was a 'return to the past', a recapturing of materialist ontology and non-teleological, mechanical causal relations (Vitzthum 1995, Chapter 2). Wallis Suchting has described these struggles in the cradle of Western science and philosophy as follows:

> Despite all the differences between Plato and Aristotle the latter carried on the work of the former in essential ways, like that of offering a metaphysical

'foundation' for the sciences and a teleological view of the world. Christianity took up elements of Platonic thought . . . but, its philosophical high-point, in Thomism, mainly appropriated Aristotle. Atomism carried on a basically marginal existence . . . till it was recuperated by Galileo.

(Suchting 1994, p. 45)

Atomism and the Scientific Revolution

The scientific revolution of the seventeenth century occurred in a Europe whose cultural, scholarly and religious life was permeated by Aristotelian philosophy, by convictions about ontology, epistemology, ethics and theology that were informed and judged by the texts of Aristotle.[7] Neo-Aristotelian scholasticism, although not monolithic in its interpretation of Aristotle,[8] dominated medieval and Renaissance universities.[9] Scholastic philosophy was intimately connected with the Catholic Church, but it also held sway in Protestant seminaries and universities (Dillenberger 1961, Chapter 2). As one commentator has observed:

The Middle Ages mean simply the absolute reign of the Christian religion and of the Church. Scholastic philosophy could not be anything else than the product of thought in the service of the reigning *Credo*, and under the supervision of ecclesiastical authority.

(De Wulf 1903/1956, p. 53)

In scholastic ontology, things were constituted by form and by matter; this was the doctrine or principle of hylomorphism, and it was fundamental to the Aristotelian tradition. Fredrick Copleston has rightly noted that Aquinas, the greatest of the scholastics, 'took over the Aristotelian analysis of substance' (Copleston 1955, p. 83) and:

According to Aquinas, therefore, every material thing or substance is composed of a substantial form and first matter. Neither principle is itself a thing or substance; the two together are the component principles of a substance. And it is only of the substance that we can properly say that it exists. 'Matter cannot be said to be; it is the substance itself which exists'.

(Copleston 1955, p. 90)

It was the 'new science' that led eventually to the unravelling of this settled, medieval, philosophical–theological worldview. This began with the publication, in 1543, of Copernicus's astronomical work *On the Revolution of the Heavenly Spheres* (Copernicus 1543/1952).[10] However, it was almost a century later that the unravelling took dramatic shape, with the publication in 1633 of Galileo's *Dialogues Concerning the Two Chief World Systems* (Galileo 1633/1953) and, 50 years later, Newton's *Principia Mathematica*

(Newton 1713/1934). These last two books, separated by a mere 50 years, embodied the intellectual core of the scientific revolution; they constituted the Galilean–Newtonian Paradigm, a GNP far more influential than any economic GNP has ever been.

The new science established the Copernican, heliocentric account of the solar system, which removed humans from their religiously and culturally privileged place in the centre of the universe; it introduced a mechanical and lawful account of natural processes; it challenged and, in many places, overthrew the long-dominant Aristotelian philosophical system that was, among other things, intimately tied up with Roman Catholic theology and ethics; and, famously, the GNP caused a reassessment of the role of religious authority in the determination of claims about both the natural and social worlds.[11]

The new science (natural philosophy) of Galileo, Descartes, Huygens, Boyle and Newton caused a massive change, not just in science, but also in European philosophy, that had enduring repercussions for religion, ethics, politics and culture. As was outlined in Chapter 2, early modern philosophers – Francis Bacon, Thomas Hobbes, John Locke, David Hume, George Berkeley, René Descartes, Gottfried Leibniz, up to Immanuel Kant – were all engaged with, and reacting to, the breakthroughs of early modern science,[12] as, of course, were the later philosophers of the French, English, German and Scottish Enlightenment; seventeenth-century science was the seed that bore eighteenth-century philosophical and worldview fruit. With the inevitable exceptions and qualifications required when talking of any large-scale transformation or revolution in thought, it can be said that all the major natural philosophers of the time rejected Aristotelianism in their scientific practice, their theorising and their enunciated philosophy. Overwhelmingly, the new philosophy to which they turned was corpuscularian, mechanical and realist – it has rightly been called the 'Mechanical World View' (Dijksterhuis 1961/1986, Westfall 1971).

In this new worldview, there was simply no place for the entities that Aristotelianism utilised to explain events in the world: hylomorphism, immaterial substances, unfolding natures and potentialities, substantial forms, teleological processes and final causes were all banished from the philosophical firmament. Much can be said about atomism and the new science, but, for current purposes, it suffices to repeat Craig Dilworth's observation that:

> The metaphysics underlying the Scientific Revolution was that of early Greek atomism. . . . It is with atomism that one obtains the notion of a physical reality underlying the phenomena, a reality in which uniform causal relations obtain. . . . What made the Scientific Revolution truly distinct, and Galileo . . . its father, was that for the first time this empirical methodology [of Archimedes] was given an ontological underpinning.
>
> (Dilworth 1996/2006, p. 201)

The Catholic Church's Condemnation of Atomism

The rise and fall of atomistic philosophy provides an example of the intertwining of science, philosophy and culture – an intertwining that students of the humanities can learn about and science students can appreciate. Many of the major seventeenth-century contributors to the new science – Galileo, Descartes, Boyle, Newton – were Christian believers, although in somewhat tense relations with their respective established churches (Roman Catholic for the first two, Anglican for the second two). Some believers rejected the new science; some wanted the new science, but not its associated metaphysics; and some, such as Joseph Priestley, embraced both the new science and its atomistic metaphysics and adjusted their religious ontology accordingly. When the seventeenth-century natural philosophers and the Enlightenment philosophers of the eighteenth century stressed the materialism, mechanism and determinism of the new science, they brought upon themselves the ire of most contemporary religious figures, who saw the emerging new worldview as anti-Christian and atheistic.[13] The historian Richard Westfall summarises the general situation well:

> Natural science rested on the concept of natural order, and the line that separated the concepts of natural order and material determinism was not inviolable. The mechanical idea of nature, which accompanied the rise of modern science in the 17th century, contradicted the assertion of miracles and questioned the reality of divine providence. Science, moreover, contained its own criteria of truth, which not only repudiated the primacy of ancient philosophers but also implied doubt as to the Bible's authority and regarded the attitude of faith enjoined by the Christian religion with suspicion.
>
> (Westfall 1973, pp. 2–3)

And Westfall goes in to say:

> Every one of the problems could be resolved in a variety of ways to reconcile science with religion. But the mere fact of reconciliation meant some change from the pattern of traditional Christianity.
>
> (Ibid.)

These 'grand historic' reconciliations between science and religion are repeated at the personal level for many science students; analysis and knowledge of the historical reconcilations can only be beneficial for students and teachers.[14]

Although Galileo was, in 1615, warned not to hold or teach the Copernican doctrine of a moving Earth, it was only after *The Assayer*, with its endorsement of atomism, was published in 1623 that he faced serious theological charges (Redondi 1988). There was a move by opponents, from general disquiet to specific repudiation.

Atomism presented particular and grievous problems for Christian belief, but the most basic and important one was the central Roman Catholic, Greek

Orthodox and Eastern Uniate teaching on Christ's presence in the Eucharist, or the doctrine of transubstantiation. The Eucharist was the sacramental heart of the Catholic mass, and the mass was and is the devotional heart of the Church. Belief in the Real Presence of Christ, brought into being by the priest's consecration of the communion host, has underwritten devotional practice and doctrinal authority for centuries. In centuries past, denial of the Real Presence was a capital offence; it was a litmus test in the Inquisition, where failure to affirm the belief meant a horrible and painful death.

Scholastic philosophy, with its Aristotelian categories of substance, accidents and qualities, could bring a modicum of intelligibility to this central mystery of faith, as it could also bring a modicum of intelligibility to doctrines such as the Incarnation, the Trinity and immortality of the soul. Scholasticism held that, at consecration, the substance of bread changed to the substance of Christ's body, but the accidents remained that of bread. So Christ became truly present, even though no sensible, observable change was apparent.

Thomas Aquinas (1225–1274), in his prodigious *Summa Theologica* (Third Part, 'Treatise on the Sacraments', Question 75, Article 4), formulated the orthodox doctrine as follows:

> Since Christ's true body is in this sacrament, and since it does not begin to be there by local motion, nor is it contained therein as in a place, as is evident from what was stated above . . . it must be said then that it begins to be there by conversion of the substance of bread into itself. . . . And this is done by Divine power in this sacrament; for the whole substance of the bread is changed into the whole substance of Christ's body, and the whole substance of the wine into the whole substance of Christ's blood. Hence this is not a formal, but a substantial conversion; nor is it a kind of natural movement: but, with a name of its own, it can be called 'transubstantiation'.
>
> (Aquinas 1270/1920).

This Thomist formulation, along with the Aristotelian philosophical apparatus required for its interpretation, was affirmed as defining Catholic orthodoxy at the Council of Trent in 1551.

John Hedley Brooke, a historian sympathetic to the positive contribution of religion to science, recognised the problem that atomism posed, 'especially for the Roman Catholic Church, which took a distinctive view of the presence of Christ at the celebration of the Eucharist' (Brooke 1991, p. 141). He writes:

> With an Aristotelian theory of matter and form, it was possible to understand how the bread and wine could retain their sensible properties while their substance was miraculously turned into the body and blood of Christ. . . . But if, as the mechanical philosophers argued, the sensible properties were dependent on an ulterior configuration of particles, then any alteration to that internal structure

> would have discernible effects. The bread and wine would no longer appear as bread and wine if a real change had occurred.
>
> (Brooke 1991, p. 142)

The atomists held, on *philosophical* grounds, that all legitimate explanation had ultimately to be in terms of the properties of atoms and of their movements and interactions. Their science was constrained by their philosophy. Clearly, the nineteenth-century addition of forces and fields to the ontology of science was not done on philosophical grounds, but on scientific grounds; it seemed that only recourse to the latter entities enabled consistent scientific explanation and progress. Some of this subsequent history of atomism was discussed in Chapter 9.

Philosophy as the 'Handmaiden' of Religion and of Politics

The history of philosophy has been a long dialogue with the history of science; both grow and learn from each other, as was well recognised by Einstein:

> The reciprocal relationship of epistemology and science is of noteworthy kind. They are dependent upon each other. Epistemology without contact with science becomes an empty scheme. Science without epistemology is – insofar as it is thinkable at all – primitive and muddled.
>
> (Einstein 1949, p. 683)

However, also accompanying the dialogue from the beginning has been a third partner: mostly religion or politics. A constant issue has been the right of the third partner to intrude in, correct and direct the primary dialogue. Contemporary debate in the US and Islamic countries about the status of evolution and what science does or does not say about human origins, the soul, the authority of revelation and so forth is a clear case of the entry of a third partner into the philosophy/science dialogue. Likewise, of course, was the intrusion of the Soviet state into the formulation of philosophy consistent with Lysenkoist genetics, and the contemporary strait-jacketing of Chinese science into dialectical garb by the Chinese Communist Party (Guo 2014).

The Roman Catholic Tradition

The Roman Catholic Church has been guided by the medieval view that 'philosophy was the handmaiden' of theology; philosophy was to be subservient to religious and theological purposes. This was the import of the sixteenth-century Tridentine decrees and curial decisions right through to the twentieth century. Pope Leo XIII (1810–1903) promulgated his encyclical *Aeterni Patris*, which gave the name *philosophia perennis* (perennial philosophy) to Thomism and directed Catholic educational institutions to base their philosophical and theological instruction upon it. In 1914, Pius X (1835–1914) issued his *Doctoris Angelici* decree, stating that:

> We desired that all teachers of philosophy and sacred theology should be warned that if they deviate so much as an iota from Aquinas, especially in metaphysics, they exposed themselves to grave risk.
>
> (Weisheipl 1968, p. 180)

It was only in the final years of the twentieth century, with Pope John Paul II's 1998 encyclical, *Fides et ratio*, that the Catholic Church relaxed its attachment to Thomism as official Church philosophy.[15]

The Thomist tradition had enormous cultural and personal impact in Catholic Europe (especially Ireland, Portugal, Poland, Spain, Italy), Latin America, the Philippines and elsewhere. For centuries, Thomism was marshalled to support Church teaching on a wide spectrum of sexual matters, including contraception, masturbation, bestiality and homosexuality. Where the Church exercised political power and influence, these teachings transferred into national law, with the supposed immoral acts becoming illegal acts and punishable by the state, and not just for Roman Catholics, but for all citizens. In all these cases, the reason for condemnation was that the activity was 'unnatural', this whole conceptualisation coming from Aristotle's understanding of objects and actions as having natures that, left alone, unfolded 'naturally' and, when interfered with, unfolded 'violently' or 'unnaturally'.[16] Science played a role in the demise of the power and scope of Thomism and, consequently, the moral and legal edifice built upon it. This is a clear example of the impact of science on philosophy and culture and the deleterious effect of third-party intrusion into the philosophy–science dialogue.[17]

The Islamic Tradition

The same dynamics have played out in the Islamic world, where the medieval view of philosophy as the servant of the Koran still holds.[18] It is very difficult, if not impossible, for a Muslim to entertain or commit to any philosophical system that cannot be reconciled with the assumed ontology, epistemology, politics and ethics of the Koran. The project of 'Islamisation of knowledge' is widely accepted as simply a part of Islam and of being a Muslim. Its purpose is to counter the humanistic and secular foundation of Western education and culture, which it sees as based on five core principles:

1. the sovereignty of man, as though supreme (humanism);
2. basing all knowledge on human reasoning and experience (empiricism);
3. unrestricted freedom of thought and expression (libertarianism);
4. unwillingness to accept 'spiritual' truths (naturalism);
5. individualism, relativism and materialism.

A representative Islamic appraisal of the scientific revolution is Seyyed Nasr's claim that the new science of Galileo and Newton had tragic consequences for the West, because it marked:

> The first occasion in human history when a human collectivity completely replaced the religious understanding of the order of nature for one that was not only nonreligious but that also challenged some of the most basic tenets of the religious perspective.
>
> (Nasr 1996, p. 130)

Nasr repeats Western religious and Romantic criticisms of the new science when he writes:

> Henceforth as long as only the quantitative face of nature was considered as real, and the new science was seen as the only science of nature, the religious meaning of the order of nature was irrelevant, at best an emotional and poetic response to 'matter in motion'.
>
> (Nasr 1996, p. 143)

These and other considerations have led to agitation for 'Islamic Science' (Hoodbhoy 1991), a programme that has received state support in Pakistan and some other Islamic countries. For such science, metaphysics is outside and above science; the external Koran-based metaphysics judges science and determines the acceptable ontology and epistemology of science and, of course, it dictates the content and ethos of school science.

Other Traditions

Third-party intervention is especially fraught when the party is tied to political and institutional power, as Thomism has been with the Roman Catholic Church, Islam in Muslim states, Marxism in the Soviet Union and notionally in China,[19] Hinduism in different Indian states and at different times in the national government[20] and National Socialism in Hitler's Germany.[21] The same situation pertains when custodians of traditional belief systems control what can be thought and taught in traditional indigenous cultures. In such cultures, the free exchange and dialogue of philosophy and science simply cannot be practised, certain subjects are off limits, and certain methodologies are demanded while others are proscribed.[22]

Educational Responses

In all of the above cases, local science and philosophy were made to answer to the dominant, institutionalised religious or political worldview, and educational bodies were forced to accept such 'direction from above' as being in the interest of the nation, religion, culture or economic advancement. The

John Scopes 'Monkey Trial' of 1925 in Tennessee is perhaps the best-known occurrence of this intrusion of outside metaphysics into science education (Larson 1997). In the US, the Scopes trial has been rerun at regular intervals – the Little Rock, Arkansas, trial of 1981 and the Dover, Pennsylvania, trial of 2006 are two recent well-publicised cases. These stand out as intrusions, but such intervention or control is simply institutionalised in many countries, especially Islamic ones. Evolution and other religion-unfriendly topics are not on the curriculum; they are not allowed to be taught. There is no apparent or dramatic outside intrusion; it is just the normal state of affairs.

This cultural–political circumstance poses acute questions for the classroom science teacher: should teachers try to teach current best science and foster independence of thought in their students, or should they be functionaries of whatever the dominant ideological power might be? Can education be legitimately shaped from within, or must it always just take the shape of the last ideological or political foot that trod upon it? These are matters requiring a thoughtful and informed philosophy of education – unfortunately, something mostly ignored in contemporary science-teacher education, where not only philosophy of education, but most foundational subjects have been progressively removed and replaced with training in pedagogical technique and classroom management; the 'apprenticeship' model of teacher education allows little opportunity for 'reflection on principles' or for understanding the history and philosophy of the discipline being taught.[23]

Science and the Spirit World

The world's major religions have had an ongoing engagement with science, investigating how their own ontological, epistemological and ethical commitments – their worldviews – are to be reconciled with both scientific findings and scientific worldviews. Sometimes, the engagement has been mutually productive; at other times, it has been destructive of one or other of the participants. Religion is the most publicly discussed and debated aspect of the science and worldview interaction, and the one that most often occupies politicians and educationalists, from their framing of national constitutions, through Supreme Court rulings, writing national and provincial curricula, arguments about multicultural and indigenous science and textbook selection, through to classroom teaching and teachers' interactions with their students. The arguments and adjustments between Christianity and science – over creation, evolution, providence, miracles, revelation – have been long debated[24] and will be the focus of this chapter. Debates about the teaching of evolution are legend and make their way into newspaper columns somewhere around the world on an almost daily basis. This section of the chapter will deal with just one of the many issues and debates that have arisen in the field: the putative existence and powers of spiritual agencies, spirits, ghosts, poltergeists and angels; inhabitants of what John Wesley, the founder of Methodism, called the 'Invisible World'.

Abrahamic Religions

Belief in a spirit-filled world is fundamental to the Judaeo–Christian–Islamic tradition. Jewish society simply took over the heavily populated world of demons that the Mesopotamian and Hellenic worlds also recognised, with their ontology of beings intermediate between gods and men: these were the *daimones*. The Judaeo–Christian explanation of this realm of troublemakers and evil-inducers was, of course, the expulsion from heaven of Satan and his fallen angels (Genesis 6:1–4). The heavens were heavily populated: in the prophet Daniel's vision, 'A thousand thousand waited on him, ten thousand times ten thousand stood before him' (Daniel 7:10). Jinn, or spirits and angels, were an integral part of the Judaic tradition; everyone in pre-Islamic Arabia believed in them; they lived in a world unseen to humans; they ate and drank and procreated; some were righteous, whereas others were evil. Illness, unusual events, misfortunes, catastrophes and so on were attributed to this host of otherworldly ne'er-do-wells. Belief in such a rich, spirit-populated world, an 'invisible world', is a requirement for the world's 1.5 billion Muslims: belief in angels is the fourth of Islam's six Articles of Faith.

The New Testament and the early Christian Church, being a sect of Judaism, simply carried on belief in the reality and powers of demons, or 'unclean spirits' as they are also called. These demons were responsible for false teaching (1 Timothy 4:1); they performed wonders (Apocalypse 16:14); they rule the kingdom of darkness (Ephesians 1:21, 3:10); and so on. Of particular account in New Testament demonology is the widespread and frequent occurrence of possession of people by the devil or evil spirits. This continued a Judaic and Mesopotamian belief in diabolical possession, one that routinely attributed psychic illness (as now understood) to such a cause (Matthew 8:16, 12:27; Mark 1:34; Luke 7:21, 11:19; Acts 19:13–16). The apostles exorcised evil spirits where they could, with the most graphic instance being the exorcism in the Gerasa cemetery, where the demons fled the person and possessed the herd of swine that they then drove to their death in the Sea of Galilee. Converts such as Paul also had such powers and exercised them effectively, such as when he drove the evil spirit from the girl from Philippi (Acts 16:16). Sometimes, they were not successful, as with the boy now seen to be most probably an epileptic (Matthew 17:14–21; Mark 9:14–29; Luke 9:37–43).

As one Catholic commentator, John L. McKenzie (from whom the foregoing textual references are taken) has written: 'The belief in heavenly beings thus runs through the entire Bible and exhibits consistency' (McKenzie 1966, p. 32). McKenzie further adds:

> But while the use of popular imagery should be understood to lie behind many details of the New Testament concept of demons, the Church has always taught the existence of personal evil spirits, insisting that they are malicious through their own will and not through their creation.
>
> (McKenzie 1966, p. 194)

The Protestant tradition holds comparable views. Martin Luther maintained that demons lived everywhere, but were especially common in Germany; John Wesley wrote in his *Journal* in 1768 that: 'The giving up of witchcraft is in effect the giving up the Bible.' He regarded witchcraft as 'one great proof of the invisible world'.

It is hardly surprising that, 500 years later, half of all Americans tell pollsters that they believe in the Devil's existence, and 10 per cent claim to have communicated with him (Sagan 1997, p. 123). The extent of such belief has been more recently documented in the findings of the large-scale 2008 Pew Report on religious belief and practice in the US.[25] This survey of 35,000 US adults, most of whom would have completed the high-school science requirement, found that belief in some form of God was nearly unanimous (92 per cent), and that this God was not the remote, untouching God of eighteenth-century Deists, but a God who was actively engaged in the affairs of people and of processes in the world. Nearly eight in ten American adults (79 per cent) agree that miracles still occur today, as in ancient times. Similar patterns exist with respect to beliefs about the existence of angels and demons. Nearly seven in ten Americans (68 per cent) believe that angels and demons are active in the world. Majorities of Jehovah's Witnesses (78 per cent), members of evangelical (61 per cent) and historically black (59 per cent) Protestant churches and Mormons (59 per cent) are *completely* convinced of the existence of angels and demons.

The whole constellation of traditional religious beliefs, especially those affirming an active, ongoing engagement of God, angels and spirits with human affairs, requires that the world, including human beings, be constituted in certain ways; that the world has a certain ontology, and that the human beings are so constituted that they can know of and interact with these supernatural agencies. All of this amounts, in part, to a religious worldview, a view about how the world and human beings need to be constituted so as to enable, or ground, religious belief, experience and practice. As is well known, Thomas Aquinas, the imposing intellect of the Middle Ages, devoted energy to determining the properties and numbers of angels, and 'Whether the Angels Differ in Species?' (Aquinas 1270/1920, Part 1, Question 50, Article 4). Henry Gill, a Catholic priest, philosopher and physics lecturer, gave succinct expression to the kind of worldview held by many religious believers:

> It will be useful to recall briefly the Catholic teaching as to the existence of spirits. The Scripture is full of references to both good and bad spirits. There are good and bad angels. Each of us has a Guardian Angel, whose presence, alas, we often forget. Angels, as the Catechism tells us, have been sent as messengers from God to man.
>
> (Gill 1944, pp. 127–28)

Traditional Societies

In traditional or indigenous cultures, convictions about the 'invisible world' and interactions between this supernatural world and the everyday world are usually bolstered with animist beliefs, where plants and natural objects are endowed with intelligences and spiritual attributes, and where natural processes can be swayed by rituals, incantations, charms, potions, magic, sorcery and spells. In most such cultures, spirits are everywhere and have immense powers; they feature in traditional stories, legends and myths and underwrite a wide variety of social and especially medical practice.

Papua New Guinea (PNG) is a representative case. In the early months of 2013, a series of horrific, gruesome, sorcery-related murders were committed. In January, outside Mt Hagen, the capital of the Western Highlands, a 20-year-old mother was accused of sorcery, doused in petrol and burned alive atop a pile of rubbish and car tyres. She supposedly had used her powers as a witch to kill a boy who had been admitted to hospital with chest pains. In March, a Highlands man ate his new-born son in order to bolster his sorcery powers. The same month, in the Southern Highlands, six supposed witches were tortured with hot irons, and one was roasted to death. In April, in Bougainville, two elderly women, accused of being witches and causing the death of a schoolteacher, were tortured for three days and then beheaded in front of a large mob that included police officers. In just one Highland province, Simbu, there are 150 sorcery-inspired and justified attacks per year.[26] At the same time, the PNG government released a report on the AIDS epidemic in the country, detailing the prevalence, and uselessness, of traditional treatments, such as having sufferers sit atop huts inside which are burned 'special fires', in expectation that the rising smoke would carry off the evil spirits inhabiting the person and causing the sickness. A long-time PNG Roman Catholic priest, Philip Gibbs, touched on the major issue of scientific and biblical worldviews when he described PNG culture as having a 'pre-Enlightenment, or Biblical, worldview. . . . They don't believe in coincidence or accidents. When something bad happens, they don't ask what did it but who did it' (Elliot 2013, p. 18).

The reality and efficacy of sorcery are recognised in the 1971 Sorcery Act. A 1977 PNG Law Commission study on *Sorcery in PNG* concluded:

> We have written some general ideas about sorcery we know from our own experience as Papua New Guineans. In order to get a balanced view of sorcery we would like to say that sorcery is very much a matter of the innermost belief of the people. Fear of, or the practice of sorcery or various occults is a world-wide phenomenon. Sorcery or black magic exists in Europe, in Asia, in Africa and in North and South America as well as the Pacific.
>
> Major world religions claim the reality of forces or personalities greater than the human and animal powers. Whether these powers or personalities can be shown to exist is often quite irrelevant to the belief. From these beliefs many practices and procedures follow.
>
> (Narokobi 1977, p. 19)

The 2013 revision of the legal code is moving to deny the reality of such powers and make supposed *Sanguma* bashings, torture and killings criminal offences. The situation with PNG traditional society is repeated in sometimes more and other times less extreme forms in many other traditional societies, and also other societies where the spirit world looms large, and where thousands of years of tradition, folklore and superstition are embedded. Although some NOMA-aligned and postmodernist commentators maintain that science cannot disprove the existence of such spirits, the revised PNG legal code appeals to modern science to assert that spirits do not exist, and so cannot be used as a defence in homicide or assault cases; purveyors of 'spirit medicine' and cures will be charged with fraud. Revision of the Sorcery Act requires rejection of the 'science is neutral' position.[27]

Of course, these beliefs and practices are not just those of 'traditional' societies: the Vatican's official exorcist, Father Gabriele Amorth, who has conducted 70,000 exorcisms, claimed that many paedophilia cases were the direct work of devils who possessed or otherwise influenced the offending priests (Amorth 2010).

Education and the Spirit World

Despite being everywhere, being endowed with amazing powers and being credited with causing tsunamis, AIDS, schizophrenia, adultery and much else, such angels and spirits do not show up in laboratories or scientific texts; they have not gained a place in the scientific understanding of the natural, social or personal worlds. No spirits have been identified by science as having any causal interaction with the world, and those who supposedly had such interaction have been in full retreat as alternative scientific explanations become established. This gives rise to a certain disconnect. Such claims are then either discordant with, or orthogonal to, the worldview and conduct of science. Denying the efficacy or existence of 'bad' spirits or devils involves proto-science. The basic claim is that, 'there is no evidence' for such possession by bad spirits, and that the evidence (paedophilia or children dying) can be accounted for by other (natural) causes. This basic claim moves discussion into the field of science and evidence appraisal. However, once that move is made, then why not extend the examination to the efficacy or existence of 'good' spirits and angels?

The educational question is what to do about such beliefs? Should nothing be done, and the cultural status quo be retained unaltered? Should students be encouraged to believe just in good spirits and not in bad ones? And, if good spirits, to believe for ontological reasons (there actually are such things) or for instrumental reasons (such belief is harmless, it encourages good behaviour and is part of the cultural or religious tradition)? Or not believe in spirits at all? The last was the choice of Joseph Priestley, the famed eighteenth-century English scientist, historian, philosopher, theologian and Dissenting Church minister:

> The notion of madness being occasioned by evil spirits disordering the minds of men, though it was the belief of heathens, of the Jews in our Savior's time, and

> of the apostles themselves, is highly improbable; since the facts may be accounted for in a much more natural way.
>
> (Rutt 1817–1832/1972, Vol. 7, p. 309)

For Priestley, Jesus was simply mistaken when he attributed the cure of madness to driving out evil spirits, because subsequent science and philosophy had shown there were no such things to be driven out.[28]

Traditional Non-Western Metaphysics

Olugbemiro Jegede, a leading African science educator, elaborates the worldviews that 'all African communities' have in common and that science education must be 'rooted in'. These are:

1. the belief in a separate being whose spiritual powers radiate through gods (of thunder, fire, iron) and ancestors;
2. reincarnation and the continuation of life after death;
3. the human as the centre of the universe in traditional African thought;
4. the theory of causality (Jegede 1989, p. 193).

Jegede cites the work of thirteen anthropologists and educators who 'now confirm the position that the African [view of nature] is anthropomorphic as opposed to the mechanistic view of nature of Western science'. In a later publication, Jegede provides the table of 'some distinguishing characteristics between the foundation of African and Western cultures', shown in Table 10.1 (Jegede 1997, p. 7)

Table 10.1 represents a clear dichotomy between African indigenous science and orthodox science. The educational and philosophical task is to work out how to deal with the dichotomy: retain the first, or the second, or both? And, for each choice, what is the intellectual, cultural and social cost? And is it worth paying? It is Jegede's view that:

Table 10.1 Traditional African and Western Science

Traditional African Science	*Western Science*
Anthropomorphic	Mechanistic, exact and hypothesis-driven
Monistic–vitalistic and metaphysical	Seeks empirical laws, principles, generalisation and theories
Based on cosmology interwoven with traditional religion	Public property, divorced from religion
Orally communicated	Primarily documented via print
The elders' repository of knowledge is truth not to be challenged	Truth is tentative and challengeable by all
Learning is a communal activity	Learning is an individual enterprise

> From this can be seen the need to design science education that satisfactorily meets with the needs of Africa in such a way that the African view of nature, socio-cultural factors, and the logical dialectical reasoning embedded in African metaphysics are catered for within a changing global community.
>
> (Jegede 1997, p. 15)

To deal intelligently with these issues requires more than the usual teacher-education curriculum focused on classroom-management skills and learning theory; the resolution of the issues requires informed historical and philosophical input, something conspicuously lacking when the Portland School District adopted the *The Portland African–American Baseline Essays* as a guide to its science programme (Adams 1986, Martel 1991).

This is not the place to canvass the myriad non-Western worldviews that activate contemporary cultures and inform their sciences.[29] Nor is it the place to canvass the detailed and rich empirical knowledge of animal life, astronomy, horticulture and technology that traditional societies possess. Rather, it is important to concentrate upon the worldview, or 'theoretical' aspects, of traditional belief systems and how to recognise and deal with them in science teaching. The observations of Jegede are sufficient to illustrate the threads of the argument to be advanced, namely that some core epistemological and ontological assumptions of Western science are in objective conflict with core assumptions of some traditional belief systems. If this is so, then thoughtful educational responses are required.

The matters of ontological contention are the following:

1 Is the world constituted in such a way as to serve human interests?
2 Are processes in the world teleological? That is, do events and behaviours occur in order to bring about some fitting end state?
3 Are inanimate and nonhuman animate processes activated and controlled by spiritual influences?

The Western scientific tradition, after centuries of investigation and tumultuous debate, answers 'no' to each of the above questions, whereas many traditional belief systems affirm some or all of the propositions. One can recognise among the pre-Socratic philosophers the slow, awkward attempts to distance their thought about the world from the mythical worldviews that were characterised by the above anthropomorphic, animistic and teleological dimensions. Western science has slowly continued this process of jettisoning these features.

The basic matters of epistemological contention are the following:

1 Does knowledge come from the observation of things as they are in their natural states?
2 Are knowledge claims validated by successful predictions?

3 Do particular classes or authority figures define knowledge or become the custodians of knowledge?
4 Is knowledge a fixed and unchanging system?

As with the ontological questions, the Western scientific tradition answers 'no' to each of these questions, whereas traditional societies affirm some or all of them. Of course, there is some debate about these questions of natural states, prediction, the institutionalisation of knowledge, and accretion versus revolutions in knowledge. However, even with more nuanced elaboration, the conflict between scientific ontology and epistemology and numerous traditional ontologies and epistemologies is still apparent. This was the view of Robin Horton, in his classic study of African and Western science. After outlining many points of similarity between African and Western science, he concluded by drawing attention to deep differences. For Horton:

> The key difference is a very simple one. It is that in traditional cultures there is no developed awareness of alternatives to the established body of theoretical tenets; whereas in scientifically orientated cultures, such an awareness is highly developed. It is this difference we refer to when we say that traditional cultures are 'closed' and scientifically oriented cultures are 'open'.
>
> (Horton 1971, p. 153)

One of the reasons for concern about teaching Western science in traditional societies is that this conflict very quickly spills over into other domains. Almost all commentators make the observation that traditional science is much more integrated with other important cultural systems than is usually apparent in the West. Traditional science is connected with religion, with health, with politics, with social structure and cultural customs. The fear is that Western science will not only subvert traditional science, but that it will, as a consequence, subvert a range of other significant social institutions and beliefs and contribute to the destruction of traditional culture.

This fear of Western science's possible disruption of culture and traditional institutions is often misplaced and, indeed, often indicates a paternalist attitude, an assumption that other cultures are so feeble that they cannot make intelligent and sensible decisions about what accommodations to make and not to make in the light of modern science. Members of the Papua New Guinea Law Commission do not cease to be members of PNG culture by virtue of declaring that the evil spirits energising sorcery do not exist. Papua New Guinea adjusts, grows and modernises. South Africans did not cease to be members of their culture when they affirmed that viruses, not evil spirits, were responsible for the crippling AIDS epidemic in the country.

Multicultural Science Education

Examples of spirit-laden cultures and traditions holding decidedly non-Western metaphysical and epistemological commitments have been given above. The teaching of science in such cultures and societies raises important matters about the purposes of science education and the distinction between *understanding* science and *believing* science. Some maintain that science education should leave cultural beliefs untouched; that students should simply leave their culture's worldview (ontology, epistemology, metaphysics, authority structure, religion) at the classroom door, enter inside to learn the instrumentally understood content of science, then go back outside and become again fully believing participants in their culture. This is close to advocating an anthropological approach to learning science. Just as anthropologists can be expected to learn *about* the beliefs and practices of different societies without any expectation that they adopt or come to believe them, some say that students can learn science in the same way – a sort of 'spectator' learning, where one learns but does not believe or internalise.[30] It is the learning that anthropologists and historians have when studying other cultures and periods: they learn what others have believed, without any requirement for themselves to believe.

Border Crossing

Glen Aikenhead, in a much-cited paper, has advocated such a strategy, calling it 'border crossing' (Aikenhead 1996). Just as tourists, when they cross borders, do not lose their cultural identity, even though they temporarily adopt foreign customs about driving, eating, dressing and language, so also science students should not lose their cultural identity (as a traditional Roman Catholic, a fundamentalist Christian, an Intelligent Designer, a PNG highlander, and so on), just because the science laboratory has no place for their own rich beliefs. The students are tourists in 'scienceland', not immigrants. This is a form of pedagogical NOMA; it gets its intellectual sustenance from Kuhnian–Feyerabendian incommensurability claims served up with liberal doses of social constructivism; it is profoundly at odds with the Enlightenment tradition that hoped for the internalisation of the scientific outlook.

The contrast between the aspirations of the Enlightenment philosophers and contemporary 'border crossing' science educators is profound and speaks to a major divergence in each one's appreciation of science. This indeed is the case. Consider the claim that educators have to learn,

> how to deprivilege science in education and to free our children from the 'regime of truth' that prevents them from learning to apply the current cornucopia of simultaneous but different forms of human knowledge with the aim to solve the problems they encounter today and tomorrow.
>
> (Van Eijck & Roth 2007, p. 944)

And the assertion that 'the social studies of science' reveal science as: 'mechanistic, materialist, reductionist, empirical, rational, decontextualised, mathematically idealised, communal, ideological, masculine, elitist, competitive, exploitive, impersonal, and violent' (Aikenhead 1997, p. 220).

This claim is more than puzzling. Is this meant to describe the work of Galileo? Newton? Huygens? Priestley? Darwin? Mendel? Faraday? Mach? Thompson? Lorentz? Maxwell? Rutherford? Planck? Einstein? Bohr? Curie? Does it describe the work of Edward Jenner in developing the smallpox vaccine? Jonas Salk and Albert Sabin in developing the polio vaccine? We are not told whose science warrants the description. It is clearly a composite or collage that requires unpicking, but this is not done. From Aikenhead's description, it is doubtful whether science should even be in the curriculum; it certainly should be rated X, with even border crossing being dangerous.

Other prominent and influential science educators share Aikenhead's unfavourable estimation of science. Consider, for instance, claims made in a contribution to a current major science-education handbook:

> One of the first places where critical inquirers might look for oppression is positivist (or modernist) science ... modernist science is committed to expansionism or growth ... modernist science is committed to the production of profit and measurement ... modernist science is committed to the preservation of bureaucratic structures. ... Science is a force of domination not because of its intrinsic truthfulness, but because of the social authority (power) that it brings with it.
>
> (Steinberg & Kincheloe 2012, pp. 1487–88)

It is not clear what is being asserted here. Such statements demonstrate the need for science educators to be well informed, careful and considered in their approach to HPS; casual reading hardly suffices. The above accounts, apart from being confused and contradictory, cannot be sustained. Taking the subject out of the picture and relying on measuring instruments (rulers, scales, thermometers, barometers, clocks) instead of subjective appraisals of length, weight, temperature, pressure and duration; utilising mathematics; introducing idealisations and abstractions; valuing objective evidence; being public, communal, publishing, criticising and debating – are all the things that enabled the scientific revolution to occur in seventeenth-century Europe and progress to its current international status. And, of course, Copernicus and Galileo had no social authority enforcing their heliocentricism, on the contrary; whereas Lysenko had all the oppressive, overwhelming authority of Stalin behind his non-Mendelian genetics, ultimately this authority counted for nothing. This understanding is missing in the unsupported criticisms of science given above.

Naturalism

The conduct of Western science presupposes at least methodological naturalism (MN). This is the view that, when doing science, whatever occurs in the world is to be explained by natural mechanisms and entities, and that these entities and mechanisms are the ones either revealed by science or in principle discoverable by science. This methodological presupposition does not rule out miracles or divine interventions or other non-scientific causes; it just means that such processes cannot be appealed to while seeking scientific explanations. There has been, historically, a transition from a more open, mixed or relaxed methodology to having MN function as a defining principle of scientific investigation. As Robert Pennock states the matter:

> Science has completely abandoned appeal to the supernatural. In large part this is simply the result of consistent failure of a wide array of specific 'supernatural theories' in competition with specific natural alternatives.
>
> (Pennock 1999, p. 282)

The US National Academy of Sciences affirmed just such a position:

> Science is limited to explaining the natural world through natural causes. Science can say nothing about the supernatural. Whether God exists or not is a question about which science is neutral.
>
> (NAS 1998, p. 58)

This seems like a nice division of territory, a live-and-let-live outcome, but this ignores the obvious question of whether science can say anything about putative evidences for God's existence. Many such evidences (design, efficacy of prayer, mystical experience) are in the purview of science, and science need not be neutral about them.

A stricter version of naturalism is ontological naturalism (ON), which is sometimes called metaphysical naturalism. This is the view that there is a scientific explanation for all events, that supernatural explanations (e.g. divine interventions, miracles) simply do not occur. Many see ON as pure dogmatism, and it can be if it is held in advance as a philosophical principle – who is to say in advance of evidence how the world works or what is in it? But it can be held on less dogmatic, two-step grounds:

1 Thus far, no credible evidence has been advanced for the existence of any putative non-natural entity, or entity not within the scientific realm.

Many, of course, reject (1), and that is a whole separate argument. However, some accept (1) and nevertheless say that ON does not follow from it, or only follows dogmatically, as no one knows what evidence might turn up. But the non-dogmatic holder of ON can add a second step to their argument:

2 Do not believe things for which there is no evidence.

If (2) is granted, then ON does indeed follow. Then, the dogmatism claim moves back to belief in (2) rather than belief in ON. However, belief in (2) need not be dogmatic; it can be the 'default' position, and its opposite, namely the holding of beliefs for which there is no evidence, is dogmatic. The latter beliefs can be entertained, that in part is what hypothesis creation is about, but they cannot be held without evidence. This was, in essence, Bertrand Russell's 'tea-pot' argument:

> I ought to call myself an agnostic; but, for all practical purposes, I am an atheist. I do not think the existence of the Christian God any more probable than the existence of the Gods of Olympus or Valhalla. To take another illustration: nobody can prove that there is not between the Earth and Mars a china teapot revolving in an elliptical orbit, but nobody thinks this sufficiently likely to be taken into account in practice. I think the Christian God just as unlikely.
>
> (Russell 1958)

Both science-informed methodological and ontological naturalists admit the existence of whatever kinds of entity (e.g. atoms, fields, forces, quarks, bosons, fermions, dark matter) science postulates or reveals as having regular causal relations with the rest of nature. However, ontological naturalists do not admit the existence of spiritual or divine entities, or any kind of entity that does not enter into scientifically demonstrated, lawful and causal relations with nature.

Although often confused, there is a difference between realism and naturalism (including materialism). Realism simply asserts that there is a world independent of human thought. Such an independent world might include spirits, minds, universals, mathematical objects, forms or any other independent existent. Realism neither rules in nor rules out any particular kind of putatively existing being. A theological realist about angels believes that angels exist; a theological instrumentalist believes that the word 'angel' is shorthand for 'makes people behave' or 'strengthens our cultural bonds'. Naturalism is a subspecies of realism; it asserts that the only existing things are the things that science postulates and incorporates into successful and mature theories; materialism, in turn, is a subspecies of naturalism. Traditional religious believers must reject ON, but, of course, religious scientists routinely adopt MN in the laboratory; to do otherwise would put them outside the scientific enterprise.[31]

Materialists are a subspecies of ontological naturalists, but they are less relaxed about what can exist. Basic or 'old-fashioned' materialists grant existence only to material, physical, 'three-dimensional' objects, the kinds of thing that can be tripped over. They reject the postulation of non-material scientific entities, believing that such postulation is a failure of scientific nerve, and it is the slippery slope to idealism. This is clearly as much an a priori metaphysical position as it is a deduction from scientific practice. Emergent materialism is a more sophisticated version, where the world is seen as material, but stratified. The properties of material aggregations are greater than, and different from, the properties of the building blocks. So cells have different kinds of property to molecules, brains have different properties to neurons, societies have different properties to individuals and so on. For emergent materialists, the world is changing and evolving, and new properties emerge as material formations become more complex; for this reason, emergent materialism is in principle anti-reductionist.[32]

Scientism

Scientism is a subspecies of naturalism; many hold the latter position, without commiting to the former.[33] Like positivism, scientism has had a bad press in social science, in 'critical' and postmodernist philosophy, and especially in constructivist science education. In these circles, 'scientism' is regarded as a synonym for reduction-ism, closedminded-ism, shallow-ism, cultural imperialism and most other -isms with which no sensible, well-educated, sensitive person would wish to be associated. But is scientism so bad, or so obviously beyond the intellectual pale?

As outlined in Chapter 2, the beginning of scientism can be seen in the once-revolutionary claim of Newton, Condorcet and the early Enlightenment philosophers that the methods and outlook of the new science should be applied outside the laboratory; they should be harnessed in understanding and solving other pressing social and cultural problems, including ones associated with superstitions and the exercise of unjustified ecclesial and feudal powers. Three hundred years later, as documented in Chapter 1, this 'proto-scientism' has been repeated by the AAAS, which maintained that:

> The scientifically literate person is one who is aware that science, mathematics, and technology are interdependent human enterprises with strengths and limitations; understands key concepts and principles of science; is familiar with the natural world and recognises both its diversity and unity; *and uses scientific knowledge and scientific ways of thinking for individual and social purposes.*
>
> (AAAS 1989, p. 4; italics added)

In its *Benchmarks for Science Literacy*, the AAAS says that science education has to 'prepare students to make their way in the real world, a world in which problems abound – in the home, in the workplace, in the community, on the planet' (AAAS 1993, p. 282). The unique contribution of the science programme to this more general problem-solving and society-improving educational goal is the cultivation and refinement of scientific habits of mind. This is where the move from proto-scientism to scientism begins.

Scientism is the view that only the methods of natural science are capable of providing knowledge of the natural, social and personal worlds; there are no other routes to such knowledge. Listening to gurus, holding Ouija boards, invoking mediums, remembering dreams, reading sacred texts or consulting astrologers simply gives no knowledge of nature (earthquakes), social circumstances (collapse of economies), public events (the outbreak of war) or personal physical episodes (sudden illness or death), or even personal psychic episodes (delusions, emotional states and so on). Such sources might provoke hypotheses or ideas to be tested, but they do not provide knowledge. Thus stated, scientism is not as 'beyond the pale' as it is usually taken to be.

For at least a century, one challenge for scientism has been the possibility of a scientific social science. An illustrious tradition – Nicolas Condorcet, Denis Diderot, Jean le Rond d'Alembert, Auguste Comte, Otto Neurath, Emile

Durkheim – thought social science had to be scientific, but this tradition has been dismissively charged with 'aping the sciences' (von Hayek 1952), and various kinds of non-scientific social science have evolved – Verstehen, hermeneutical, humanistic and so on. This is not the occasion to settle this argument, but it is by no means obvious that the non-scientific social sciences have shed light on, much less explained, major historical, economic or social structures or events – the global financial crisis, the Arab Spring, the invasion of Iraq and so on.

Mario Bunge, a defender of scientism, has illustrated his position using the accompanying 'ecology of progressive science' diagram (Bunge 2010) (Figure 10.1). For Bunge, science requires, and can only flourish in, social and intellectual environs characterised by the following political, ethical and philosophical commitments.

- *Humanism*: scientists need to promote human welfare, not misery, business advancement or political advantage. The latter purposes more easily lead to corruption of science (witness Nazi Germany, Stalinist Russia or current 'big business' science). There can and should be applied science, but it ought be for human welfare and improvement.
- *Systemism*: scientists need to recognise that there are no isolated events, mechanisms or problems in the world. Structures and events are parts of systematic, causal wholes, or, as John Donne famously wrote, 'no man is an island'. Thus, good science generates cross-disciplinary research

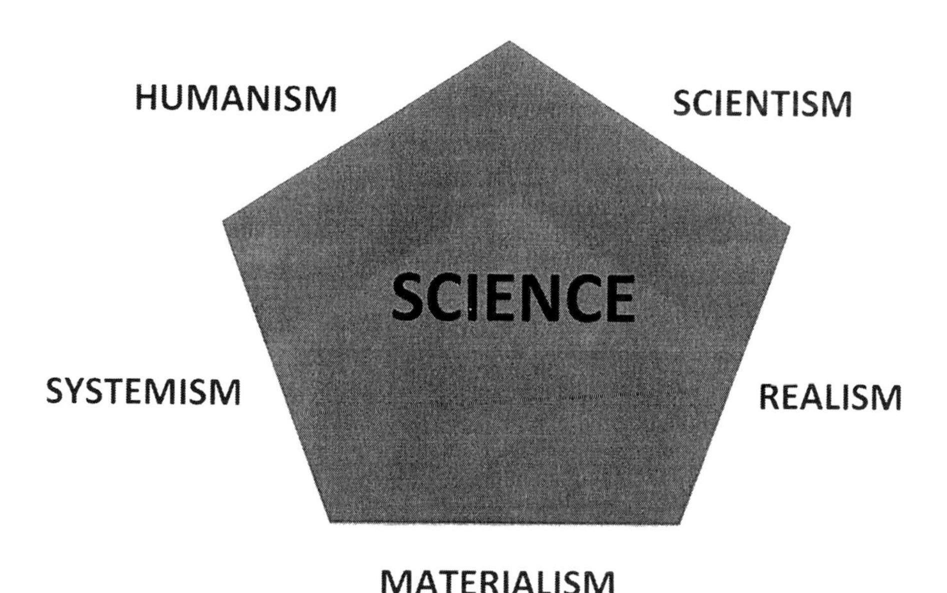

Figure 10.1 Bunge's Ecology of Science

fields: geophysics, astrophysics, biochemistry, social psychology, molecular biology, psycholinguists and so on. And it rules out such hybrids as astropsychology or creation science.
- *Materialism*: scientists need to seek for causes and explanations in the kinds of thing that are within the accepted ontology of science. *Naturalism* can satisfy this requirement, but evocation of *spiritualism* or *supernaturalism* or *traditionalism* violates it. To the degree that a society believes that the gods or spirits are responsible for earthquakes, then money for geophysical research will be limited.
- *Realism*: scientists need to recognise that there is an external world independent of human consciousness or experience, and that science attempts to provide knowledge of such a world. The external world passes judgement on scientific efforts to understand it.
- *Scientism*: scientists need to hold the conviction that scientific methods are applicable outside the laboratory and are the only way in which knowledge of the world and society is attained. Without this commitment, social and cultural problems are addressed in wholly ineffective ways; praying for the end of Middle East conflict can be a nice cultural engagement, but it can shed no light on the conflict or its remediation.

For Bunge, to the degree that one or more of these elements is missing from the ecology or cultural envirnons of science, the science is constrained, compromised, misdirected, or becomes pseudoscience.

Compatibility of Science and Religion

The compatibility or otherwise of science and religion is an enduring issue for educators; it bears on the interaction of schools with their society and on school life, curricula and classroom teaching. When considering the question, we need to distinguish between a number of sometimes-conflated issues.[34]

First, do religious claims and understandings have to be adjusted to fit proven scientific facts and theories? There really is no longer any serious debate on this issue; sensible believers and informed theologians acknowledge that religious claims need to be modified or given a non-literal interpretation to fit with proven or even highly probable scientific claims. Joseph Priestley, the eighteenth-century enlightened believer, told the story of one of his congregation:

> A good old woman, who, on being asked whether she believed the literal truth of Jonah being swallowed by the whale, replied, yes; and added, that if the Scriptures had said that Jonah swallowed the whale, she would have believed it too.

Priestley thought that such convictions simply indicated that the term 'belief' was being misused in the context: 'How a man can be said to *believe* what is, in the nature of things, *impossible*, on any authority, I cannot conceive' (Rutt 1817–1832/1972, Vol.6, p. 33).

Since Saint Augustine, all serious thinkers on the topic agree with Priestley. There has been debate about just what degree of proof a factual scientific claim needs to have before it triggers a revision in a competing factual religious claim – Augustine thought revision was needed only in the face of absolutely proven 'scientific' claims. The details of this debate do not bear on the present argument.[35]

Second, can religious believers be scientists? Again, at one level, there is no debate on this matter. As a simple matter of anthropological and psychological fact, there have been and are countless believers of all religious stripes who are scientists. John Polkinghorne, an Anglican priest, could be picked out as an exemplar of a research physicist and believer (Polkinghorne 1991, 1996). Many such individuals can be found contributing to journals such as *Zygon: Journal of Religion & Science*. For just one compilation of contemporary Christian scientists, see Mott (1991). There are comparable compilations of Hindu, Islamic, Buddhist, Mormon and Jewish scientists. These lists are relevant to the question of the psychological compatibility between scientific and religious beliefs, but in themselves the lists do not bear on the philosophical or rational compatibility of science and religious belief; that such lists can be compiled might incline a person to the compatibility postion, but extra argument is required. Some scientists are astrologers, others channel spirits, some think they are Napoleon reincarnated, some are racist and others are sexist, and so on for a whole spectrum of beliefs that, as a matter of fact, have been held by scientists.

No one doubts that science, as a matter of anthropological fact, is compatible with any number of belief systems – recall that the Nobel laureates Philipp Lenard and Johannes Stark were both Nazi ideologues. Scientists are humans, and humans notoriously can believe all sorts of things at the same time, but such psychological compatibility has no bearing on the rationality or reasonableness of their beliefs, nor on the philosophical compatibility between science and belief systems; much less are the latter rational simply in virtue of one or more scientists believing them. The rationality is a logical or normative matter. The philosophically interesting questions are whether a scientist can be a *rational* religious believer (or astrologer, diviner, re-incarnationer, racist, sexist, Nazi, etc.) and what are the arguments for so being.

Third, is religion compatible with the metaphysics and worldview of science? Where there is incompatibility between scientific and religious metaphysics and worldviews – as in the case of atomism and traditional Roman Catholic doctrine developed above – the options usually taken to reconcile the differences are to claim that:

1 Science has no metaphysics; it deals just with appearances and makes no claims about reality. This is the option made famous by the Catholic positivist Pierre Duhem.[36] It is the claim made by many fundamentalists who say, specifically of evolution, that 'it is just a theory' (Ben-Ari 2005).
2 The metaphysics of science is false, at least any such purported metaphysics that is inconsistent with religious beliefs. This is the option

advocated by the scholastic tradition discussed above, by Claude resmontant and Seyyed Nasr, who are quoted above, and by philosophical theologians such as Alvin Plantinga (2011), E.L. Mascall (1956) and numerous others.

3 There can be parallel, equally valid, metaphysics. This is an old option given recent prominence by Stephen Gould in his NOMA formulation (Gould 1999). Gould's much-repeated claim was that:

> The magisterium of science covers the empirical realm: what the Universe is made of (fact) and why does it work in this way (theory). The magisterium of religion extends over questions of ultimate meaning and moral value. These two magisteria do not overlap, nor do they encompass all inquiry (consider, for example, the magisterium of art and the meaning of beauty).
> (Gould 1999, p. 6)

The problem for NOMA is that, apart from classical deists for whom God stays remote in His heaven and has no dealings with His creation, the core conviction of religious traditions is that the two realms overlap: that the supernatural has engagement with the natural; that God engages with His creation; that certain texts (Torah, Bible, Koran, Book of Mormon, Sikh scriptures) are divinely inspired, if not divinely written; miracles occur; prayers are answered; and so on. If something is claimed to happen in the world, then its causation can be investigated by science. There may be natural causes identified for the event, as have routinely been established for momentous events such as tsunamis and earthquakes and medical episodes such as epilepsy and AIDS, where, previously, unnatural causes were invoked, or the current failure of science to identify a cause can be acknowledged. However, the latter does not entail the truth or even the reasonableness of the supernatural causal hypothesis. Evidence needs to be adduced for the latter and indication given of the conditions for its acceptance. These two requirements are open to ordinary scientific investigation.[37]

Conclusion

As well as enabling our social and technical lives, science has contributed immensely to our philosophical and cultural tradition – this is part of the 'flesh' of science; however, too often science teaching presents just the 'bare bones' of laws, formulae and problems, the 'final products' of science. This is one reason why, notoriously, advanced 'technical' science is so often associated with religious and ideological fundamentalism and bigotry.[38] The cultural flesh of science should be part of any serious science programme.

In a good liberal education, science students, and hopefully other students as well, will learn about the philosophical dimensions of science, beginning with routine matters such as conceptual analysis, epistemology, values and so on. They will also learn about the metaphysical, especially ontological, dimensions of science, some of which have been discussed above. They should also be introduced to, and hopefully make decisions about, the constitution and applicability of the scientific outlook, habit of mind or temper. They should entertain questions such as: Is a scientific outlook required for the solution of social and ideological problems? By reading about any number of courageous scientists, beginning with Galileo and moving through Joseph Priestley and on to Andrei Sakharov (Sakharov 1968), students can be introduced to the issue of the social and cultural requirements for the pursuit of science, the issue that so animated the Enlightenment scientists, philosophers and social reformers.

In particular, students might think through the Enlightenment tradition's claims that, on purely epistemological grounds, science, and more generally the pursuit of truth in all human domains, requires the legal protection of free speech, freedom of the press and support for diversity, unhindered scholarly publication and freedom of association. Students, in essays, debates, mock trials or drama, can entertain questions such as: Does the promotion and spread of science entail a liberal, secular, democratic, non-authoritarian state? An engaging question in China, Pakistan, Saudi Arabia and many other countries.

All of this makes science classes more intellectually engaging, it promotes 'minds-on' science learning and it enables diverse subjects in a school curriculum (history, mathematics, technology, religion) to be related. The introduction of history and philosophy to science lessons enables students to better understand the science and the scientific methodology they are learning, to better appreciate the role of science in the formation of the modern world and contemporary worldviews, and perhaps to obtain the knowledge and enthusiasm to support science and the spread of the 'scientific habit of mind'.

Undoubtedly, such an education has an impact on, and contributes to, the worldviews of students. So it is worth noting Frederick Copleston's caution:

> It must be recognized, I think, that the creation of worldviews is none the less a pretty risky procedure. There is, for example, the risk of making unexamined or uncritized presuppositions in a desire to get on with the painting of the picture. Again, there are the risks of over-hastily adopting desired conclusions, and also of allowing one's judgements to be determined by personal prejudices or psychological factors.
>
> (Copleston 1991, p. 71)

In the liberal tradition, science teachers are not so much creating worldviews but rather encouraging students to identify and then to begin to analyse and appraise aspects of worldviews. For educators, it is the student's enquiry and thinking that are important. A good science teacher can agree with Bertrand Russell, who famously said in 1916, at the height of the Great War, when he criticised the use of schools by both sides for nationalist indoctrination:

> Education would not aim at making them [students] belong to this party or that, but at enabling them to choose intelligently between the parties; it would

aim at making them able to think, not at making them think what their teachers think.

(Egner & Denonn 1961, pp. 401–402)

Notes

1 This chapter is dependent on research published in Matthews (2009a).
2 A classic account of the history of these interactions is J.D. Bernal's four-volume study, *Science in History* (Bernal 1965). See also Crombie (1994), Dewitt (2004) and Randall (1962).
3 Richard Attenborough's 2013 television documentary series, *The Galapagos Islands*, is promoted as: 'The islands that transformed our view of life on Earth.'
4 Learning that *homo sapiens* shares 98.4 per cent of its genes with pygmy chimpanzees can change a person's views of their relationship to the animal world.
5 Of the voluminous literature on Darwinism and worldviews, see especially: Dennett (1995), Greene (1981), McMullin (1985) and Ruse (1989).
6 Benjamin Farrington's *Science and Politics in the Ancient World* (1939) is a classic treatment of these themes.
7 An informative guide to the vast literature and debates about the scientific revolution (including whether to capitalise the term) is H. Floris Cohen's *The Scientific Revolution* (Cohen 1994).
8 The varieties of medieval and renaissance Aristotelianism arose from efforts to accommodate ever new developments and discoveries in natural philosophy. See Blum (2012) and Schmitt (1983).
9 A classic work on the doctrines and history of scholastic philosophy is De Wulf (1903/1956). See also Volumes 2 and 3 of Frederick Copleston's *History of Philosophy* (Copleston 1950).
10 For the background, context and impact of Copernicus, see Blumenberg (1987), Gingerich (1975, 1993) and Grant (2004).
11 A classic discussion is Dijksterhuis's *The Mechanization of the World Picture* (1961/1986). On the wider impact of the Galilean–Newtonian method, see Butts and Davis (1970), Cohen (1980), McMullin (1967) and Shank (2008).
12 Unfortunately, these early modern philosophers are frequently studied in isolation from the contemporary science with which they were engaged; early modern philosophy is presented to students as a drawn-out soliloquy, not the dialogue and debate with early modern science that it was. This theme, with texts, is developed in Matthews (1989).
13 See, for instance, Brooke (1991, Chapter V), Israel (2001) and Porter (2000).
14 On this much-researched topic, see at least the following: Blancke *et al.* (2012), Lawson and Worsnop (1992), Yasri *et al.* (2013), Martin-Hansen (2008), Sinatra and Nadelson (2011), Smith and Siegel (2004) and Taber *et al.* (2011).
15 On John Paul II's encyclical and how it reviewed and revised the status of Thomism, see Ernst (2006).
16 An example of this reasoning and mindset is Aquinas's view that sexual intercourse was 'naturally ordained for procreation' (*Sentences* 4.31.2.2), so that even indulging in coitus for reasons of health (a good purpose) nevertheless rendered the act unnatural and thus sinful, as it was not done for its primary end. At the time, and right through to the present, such reasoning convinced millions of sensible people. Most now see it as plainly ridiculous. For the history, see Noonan (1965, Chapter 8).
17 The relationship of Thomism to science is a complex matter. Thomism has made adjustments and, in many quarters, is still thriving (Ashley 1991, Lamont 2009). The philosophy journal *New Scholasticism* was published from 1927 to 1989, *The Thomist* journal has been published continuously since 1939, and *The Modern Schoolman* has been published continuously since 1925. And, of course, numerous 'scholastic' philosophy journals are still published in languages other than English.

18 On the tensions and accommodations between science and Islam, see at least: Edis (2007), Edis and BouJaoude (2014) and Hoodbhoy (1991).
19 For 'official' philosophy in the Soviet Union, and its contested relationship to science, see Graham (1973); for Maoism and Chinese philosophy and science, see Chan (1969) and Guo (2014).
20 See Nanda (2003).
21 See Beyerchen (1977, 1992) and Cornwell (2003).
22 This ultimate control was exercised in the Roman Catholic Church through the requirement of all writers of history, theology and philosophy to obtain the *Nihil obstat* certificate from the diocesesian censor and then the *imprimatur* from the bishop that allows publication. Communist regimes have their own equivalents.
23 On philosophy of education in science education, see Schulz (2009, 2014); on the larger issue of educational foundations, see contributions to Tozer *et al.* (1990).
24 Among a veritable library of relevant books, see Barbour (1966), Brooke (1991), Haught (1995), Jaki (1978), Mascall (1956) and contributions to Lindberg and Numbers (1986).
25 The survey was conducted between May and August 2007 and published in June 2008 in the Pew Report at http://religions.pewforum.org/pdf/report-religious-landscape-study-full.pdf
26 See accounts and interviews in Elliot (2013).
27 Postmodernists and NOMA advocates can, of course, claim that the PNG legislators are mistaken in their claims about the reach of science. The first appeal against a murder conviction will bring out philosophers on both sides of the courtroom, just as was done in the 1981 Little Rock, Arkansas, case, where Judge Overton endorsed the philosopher Michael Ruse's argument that Creationism could be shown not to be science. Larry Laudan subsequently argued that there could be no such demonstration. Teaching Creationism in schools was thus deemed illegal, just as sorcery-informed murder in PNG will be deemed illegal.
28 In passing, it is worth noting that every account of Priestley's life shows that a calm, considerate and deeply 'spiritual' life is possible without any belief in spirits.
29 A useful introduction to the literature is Selin (1997), and numerous websites are devoted to this subject.
30 Jegede (1997) calls this 'collateral learning'. The position is argued in Ogunniyi (1988) and by others, although the usual option is to 'skirt around' it – to muddy the argumentative waters.
31 On naturalism see: Devitt (1998), Fishman and Boudry (2013), French *et al.* (1995), Mahner (2012, 2014), Nagel (1956), Rosenberg (2011) and Wagner and Warner (1993).
32 On emergent materialism, see Broad (1925), Bunge (1977, 1981, 2003, 2012) and Sellars (1932).
33 Two proponents of scientism are Mario Bunge (Bunge 2010) and Alex Rosenberg (Rosenberg 2011). Tom Sorell articulates the anti-scientism argument (Sorell 1991).
34 Different taxonomies or ways of classifying science–religion relationships are developed in Barbour (1990), Haught (1995) and Polkinghorne (1986). These, and other related matters, are discussed in Reiss (2014) and Yasri *et al.* (2013).
35 For the arguments and literature, see McMullin (2005).
36 See extensive discussion and bibliography in Martin (1991).
37 The anti-NOMA view that science can test supernatural claims is argued by many, including Boudry *et al.* (2012), Fishman (2009), Slezak (2012) and Stenger (2007).
38 That there is no connection between advanced technology and advanced thinking was sadly demonstrated when scores of spectators at the Papua New Guinea witch burning described above captured the event on their mobile-phone cameras and uploaded the burning on to the Internet.

References

AAAS (American Association for the Advancement of Science): 1989, *Project 2061: Science for All Americans*, AAAS, Washington, DC (also published by Oxford University Press, 1990).

AAAS (American Association for the Advancement of Science): 1993, *Benchmarks for Science Literacy*, Oxford University Press, New York.

Adams III, H.H.: 1986, 'African and African-American Contributions to Science and Technology'. In *The Portland African-American Baseline Essays*, Portland Public Schools, Portland, OR.

Aikenhead, G.S.: 1996, 'Science Education: Border Crossing Into the Subculture of Science', *Studies in Science Education* 27(1), 1–52.

Aikenhead, G.S.: 1997, 'Toward a First Nations Cross-cultural Science and Technology Curriculum', *Science Education* 81(2), 217–238.

Allen, R.E. (ed.): 1966, *Greek Philosophy. Thales to Aristotle*, The Free Press, New York.

Amorth, G.: 2010, *The Memoirs of an Exorcist*, Ediciones Urano, Rome.

Aquinas, T.: 1270/1920, *Summa Theologica*, trans. English Dominican Province, Burns, Oates & Washbourne, London (online edition, 2008).

Ashley, B.M.: 1991, 'The River Forest School and the Philosophy of Nature Today'. In R.J. Long (ed.) *Philosophy and the God of Abraham. Essays in Memory of James A. Weisheipl, OP*, Pontifical Institute of Medieval Studies, Toronto, pp. 1–15.

Barbour, I.G.: 1966, *Issues in Science and Religion*, SCM Press, London.

Barbour, I.G.: 1990, *Religion in an Age of Science*, SCM Press, London.

Ben-Ari, M.: 2005, *Just a Theory: Exploring the Nature of Science*, Prometheus Books, Amherst, NY.

Bernal, J.D.: 1965, *Science in History*, 4 volumes, 3rd edn, C.A. Watts, London.

Beyerchen, A.D.: 1977, *Scientists Under Hitler: Politics and the Physics Community in the Third Reich*, Yale University Press, New Haven, CT.

Beyerchen, A.D.: 1992, 'What We Know About Nazism and Science', *Social Research* 59, 615–641.

Blancke, S., de Smedt, J., de Cruz, H., Boudry, M. and Braeckman, J.: 2012, 'The Implications of the Cognitive Sciences for the Relation Between Religion and Science Education: The Case of Evolutionary Theory', *Science & Education* 21(8), 1167–1184.

Blum, P.R.: 2012, *Studies on Early Modern Aristotelianism*, Brill, Leiden, The Netherlands.

Blumenberg, H.: 1987, *The Genesis of the Copernican World*, MIT Press, Cambridge, MA.

Boudry, M., Blancke, S. and Braeckman, J.: 2012, 'Grist to the Mill of Anti-evolutionism: The Failed Strategy of Ruling the Supernatural Out of Science by Philosophical Fiat', *Science & Education* 21, 1151–1165.

Broad, C.D.: 1925, *The Mind and Its Place in Nature*, Harcourt Brace, New York.

Brooke, J.H.: 1991, *Science and Religion: Some Historical Perspectives*, Cambridge University Press, Cambridge, UK.

Bunge, M.: 1977, *Treatise on Basic Philosophy. Vol.3, The Furniture of the World*, Reidel, Dordrecht, The Netherlands.

Bunge, M.: 1981, *Scientific Materialism*, Reidel, Dordrecht, The Netherlands.

Bunge, M.: 2003, *Emergence and Convergence*, University of Toronto Press, Toronto.

Bunge, M.: 2010, *Matter and Mind: A Philosophical Inquiry*, Springer, Dordrecht, The Netherlands.

Bunge, M.: 2012, *Evaluating Philosophies, Boston Studies in the Philosophy of Science*, Vol.295, Springer, Dordrecht, The Netherlands.

Butts, R.E. and Davis, J.W. (eds): 1970, *The Methodological Heritage of Newton*, University of Toronto Press, Toronto.

Chan, W.-T.: 1969, *A Source Book in Chinese Philosophy*, Princeton University Press, Princeton, NJ.

Cohen, H.F.: 1994, *The Scientific Revolution: A Historiographical Inquiry*, University of Chicago Press, Chicago, IL.

Cohen, I.B.: 1980, *The Newtonian Revolution*, Cambridge University Press, Cambridge, UK.
Copernicus, N.: 1543/1952, *On the Revolutions of the Heavenly Spheres* (trans. C.G. Wallis), Encyclopædia Britannica, Chicago. IL.
Copleston, F.C.: 1950, *A History of Philosophy*, 8 volumes, Doubleday, New York.
Copleston, F.C.: 1955, *Aquinas*, Penguin Books, Harmondsworth, UK.
Copleston, F.C.: 1991, 'Ayer and World Views'. In A. Phillips Griffiths (ed.) *A.J. Ayer: Memorial Essays*, Cambridge University Press, Cambridge, UK, pp. 63–75.
Cornwell, J.: 2003, *Hitler's Scientists: Science, War and the Devil's Pact*, Penguin, London.
Crombie, A.C.: 1994, *Styles of Scientific Thinking in the European Tradition*, 3 volumes, Duckworth, London.
De Wulf, M.: 1903/1956, *An Introduction to Scholastic Philosophy: Medieval and Modern* (trans. P. Coffey), Dover Publications, New York.
Dennett, D.C.: 1995, *Darwin's Dangerous Idea: Evolution and the Meanings of Life*, Allen Lane, Penguin, London.
Devitt, M.: 1998, 'Naturalism and the A Priori', *Philosophical Studies* 92, 45–65.
Dewitt, R.: 2004, *Worldviews: An Introduction to the History and Philosophy of Science*, Blackwell Publishing, Oxford, UK.
Dijksterhuis, E.J.: 1961/1986, *The Mechanization of the World Picture*, Princeton University Press, Princeton, NJ.
Dillenberger, J.: 1961, *Protestant Thought & Natural Science: A Historical Study*, Collins, London.
Dilworth, C.: 1996/2006, *The Metaphysics of Science. An Account of Modern Science in Terms of Principles, Laws and Theories*, Kluwer Academic Publishers, Dordrecht, The Netherlands (2nd edition, 2006).
Edis, T. and BouJaoude, S.: 2014, 'Rejecting Materialism: Responses to Modern Science in the Muslim Middle East'. In M.R. Matthews (ed.) *International Handbook of Research in History, Philosophy and Science Teaching*, Springer, Dordrecht, The Netherlands, pp. 1663–1691.
Edis, T.: 2007, *An Illusion of Harmony: Science and Religion in Islam*, Prometheus Books, Amherst, NY.
Egner, R.E. and Denonn, L.E. (eds): 1961, *The Basic Writings of Bertrand Russell*, George Allen & Unwin, London.
Einstein, A.: 1949, 'Remarks to the Essays Appearing in this Collective Volume'. In P.A. Schilpp (ed.) *Albert Einstein: Philosopher-Scientist*, Tudor Publishing Company, New York, pp. 663–688.
Elliot, T.: 2013, 'Witch-Hunt', *Sydney Morning Herald*, Good Weekend, 20 April, pp. 16–21.
Ernst, H.E.: 2006, 'New Horizons in Catholic Philosophical Theology: *Fides et Ratio* and the Changed Status of Thomism', *The Heythrop Journal* 47(1), 26–37.
Farrington, B.: 1939, *Science and Politics in the Ancient World*, George Allen & Unwin, London.
Fishman, Y.I.: 2009, 'Can Science Test Supernatural Worldviews?' *Science & Education* 18(6–7), 813–837.
Fishman, Y.I. and Boudry, M.: 2013, 'Does Science Presuppose Naturalism (or, Indeed, Anything at All)? *Science & Education* 22(5), 921–949.
French, P.A., Uehling, T.E. and Wettstein, H.K. (eds): 1995, *Philosophical Naturalism*, University of Notre Dame Press, Notre Dame, IN.
Galileo, G.: 1633/1953, *Dialogue Concerning the Two Chief World Systems*, (trans. S. Drake), University of California Press, Berkeley, CA (2nd revised edition, 1967).
Gauch Jr, H.G.: 2009, 'Science, Worldviews and Education', *Science & Education* 18(6–7), 667–695.
Gill, H.V.: 1944, *Fact and Fiction in Modern Science*, M.H. Gill, Dublin.

Gingerich, O. (ed.): 1975, *The Nature of Scientific Discovery: A Symposium Commemorating the 500th Anniversary of the Birth of Nicolaus Copernicus*, Smithsonian Institution Press, Washington, DC.

Gingerich, O.: 1993, *The Eye of Heaven: Ptolemy, Copernicus, Kepler*, American Institute of Physics, New York.

Gould, S.J.: 1999, *Rock of Ages: Science and Religion in the Fullness of Life*, Ballantine Books, New York.

Graham, L.R.: 1973, *Science and Philosophy in the Soviet Union*, Alfred A. Knopf, New York.

Grant, E.: 2004, *Science and Religion, 400 bc to ad 1550. From Aristotle to Copernicus*, Johns Hopkins University Press, Baltimore, MD.

Greene, J.C.: 1981, *Science, Ideology and World View: Essays in the History of Evolutionary Ideas*, University of California Press, Berkeley, CA.

Guo, Y.: 2014, 'The Philosophy of Science and Technology in China: Political and Ideological Influences', *Science & Education* 23.

Haught, J.F.: 1995, *Science and Religion: From Conflict to Conversation*, Paulist Press, New York.

Hoodbhoy, P.: 1991, *Islam and Science: Religious Orthodoxy and the Battle for Rationality*, Zed Books, London.

Horton, R.: 1971, 'African Traditional Thought and Western Science'. In M.F.D. Young (ed.) *Knowledge and Control*, Collier-Macmillan, London, pp. 208–266.

Israel, J.: 2001, *Radical Enlightenment: Philosophy and the Making of Modernity 1650–1750*, Oxford University Press, Oxford, UK.

Jaki, S.L.: 1978, *The Road of Science and the Ways to God*, University of Chicago Press, Chicago, IL.

Jegede, O.J.: 1989, 'Toward a Philosophical Basis for Science Education of the 1990s: An African View-Point'. In D.E. Herget (ed.) *The History and Philosophy of Science in Science Teaching*, Florida State University, Tallahassee, FL., pp. 185–198.

Jegede, O.J.: 1997, 'School Science and the Development of Scientific Culture: A Review of Contemporary Science Education in Africa', *International Journal of Science Education* 19(1), 1–20.

Lamont, J.: 2009 'The Fall and Rise of Aristotelian Metaphysics in the Philosophy of Science', *Science & Education* 18(6–7), 861–884.

Larson, E.J.: 1997, *Summer for the Gods. The Scopes Trial and America's Continuing Debate Over Science and Religion*, Basic Books, New York.

Lawson, A.E. and Worsnop, W.A.: 1992, 'Learning About Evolution and Rejecting a Belief in Special Creation: Effects of Reflective Reasoning Skill, Prior Knowledge, Prior Belief and Religious Commitment', *Journal of Research in Science Teaching* 29(2), 143–166.

Lindberg, D.C. and Numbers, R.L. (eds): 1986, *God and Nature: Historical Essays on the Encounter between Christianity and Science*, University of California Press, Berkeley, CA.

McKenzie, J.L.: 1966, *Dictionary of the Bible*, Geoffrey Chapman, London.

McMullin, E. (ed.): 1967, *Galileo Man of Science*, Basic Books, New York.

McMullin, E.: 1985, 'Introduction: Evolution and Creation'. In E. McMullin (ed.) *Evolution and Creation*, University of Notre Dame Press, Notre Dame, IN, pp. 1–58.

McMullin, E.: 2005, 'Galileo's Theological Venture'. In E. McMullin (ed.) *The Church and Galileo*, University of Notre Dame Press, Notre Dame, IN, pp. 88–116.

Mahner, M.: 2012, 'The Role of Metaphysical Naturalism in Science', *Science & Education* 21(10), 1437–1459.

Mahner, M.: 2014, 'Science, Religion, and Naturalism: Metaphysical and Methodological Incompatibilities'. In M.R. Matthews (ed.) *International Handbook of Research in History, Philosophy and Science Teaching*, Springer, Dordrecht, The Netherlands, pp. 1793–1835.

Martel, E.: 1991, 'How Valid Are the Portland Baseline Essays?' *Educational Leadership* Dec.–Jan., 20–23.
Martin, R.N.D.: 1991, *Pierre Duhem: Philosophy and History in the Work of a Believing Physicist*, Open Court, LaSalle, IL.
Martin-Hansen, L.M.: 2008, 'First-Year College Students' Conflict with Religion and Science', *Science & Education* 17(4), 317–357.
Mascall, E.L.: 1956, *Christian Theology and Natural Science: Some Questions in Their Relations*, Longmans, Green, London.
Matthews, M.R. (ed.): 1989, *The Scientific Background to Modern Philosophy*, Hackett Publishing, Indianapolis, IN.
Matthews, M.R.: 2009a, 'Teaching the Philosophical and Worldview Components of Science', *Science & Education* 18(6–7), 697–728.
Matthews, M.R. (ed.): 2009b, *Science, Worldviews and Education*, Springer, Dordrecht, The Netherlands.
Montalenti, G.: 1974, 'From Aristotle to Democritus via Darwin: A Short Survey of a Long Historical and Logical Journey'. In F.J. Ayala and T. Dobzhansky (eds) *Studies in the Philosophy of Biology: Reduction and Related Problems*, University of California Press, Berkeley, CA, pp. 4–19.
Mott, N. (ed.): 1991, *Can Scientists Believe?* James & James, London.
Nagel, E.: 1956, 'Naturalism Reconsidered'. In his *Logic without Metaphysics*, Freepress, Glencoe, IL, Chapter 1.
Nanda, M.: 2003, *Prophets Facing Backward. Postmodern Critiques of Science and Hindu Nationalism in India*, Rutgers University Press, New Brunswick, NJ.
Narokobi, B. (ed.): 1977, 'Occasional Paper No.4, Sorcery', Papua New Guinea Law Commission, Port Moresby.
NAS (National Academy of Science): 1998, *Teaching About Evolution and the Nature of Science*, National Academy Press, Washington, DC.
Nasr, S.H.: 1996, *Religion and the Order of Nature*, Oxford University Press, Oxford, UK.
Newton, I.: 1713/1934, *Principia Mathematica*, 2nd edn (trans. Florian Cajori), University of California Press, Berkeley, CA (1st edition, 1687).
Noonan, J.T.: 1965, *Contraception: A History of Its Treatment by Catholic Theologians and Canonists*, Mentor-Omega Books, New York.
Ogunniyi, M.B.:1988, 'Adapting Western Science to Traditional African Culture', *International Journal of Science Education* 10(1), 1–9.
Pennock, R.T.: 1999, *Tower of Babel: The Evidence Against the new Creationism*, MIT Press, Cambridge, MA.
Plantinga, A.: 2011, *Where the Conflict Really Lies. Science, Religion and Naturalism*, Oxford University Press, New York.
Polkinghorne, J.: 1996, *The Faith of a Physicist: Reflections of a Bottom-up Thinker.* Fortress Press, Minneapolis, MN.
Polkinghorne, J.C.: 1986, *One World: The Interaction of Science and Theology*, SPCK, London.
Polkinghorne, J.C.: 1991, *Reason and Reality: The Relationship between Science and Theology*, SPCK, London.
Popper, K.R.: 1963, *Conjectures and Refutations: The Growth of Scientific Knowledge*, Routledge & Kegan Paul, London.
Porter, R.: 2000, *The Enlightenment: Britain and the Creation of the Modern World*, Penguin Books, London.
Randall Jr, J.H.: 1962, *The Career of Philosophy*, Columbia University Press, New York.
Redondi, P.: 1988, *Galileo Heretic*, Allen Lane, London.
Reiss, M.: 2014, 'What Significance Does Christianity Have for Science Education?'. In M.R. Matthews (ed.) *International Handbook of Research in History, Philosophy and Science Teaching*, Springer, Dordrecht, The Netherlands, pp. 1637–1662.
Rohrlich, F.: 1987, *From Paradox to Reality: Our Basic Concepts of the Physical World*, Cambridge University Press, Cambridge, UK.

Rosenberg, A.: 2011, *The Atheist's Guide to Reality: Enjoying Life Without Illusions*, W.W. Norton, New York.
Ruse, M.: 1989, *The Darwinian Paradigm. Essays on Its History, Philosophy, and Religious Implications*, Routledge, London.
Russell, B.: 1958, 'Letter to Mr Major', in *Dear Bertrand Russell: A Selection of his Correspondence with the General Public, 1950–1968*, Allen & Unwin, London, 1969.
Rutt, J.T. (ed.): 1817–1832/1972, *The Theological and Miscellaneous Works of Joseph Priestley*, 25 volumes, J. Johnson, London (Kraus Reprint, New York, 1972).
Sagan, C.: 1997, *The Demon-Haunted World: Science as a Candle in the Dark*, Headline Book, London.
Sakharov, A.D.: 1968, *Progress, Coexistence and Intellectual Freedom*, W.W. Norton, New York.
Schmitt, C.B.: 1983, *Aristotle and the Renaissance*, Harvard University Press, Cambridge, MA.
Schulz, R.M.: 2009, 'Reforming Science Education: Part I. The Search for a Philosophy of Science Education', *Science & Education* 18 (3–4), 225–249.
Schulz, R.M.: 2014, 'Philosophy of Education and Science Education: A Vital but Underdeveloped Relationship'. In M.R. Matthews (ed.) *International Handbook of Research in History, Philosophy and Science Teaching*, Springer, Dordrecht, The Netherlands, pp. 1259–1315.
Selin, H. (ed.): 1997, *Encyclopedia of the History of Science, Technology, and Medicine in Non-Western Cultures*, Kluwer Academic Publishers, Dordrecht, The Netherlands.
Sellars, R.W.: 1932, *The Philosophy of Physical Realism*, Macmillan, New York.
Shank, J.B.: 2008, *The Newton Wars and the Beginning of the French Enlightenment*, University of Chicago Press, Chicago, IL.
Sinatra, G.M. and Nadelson, L.: 2011, 'Science and Religion: Ontologically Different Epistemologies'. In R.S. Taylor and M. Ferrari (eds) *Epistemology and Science Education: Understanding the Evolution vs. Intelligent Design Controversy*, Routledge, New York, pp. 173–193.
Slezak, P.: 2012, 'Review of Michael Ruse Science and Spirituality: Making Room for Faith in the Age of Science', *Science & Education* 21, 403–413.
Smith, M.U. and Siegel, H.: 2004, 'Knowing, Believing and Understanding: What Goals for Science Education?' *Science & Education* 13, 553–582.
Sorell, T.: 1991, *Scientism: Philosophy and the Infatuation with Science*, Routledge, London.
Steinberg, S.R. and Kincheloe, J.: 2012, 'Employing the Bricolage as Critical Research in Science Education'. In B. Fraser, K. Tobin and C. McRobbie (eds) *International Handbook of Science Education*, 2nd edn, Springer, Dordrecht, The Netherlands, pp. 1485–1500.
Stenger, V.J.: 2007, *God: The Failed Hypothesis: How Science Shows That God Does not Exist*, Prometheus Books, Amherst, NY.
Suchting, W.A.: 1994, 'Notes on the Cultural Significance of the Sciences', *Science & Education* 3(1), 1–56.
Taber, K.S., Billingsley, B., Riga, F. and Newdick, H.: 2011, 'Secondary Students' Responses to Perceptions of the Relationship Between Science and Religion: Stances Identified From an Interview Study', *Science Education* 95(6), 1000–1025.
Tozer, S., Anderson, T.H. and Armbruster, B.B. (eds): 1990, *Foundational Studies in Teacher Education: A Reexamination*, Teachers College Press, New York.
Van Eijck, M. and Roth W.-M.: 2007, 'Keeping the Local Local: Recalibrating the Status of Science and Traditional Ecological Knowledge (TEK) in Education', *Science Education* 91, 926–947.
Vitzthum, R.C.: 1995, *Materialism: An Affirmative History and Definition*, Prometheus, Amherst, NY.
von Hayek, F.: 1952, *The Counter-Revolution of Science*, Free Press, Glencoe, IL.
Wagner, S. and Warner, R. (eds): 1993, *Naturalism: A Critical Appraisal*, University of Notre Dame Press, Notre Dame, IN.

Weisheipl, J.A.: 1968, 'The Revival of Thomism as a Christian Philosophy'. In R.M. McInerny (ed.) *New Themes in Christian Philosophy*, University of Notre Dame Press, South Bend, IN, pp. 164–185.

Westfall, R.S.: 1971, *The Construction of Modern Science: Mechanisms and Mechanics*, Cambridge University Press, Cambridge, UK.

Westfall, R.S.: 1973, *Science and Religion in Seventeenth-Century England*, University of Michigan Press, Ann Arbor, MI.

Yasri, P., Arthur, S., Smith, M.U. and Mancy, R.: 2013, 'Relating Science and Religion: An Ontology of Taxonomies and Development of a Research Tool for Identifying Individual Views', *Science & Education* 22(10), 2679–2707.

Chapter 11

The Nature of Science and Science Teaching[1]

There has been a long tradition in education that has advocated the cultural, educational, personal and scientific benefits of infusing HPS into science classes, curricula and teacher education, or, in current terms, of bringing the NOS into classrooms, curricula and teacher education. This might be called the normative NOS tradition, the tradition that argues that, for a range of personal, cultural and disciplinary purposes, students learning science should also learn about science, in particular its philosophical or methodological distinctiveness. Joseph Priestley, in the eighteenth century, could be thought of as the founder of this tradition, as shown in Chapter 7. He wrote the first ever books on the history of electricity and the history of optics, so that natural philosophers could learn from the successes and failures of those who went before them. In the nineteenth century, the central figures in this tradition were William Whewell (Whewell 1855), Thomas Huxley (Huxley 1868/1964) and Ernst Mach (Mach 1886/1986). In the early decades of the twentieth century, John Dewey (Dewey 1910), in the US, and Frederick Westaway (Westaway 1929) and Eric Holmyard (Holmyard 1924), in the UK, were central figures. In the North American world, the tradition was continued, in the 1940s, by Joseph Schwab (Schwab 1949); in the 1960s, by Leo Klopfer (Klopfer 1969) and James Robinson (Robinson 1968); and in the 1970s, by Jim Rutherford (Rutherford 1972, 2001), Gerald Holton (Holton 1975, 1978), Robert Cohen (Cohen 1975) and Michael Martin (Martin 1972, 1974).

In the past three decades, a number of science educators have extended this normative tradition. Perhaps the best known are Derek Hodson (1986, 1988, 2008, 2009, 2014), Richard Duschl (1985, 1990, 2004) and Mansoor Niaz (2009, 2010), and there have been many others.[2] As outlined in the Preface to this book, the IHPST Group, through its conferences held biennially since 1989 and associated journal, *Science & Education*, has contributed significantly to the tradition. Since its inception in 1992, the latter journal has published more than 800 research articles on 'History, Philosophy and Science Teaching', and hundreds of papers have been presented to IHPST international and regional conferences.

As well as advocacy or normative work, there has been, more recently, a steady growth of *empirical* NOS research. This has focused less on why students should learn NOS, but more on how they learn and if they have learned the subject. This empirical tradition has studied questions such as: Can NOS be effectively taught in elementary school? How is NOS best learned? What are the different outcomes between explicit or implicit NOS instruction? What NOS views are held by scientists, teachers and representative historians and philosophers? What, if any, is the connection between learning NOS and learning science content? What 'long term' gains and transferability there might be from NOS learning? How can valid, reliable and efficient NOS tests be developed? And so on.[3] The work of Norman Lederman (2004, 2007, Lederman *et al.* 2014), Fouad Abd-el-Khalick (2005), William McComas (1998a, 2014), Keith Taber (2009, 2014) and their research teams has had a particular impact.

From the beginning, there has naturally been ambiguity and sometimes tension over just how wide to draw the NOS net for educational purposes. Originally, NOS was identified with philosophy of science. There was a concentration on the epistemology, methodology, ontology and ethics of science; on learning how evidence related to theory appraisal, what determined theory choice, what were the characteristics of a competent experiment and so on. The view was that this cluster constituted the distinctive, defining feature of science. The NOS net was widened to include history of science, alongside philosophy of science, on the grounds that the latter needed the former, and that to learn about science required knowing something of its history and the actual processes of scientific discovery and theory acceptance – 'Philosophy of science without history of science is empty; history of science without philosophy of science is blind' (Lakatos 1978, p. 102). With renewed interest in large-scale (industrial) and small-scale (laboratory) sociology of science, the NOS net was further widened to include sociology, and then psychology of science. At this point, NOS basically became 'science studies', with philosophy and epistemology no longer accorded central status.[4] Both the normative and the empirical traditions of NOS research will adjust to how tight or relaxed is the NOS definition adopted.

Science is a human, and thus historically embedded, truth-seeking enterprise that has many features: cognitive, social, commercial, cultural, political, structural, ethical, financial, psychological, etc. All of these features are worthy of study by science students, as well as by disciplinary specialists, and different ones come into clearer focus when considering different sciences, and when considering different aspects of the history, achievements and practice of the different sciences. Some of the features are shared to a large degree with other knowledge-acquiring enterprises, some are shared to a limited degree, and some are not shared at all. Given these characteristics of science, it is useful to understand NOS, not as some list of necessary and sufficient conditions for a practice to be scientific, but rather as something that, following Wittgenstein's terminology, identifies a 'family resemblance' of features that warrant different enterprises being called scientific. Truth seeking must be retained as a defining goal of science in order to give any limits to its characterisation; whether science is successful and just what 'truth' means are subsidiary issues.[5]

This chapter will recommend a change of terminology and research focus from the essentialist and epistemologically focused NOS to a more relaxed, contextual and heterogeneous 'features of science' (FOS). Such a change of terminology and focus avoids the following philosophical and educational pitfalls that have been associated with much of recent NOS research:

1 the confused aggregation of epistemological, sociological, psychological, ethical, commercial and philosophical features into a single NOS list;
2 argument about just how many items should be included in an NOS list: seven for the Lederman group, ten for the McComas group, still other numbers for other groups;
3 the privileging of one side of what are contentious and much-debated arguments about the methodology or 'nature' of science;
4 the assumption of particular solutions of the demarcation dispute;
5 the assumption that NOS learning can be judged and assessed by students' capacity to identify some number of declarative statements about NOS.

William Whewell: A Precursor to Contemporary NOS Debates

William Whewell (1794–1866), the formidable English scientist, philosopher, historian, theologian and moralist, gave a lecture in Leeds in 1854 to the Royal Institution of Great Britain, entitled 'On the Influence of the History of Science Upon Intellectual Education' (Whewell 1855). He prepared the ground for his particular argument by saying:

> As the best sciences which the ancient world framed supplied the best elements of intellectual education up to modern times, so the grand step by which, in modern times, science has sprung up into a magnitude and majesty far superior to her ancient dimensions, should exercise its influence upon modern education, and contribute its proper result to modern intellectual culture.
>
> (Whewell 1855, p. 242)

In the lecture, he provided passionate argument for the inclusion of NOS (as it is now called) into all liberal education, saying:

> In the History of Science we see the infinite variety of nature; of mental, no less than bodily nature; of the intellectual as well as of the sensible world. . . . The history of science . . . may do, and carefully studied, must do, much to promote that due apprehension and appreciation of inductive discovery; and inductive discovery, now that the process has been going on with immense vigour in the nations of Europe for the last three hundred years, ought, we venture to say, to form a distinct and prominent part of the intellectual education of the youth of those nations.
>
> (Whewell 1855, pp. 248–249)

Whewell believed that the history of science was indispensable for understanding 'intellectual culture' more generally, by which he meant the processes of knowledge creation or epistemology. One hundred and more years before Karl Popper, Imre Lakatos and Thomas Kuhn made the view popular, Whewell argued that philosophy of science has to be informed by history of science. Whewell's point is worth drawing attention to, as so much NOS discussion in science education goes on in direct violation of it. NOS is frequently taught without reference to history and is not informed by history. Unfortunately, many teachers wishing to convey something of NOS do so by having students 'reflect on', 'brainstorm' or 'discuss' just their own classroom activities or investigations. This has only limited value. Whatever lessons might be learned depend on extrapolating from classroom science to 'big picture', historically and socially embedded science. Caution and caveats are required in making any such extrapolation. Without such humility, the exercise cultivates hubris and promotes narcissism: 'I will tell you about science and its aims, methods and values by reflecting on what I do in the classroom.'

Whewell expressed two concerns that have occupied much contemporary NOS research when he went on to ask: 'How is such a culture to be effected? And also, how are we to judge whether it has been effected?' (Whewell 1855, p. 249).

Whewell was, in contemporary terms, asking: How can NOS best be taught? And, how can NOS learning best be assessed? Educators and researchers are still asking and answering these questions.

Current NOS Research

As mentioned above, contemporary NOS research is either normative or empirical, with, of course, some overlap, as people do not usually research topics unless they think they are important. The basic challenge for normative NOS research is that it needs be well informed by historical and philosophical studies of science. This challenge has been more or less well met, depending on the researcher's own training or grasp of the fields. Science educators have typically taken a broad, relaxed and not overly examined view of the nature of science; the tendency has been to 'go with the HPS flow' or, less kindly, to embrace whatever HPS position is fashionable at the time: logical empiricism in the 1960s (Matthews 1997), Kuhnianism in the 1970s and 1980s (Matthews 2004) and constructivism in the 1980s and 1990s (Matthews 2000b);[6] more recently, many educators are embracing various postmodernist and sociocultural versions of HPS, as the following attests:

> Recent scholarship in science studies has opened the way for more thoughtful science education discourses that consider critical, historical, political and sociocultural views of scientific knowledge, and practice . . . Increased attention to the problematic nature of traditional western science's claims to objectivity and universal truth has created an educational space where taken-for-granted meanings

> are increasingly challenged, enriched and rejected. . . . Thus, science's long accepted claim to epistemological superiority has now become bound to the consideration of cultural codes, social interests, and economic imperatives.
>
> (Bazzul & Sykes 2011, p. 268)

Because reasonable training in history or philosophy of science is rare in the science-education community, many of the published claims made about NOS, such as those in the preceding quotation, are false, ill informed or mythical.[7] This, of course, has consequences for teacher education when NOS is included in the programme, and it has flow-on consequences for classroom teaching when teachers attempt to promote NOS knowledge in schools. Consider the following confident claim made in a recent article:

> People with more sophisticated epistemological beliefs about science reject the objective truth, recognize multiple realities and consider science knowledge as human construction. These sophisticated epistemological perspectives are promoted in the US science education reform documents as both learning goals and teaching approaches.
>
> (Kang 2008, p. 480)

Here, of the three NOS claims, one, 'scientific knowledge is a human construction', is quite true but verges on tautological (could it be a tree's construction? A polar bear's construction?). The other two claims are simply false, or at best highly contested; nevertheless, they are labelled 'sophisticated'. This pattern of announcing false or highly contentious NOS claims as 'sophisticated' is widespread. Consider:

> The constructivist mode of learning may be associated with teachers having sophisticated epistemologies, and an orientation to the traditional/transmissive conception may be reflective of teachers holding naive epistemologies associated with omniscient authority and certain knowledge.
>
> (Chan & Elliot 2004, p. 819)

This misdirection flows over into empirical research, where some NOS tests are scored so that realism is the 'immature' postion and is indicative of inadequate NOS understanding and of the need for teachers to 'try again' to get their students to believe the more 'sophisticated' position (Matthews 1998). Clearly, the NOS position taken by researchers bears upon the validity of test instruments and of informed assessment of NOS learning.[8]

One influential group in current NOS research is that associated with Norman Lederman.[9] The group's definition of NOS is characteristically catholic: 'Typically, NOS refers to the epistemology and sociology of science, science as a way of knowing, or the values and beliefs inherent to scientific knowledge and its development' (Lederman *et al.* 2002, p. 498). It is noteworthy that, in this definition, both epistemological *and* sociological aspects of science are brought into the NOS tent. This necessitates a very large tent. Sociology of science will include politics, commerce, professional structures, employment and whatever else those studying the functioning of science see as important. This stretching is an important reason, advanced below, for moving from NOS terminology to FOS. Restricting NOS to epistemological or methodological characteristics of science presents problems for identifying any core, essential 'nature' of science; once sociology is allowed in, and then psychology of science is also brought inside the NOS tent, then all attempts at delimiting any 'nature' need to be abandoned.

The Lederman group maintains that, 'no consensus presently exists among philosophers of science, historians of science, scientists, and science educators on a specific definition for NOS' (Lederman 2004, p. 303). Although recognising no across-the-board consensus on NOS, the group does claim that there is sufficient consensus on central matters for the purposes of NOS instruction in K-12 classes. The group has elaborated and defended seven elements of NOS that they believe fulfil the criteria of:

1 accessibility to school students;
2 wide enough agreement among historians and philosophers; and
3 being useful for citizens to know.

The seven elements are as follows:[10]

1 The *empirical nature of science*, where they recognised that, although science is empirical, scientists do not have direct access to most natural phenomena. It is claimed that:

> Students should be able to distinguish between observation and inference. . . . An understanding of the crucial distinction between observation and inference is a precursor to making sense of a multitude of inferential and theoretical entities and terms that inhabit the worlds of science.
>
> (Lederman *et al.* 2002, p. 500)

2 *Scientific theories are different from scientific laws*, where they hold that:

> Laws are descriptive statements of relationships among observable phenomena. . . . Theories by contrast are inferred explanations for observed phenomena or regularities in those phenomena. . . . Theories and laws are different kinds of knowledge and one does not become the other.
>
> (Lederman *et al.* 2002, p. 500)

3 The *creative and imaginative nature of scientific knowledge*, where they hold that:

> Science is empirical. . . . Nonetheless, generating scientific knowledge also involves human imagination and creativity. Science . . . is not a lifeless,

> entirely rational and orderly activity. . . . Scientific entities, such as atoms and species are functional theoretical models rather than copies of reality.
>
> (Lederman *et al.* 2002, p. 500)

4 The *theory-laden nature of scientific knowledge*, where it is held that:

> Scientists' theoretical and disciplinary commitments, beliefs, prior knowledge, training, experiences, and expectations actually influence their work. All these background factors form a mindset that affects the problems scientists investigate and how they conduct their investigations.
>
> (Lederman *et al.* 2002, p. 501)

5 The *social and cultural embeddedness of scientific knowledge*, where it is held that:

> Science as a human enterprise is practised in the context of a larger culture and its practitioners are the product of that culture. Science, it follows, affects and is affected by the various elements and intellectual spheres of the culture in which it is embedded.
>
> (Lederman *et al.* 2002, p. 501)

6 The *myth of scientific method*, where it is held that:

> There is no single scientific method that would guarantee the development of infallible knowledge . . . and no single sequence of activities . . . that will unerringly lead [scientists] to functional or valid solutions or answers.
>
> (Lederman *et al.* 2002, p. 502)

7 The *tentative nature of scientific knowledge*, where it is maintained that:

> Scientific knowledge, although reliable and durable, is never absolute or certain. This knowledge, including facts, theories, and laws, is subject to change
>
> (Lederman *et al.* 2002, p. 502).

This list has functioned widely in science education as an NOS checklist and it informs the group's hugely popular series of Views of Nature of Science (VNOS) tests, which are used in hundreds of published research papers to measure effectiveness of NOS teaching (Lederman *et al.* 2014) and degrees of NOS understanding.[11] The positive side of the list is that it puts NOS into classrooms, it provides researchers with an instrument for measurement of NOS learning and it can give teachers and students some NOS matters to think through and become more knowledgeable about. The negative side of the list is that, despite the wishes of its creators, it can function as a mantra, as a catechism, as yet another something to be learned. Instead of teachers and students reading, analysing and coming to their own views about NOS matters, the list often short-circuits all of this. And, in as much as it does so, it is directly antithetical to the very goals of thoughtfulness and critical thinking that most consider the reason for having NOS (or HPS) in the curriculum.

The Contribution of HPS

The seven features of science, or NOS elements, can usefully be philosophically and historically refined and developed in order to better achieve NOS outcomes for teachers and students. This is not just the obvious point that, when seven matters of considerable philosophical subtlety, and with long traditions of debate behind them, are dealt with in a few pages, then they will need to be further elaborated. Rather, it is the more serious claim that, at crucial points, there is ambiguity that mitigates the usefulness of items on the list as curricular objectives, assessment criteria and goals of science-teacher education courses.

Empirical Base

For instance, consider the first item on the list – the empirical basis of science. This subject has been partially discussed in Chapter 6 when elaborating the meaning of 'observation'. However, further elaboration is warranted, as the distinction as stated raises question about: first, the ontological status of theoretical entities in science and, second, the role of abstraction and idealisation in science.

First, in discussing the empirical nature of science, it is maintained that there is wide enough agreement on the 'existence of an objective reality, for example, as compared to phenomenal realities' (Lederman 2004, p. 303). This is quite so, but the serious subject of debate among philosophers is not the reality of the world, but the reality of explanatory entities proposed in scientific theories. This debate has been between realists, on the one hand, and empiricists, constructivists and instrumentalists, on the other; it has gone on since Aristotle's time.

Something of the history and philosophical dimensions of the debate have been outlined in Chapter 9. Throughout the 2,700 years since Anaximander's postulation of crystalline spheres to carry the planets, it has not been the existence of the world that has been doubted – Bellarmine, Berkeley, Mach and Bohr did not doubt the existence of objects, just the unseen entities and mechanisms that the science of their time was postulating to explain the visible, macro or phenomenal behaviour of the objects. This whole history is removed from science-education discussion when the first element in the Lederman list simply says that 'science has an empirical base'. Well yes, it does, but the issue is more complex, and, as with many things, the devil is in the detail. It might be said that students cannot comprehend the detail, but this is an empirical matter; certainly, teachers can and should comprehend the detail, and truly useful NOS lists will lead them to these details.

The Lederman group are realists about the world, but it is very unclear whether they are realists about science's theoretical entities – the very issue on which the realist/instrumentalist (constructivist) debate has hinged. It is not the reality of the world that teachers need guidance about, it is the reality or otherwise of entities postulated in scientific theories. Lederman asks: 'can it be said that a student truly understands the concept of a gene if he/she does not realize that a "gene" is a construct invented to explain experimental results?' (Lederman 2004, p. 314). And he repeats the question by asking: 'Does the student who views genes as possessing physical existence analogous to pearls on a necklace possess an in-depth understanding of the concept?' (ibid.) and:

> Does the student who is unaware that the atom (as pictured in books) is a scientific model used to explain the behavior of matter and that it has not been directly observed have an in-depth understanding of the atom?
>
> (Ibid.)

With some philosophical sensibility, it can be seen that these questions mask a serious and misleading ambiguity concerning the existence of genes and atoms. At first reading, the questions seem to suggest an instrumentalist, non-realist view of these central explanatory entities; they appear to 'in principle' not exist, but be merely a human 'construct'. What if the student thinks of genes, not as pearls on a necklace, but as links in a necklace chain: is this sufficient sophistication to rate as high NOS understanding? Or, what if a student thinks of atoms, not as pictured in the textbook, but as some sort of micro particle: is this sufficient to rate as high NOS understanding? The crucial NOS issue is whether genes and atoms exist at all, exist in principle, not whether any particular picture of them is correct. Once we grant in principle existence, we can be reasonably relaxed about any particular picture; this is just a matter for good science education to fill in. However, Lederman is silent about whether it is in principle existence or just some particular existence – pearl-like genes, or red and green atoms – that is being denied.

The same ambiguity can be seen when another member of the group, Fouad Abd-El-Khalick, recognises that, 'The world of science is inhabited by a multitude of theoretical entities, such as atoms, photons, magnetic fields, and gravitational forces to name only a few.' All realists recognise that the entities listed are both theoretical and central to science, but Abd-El-Khalick proceeds to say that these are 'functional theoretical models rather than faithful copies of "reality"' (Abd-El-Khalick 2004, pp. 409–410). Here, again, is the crucial ambiguity. One wonders why 'reality' was put in scare quotes, as this introduces some element of doubt about reality itself, but this can be left aside for the moment, as Abd-El-Khalick is a realist about reality. More importantly, however, functional theoretical models can either have a reference (denote something existing) or merely link observables in a, usually, mathematical way that has no ontological import. Abd-El-Khalick's claim is ambiguous at the crucial point of whether the listed theoretical entities are non-existing-in-principle 'functional theoretical models', by virtue of them not being 'faithful copies of reality' or by virtue of their very nature.

This is a rephrasing of the long-discussed distinction between hypothetical constructs (which, in principle, can have existence, although they may, as a matter of fact, not exist or not have the properties attributed to them) and intervening variables (which, in principle, have no existence, but merely link observables).[12] In the nineteenth century, caloric and Neptune were hypothetical constructs; one turned out to have existence, the other did not. The notion of 'average-family number' when applied to societies functions as an intervening variable: there is no suggestion that any particular family has 3.7 members; the latter is not meant to copy, faithfully or otherwise, any particular reality. The crucial question is whether atoms, photons, magnetic fields and gravitational forces are like average-family numbers? Bellarmine, Berkeley, Mach and Bohr would say 'yes'; it is simply unclear if Abd-El-Khalick agrees with them or not. If attention had been paid to spelling out the meaning of 'functional theoretical model', this ambiguity would be removed, and, more importantly, teachers and students could be introduced to a long and rich philosophical conversation in the history of science.

At a surface reading, it would seem that the Lederman group are empiricists and constructivists about theoretical entities in science. If so, this is a mistake, and it is not the message about NOS that science teachers should convey. The mistake is not so much the affirmation of one philosophical side, constructivism, in this debate, but rather giving the impression that there is no debate or no alternative position that can and has been adopted – the realist position. Once again, a concentration on learning the NOS list rather than open discussion and enquiry about FOS leads to this mistake.

The second problem with the Lederman group's 'empirical basis' characterisation is that it disguises, if not completely distorts, the non-empirical component of science. The very process of abstraction, and idealisation, is the beginning of modern science. It is an ability to see the forest, and not just the trees. Consider Galileo's 'thousands of swings' of the pendulum. As has been detailed in Chapter 6 of this book, Galileo clearly saw no such thing; it is a claim about what he would see, if the impediments to pendulum motion were removed. Similarly, Newton did not see inertial bodies continuing to move in a straight line, indefinitely. This is what he would have seen, if all resistance were removed. Fermi and Bernardini, in their biography of Galileo, emphasise this innovation:

> In formulating the 'Law of Inertia' the abstraction consisted of imagining the motion of a body on which no force was acting and which, in particular, would be free of any sort of friction. This abstraction was not easy, because it was friction itself that for thousands of years had kept hidden the simplicity and validity of the laws of motion. In other words, friction is an essential element in all human experience; our intuition is dominated by friction; men can move around because

of friction; because of friction they can grasp objects with their hands, they can weave fabrics, build cars, houses, etc. To see the essence of motion beyond the complications of friction indeed required a great insight.

(Fermi & Bernardini 1961, p. 116)

The point of this drawn-out discussion of the first item on the Lederman list is to indicate that such a claim about the empirical basis, and the role of inference, needs to be elaborated at a much more sophisticated level, in order to both be useful and to avoid massive misunderstandings of the scientific endeavour. Further, with just the slightest HPS-informed elaboration, the more or less uncontroversial and mundane claim – that science has an empirical base – can be transformed into an engaging enquiry that can link teachers and students with a central philosophical argument in the history of philosophy, namely realist or instrumentalist interpretation of scientific theory, a debate where the greatest minds can be found on either side. It is not a simple, 'open-and-shut' matter that can be reduced to a declarative list.

The same kind of argument can be mounted against each of the other items on the Lederman list. A general point is that such necessary elaboration depends upon teachers having some competence, or at least familiarity, with HPS, and, notoriously, such training is absent from teacher education programmes.

Subjectivity

The fourth claim is that, 'Scientific knowledge is subjective or theory-laden'. Again, the claim is ambiguous: one can say both 'yes' and 'no'. First, to acknowledge that some claim is theory-laden is not equivalent to saying it is subjective in the usual psychological meaning of the term. A theory-laden description is as good and reliable as the theory embodied in the description; of itself, it has no connection with subjectivity. However, the meaning of 'subjectivity' being used by the Lederman group is ambiguous. For instance, Lederman says that, 'I am not advocating that scientists be subjective' (Lederman 2004, p. 306). Here, 'subjective' must be the everyday psychological sense of the term, and assuredly scientists should avoid this subjectivity. Previously, however, we have been dealing with what one might call 'philosophical subjectivity', as it has been stated that subjectivity is equivalent to theory-ladenness, and that 'subjectivity is unavoidable' (ibid.). Clearly, all science is theory-laden, as Lederman rightly points out; but, if so, then scientists have to be subjective (as in philosophical subjectivity), whether it is advocated or not advocated, but this is entirely different from psychological subjectivity.

The long history of modern science is an effort to take out, or minimise, the psychological subjectivity in measurement and explanation – beginning with the earliest use of measuring instruments in order to get intersubjective agreement about weight, length, time, etc. As mentioned in Chapter 6, Galileo created the *pulsilogium* so as to be able to objectively measure pulse rate for medical diagnosis. The entirely subjective judgements of 'fast', 'medium' or 'slow' pulse were replaced by the length of a pendulum beating in time with the patient's pulse. This length could be seen by all and measured; pulse rate was no longer an entirely internal, subjective matter. This process and concern repeat themselves in the development of all measuring instruments in science (ammeters, voltmeters, spring balances, etc.) and in social science (intelligence tests, personality tests, wellness measures, etc.). The force of the fourth claim trades entirely upon an ambiguity, which is unfortunate in something so widely used as a checklist of NOS understanding.

Culture

The group's fifth claim is that science is embedded in culture; that it 'affects and is affected by the various elements and intellectual spheres of the culture in which it is embedded' (Lederman 2004, p. 306). It is important that this be recognised, but again the devil is in the detail, and the detail is not provided. As has been outlined in Chapter 10, we know that the cultures of Nazism (Beyerchen 1977), Stalinism (Birstein 2001, Graham 1973), Islam (Hoodbhoy 1991) and Hinduism (Nanda 2003), to take just some examples, dramatically affected scientific investigation wherever they were powerful enough so to do. And, of course, the impact, for good and bad, of Christian culture, beliefs and authorities on science is well documented (Lindberg & Numbers 1986, Reiss 2014). Clearly, indigenous sciences are affected by the worldviews and social structures of the traditional societies in which they are practised; to be so affected constitutes the meaning of indigenous science.

All commentators on the European scientific revolution recognise that the blossoming of the new science of Galileo, Huygens, Newton, Boyle and others was dependent on, though not caused by, the social and cultural circumstances of seventeenth-century Europe. Whole research programmes have been dedicated to cataloguing these cultural contributions.[13] Running counter to this, scholars have tried to identify the absence of such circumstances in China at the time, to account for why there was no comparable scientific revolution in China (Needham & Ling 1954–1965). In a famous and contentious study, Paul Forman attempted to provide a causal link between the culture of Weimar Germany and the creation of indeterminate quantum theory (Forman 1971).

The sociological and historical facts of the matter are not in dispute – science depends upon technology, mathematics, communications, money, education, philosophy, and culture more broadly – and it is useful for students and teachers to be reminded of all this and to be given examples. However, for this fact to be truly useful, and not just a sort of anthropological observation, teachers (and their pupils) need to be engaged in or enquire about issues such as: separating benign from adverse effects of culture; distinguishing good from bad science; identifying internal and external factors in scientific development; trying to determine just how analogous are Western and indigenous science; being clear about how any of this information bears upon the truth of what is claimed or the progressiveness of the science so affected; and so on. However, the Lederman group is silent on these ultimately normative matters.

We are told just that, although Western science dominates North American schools, there 'exist other analogous sciences (e.g., indigenous science) in other parts of the world' (Lederman 2004, p. 307). The ambiguity here over 'analogous' means that this item on the list gives no direction to teachers, either in cultures that are resistant to Western science or in multicultural situations. It is a too-easy step to move from this anthropological claim to the educational conclusion that, where other analogous sciences exist, then they should be taught.[14]

The Lederman group does, correctly, say that, to teach NOS means, among other things, identifying the 'values and beliefs inherent to scientific knowledge and its development' (Lederman 2004, p. 303), but there is little, if any, elaboration of just what these values are. This can be a good thing, if teachers and students are meant to work out their own answer, but the list is meant to function as a characterisation of the nature of science, and it is meant to be useful to teachers, but, as has been argued, the usefulness depends on HPS-informed elaboration. As has been shown in Chapter 5, there is considerable HPS literature on the role of internal and external values and cognitive and ethical values in science, and teachers can, with benefit, be introduced to these discussions.

The history and sociology of science show the influence of personal, social, sexual and cultural interests on the development of science. The recognition of such influences is an important component of good science education. Often, it is only from the perspective of history, or of another culture, that these assumptions become apparent. Few have expressed this idea better than Ernst Mach, who, in his 1883 history of mechanics, said:

> The historical investigation of the development of a science is most needful, lest the principles treasured up in it become a system of half-understood precepts, or worse, a system of prejudices. Historical investigation not only promotes the understanding of that which now is, but also brings new possibilities before us.
> (Mach 1883/1960, p. 316)

The fact that scientists carry cultural baggage does not imply subjectivity in their discoveries, or that their work is intellectually compromised. Over the centuries, scientists from diverse cultural, racial and religious milieux have built upon the work of scientists from other cultures and earlier centuries. Today, 'Western' science is contributed to by scientists from all corners of the globe; indeed, some non-Western cultures are putting more effort into, and having more success with, Western science education than the UK and the US. This fact accords with structuralist or objectivist theories of science, theories that maintain that science has an independence from any particular scientist's experience, and that, in general, it is the state of science that determines the experience of scientists, rather than the experience of scientists determining the state of science (Chalmers 1990).

Features of Science

It has been argued in the foregoing that the seven items on the Lederman group's NOS list could better be thought of as different FOS to be elaborated, discussed and enquired about, rather than NOS items to be learned and assessed. Each of these features has been richly written about by philosophers, historians and others – as has been indicated above for just three items on the list. However, if they are FOS, then there is no good reason why just those seven features are picked out, and not others of the numerous features – epistemological, historical, psychological, social, technological, economic, etc. – that can be said to characterise scientific endeavour, and that also meet the three criteria of accessibility, consensus and usefulness that the Lederman group additionally utilises to reduce NOS matters to classroom size.

The group recognises that many other things can be added to the above list; it does not regard it as a closed list. But the in principle openness is disguised by the essentialist terminology of the NOS. Among philosophers, NOS discussion and debate have traditionally revolved around investigations of the epistemological, methodological, ontological and ethical commitments of science. However, there are illuminating, non-philosophical studies of science, such as conducted by historians, cognitive psychologists, sociologists, economists, anthropologists and numerous other disciplines, all of which together can be labelled 'science studies'. This term encompasses the complete academic spectrum, and all components have useful things to say about different features of science. The following are just some of the additional features, topics, issues or questions that can usefully engage science teachers and students when learning about science, or when teaching NOS. These features have been highlighted in Chapters 5–7 of this book.

Item 8: Experimentation

As was shown in Chapter 6, the long-standing Aristotelian injunction about not interfering with nature if we want to understand her was rejected first by Galileo, with his famous inclined plane experiments, conducted so as to understand the phenomenon of free fall, then progressively by the other foundation figures of early modern science, most notably Newton, with his prism manipulations in optics, where light was bent, 'forced' apart and reunited. It was this newly introduced experimentalism that occasioned Kant to remark that:

> When Galileo caused balls, the weights of which he had himself previously determined, to roll down an inclined plane; when Torricelli made the air carry a weight which he had calculated beforehand to be equal to that of a definite

> volume of water . . . a light broke upon all students of nature. They learned that reason has insight only into that which it produces after a plan of its own, and that it must not allow itself to be kept, as it were, in nature's leading-strings.
>
> (Kant 1787/1933, p. 20)

As Kant was writing this, Priestley, as shown in Chapter 7, was practising it in his pneumatic chemistry and his investigations of what would be called photosynthesis.

Historians and philosophers have written a great deal on this topic, with some maintaining that it was experimentation and manipulation of nature that marked out the scientific revolution. And the topic can connect immediately with a more sophisticated understanding of school laboratory work and student experimentation (Chang 2010, Hodson 1993, 1996, Jenkins 1999). These are good reasons for having experimentation on any NOS list, and it can easily be added without debate about what to take off, if one moves from NOS thinking to FOS thinking. Experiments are a 'hallmark' of school science and, with HPS-informed teachers, can be the occasion for elaborating an important feature of science.

Item 9: Idealisation

As has also been documented in Chapter 6, and at greater length in Matthews (2000a), Galileo was the first to build idealisation into the investigation of nature, and it was this methodological move that enabled his new science to emerge from its medieval and Renaissance milieu. What Galileo recognised was that nature's laws were not obvious in nature; they were not given in immediate experience; the laws applied only to idealised circumstances. This employment of idealisation was also in flat contradiction to the long, empiricist Aristotelian tradition, whereby 'science' was to be about the world as seen and experienced. As Aristotle maintained: 'If we cannot believe our eyes what should we believe?' In contrast, Galileo, immediately after proving his famous Law of Parabolic Motion, says:

> I grant that these conclusions proved in the abstract will be different when applied in the concrete and will be fallacious to this extent, that neither will the horizontal motion be uniform nor the natural acceleration be in the ratio assumed, nor the path of the projectile a parabola.
>
> (Galileo 1638/1954, p. 251)

Of crucial importance was the fact that idealisation, and only idealisation, gave specific direction to experimentation, so that students of nature could mould nature 'after a plan of its own', in Kant's famous words. The decades and centuries of classical mechanics, begun by Galileo, were a long process of transforming nature in the image of theory; that is what an experiment was: controlling all variables identified by theory as being irrelevant, and varying the one held responsible for the phenomenon.

The historian William Brock, in discussing Joseph Priestley's procedures, has well stated this matter:

> When science idealizes, it leaves anomalies for later followers to add explanations such as 'side reactions', the presence of impurities, altered physical conditions, etc. But as examples from the past repeatedly show . . . simplification is a necessary feature of scientific progress and the first step towards advancing knowledge.
>
> (Brock 2008, p. 78)

Without idealisation, there would be no modern science. Thus, there are strong claims for idealisation to be added to any NOS list, and once more this can easily be done, if one moves from essentialist NOS thinking to more flexible FOS thinking.

Item 10: Models

The ubiquity of models in the history and current practice of science is widely recognised, indeed it is difficult to think of science without models: the 'billiard ball', 'plum-pudding' and 'solar system' models of the atom, the electron orbit model for the periodic table, the 'lattice' model of salt structure, the fluid-flow model of electricity, the double-helix model of the chromosome, the 'survival of the fittest' model for population expansion in ecosystems, the particle model of light, the 'big bang' model in cosmology, the 'three-body' model for Sun–Earth–Moon interaction, full dinosaur models from bone fragments in palaeontology, the plate-tectonic model in geophysics, the scores of mathematical models in hereditary and population studies, the thousands of mathematical models in economics, engineering, and so on. Any ten pages of a science textbook might be expected to contain twice that number of models, many in full glossy colour, with state-of-the-art graphics.

In the past half-century, historians and philosophers of science have devoted considerable time to documenting and understanding the role of models in science and social science. These studies have led scholars to examine model-related topics, such as the nature of scientific theory, the status of hypothesis, the role of metaphor and analogy in scientific explanation, thought experiments in science and the centrality of idealisation for the articulation, application and testing of models. Mary Hesse's *Models and Analogies in Science* (Hesse 1966) was of particular importance.

What gave impetus to model-related education research was the work done on models by psychologists and cognitive scientists (Gentner & Stevens 1983), with Philip Johnson-Laird's book *Mental Models* (1983) being enormously influential. He, and associates, provided an explanation for the ubiquity of models in science when they detailed how models were ubiquitous, not just in science, but *across the brain* in mental life itself. Johnson-Laird wrote that:

> It is now plausible to suppose that mental models play a central and unifying role in representing objects, states of affairs, sequences of events, the way the world

> is, and the social and psychological actions of daily life. They enable individuals to make inferences and predictions, to understand phenomena, to decide what action to take and to control its execution, and above all to experience events by proxy; they allow language to be used to create representations comparable to those deriving from direct acquaintance with the world; and they relate words to the world by way of conception and perception.
>
> (Johnson-Laird 1983, p. 397)

Johnson-Laird here skirts around the central epistemic question. Although recognising the ubiquity of models in human reasoning is important, and an accomplishment for psychology, this recognition just makes more important the need to combine epistemology with learning theory, the more so for the learning of science. To realise that every individual and culture have mental models of their natural and social world that they utilise in interpreting perceptions and framing actions is one thing, and working out just how these models function is a legitimate *psychological* and *anthropological* study, but what educators, philosophers and scientists are interested in is what mental models more accurately reflect or capture the world and its processes, which models are conducive to development of knowledge, as distinct from opinion, or even very useful opinion. Psychologists need not do this, as the study of learning is neutral with respect to the truthfulness of what is learned: the mental processes whereby astrology is learned will be no different from how astronomy is learned; learning about psychic auras will involve the same mechanisms as learning about arteries; the learning of falsehoods will be no different from the learning of truths. Learning theory is epistemologically blind, but education cannot be: the task of education is to promote truth, not ignorance, to encourage rationality, not irrationality. It might be comforting to say that models are involved in all reasoning, but this is premature comfort. It is the situation that Hegel spoke of, when he said, 'at night, all cows are black'. In the morning, the task of separating black cows from white ones still needs to be undertaken.

If models are seen as an important feature of science, then a competent HPS-informed teacher can provide rich materials and questions for class discussion on the topic: How do models relate to the world they model? Is learning the properties of models the same as learning about the world?. As with so many FOS questions, there is no uncontested answer, just better informed and better argued answers.[15]

Once experimentation, idealisation and models are accepted on to an NOS list, then the list can simply be extended to include any number of other important and engaging features of science, of which the following are just some:

11 values
12 mathematisation
13 technology
14 explanation

15 worldviews and religion
16 rationality
17 feminism
18 realism and constructivism.

Different ones of these features can be elaborated as the curricular occasion allows, as significant, topical, science-related episodes are reported in the media or as teachers themselves have the interest.

Goals of FOS Teaching

Teachers should have modest goals when teaching about science – either FOS or NOS. Pleasingly, in the opening page of the AAAS *Benchmarks* document, it was stated that: 'Little is gained by presenting these beliefs to students as dogma. For one thing, such beliefs are subtle' (AAAS 1993, p. 5). The same points are made in the UK *Perspectives on Science* course, where it is repeatedly stated that students will gain appreciation of NOS positions and issues, and competence in NOS thinking, rather than declarative knowledge of NOS. It is important to stress these points: First, FOS claims should not be presented as dogma – to do so is to confuse education with indoctrination; and second, most, if not all, statements about FOS are subtle, and recognition of this subtlety simply depends upon having HPS awareness. Both these points have implications for the very vexed and much-written-up topic of the assessment of NOS learning.

It is unrealistic to expect students, or trainee teachers, to become competent historians, sociologists or philosophers of science. We should have limited aims in introducing NOS or FOS questions in the classroom. Teachers should aim for a more complex understanding of science, not a total, or even a very complex, understanding. Fortunately, philosophy does not have to be artificially imported into the science classroom, as it is not far below the surface in any lesson or textbook. At a most basic level, any text or scientific discussion will contain terms such as 'law', 'theory', 'model', 'explanation', 'cause', 'truth', 'knowledge', 'hypothesis', 'confirmation', 'observation', 'evidence', 'idealisation', 'time', 'space', 'fields', 'species'. Philosophy begins as soon as these common and ubiquitous terms are explained, amplified and discussed.

There is no need to overwhelm students with 'cutting-edge' philosophical questions. They have to crawl before they can walk, and walk before they can run. This is no more than common-sensical pedagogical practice. There are numerous low-level philosophical questions that are legitimate FOS queries: What is a scientific explanation? What is a controlled experiment? What is a crucial experiment, and are there any? How do models function in science? How much confirmation does a hypothesis require before it is established? Are there ways of evaluating the worth of competing research programmes? Did Newton's religious belief affect his science? Was Darwin's 'damaged book' analogy a competent reply to critics who pointed to all the evidence that contradicted his evolutionary theory? Was Planck culpable for remaining in Nazi Germany and continuing his scientific research during the war? And so on.

Likewise, history is unavoidable. Texts are replete with names such as Aristotle, Copernicus, Kepler, Galileo, Huygens, Newton, Boyle, Hooke, Darwin, Mendel, Faraday, Volta, Lavoisier, Priestley, Dalton, Rutherford, Mach, Curie, Bohr, Heisenberg, Einstein and others. History 'lite' begins when teachers, as Westaway was quoted earlier, 'talk to [students] about the personal equations, the lives, and the work' of such figures and encourage students to do their own research on the life, work, times and impact of these scientists.[16] History 'full strength' begins when the experiments and debates of these figures are reproduced in the classroom, when 'historical–investigative' teaching is practised, and when connections with other subjects are made.

Conclusion

Each day, newspapers, TVs and the Internet feature socio-scientific controversies and debates about genetics, agri-business, climate change, GM crops, global warming and so on. If understanding FOS is embraced as a curricular goal, then well-prepared teachers should be able to elaborate a little on these matters as they occur and facilitate useful classroom discussion that draws out appropriate FOS items from the individual controversies. Advancing the discussion and understanding of each issue is one goal, but seeing the appropriate FOS components of each debate is a second, important classroom goal. Things such as experiment, models, objectivity, values, cognitive and non-cognitive values and so on can all be elaborated. The change of focus from NOS to FOS greatly facilitates this orientation; something topical and important may not be on an NOS list, but it can be elaborated and taught. NOS research has concentrated on the nature of scientific knowledge; FOS includes this, but is also concerned with the processes, institutions and cultural and social contexts in which this knowledge is produced. As argued elsewhere, some caution is needed:

> Science educators should be modest when urging substantive positions in the history and philosophy of science, or in epistemology. . . . Modesty does not entail vapid fence-sitting, but it does entail the recognition that there are usually two, if not more, sides to most serious intellectual questions. And this recognition needs to be intelligently and sensitively translated into classroom practice.
>
> (Matthews 1998, pp. 169–170)

Notes

1 This chapter draws on research published in Matthews (2012).
2 See at least: Arons (1988), Jung (1994, 2012), Norris (1985, 1997), Schulz (2009, 2014), Stenhouse (1985) and Stinner (1989), and the seventy-six chapters in Matthews (2014).
3 See contributions to the special issues of *Science & Education* (Vol.6, No.4 1997, Vol.7, No.6 1998), McComas (1998b), Flick and Lederman (2004) and Khine (2012).

See also the literature reviews in Abd-El-Khalick and Lederman (2000) and Lederman (2007), and the bibliography in Bell *et al.* (2001).

4 The history of NOS research is extensively documented in Duschl and Grandy (2013), Hodson (2014) and Lederman *et al.* (2014).

5 This point has been persuasively argued by Gürol Irzik and Robert Nola (2011, 2014).

6 In one recent article, titled 'The Interplay Between Philosophy of Science and the Practice of Science Education', it is recognised that, 'The science education community has witnessed a paradigm shift from logical positivism or empiricism to constructivism in recent decades' (Tsai 2003).

7 William McComas, in one publication, 'presents and discusses fifteen widely-held, yet incorrect ideas about the nature of science' (McComas 1998b, p. 53).

8 For accounts of instruments used for NOS assessment from the 1950s to the present, see Hodson (2014) and Lederman *et al.* (2014).

9 Norman Lederman, now Professor of Science Education at the Chicago Institute of Technology, was formerly at Oregon State University. His original Oregon State students included Fouad Abd-El-Khalick, Renee Schwartz, Valarie Akerson and Randy Bell – all of whom have published widely in this field.

10 The list is articulated and defended in, among other places: Lederman *et al.* (2002, pp. 499–502), Lederman (2004, pp. 303–308) and Schwartz and Lederman (2008, pp. 745–762).

11 See at least: Flick and Lederman (2004, Chapter IV), Schwartz and Lederman (2008) and Chen (2006).

12 A classic discussion of the difference between hypothetical constructs (that in principle have existence) and intervening variables (that in principle do not have existence) is Meehl and MacCorquodale (1948). Clarity on this issue is of absolute importance in social science: Is, for instance, 'intelligence' to be understood as a hypothetical construct or an intervening variable? Rivers of ink have been spilled because researchers have not clarified the kind of thing they are looking for.

13 The classic statement of this 'causal' position is Boris Hessen's 1931 'The Social and Economic Roots of Newton's *Principia*' (Hessen 1931). For Hessen's text and commentary, see Freudenthal and McLaughlin (2009).

14 This topic has been discussed in Chapter 10; see also Nola and Irzik (2005).

15 A number of informative studies can be seen in the special issue of *Science & Education* devoted to the subject – 'Models in Science and in Science Education' (2007, Vol.16 nos.7–8). The whole topic of 'models in science education' is reviewed in Passmore *et al.* (2014).

16 In the 20 years since the first edition of this book, such student research has been transformed by the ubiquity of Wikipedia and countless web-based sources of original works.

References

AAAS (American Association for the Advancement of Science): 1993, *Benchmarks for Science Literacy*, Oxford University Press, New York.

Abd-El-Khalick, F.: 2004, 'Over and Over Again: College Students' Views of Nature of Science'. In L.B. Flick and N.G. Lederman (eds) *Scientific Inquiry and Nature of Science: Implications for Teaching, Learning, and Teacher Education*, Kluwer Academic Publishers, Dordrecht, The Netherlands, pp. 389–425.

Abd-El-Khalick, F.: 2005, 'Developing Deeper Understanding of Nature of Science: The Impact of a Philosophy of Science Course on Preservice Science Teachers' Views and Instructional Planning', *International Journal of Science Education* 27(1), 15–42.

Abd-El-Khalick, F. and Lederman, N.G.: 2000, 'Improving Science Teachers' Conceptions of the Nature of Science: A Critical Review of the Literature', *International Journal of Science Education* 22(7), 665–701.

Arons, A.B.: 1988, 'Historical and Philosophical Perspectives Attainable in Introductory Physics Courses', *Educational Philosophy and Theory* 20(2), 13–23.
Bazzul, J. and Sykes, H.: 2011, 'The Secret Identity of a Biology Textbook: Straight and Naturally Sexed', *Cultural Studies of Science Education*, 6, 265–286.
Bell, R.L., Abd-el-Khalick, F., Lederman, N.G., McComas, W.F. and Matthews, M.R.: 2001, 'The Nature of Science and Science Education: A Bibliography', *Science & Education* 10(1–2), 187–204.
Beyerchen, A.D.: 1977, *Scientists Under Hitler: Politics and the Physics Community in the Third Reich*, Yale University Press, New Haven, CT.
Birstein, V.J.: 2001, *The Perversion of Knowledge: The True Story of Soviet Science*, Westview, Cambridge, MA.
Brock, W.H.: 2008, 'Joseph Priestley, Enlightened Experimentalist'. In I. Rivers and D.L. Wykes (eds) *Joseph Priestley: Scientist, Philosopher, and Theologian*, Oxford University Press, Oxford, UK, pp. 49–79.
Chalmers, A.F.: 1990, *Science and Its Fabrication*, Open University Press, Milton Keynes, UK.
Chan, K.W. and Elliott, R.G.: 2004, 'Relational Analysis of Personal Epistemology and Conceptions About Teaching and Learning', *Teaching and Teacher Education* 20(8), 817–831.
Chang, H.: 2010. 'How Historical Experiments Can Improve Scientific Knowledge and Science Education: The Cases of Boiling Water and Electrochemistry', *Science & Education* 20(3–4), 317–341.
Chen, S.: 2006, 'Development of an Instrument to Assess Views on Nature of Science and Attitudes Towards Teaching Science', *Science Education* 90(5), 803–819.
Cohen, R.S.: 1975, *Physical Science*, Holt, Rinehart & Winston, New York.
Dewey, J.: 1910, 'Science as Subject-Matter and as Method', *Science* 31, 121–127. Reproduced in *Science & Education* 1995, 4(4), 391–398.
Duschl, R. and Grandy, R.: 2013, 'Two Views About Explicitly Teaching Nature of Science', *Science & Education* 22(9), 2109–2139.
Duschl, R.A.: 1985, 'Science Education and Philosophy of Science, Twenty-five Years of Mutually Exclusive Development', *School Science and Mathematics* 87(7), 541–555.
Duschl, R.A.: 1990, *Restructuring Science Education: The Importance of Theories and Their Development*, Teachers College Press, New York.
Duschl, R.A.: 2004, 'Relating History of Science to Learning and Teaching Science: Using and Abusing'. In L.B. Flick and N.G. Lederman (eds) *Scientific Inquiry and Nature of Science: Implications for Teaching, Learning, and Teacher Education*, Kluwer Academic Publishers, Dordrecht, The Netherlands, pp. 319–330.
Fermi, L. and Bernadini, G.: 1961, *Galileo and the Scientific Revolution*, Basic Books, New York.
Flick, L.B. and Lederman, N.G. (eds): 2004, *Scientific Inquiry and Nature of Science: Implications for Teaching, Learning and Teacher Education*, Kluwer, Dordrecht, The Netherlands.
Forman, P.: 1971, 'Weimar Culture, Causality and Quantum Theory, 1918–1927: Adaptation by German Physicists and Mathematicans to a Hostile Intellectual Environment'. In R. McCormmach (ed.) *Historical Studies in the Physical Sciences*, No.3, University of Pennsylvania Press, Philadelphia, pp. 1–116.
Freudenthal, G. and McLaughlin, P. (eds): 2009, *The Social and Economic Roots of the Scientific Revolution. Texts by Boris Hessen and Henryk Grossmann*, Springer, Dordrecht, The Netherlands.
Galileo, G.: 1638/1954, *Dialogues Concerning Two New Sciences* (trans. H. Crew and A. de Salvio), Dover Publications, New York (originally published 1914).
Gentner, D. and Stevens, A. (eds): 1983, *Mental Models*, Lawrence Earlbaum, Hillsdale, NJ.
Graham, L.R.: 1973, *Science and Philosophy in the Soviet Union*, Alfred A. Knopf, New York.

Hesse, M.B.: 1966, *Models and Analogies in Science*, University of Notre Dame Press, South Bend, IN.
Hessen, B.M.: 1931, 'The Social and Economic Roots of Newton's *Principia*'. In *Science at the Crossroads*, Kniga, London. Reprinted in G. Basalla (ed.) *The Rise of Modern Science: External or Internal Factors?* D.C. Heath, New York, 1968, pp. 31–38.
Hodson, D.: 1986, 'Philosophy of Science and the Science Curriculum', *Journal of Philosophy of Education* 20, 241–251. Reprinted in M.R. Matthews (ed.) *History, Philosophy and Science Teaching: Selected Readings*, OISE Press, Toronto, 1991, pp. 19–32.
Hodson, D.: 1988, 'Toward a Philosophically More Valid Science Curriculum', *Science Education* 72, 19–40.
Hodson, D.: 1993, 'Re-Thinking Old Ways: Towards a More Critical Approach to Practical Work in Science', *Studies in Science Education* 22, 85–142.
Hodson, D.: 1996, 'Laboratory Work as Scientific Method: Three Decades of Confusion and Distortion', *Journal of Curriculum Studies* 28, 115–135.
Hodson, D.: 2008, *Towards Scientific Literacy: A Teachers' Guide to the History, Philosophy and Sociology of Science*, Sense Publishers, Rotterdam, The Netherlands.
Hodson, D.: 2009, *Teaching and Learning About Science: Language, Theories, Methods, History, Traditions and Values*, Sense Publishers, Rotterdam, The Netherlands.
Hodson, D.: 2014, 'Nature of Science in the Science Curriculum: Origin, Development and Shifting Emphases'. In M.R. Matthews (ed.) *International Handbook of Research in History, Philosophy and Science Teaching*, Springer, Dordrecht, The Netherlands, pp. 911–970.
Holmyard, E.J.: 1924, *The Teaching of Science*, Bell, London.
Holton, G.: 1975, 'Science, Science Teaching and Rationality'. In S. Hook, P. Kurtz and M. Todorovich (eds) *The Philosophy of the Curriculum*, Prometheus Books, Buffalo, NY, pp. 101–118.
Holton, G.: 1978, 'On the Educational Philosophy of the Project Physics Course'. In his *The Scientific Imagination: Case Studies*, Cambridge University Press, Cambridge, UK, pp. 284–298.
Hoodbhoy, P.: 1991, *Islam and Science: Religious Orthodoxy and the Battle for Rationality*, Zed Books, London.
Huxley, T.H.: 1868/1964, 'A Liberal Education; and Where to Find It'. In his *Science & Education*, Appleton, New York, 1897 (originally published 1885). Reprinted, with Introduction by C. Winick, Citadel Press, New York, 1964, pp. 72–100.
Irzik, G. and Nola, R.: 2011, 'A Family Resemblance Approach to the Nature of Science for Science Education', *Science & Education* 20(7–8), 591–607.
Irzik, G. and Nola, R.: 2014, 'New Directions in Nature of Science Research'. In M.R. Matthews (ed.) *International Handbook of Research in History, Philosophy and Science Teaching*, Springer, Dordrecht, The Netherlands, pp. 999–1021.
Jenkins, E.W.: 1999, 'Practical Work in School Science: Some Questions to be Answered'. In J. Leach and A.C. Paulsen (eds) *Practical Work in Science Education: Recent Research Studies*, Kluwer Academic Publishers, Dordrecht, The Netherlands, pp. 19–32.
Johnson-Laird, P.N.: 1983, *Mental Models*, Harvard University Press, Cambridge, MA.
Jung, W.: 1994, 'Preparing Students for Change: The Contribution of the History of Physics to Physics Teaching', *Science & Education* 3(2), 99–130.
Jung, W.: 2012, 'Philosophy of Science and Education', *Science & Education* 21(8), 1055–1083.
Kang, N.H.: 2008, 'Learning to Teach Science: Personal Epistemologies, Teaching Goals, and Practices of Teaching', *Teaching and Teacher Education* 24, 478–498.
Kant, I.: 1787/1933, *Critique of Pure Reason*, 2nd edn (trans. N.K. Smith), Macmillan, London (1st edition, 1781).
Khine, M.S. (ed.): 2012, *Advances in Nature of Science Research: Concepts and Methodologies*, Springer, Dordrecht, The Netherlands.

Klopfer, L.E.: 1969, 'The Teaching of Science and the History of Science', *Journal of Research in Science Teaching* 6, 87–95.
Lakatos, I.: 1978, 'History of Science and Its Rational Reconstructions'. In J. Worrall and G. Currie (eds) *The Methodology of Scientific Research Programmes*, Vol.I, Cambridge University Press, Cambridge, UK, pp. 102–138 (originally published 1971).
Lederman, N.G.: 2004, 'Syntax of Nature of Science Within Inquiry and Science Instruction'. In L.B. Flick and N.G. Lederman (eds) *Scientific Inquiry and Nature of Science*, Kluwer Academic Publishers, Dordrecht, The Netherlands, pp. 301–317.
Lederman, N.G.: 2007, 'Nature of Science: Past, Present and Future'. In S.K. Bell and N.G. Lederman (eds) *Handbook of Research on Science Education*, Lawrence Erlbaum, Mahwah, NJ, pp. 831–879.
Lederman, N., Abd-el-Khalick, F., Bell, R.L. and Schwartz, R.S.: 2002, 'Views of Nature of Science Questionnaire: Towards Valid and Meaningful Assessment of Learners' Conceptions of the Nature of Science', *Journal of Research in Science Teaching* 39, 497–521.
Lederman, N.G, Bartos, S.A. and Lederman, J.: 2014, 'The Development, Use, and Interpretation of Nature of Science Assessments'. In M.R. Matthews (ed.) *International Handbook of Research in History, Philosophy and Science Teaching*, Springer, Dordrecht, The Netherlands, pp. 971–997.
Lindberg, D.C. and Numbers, R.L. (eds): 1986, *God and Nature: Historical Essays on the Encounter between Christianity and Science*, University of California Press, Berkeley, CA.
McComas, W.F. (ed.): 1998a, *The Nature of Science in Science Education: Rationales and Strategies*, Kluwer Academic Publishers, Dordrecht, The Netherlands.
McComas, W.F.: 1998b, 'The Principal Elements of the Nature of Science: Dispelling the Myths'. In W.F. McComas (ed.) *The Nature of Science in Science Education: Rationales and Strategies*, Kluwer Academic Publishers, Dordrecht, The Netherlands, pp. 53–70.
McComas, W.F.: 2014, 'Nature of Science in the Science Curriculum and in Teacher Education Programmes in the United States'. In M.R. Matthews (ed.) *International Handbook of Research in History, Philosophy and Science Teaching*, Springer, Dordrecht, The Netherlands, pp. 1993–2023.
Mach, E.: 1883/1960, *The Science of Mechanics*, Open Court Publishing, LaSalle, IL.
Mach, E.: 1886/1986, 'On Instruction in the Classics and the Sciences'. In his *Popular Scientific Lectures*, Open Court Publishing, LaSalle, IL, pp. 338–374.
Martin, M.: 1972, *Concepts of Science Education: A Philosophical Analysis*, Scott, Foresman, New York (reprinted University Press of America, 1985).
Martin, M.: 1974, 'The Relevance of Philosophy of Science for Science Education', *Boston Studies in Philosophy of Science* 32, 293–300.
Matthews, M.R.: 1997, 'James T. Robinson's Account of Philosophy of Science and Science Teaching: Some Lessons for Today from the 1960s', *Science Education* 81(3), 295–315.
Matthews, M.R.: 1998, 'In Defence of Modest Goals for Teaching About the Nature of Science', *Journal of Research in Science Teaching* 35(2), 161–174.
Matthews, M.R.: 2000a, *Time for Science Education: How Teaching the History and Philosophy of Pendulum Motion Can Contribute to Science Literacy*, Kluwer Academic Publishers, New York.
Matthews, M.R.: 2000b, 'Constructivism in Science and Mathematics Education'. In D.C. Phillips (ed.) *National Society for the Study of Education 99th Yearbook*, National Society for the Study of Education, Chicago, IL, pp. 161–192.
Matthews, M.R.: 2004, 'Thomas Kuhn and Science Education: What Lessons can be Learnt?', *Science Education* 88(1), 90–118.
Matthews, M.R.: 2012, 'Changing the Focus: From Nature of Science (NOS) to Features of Science (FOS)'. In M.S. Khine (ed.) *Advances in Nature of Science Research*, Springer, Dordrecht, The Netherlands, pp. 3–26.
Matthews, M.R. (ed.): 2014, *International Handbook of Research in History, Philosophy and Science Teaching*, 3 volumes, Springer, Dordrecht, The Netherlands.

Meehl, P. and MacCorquodale, K.: 1948, 'On a Distinction Between Hypothetical Constructs and Intervening Variables', *Psychological Review* 55, 95–107.
Nanda, M.: 2003, *Prophets Facing Backward. Postmodern Critiques of Science and Hindu Nationalism in India*, Rutgers University Press, New Brunswick, NJ.
Needham, J. and Ling, W.: 1954–1965, *Science and Civilisation in China*, Vols.1–4, Cambridge University Press, Cambridge, UK.
Niaz, M.: 2009, *Critical Appraisal of Physical Science as a Human Enterprise: Dynamics of Scientific Progress*, Springer, Dordrecht, The Netherlands.
Niaz, M.: 2010, *Innovating Science Teacher Education: A History and Philosophy of Science Perspective*, Routledge New York.
Nola, R. and Irzik, G.: 2005, *Philosophy, Science, Education and Culture*, Springer, Dordrecht, The Netherlands.
Norris, S.P.: 1985, 'The Philosophical Basis of Observation in Science and Science Education', *Journal of Research in Science Teaching* 22(9), 817–833.
Norris, S.P.: 1997, 'Intellectual Independence for Nonscientists and Other Content-Transcendent Goals of Science Education', *Science Education* 81(2), 239–258.
Passmore, C., Svoboda-Gouvea, J. and Giere, R.: 2014, 'Models in Science and in Learning Science: Focusing Scientific Practice on Sense-making'. In M.R. Matthews (ed.) *International Handbook of Research in History, Philosophy and Science Teaching*, Springer, Dordrecht, The Netherlands, pp. 1171–1202.
Reiss, M.: 2014, 'What Significance Does Christianity Have for Science Education?'. In M.R. Matthews (ed.) *International Handbook of Research in History, Philosophy and Science Teaching*, Springer, Dordrecht, The Netherlands, pp. 1637–1662.
Robinson, J.T.: 1968, *The Nature of Science and Science Teaching*, Wadsworth, Belmont, CA.
Rutherford, F.J.: 1972, 'A Humanistic Approach to Science Teaching', *National Association of Secondary School Principals Bulletin* 56(361), 53–63.
Rutherford, F.J.: 2001, 'Fostering the History of Science in American Science Education: The Role of Project 2061', *Science & Education* 10(6), 569–580.
Schulz, R.M.: 2009, 'Reforming Science Education: Part I. The Search for a Philosophy of Science Education', *Science & Education* 18(3–4), 225–249.
Schulz, R.M.: 2014, 'Philosophy of Education and Science Education: A Vital but Underdeveloped Relationship'. In M.R. Matthews (ed.) *International Handbook of Research in History, Philosophy and Science Teaching*, Springer, Dordrecht, The Netherlands, pp. 1259–1315.
Schwab, J.J.: 1949, 'The Nature of Scientific Knowledge as Related to Liberal Education', *Journal of General Education* 3, 245–266. Reproduced in I. Westbury and N.J. Wilkof (eds) *Joseph J. Schwab: Science, Curriclum, and Liberal Education*, University of Chicago Press, Chicago, IL, 1978, pp. 68–104.
Schwartz, R. and Lederman, N.: 2008, 'What Scientists Say: Scientists' Views of Nature of Science and Relation to Science Context', *International Journal of Science Education* 30(6), 727–771.
Stenhouse, D.: 1985, *Active Philosophy in Education and Science*, Allen & Unwin, London.
Stinner, A.: 1989, 'The Teaching of Physics and the Contexts of Inquiry: From Aristotle to Einstein', *Science Education* 73(5), 591–605.
Taber, K.S.: 2009, *Progressing Science Education: Constructing the Scientific Research Programme into the Contingent Nature of Learning Science*, Springer, Dordrecht, The Netherlands.
Taber, K.S.: 2014, 'Methodological Issues in Science Education Research: A Perspective From the Philosophy of Science'. In M.R. Matthews (ed.) *International Handbook of Research in History, Philosophy and Science Teaching*, Springer, Dordrecht, The Netherlands, pp. 1839–1893.
Tsai, C.-C.: 2003, 'The Interplay between Philosophy of Science and the Practice of Science Education', *Curriculum and Teaching* 18, 27–43.

Westaway, F.W.: 1929, *Science Teaching*, Blackie, London.
Whewell, W.: 1855, 'On the Influence of the History of Science Upon Intellectual Education', *Lectures on Education Delivered at the Royal Institution on Great Britain*, J.W.Parker, London.

Chapter 12

Philosophy and Teacher Education[1]

This chapter will argue that, for a science teacher to be a well-prepared educator (as distinct from just coach, instructor or merely the teacher of subject matter) they need, to the appropriate degree, subject-matter competence, foundational training – especially philosophical – and knowledge of HPS, and, finally, that they cultivate a defensible philosophy of education. All of this cannot happen at once, but beginnings can be made in each area in teacher-training courses, and they can subsequently be built upon.

Everyone agrees that intelligent, knowledgeable and engaging teachers – who know their subject, are interested in children, have an interest and competence in teaching, are familiar with new web and other technologies and teach engagingly – are crucial for good education. Furthermore, teachers increasingly have to do more than just teach and enthuse others with their subject: they need to develop local curricula or to interpret national or provincial curricula for local use, they take part in school governance and in policy-making that bears upon what subjects, at what level, are taught to what students in their school, and so on. And, of course, they need to relate to students, parents and administrators in a whole spectrum of matters beyond the classroom. Good, well-prepared teachers are necessary for these complex and important tasks; teachers do many things, on many fronts, with many people. There has, however, been less agreement on how best to prepare such teachers (Yager & Penick 1990). What is needed is some sense of coherent purpose that provides a guide and standard for their engagements. Charles Silberman, writing 40 years ago on the then crisis in American education, well identified this:

> The central task of teacher education, therefore, is to provide teachers with a sense of purpose, or, if you will, with a philosophy of education. This means developing teachers' ability and desire to think seriously, deeply, and continuously about the purposes and consequences of what they do – about the ways in which their curriculum and teaching methods, classroom and school organization, testing and grading procedures, affect purpose and are affected by it.
>
> (Silberman 1970, p. 472)

Science teachers, as with all teachers, do need to think 'seriously and deeply' about the subject they teach; this is another way of saying that all teachers need to think about the history and philosophy of their subject. Teachers need an educational compass, but there are many available, all pointing in different directions. Philosophy is necessary to craft an educational compass that points to educational north, can contribute to a teacher's identity as a genuine educator and their social standing and be a guide in their myriad professional decision-making. How the creation and correction of such a compass fit into teacher training has been long debated, with contributions from entrenched professional and academic interests, and with political and economic expediency looming over most policy decisions.[2]

In 1986, two reports on teacher education were published in the US that galvanised debate on the subject – the Carnegie Foundation's *A Nation Prepared* and the Holmes Group's *Tomorrow's Teachers* (Fraser 1992). There is a range of views about the best organisation of teacher training programmes. Some countries require none, or only minimal training: just take interested science graduates or folk working in science-based professions and put them into schools in some kind of apprenticeship role. This has happened in the US, with different of the 'Teach for America' programmes, and in the UK, where the government has moved to partly bypass university teacher training in favour of a nineteenth-century, in-school-apprenticeship mode of training (Hirst 2008). However, still the more usual arrangement is to require formalised teacher training (Darling-Hammond 1999).

Where university education studies are required for prospective teachers, their content has been contentious. Such studies have traditionally consisted of both theoretical or foundation studies (typically philosophy, sociology and psychology of education) and applied or pedagogic studies (typically curriculum studies, teaching methods and practice teaching). The trend in the past four decades has been to emphasise the latter at the expense of the former.[3] Paul Hirst observes of the requirements of the UK 1994 Education Act that:

> These restrictions, insisting on the direct practical relevance of all education courses, have led to the near demise of all courses concerned specifically with the disciplines of educational theory within British universities.
>
> (Hirst 2008, p. 309)

So, for instance, a recent seventy-six-chapter *Handbook of Research on Teachers and Teaching* (Saha & Dworkin 2009) has no entry for 'foundation studies' or indeed for any specific foundation discipline. This chapter will suggest ways in which HPS-informed philosophy and psychology programmes can substitute for, or enrich, the usual foundation offerings and greatly diminish the 'irrelevance' factor that students frequently bemoan. Further HPS can enrich the standard methods or curriculum and instruction courses. These can be understood as Trojan-horse ways of getting required philosophy into teacher education.

Philosophy of Education

Philosophical questions arise for all teachers. Some of these arise at an individual teacher–student level (what is and is not appropriate discipline?), some at classroom level (what should be the aim of maths instruction?), some at school level (should classes be organised on mixed-ability or graded-ability lines?) and some at system level (should governments fund private schooling, and, if so, on what basis?). Some of the quite ubiquitous questions that teachers deal with and that require degrees of philosophical input are: How to elaborate and defend specific educational aims? How to demarcate indoctrination from education and determine the legitimate scope of the former? What are the mutual requirements of respect between teachers and students? What are the limits on the professional independence of teachers? In teaching 'controversial issues', should teachers make known their own positions? On these issues, should teachers seek to convert students to their own opinions? What are the grounds for inclusion and exclusion of topics in curricula? What is the legitimate authority of state, church, business, school boards, parents and other stakeholders in curriculum construction, assessment and textbook selection? What are the legitimate versus illegitimate claims of culture and tradition on educational processes? How can schools resolve competing purposes between the transmission of culture and the reform of culture? What are the ethical and political justifications for myriad funding, staffing and class-grouping decisions?

These philosophical, normative, non-empirical questions impinge equally on all teachers, whether they are teaching mathematics, music, economics, history, literature, theology or anything else in an institutional setting. The more informed, thoughtful and nuanced a teacher's engagement with these general questions, the more professional and adequately prepared the teacher. The need for such preparation has long justified the inclusion of philosophy of education in pre-service and in-service teacher education programmes: it is clear that just applied, 'how to' or classroom-management courses give no guidance on such questions. Of course, justification for philosophy has not always led to inclusion; in education, correct arguments are always balanced against other considerations.

The foregoing questions and engagements belong to what can be called general philosophy of education, a subject with a long and distinguished past, contributed to by a roll-call of well-known philosophers and educators such as Plato, Aristotle, Aquinas, Locke, Mill, Whitehead, Russell, Dewey, Peters, Hirst and Scheffler (to name just a Western First XI). These general issues can be found fleshed out and debated in the field's major research journals such as: *Educational Philosophy and Theory*, *Studies in Philosophy and Education* and *Journal of Philosophy of Education* and *Educational Theory*; in annual conference proceedings of different national philosophy-of-education societies, especially those of the US and UK; in large handbooks, such as *The Blackwell Companion to the Philosophy of Education* (Curren 2003) and *The Oxford Handbook of Philosophy of Education* (Siegel 2009); and in numerous substantial books that have been published in the past 60 years.

A noteworthy, if depressing, feature of formal philosophy of education has been its neglect of science education and the specific philosophical considerations that arise in the teaching of science. The huge corpus of analytic philosophy of education, which dominated the professional field in the final decades of the last century, is almost devoid of questions about science education or any analysis informed by HPS. There is, for instance, no mention of these topics in the influential books of Richard Peters, books that largely defined the field (Peters 1959, 1966, 1967, 1973, 1974), or in collections such as the major three-volume anthology *Education and the Development of Reason* (Dearden *et al.* 1972), which was the flagship of analytic philosophy of education. These books and associated literatures were the staple of philosophy in teacher education programmes, including science-teacher education. The neglect of science and HPS is the more puzzling as, across the world, science curricula were being overhauled in the post-*Sputnik* era, and, at the same time, HPS had its own '*Sputni*k moment' with the publication of Kuhn's *The Structure of Scientific Revolutions* (Kuhn 1962/1970) and the consequent tsunami of scholarly studies that washed over *all* university faculties. It could have been expected that formal philosophy of education would have engaged with both the science-curricular and the scholarly HPS enterprises, but, with some exceptions, this did not happen.[4]

Peter Nidditch, the Hume scholar, did contribute the final article, titled 'Philosophy of Education and the Place of Science in the Curriculum', to the 1973 anthology *New Essays in the Philosophy of Education* (Nidditch 1973). David Stenhouse published a book on philosophy of education and science education (Stenhouse 1985). Denis Phillips, a former president of the US PES, published a number of valuable pieces in the field (Phillips 1981, 2000). Harvey Siegel, another former PES president and student of Israel Scheffler, also published important articles on issues in science education (Siegel 1978, 1979, 1989, 1993). Another exception to neglect or indifference to science by philosophers of education was the educational writing of Israel Scheffler (Scheffler 1973), who was also a significant philosopher of science (Scheffler 1963b, 1966/1982) and whose work will be elaborated below.

Philosophy and Clear Communication

The best philosophy has always been concerned with the refinement and improvement of ideas and the betterment of understanding, and so it has valued clear communication, which is a prerequisite for such advances; teacher education and graduate programmes should value and promote the same thing. The habitual Socratic 'what do you mean by . . .?' question is something that needs to be cultivated in science education, where there is an abundance of less-than-clear and sometimes completely opaque communication. Forty years ago, Paul Wagner and Christopher Lucas affirmed the point as follows:

> We shall argue that in order to move toward the broader goals of science education cited above, careful attention ought to be given to the use of philosophic analysis

> as a pedagogical resource for the teaching of science. The thesis amounts to this: Philosophical questioning should precede or accompany children's participation in science activities in the elementary grades, the place where the foundations for later scientific learning are laid down.
>
> (Wagner & Lucas 1977, p. 550)

At its simplest level, this amounts to being careful with language, knowing what words mean and trying to be clear and unambiguous in communication. Scientific thinking and practice can only advance if these basic analytic sensibilities are cultivated and valued to degrees appropriate to the maturity of students; conversely, to the extent that careless writing and jargon prevail, then scientific thinking and practice are retarded.

Being careful about communication is not a modern idea. Thomas Sprat, in his 1667 *History of the Royal Society of London, For the Improving of Natural Knowledge*, thought it worth drawing attention to 'Their Manner of Discourse':

> There is one thing more, about which the Society has been most solicitous; and that is, the manner of their Discourse. . . . They have exacted from all their members a close, naked, natural way of speaking; positive expressions; clear senses; a native easiness: bringing all things as near the mathematical plainness as they can; and preferring the language of artizans, countrymen, and merchants, before that of wits or scholars.
>
> (Sprat 1667/1966, Pt.II, Sect.xx)

Three hundred and fifty years later, much would be improved in science education if academics had 'taken on board' the Royal Society's advice to speak plainly. Consider the following:

> The dance of agency, seen asymmetrically from the human end, thus takes the form of a *dialectic of resistance and accommodation*, where resistance denotes the failure to achieve an intended capture of agency in practice, and accommodation an active human strategy of response to resistance, which can include revisions to goals and intentions as well as to the material form of the machine in question and to the human frame of gestures and social relations that surround it.
>
> (Pickering 1995, p. 22)

It is a struggle to understand what this seventy-eight-word sentence is, if anything, asserting. Or consider:

> Only a living/lived, sensible body – one that can sense itself moving because of auto-affection, one that has practical knowledge of the coordination that occurs with the sense of self-movement and sensation through external senses – can be a body of knowledge. . . . The unity and identity of the sensory phenomena are

> not realized by a synthesis of recognition in the concept, as an intellectualist approach would conceive of it, but this unity is built upon the unity of identity of the living/lived body (flesh) as a synergistic ensemble.
>
> (Roth 2011, p. 233)

The meaning here, if there is one, is less than clear. Or consider:

> If, on the other hand, we begin with the ontological assumption of difference that exists in and for itself, that is, with the recognition that $A \downarrow A$ (e.g., because different ink drops attached to different paper particles at a different moment in time), then all sameness and identity is the result of work that not only sets two things, concepts, or processes equal but also deletes the inherent and unavoidable differences that do in fact exist. This assumption is an insidious part of the phallogocentric epistemology undergirding science as the method of decomposing unitary systems into sets of variables, which never can be more than external, one-sided expressions of a superordinate unit.
>
> (Roth & Tobin, 2007, pp. 99–100)

What is being asserted in these 111 words is a mystery, and probably beyond the possibility of a translation into the language of artisans, countrymen and merchants. Such unintelligible assertions abound. Consider the following claim in a major science-education handbook:

> While different 'methodologies' are employed, bricolage cloaks itself within a critical theoretical commitment to social justice and a critical pedagogical underpinning combining theory, discourse, identity, and the political. . . . The synergism of the conversation between the research bricolage and critical theory involves an interplay between the praxis of the critical and the radical uncertainty of what is often referred to as the postmodern.
>
> (Steinberg & Kincheloe 2012, p. 1492)

The meaning of this claim, to put it mildly, is elusive – something doubly regrettable when social justice is supposedly being championed. Social justice is always best served by plain speaking, by being expressed in the language of artisans. Consider the following:

> During the last quarter of the 20th century the curriculum field underwent a paradigmatic transformation. The hegemony of positivistic inquiry and the monochromatic vision of curriculum development were dislodged through a triumvirate of political analysis, phenomenological studies, and gender-based psychoanalytic thought.
>
> (Sears 2003, p. 227)

With difficulty, the meaning can be teased out, but the writing is clichéd and jargonistic. Such writing might resonate with the choir, but it hardly advances the reputation of education faculty among citizens, teachers or administrators.

The reason for highlighting these examples is that they are merely the signs of an obfuscationist disease that is a threat to scholarship and understanding, not just in education, but across the academy. There are countless other such examples in science-education books and journals.[5] Stephen Shapin's account of this blight in the discipline of history of science is applicable to science education:

> But the problem to which it is worth drawing attention is the particular species of bad writing that is, so to speak, institutionally intentional. Initiates learn to write badly as a badge of professionalism; they resist using the vernacular because it doesn't sound smart enough; they infer from obscurity to profundity. Some things are indeed hard to say in ordinary English, but not nearly so many as academics pretend.
>
> (Shapin 2005, p. 239)

Daniel Dennett relates a conversation with Michel Foucault that nicely captures this academic blight:

> 'Michel, you're so clear in conversation; why is your written work so obscure?' To which Foucault replied, 'That's because, in order to be taken seriously by French philosophers, twenty-five percent of what you write has to be impenetrable nonsense'.
>
> (Dennett 2006, p. 405)

Dennett coins a new word for such debasing of language: *eumerdification* – a neologism best left to the reader to translate.

Nicholas Shackel, in a wide-ranging critique of postmodernist philosophising (which he takes to include poststructuralists, deconstructivists, strong programmers and feminist anti-rationalists), draws attention to the persistent use of linguistic and rhetorical ploys to gain conviction:

> Many of the philosophical doctrines purveyed by postmodernists have been roundly refuted, yet people contine to be taken in by a set of dishonest devices used in proselytizing for postmodernism. It is getting tiring to repeat refutations of the same type for each new appearance of these various manoeuvres.
>
> (Shackel 2005, p. 295)

Not everything has to, or can be, expressed in simple terms; intellectual advance does require new concepts and new or redefined words – 'surplus value', 'liberation', 'natural selection', 'inertia', 'velocity', 'acceleration', 'oxygen' and so on. However, 'What does this mean?' is a question that needs be asked of all such terms; the asking is a basic philosophic habit that warrants cultivation in teacher education and graduate education programmes. Writing such as in the above extracts, and the genres represented, would surely benefit from routine application of this basic philosophic practice, the 'slow lingering method', as David Hume called it. With enough application, this method can even overcome obscurantist disease.

Whatever problems analytic philosophy might have had, it did have the virtue of valuing clarity and direct communication, both of which are important for science and science teaching. Consider the extracts in Table 12.1 and their translation into Royal Society-like 'language of artizans, countrymen, and merchants'.

It is a useful exercise for students to collect such examples, to provide 'everyday' translations and to ascertain what is gained and what is lost in so doing.[6] Assuredly, some things will be lost, as in any translation, but the task is to see if the communicative gain outweighs the loss. There is no in-principle problem with specialised vocabularies and theoretical terms: natural science is full of them, and, as the occasion demands, they need to be learned. But, whereas natural science uses theoretical terms to simplify complex matters (lots of disparate movements can be unified as magnetic phenomena), social science, at least in the examples in Table 12.1, is using supposed theoretical terms to make simple matters more complex, to obfuscate.

The above examples, and their ilk, are the kinds of writing that moved Isaiah Berlin (1909–1997) to complain that:

> Pretentious rhetoric, deliberate or compulsive obscurity or vagueness, metaphysical patter studded with irrelevant or misleading allusions to (at best) half-understood scientific or philosophical theories or to famous names, is an old, but at present particularly prevalent, device for concealing poverty of thought or muddle, and sometimes perilously near a confidence trick.
>
> (Berlin 2000, p. 221)

Philosophy for Science Education

However, as well as *general* philosophy of education, there is a need for *disciplinary* philosophy of education, and, for science education, such philosophy is dependent upon HPS. Some of the disciplinary questions are *internal* to teaching the subject and might be called 'philosophy *for* science teaching'. This covers the following kinds of question: Is there a singular scientific method? What is the scope of science? What is a scientific explanation? Can observational statements be separated from theoretical statements? Do experimental results bear inductively, deductively or abductively upon hypotheses being tested? What are legitimate and illegitimate ways to rescue theories from contrary evidence? Other of the disciplinary questions are *external* to the subject, and might be called 'philosophy *of* science teaching'. Here, questions might be: Can science be justified as a compulsory school subject? What characterises scientific 'habits of mind' or 'scientific temper'? How might competing claims of science and religion be reconciled? Should local or indigenous knowledge be taught in place of orthodox science, or alongside it, or not taught at all? Doubtless, the same kinds of question

Table 12.1 Plain-Speaking Translations

Eduspeak	*Plain-Speak Translation*
'Since co-participation involves the negotiation of a shared language, the focus is on sustaining a dynamic system in which discursive resources are evolving in a direction that is constrained by the values of the majority culture while demonstrating respect for the habitus of participants from minority cultures, all the time guarding against the debilitation of symbolic violence' (Tobin 1998, p. 212)	Teach in a way that is sensitive to cultural values
'Making meaning is thus a dialogic process involving persons-in-conversation, and learning is seen as the process by which individuals are introduced to a culture by more skilled members. As this happens they "appropriate" the cultural tools through their involvement in the activities of this culture' (Driver *et al.* 1994, p. 7)	Students need the assistance of teachers when learning new concepts
'[Constructivism] suggests a commonality amongst school science students and research scientists as they struggle to make sense of perturbations in their respective experiential realities' (Taylor 1998, p. 1114)	Students and scientists consider adjusting their understandings when confronted with anomalies
'Speaking from the sociocultural perspective, [we] define negotiation as a process of mutual appropriation in which the teacher and students continually coopt or use each others' contribution' (Cobb 1994, p. 14)	Teachers and students can exchange ideas
'The path of constituted objectivity is in essence welcoming a variety of worldviews, and thus recognises the path of transcendental objectivity as a legitimate one, because even though one pretends to make observations in transcendental objectivity, the human praxis in which these observations are made is still a path of constituted objectivity' (Maheus *et al.* 2010, p.218)	Unbiased observation is impossible
'Activities, or rather, societally mediated motives do not get themselves realised; concrete goals are required for directing individual human subjects to realise the activities in which they participate and of which they are constitutive moments' (Roth 2007, p. 165)	People act when they have strong motives to do so

arise for teachers of other subjects – mathematics, economics, music, art, religion.

Teachers have theories of knowledge, or an epistemology, that bears on and arises from their understanding of science. This image and understanding will be conveyed to classes and will inform teachers' decision-making about textbook choice, curriculum, lesson preparation, assessment and other pedagogic matters. Teachers' epistemology is an important part of their 'pedagogical content knowledge' that Lee Shulman has identified as so important for effective teaching (Shulman 1986, 1987) and that has been discussed in Chapter 4. A teacher's PCK influences the understanding of science that students retain, long after they have forgotten the details of what has been taught in specific biology, chemistry or physics classes.

It has long been argued that HPS should be part of the education of science teachers. In the UK, the *Thompson Report* in 1918 said, 'some knowledge of the history and philosophy of science should form part of the intellectual equipment of every science teacher in a secondary school' (Thompson 1918, p. 3). The between-wars arguments advanced by Westaway and Holmyard have been mentioned in Chapter 3. A 1981 review of the place of philosophy of science in British science-teacher education said:

> This more philosophical background which is being advocated for teachers would, it is believed, enable them to handle their science teaching in a more informed and versatile manner and to be in a more effective position to help their pupils build up the coherent picture of science – appropriate to age and ability – which is so often lacking.
>
> (Manuel 1981, p. 771)

All of this amounts to advocacy of *disciplinary* philosophy of science education. To effectively teach science with understanding requires that teachers be engaged by HPS.

In the US, 40 years ago, Israel Scheffler, who had a joint appointment in philosophy and education, argued for the inclusion of philosophy of science courses in the preparation of science teachers. It was part of his wider argument for the inclusion of courses in the philosophy of the discipline in programmes that are preparing people to teach that discipline. His suggestion was that, 'philosophies-of constitute a desirable additional input in teacher preparation beyond subject-matter competence, practice in teaching, and educational methodology' (Scheffler, 1970, p. 40). He summarised his argument as follows:

> I have outlined four main efforts through which philosophies-of might contribute to education: (1) the analytic description of forms of thought represented by teaching subjects; (2) the evaluation and criticism of such forms of thought; (3) the analysis of specific materials so as to systematise and exhibit them as exemplifications of forms of thought; and (4) the interpretation of particular exemplifications in terms accessible to the novice.
>
> (Scheffler 1970, p. 40)

Each of these four contributions can only be made on the basis of historical and philosophical studies of the relevant teaching subject.[7]

A few years later, Michael Martin, a philosopher of science, argued and elaborated these points in his *Concepts of Science Education* (Martin 1972). Martin's book is infused with philosophy of science. The references include books by Agassi, Bridgman, Carnap, Feyerabend, Hanson, Hempel, Kuhn, Lakatos, Nagel, Popper, Quine, Reichenbach and Scheffler, and he uses their illustrations, arguments and analyses to explicate, in five chapters, the core topics of: Scientific Inquiry, Explanation, Definition, Observation and Goals of Science Education. These topics are part of all science curricula and teaching, and their pedagogical relevance would be obvious to trainee teachers. Unfortunately, the arguments of Scheffler, Martin and others were articulated to a mostly empty audience: science educators were not attending to history or philosophy of science (Duschl 1985), and nor were philosophers of education.

However, there were exceptions, even if the terminology of HPS was not used. Thirty years ago, Lee Shulman, with his National Teacher Assessment Project, rejected the behaviourist, managerial measures of teacher competence so long enshrined in evaluation practice. He asked about the 'missing paradigm', the command of subject matter, and the ability to make it intelligible to students. For Shulman:

> Teachers must not only be capable of defining for students the accepted truths in a domain. They must also be able to explain why a particular proposition is deemed warranted, why it is worth knowing, and how it relates to other propositions, both within the discipline and without, both in theory and in practice.
>
> (Shulman 1986, p. 9)

This was a core component of his widely endorsed PCK concept, which has been discussed in Chapter 4. But, to explain why a particular proposition is deemed warranted – for instance, a proposition about genetic inheritance, or the conservation of energy, or the valency of sodium, or the shape of the Earth – assumes an epistemology of science. Such epistemology will include standard matters of, for example, evidence, both empirical and non-empirical, as have already been discussed in this book, and considerations about testimony, as the bulk of what anyone knows in science comes by virtue of reliance on the testimony of others. Teachers are a responsible link in the chain of testimony.[8] Teachers who have thought through some basic epistemological questions, and know something of the history and sociology of science, will be much better able to explain why a proposition is deemed warranted than those who have no such training.

There is a comparable situation in the UK. Susanne Lakin and Jerry Wellington, in their study of UK science teachers' understanding of the nature of science, observed that:

> Teachers' lack of knowledge about the nature and history of science emerged strongly in the study. . . . As well as verbally recognising that their knowledge was

> patchy and their ideas not well formulated, non-verbal signals reflected an insecurity when the issues were probed in depth.
>
> (Lakin & Wellington, 1994, p. 186)

They also noted that, 'The lack of reflection was most apparent in the neglect of the cultural, moral and philosophical aspects of science.' This is a circumstance unlikely to be improved until HPS studies become accepted as a routine part of science-teacher education, and until the gulf between the science-education community and the HPS community is bridged. In Canada, the Science Council of Canada, after advocating increased attention to HPS matters in the science curriculum, said: 'Although Council does not expect children or adolescents to be trained in the philosophy of science, it does expect science educators to be trained in this area' (Science Council of Canada 1984, p. 37).

The Philosophical Health of Teacher Education

Teacher education is not in good philosophical health. Despite all of the concerns and arguments that have long been known and that have been documented in this book, competence in philosophy, and more specifically in HPS, is rare in schools of education, nor is its attainment much encouraged. In 1989, only four of fifty-five institutions providing science-teacher training in Australia offered any HPS-related course. In 1990, of the fifteen leading centres of science-teacher training in the US, only about half required a course in philosophy of science; the proportion in the remaining hundreds of centres was far lower (Loving 1991). The situation in the rest of the world is no more encouraging. Thus, teachers' grasp of HPS is largely picked up in their own science courses, and it is seldom consciously examined or refined. This is epistemology by osmosis and is less than desirable for the formation of something so influential in teaching practice and so important for professional development.

The paucity of serious HPS input into science-teacher education is depressingly well documented in Peter Fensham's book *Defining an Identity: The Evolution of Science Education as a Field of Research* (Fensham 2004). The book opens a representative and authoritative window on to international science-teacher education and the ethos of science-education graduate schools. Fensham is one of the most respected and influential science educators of the past 40 years (Cross 2003). His *Identity* book is built around his interviews with seventy-nine leading science educators from sixteen countries; they include at least sixteen past presidents of the NARST and ten to fifteen current or past editors of major international science-education research journals. The interviewees have authored or edited hundreds of books and a thousand or more research articles and have overseen the same number of doctoral students. So, although the book provides a numerically very small sample of the profession, nevertheless it is a fair sample of leading science-education academics and provides some reasonable warrant for extrapolation to the wider science-education academic community. The interviewees were asked by Fensham to respond to two questions (Fensham 2004, p.xiv):

- Tell me about two of your publications in the field that you regard as significant.
- Tell me about up to three publications by others that have had a major influence on your research work in the field.

It quickly becomes apparent that there is a major problem with 'science education as a field of research': namely, researchers in the field are poorly prepared for conducting their research. One prominent interviewee represents most when he says:

> When I began teaching more than a decade ago, I had just completed a masters degree in physics, but I did not have any background in educational psychology or methodology.
>
> (Roth 1993, p. 145)

Jay Lemke, mentioned by Fensham as a pioneering researcher into the effect of language in science learning (Fensham 2004, p.201), well recognises this problem when he writes:

> Science education researchers are not often enough formally trained in the disciplines from which socio-cultural perspectives and research methods derive. Most of us are self-taught or have learned these matters second-hand from others who are also not fully trained.
>
> (Lemke 2001, p. 303)

Fensham remarks on many occasions that the pioneer researchers came into the field either from a research position in the sciences or from senior positions in school teaching. For both paths, training in psychology, sociology, history or philosophy – the foundation disciplines essential for most serious research in education – was exceptional. He mentions Joseph Schwab, 'a biologist with philosophical background' as an exception (Fensham 2004, p. 20).[9] Among the pioneers, there were not many other exceptions that could be so named.[10]

This failure of preparation was not rectified for second-generation or younger researchers. Indeed, it has perhaps got worse, as proportionally fewer science-education researchers have the experience of scientific research that the founders of the discipline, including Fensham, had. The interviews reveal that the overwhelming educational pattern for current researchers is: an undergraduate science degree, school teaching and then a doctoral degree in science education. Fensham remarks: 'The pattern of movement, from teaching school science to curriculum development to tertiary science education was to be followed in many countries over the ensuing years' (Fensham 2004, p. 25).

Ernst von Glasersfeld's Influence

Philosophical training among the interviewees is minimal and is skewed towards constructivism. A number of interviewees cite Ernst von Glasersfeld as 'a most significant influence'. Fensham states that:

> von Glasersfeld's many writings on personal constructivism have had a very widespread influence on researchers in science education . . . In their published research he is regularly cited as a general source for constructivist learning.
>
> (Fensham 2004, p. 5)

This will be no surprise to the science-education community. Von Glasersfeld was an invited plenary speaker at the large NARST conference in 1990, and, in an unprecedented gesture, the invitation was repeated for the 1993 conference. In the 3 years from 1999 to 2001, there were forty-two citations of his work in the four major science-education journals (Niaz *et al.* 2003, p. 791).

Citation counts alone do not give the full measure of von Glasersfeld's influence; for this, one needs to see the impact on graduate students and next-generation faculty of those who initially 'converted' to radical constructivism. One prominent interviewee has written that:

> According to radical constructivism, we live forever in our own, self-constructed worlds; the world cannot ever be described apart from our frames of experience. This understanding is consistent with the view that there are as many worlds as there are knowers.
>
> (Roth 1995, p. 13)

As there are 7 billion people in the world, this amounts to saying there are 7 billion different worlds. This is plainly nonsense, but nevertheless still said and published.[11] The same interviewee goes on to state that, 'Radical constructivism forces us to abandon the traditional distinction between knowledge and beliefs. This distinction only makes sense within an objective–realist view of the world' (Roth 1995, p. 14), and, for good measure, adds, 'Through this research [sociology of science], we have come to realise that scientific rationality and special problem solving skills are parts of a myth' (Roth 1995, p. 31). All of this, and comparable epistemological and ontological views held by other early enthusiasts for radical constructivism, can be traced back to von Glasersfeld's philosophy and, ultimately, to Bishop Berkeley (Matthews 2000, pp. 171–174). From these early enthusiasts, the idealist and relativist message spread quickly and widely in science education.

Importantly, von Glasersfeld acknowledges his lack of training in philosophy and describes himself as an amateur in the field (von Glasersfeld 1995, p. 4). His philosophical position is simply Bishop Berkeley's empiricism, with some Piagetian additives. He freely admits this, saying at one point that Bishop Berkeley was the first philosopher he read, and that, '1710 was the greatest year in the history of philosophy', on account of it marking the publication of Berkeley's *Principles of Human Knowledge*. Science-education researchers may not have been so swept away by von Glasersfeld's constructivist restatement of the bishop if they had done an undergraduate philosophy course where Berkeley's empiricism is historically situated and his theory of perception, his account of mental imagery, his theory of knowledge and his critiques of Newtonian science are all routinely explicated and criticised. There is nothing novel in von Glasersfeld's philosophy.

Thomas Kuhn's Influence

The other philosopher mentioned by Fensham's interviewees as having a major influence on them was Thomas Kuhn. However, as with von Glasersfeld, the science-education community took up Kuhn's ideas in an altogether uncritical way: 'the community became a cheer-squad for Kuhn' is how Cathleen Loving and William Cobern summarised Kuhn's impact (Loving & Cobern 2000).[12] Of course, Kuhn is more cited than read; depressingly, the mere citation of Kuhn is considered to constitute an argument or to provide evidence for philosophical views. One interviewee, in a publication, writes that:

> In recent years, the rational foundations of Western science and the self-perpetuating belief in the scientific method have come into question . . . The notion of finding a truth for reality is highly questionable.
>
> (Fleer 1999, p. 119)

No evidence is adduced for this sweeping claim, except an unpaginated reference to Kuhn. This practice of having a Kuhn citation or even his name substitute for evidence or argument is widespread in education; the mere word 'Kuhn' functions as a 'get out of philosophical jail free' card, as 'the Bible' or 'the Church' or 'the Party' might in some other circles.

The extensive philosophical critiques of Kuhn's notions of paradigm, incommensurability, theory dependence of observation, intra-theoretic rationality, and so on, have gone largely unnoticed (Matthews 2004). One sympathetic appraisal of Kuhn correctly maintains that:

> Kuhn's treatment of philosophical ideas is neither systematic nor rigorous. He rarely engaged in the stock-in-trade of modern philosophers, the careful and precise analysis of the details of other philosopher's views, and when he did so the results were not encouraging.
>
> (Bird 2000, p. ix)

Abner Shimony, a physicist and philosopher, said of the key Kuhnian move of deriving methodological lessons from scientific practice that:

> His work deserves censure on this point whatever the answer might turn out to be, just because it treats central problems of methodology elliptically, ambiguously, and without the attention to details that is essential for controlled analysis.
>
> (Shimony 1976, p. 582)

Kuhn, like von Glasersfeld, admits that he is an interloper in philosophy, having never taken a course in the subject, and he candidly confesses that his treatment of philosophical issues in his famous *Structure of Scientific Revolutions* was 'irresponsible' (Conant & Haugeland 2000, p. 305). In reviewing his achievements, he regretted the 'purple passages' he wrote in *Structures*. This acknowledgement is letting himself off very lightly indeed. Unfortunately, it was often the purple passages that were taken up by many in the science-education community. By the time Kuhn regretted them and tried to close the stable door, they had bolted out into thousands of higher degrees, articles and books. And the influence was not just confined to the academy. Versions of 'Kuhnianism', especially putative incommensurability and cross-paradigm relativism, are, for instance, routinely appealed to in support of curricular decisions in multicultural science education, where, supposedly, Kuhn establishes that there is epistemological equality between all sciences, and so the choice of Western science in schools is merely political.

Impact of the 'Strong Programme'

The effect of von Glasersfeld and Kuhn on the Fensham cohort was only strengthened by the impact of the Edinburgh 'Strong Programme in the Sociology of Scientific Knowledge' (SSK), a programme that owed its origins to Thomas Kuhn. Fensham reports that, 'One book stood out as an influence about the culture of science and that was Latour and Woolgar's *Laboratory Life*' (Fensham 2004, p. 58). This book is at the extreme, idealist wing of the SSK movement. It is an attempt at an anthropological study of laboratory research on THR (thyrotropin releasing hormone), where one author, Latour, surprisingly thought it advantageous that he knew absolutely no science. The book argues that all science is 'the construction of fictions', and that scientific success is simply the ability of one group, in this case the Nobel Prize winners Schally and Guillemin, to 'extract compliance' from other scientists (Latour & Woolgar 1979/1986, p. 285). They make the outright idealist claim that THR exists only if a certain bioassay procedure is accepted. Just stating this should suffice to set off philosophical alarm bells. Their THR claim is peculiar. Acceptance or otherwise of a bioassay result might be grounds for believing or not believing in THR, but hardly grounds for it coming into and out of existence.

However, adherents of the SSK programme make the same claim about even massive bodies, such as planets, which supposedly only come into existence when discovered. This is idealism in its purest form: the world is supposedly dependent upon human minds (how dependent it might be on animal minds we are not told, nor whether there are as many worlds as there are animals having experiences). This SSK idealism is evident when another Fensham interviewee writes:

> For constructivists, observations, objects, events, data, laws, and theory do not exist independently of observers. The lawful and certain nature of natural phenomena are properties of us, those who describe, not of nature, that is described.
>
> (Staver 1998, p. 503)

Clearly, observations, data and theory do not come into existence independently of us, but to claim that *objects*, *events* and *lawful* behaviour depend on us is pure and unsupported idealism, something that King Canute dramatically demonstrated to his courtiers. Even social institutions, such as marriage or armies, that would not exist without humans do not come and go out of existence depending on how sociologists view or describe them. But a steady diet of von Glasersfeld, Kuhn and the Edinburgh School results in such claims flowing off the keyboard and moving unchecked and unexamined into science-education research literature and graduate classrooms.

Latour, Woolgar and those more deeply affected by them believe that the efforts of Galileo, Newton, Darwin, Einstein and the roll-call of contributors to the scientific tradition have not revealed even approximate truths about the world, but have revealed how to succeed in science, and this success owes nothing to any fit between scientific claims and the world; the contribution of nature is negligible. They maintain that, inside Schally and Guillemin's laboratory, and indeed inside all science laboratories, 'nothing extraordinary and nothing "scientific" was happening inside the sacred walls of these temples' (Latour & Woolgar 1979/1986, p. 141). Peter Slezak, a philosopher, provides a long, detailed and withering critique of *Laboratory Life* and its pretensions. In his estimation, the book is 'in many respects . . . completely incoherent and unintelligible' (Slezak 1994, p. 335). Stephen Cole, a sociologist, says, 'there is not a single example in the entire constructivist literature that supports this [Latour and Woolgar's] view of science' (Cole 1996, p. 278). Many other philosophers, sociologists and historians share this opinion.[13] Even Thomas Kuhn, in his 1991 Rothchild Lecture, was moved to say of the SSK programme that:

> I am among those who have found the claims of the strong program absurd: an example of deconstruction gone mad. And the more qualified sociological and historical formulations that currently strive to replace it are, in my view, scarcely more satisfactory.
>
> (Conant & Haugeland 2000, p. 110)

Despite all of this, *Laboratory Life* was nominated, in 2004, an 'important influence' by many of Fensham's seventy-nine key international science educators. Indeed, Latour is uncritically cited right up to the 2007 US NRC's *Taking Science to School* report, where his work is evoked to support 'The view of science as practice' (NRC 2007, p. 29). Abner Shimony's above comment on Kuhn is to the point here: Most human activity, including science, is a practice; being told that, although it sounds profound, is not being told very much. Football and politics are practices with their own criteria for success, and sociologists can study how such success is achieved. However, science is a distinctly different kind of practice, and this is not recognised. The critical issues are: Is the practice truth-seeking? What indicators are there of success, or of practices that are 'better than' alternative truth-seeking practices? In as much as Latour gives answers to these questions, they are the wrong answers.

Latour and the SSK programme ignore truth, reasonableness, warrant or any epistemological factor as an explanation for the success of a theory; the latter is all a matter of influence, compliance or disciplinary politics. The first kind of explanation requires that the sociologist know some science and know something about philosophy of science, minimally about theory appraisal; without this knowledge, all that can be examined are the latter factors, and the natural tendency is then to claim that these are the important or only ones. If an observer knows something of mathematics and sees a group of students all saying that '2 + 2 = 5', then he or she realises that there is something to be explained. Without knowledge of mathematics, nothing appears amiss or in need of explanation. Sociologists without scientific knowledge, as Latour proudly confesses, are in this situation.

Psychological Theorising

Leaving philosophy aside, the lack of rigorous preparation for science-education research is also evidenced by the extent that shallow psychological theory is utilised in the field. Fensham recognises this and says that, 'science educators borrow psychological theories of learning . . . for example Bruner, Gagne and Piaget' (Fensham 2004, p. 105). He goes on to say, damningly, that, 'The influence of these borrowings is better described as the lifting of slogan-like ideas from these theories' (Fensham 2004, p. 105). It is the lack of thorough training in undergraduate psychology that leaves researchers dependent upon a sloganistic interpretation of theory. Three or four years' hard slog in learning a discipline usually liberates students from dependence on slogans, but, as Fensham so clearly attests, few in the science-education research community have done this hard slog in the foundation disciplines.

Jerome Bruner lamented educators' sloganistic interpretations of his idea of discovery learning (Bruner 1974). More recently, as was mentioned in Chapter 4, the same fate befell the much-cited article co-authored by one of Fensham's interviewees, Peter Hewson: 'Accommodation of a Scientific Conception: Toward a Theory of Conceptual Change' (Posner *et al.* 1982). Ten years after its publication, two of the co-authors were moved to publish,

'A Revisionist Theory of Conceptual Change' (Strike & Posner 1992), where they pointed out that the original paper was intended to be an account of rational conceptual change; it was not a psychological theory of conceptual change, much less was it a pedagogical template for classroom teaching:

> This theory is largely an epistemological theory, not a psychological theory. It follows that it is also a normative theory. It is rooted in a conception of the kinds of things that count as good reasons.
>
> (Strike & Posner 1992, p. 150)

They later speculate on the reasons for the misunderstandings of the original paper:

> Perhaps they result from 'constructivist' epistemologies that have forgotten the social character of knowledge and that assume that people do not understand any conceptions that they have not constructed for themselves.
>
> (Strike & Posner 1992, p. 170)

At the heart of the much-cited original paper is the notion of 'good reasons'. This is a topic that the history of epistemology since Plato has engaged with, and it is a history that could have contributed to educators' deeper understanding of the original paper, except that such history forms no part of the preparation of science-education researchers. Trying to think through the difference between causes, reasons, good reasons and warrants for beliefs is a beneficial exercise and could be part of any decent teacher education programme.

Is Science Education an Autonomous Discipline?

Fensham concludes his book with the observation that science-education research has matured and has, at least in some areas, 'realised an identity' (Fensham 2004, p. 209). This is a moot point. A lot hinges on what 'having an identity' entails. A basic issue is whether science education as a field of research is autonomous, or derivative or something in between. This is a 'local' manifestation of the larger issue debated some 50 years ago about whether education itself is an autonomous discipline, or whether it is exhausted by its constituent components of philosophy, psychology, sociology, history and economics?[14] Education covers an enormous range of areas or activities: schooling, learning, teaching, funding, curricula, governments and so on. It is doubtful if there could be any just educational analysis of each of these areas; rather, there will be sociology of education, history of education, psychology of education, philosophy of education, economics of education and so on.

Two years before Fensham's book appeared, a group of thirteen international science educators argued for the autonomous position, saying:

> This emergence of science education as a scientific domain is usually associated with the establishment of what Novak . . . called an 'emergent consensus' about constructivist positions . . . considered to be the most important contribution to the last decades in science education. . . . it is necessary to build a specific science education body of knowledge.
>
> (Gil-Pérez *et al.* 2002, pp. 558, 562)

The authors do not 'intend to ignore the contributions from other fields such as educational psychology and the history of science' (Gil-Pérez *et al.* 2002, p. 560), but they say that debates in those fields, including philosophical criticism of constructivism, 'are not our debates'. This autonomous position is the default position in science education: conferences are confined largely to science educators, journals mainly publish just science-education articles that are reviewed by science educators, and graduate programmes are almost entirely occupied with science-education courses. The autonomous position can easily morph into isolationism and hyper-professionalism. Stephen Sharpin observes of the latter, that:

> One mark of hyper-professional disciplines is that they are self-referential. They conflate the professional literature with the things-in-the-world to which that literature purportedly refers; they equate literature citing with learning. . . . You can tell a self-referential discipline by the fact that no one reads its products who doesn't have to.
>
> (Shapin 2005, p. 239)

The argument of this book has been that this pursuit of autonomy or 'identity' for science education, without learning something of the foundation disciplines, is damaging to all concerned. Isolationism is as unfortunate as it is unrealistic. Science-education discussion, debate, teaching and research are concerned with questions such as: How do children learn? What do we know about the maturation processes of children's minds, and how does this impact on learning? What should determine the content of science curricula and national standards? What have been past curricula, and why were they changed? What is the nature of the discipline, science, that is to be taught? What rationales can best be advanced for teaching science? What acknowledgement needs be made of indigenous knowledge traditions? To what level should science be compulsory? Does science require a liberal, democratic state? And so on.

None of these questions can be answered, or even intelligently engaged with, without attention to the foundation disciplines of philosophy, psychology, history, sociology and, indeed, others such as economics and politics. Conversely, when the latter foundation disciplines are ignored or merely utilised at an uncritical, sloganistic level, then the field of science education as a whole is diminished. The dilemma for the profession, so well captured in the Fensham interviews, is that science educators, once appointed, have to engage with, answer questions and teach about all of the above questions, for which they are not trained. They have to teach how children learn, without having studied psychology; to teach about the nature of science or comment on issues of science and religion, without having studied history or philosophy of science; to provide rationales for science education without having studied philosophy of education; to account for changes in curricula, without having studied history; to appraise the philosophical claims of constructivism, without having studied philosophy; and so on across the whole spectrum of schools of education programmes.

The first step in dealing with these dilemmas is to acknowledge them; then be modest in what is claimed or asserted on each matter; and, finally, to do one's best to grasp the relevant foundational disciplines and literatures.

The academic field of science education cannot be entirely derivative; there are genuine theoretical, curricula and pedagogical issues *in* and *peculiar to* science education, and it is these issues that need to be illuminated by the foundation disciplines. What needs to be developed for these purposes is a philosophy *of* science education. The beginnings of this, as argued above in this chapter, lie in developing a philosophy of education coupled with knowledge and appreciation of HPS. This can be understood as philosophy *for* science education, assuredly an important thing and something to which this book has hopefully contributed.[15]

The following are some steps that might be taken to strengthen science education as a discipline:

1 For the preparation of future researchers, instead of science teachers doing higher degrees in education after their first degree in science, they could be encouraged to first do an undergraduate degree in an appropriate foundation discipline (philosophy, psychology, history, etc.) and, after that, do a PhD in education. This is good for their personal growth or education and it is ultimately beneficial to whatever research programme in which they might engage.
2 Include a suitable undergraduate or graduate foundation course in science-education graduate programmes.
3 Include foundation faculty in science-education PhD committees. The participation of psychology, philosophy, history or linguistics professors on thesis committees would contribute to raising candidate and supervisor awareness of past and current literature in the foundation discipline that bears upon the student's research.
4 Ease publication pressure on newly appointed staff, so that they can catch up on reading and scholarship in the foundation fields neglected in their own training. This might amount to getting institutions to trade off quantity for quality in appraising a new staff member's output, and giving them longer to make such contributions. Institutions should recognise that one substantial, long-shelf-life publication contributes more to the field than ten second-rate, shallow, ill-thought-out publications that simply occupy journal space. It is better for newly appointed science-education

staff to spend a semester attending a philosophy, psychology, linguistics or history course, and reading substantial books, than conducting yet another study of misconceptions, or transcribing videos of some classroom intervention for which there are no controls and which is not guided by any decent theory. Better that a few core things be done well, than a dozen peripheral things be done poorly.

5 Encourage a system of joint appointments between education and foundation disciplines. Encouragingly, this happens to a small extent between education and science disciplines; if other faculty could be cross-appointed to philosophy, HPS, psychology or sociology, this would assuredly lift the quality of scholarship and research in the field. The models of cross-appointment between philosophy and science disciplines are exemplary.

For the preparation of science teachers, the argument of this book is that, whatever else is done, the inclusion of some education-related HPS course or HPS-related education course in the programme is a necessity. Ideally, it means the creation of specific courses that pick up tangible theoretical, curricular and pedagogical topics in science teaching that teachers can identify and recognise as genuine problems in their teaching life, and then elaborate how HPS considerations can contribute to the better understanding and resolution of the issues.[16] Figure 12.1 – which is an elaboration of an informative comparable diagram in Roland Schulz (2014b) – captures the components discussed above that support the formation of well-prepared science teachers:

- *science*: undergraduate and/or postgraduate science degree, etc.;
- *history and philosophy of science*: internal curriculum-based HPS and external education-related HPS studies, etc.;
- *pedagogy*: practice teaching, educational technology, instructional theory, local curricula, assessment theory and practice, administrative matters, special-needs education, etc.;
- *philosophy of education*: aims of education, personal and social goals of education, ethical standards for classroom teaching and teacher–student interactions, and for school systems, conceptual analysis of teaching and learning, etc.;
- *education foundation subjects*: sociology of education, history of science education, psychology and cognitive science, developmental psychology, curriculum theory, etc.

Conclusion

There are many reasons why study of HPS should be part of pre-service and in-service science-teacher education programmes. Increasingly, school science courses address historical, philosophical, ethical and cultural issues occasioned by science. Teachers of such curricula obviously need knowledge of HPS. Without such knowledge, they either present truncated and partial versions

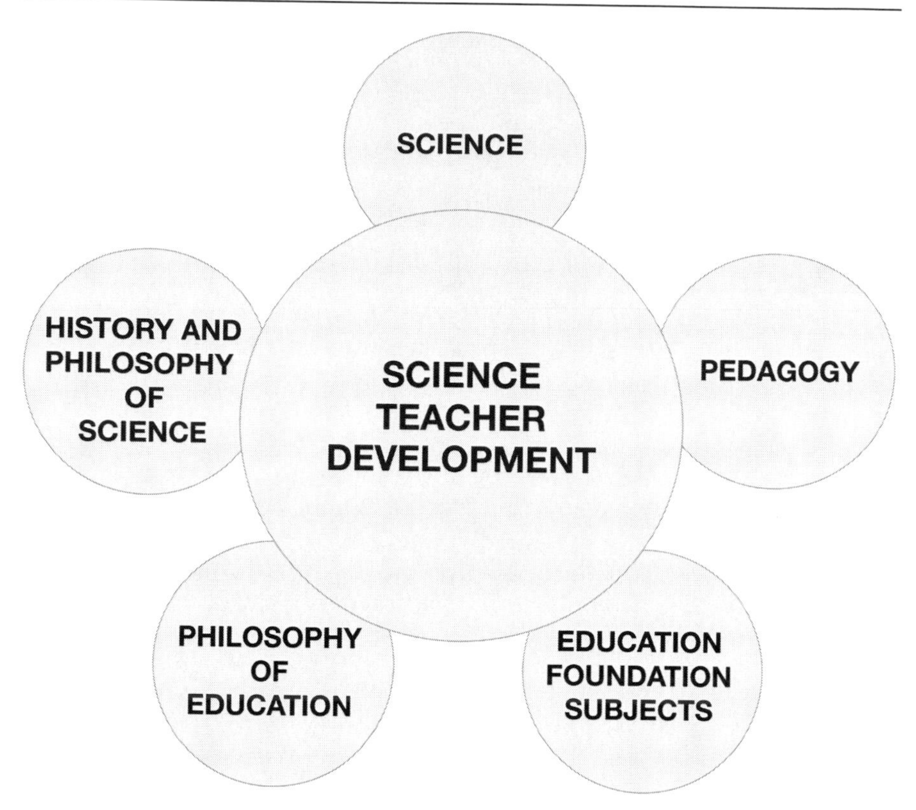

Figure 12.1 Science-Teacher Development

of the curricula, or they repeat shallow academic hearsay about the topics mentioned ('slogans', as Peter Fensham wrote). Either way, their students are done a disservice. Even where curricula do not include such 'nature of science' sections, HPS can contribute to more interesting and critical teaching of the curricular content.

Beyond these 'practical' arguments for HPS in teacher education, there are compelling 'professional' arguments. A teacher ought to know more than just what he or she teaches. As an educator, they need to know something about the body of knowledge they are teaching, something about how this knowledge has come about, how its claims are justified, what its limitations are and, importantly, what the strengths and contributions of science have been to the betterment of human understanding and life. Teachers should have an appreciation of, and value, the tradition of enquiry into which they are initiating students. HPS fosters this.

Enough has been said in this book to suggest that many of the issues in HPS are complex and contentious. The jury is still out on important matters. The art of the teacher is to judge the sophistication of his or her students and present a picture of science that is intelligible to them, without being overwhelming. Students need to find their feet, to become familiar with a tradition, before facing the 'cutting-edge' questions. The teacher may have strong opinions on various HPS issues, but the point of education is to develop the students' minds, which means giving students the knowledge and wherewithal to develop informed opinions. If HPS in science teaching becomes a catechism, then it defeats one of its major purposes. HPS in teacher training programmes can do something towards broadening the vision of teachers and having their students, not just arrive at destinations (scientific competence), but arrive with broader horizons, having travelled with a different view. In the long run, this contributes to the health of science, society and culture.

Education systems have a responsibility to identify and transmit the best of our cultural heritage. Science is one of the most important parts of this heritage; it is now part of the human patrimony. HPS allows science teachers and science-education researchers to better understand their own social and professional responsibilities as part of a great tradition, and so interest and competence in HPS need to be cultivated in teacher-education and graduate courses.

Notes

1. This chapter is partly dependent on material published in Matthews (2014).
2. See Labaree (2008) and contributions to Cochran-Smith *et al.* (2008) and Roth (1999).
3. Arguments for the place of foundation studies in teacher education can be found in the contributions to Tozer *et al.* (1990) and Waks (2008).
4. See Schulz (2014a, 2014b) for a detailed survey of the field of philosophy of education and science education.
5. For an account of postmodernism in current science education, see Mackenzie *et al.* (2014). For an account of cultural studies and science education, see McCarthy (2014).
6. See Slezak (2014) for other examples and discussion of this issue.
7. For elaboration and appraisal of Scheffler's arguments about science teaching, see Matthews (1997a); for wider appraisals see contributions to Siegel (1997).
8. The basic modern work on testimony in human affairs and science is Coady (1992).
9. For the life and achievements of Schwab, see DeBoer (2014).
10. James Robinson would be one (Robinson 1968, 1969). For an account of his life and work, see Matthews (1997b).
11. When such claims are questioned, there is the inevitable 'I did not mean what I wrote' response. This means that people should write more carefully. For examples of this common ploy, see Shackel (2005).
12. Loving and Cobern list thirty articles in the *Journal of Research in Science Teaching and Science Education* in the 13 years from 1985 to 1997 that cite Kuhn and Kuhnian themes.
13. See critiques of Latour and SSK by philosophers Mario Bunge (1991, 1992), Susan Haack (1996) and Nicholas Shackel (2005), sociologist Stephen Cole (1992, 1996) and historian Margaret Jacob (1998).
14. See Scheffler (1963a) and contributions to Tibble (1966).
15. Roland Schulz has argued that, in addition, what is required for the field is a philosophy of science education (Schulz 2009, 2014a, 2014b).
16. Details of such HPS and NOS courses in teacher education programmes can be seen in Matthews (1990), McComas (1998), Rosa and Martins (2009), Schulz (2014b) and Sullenger and Turner (1998).

References

Berlin, I.: 2000, *The Power of Ideas*, H. Hardy (ed.), Chatto & Windus, London.

Bird, A.: 2000, *Thomas Kuhn*, Princeton University Press, Princeton, NJ.

Bruner, J.S.: 1974, 'Some Elements of Discovery'. In his *Relevance of Education*, Penguin, Harmondsworth, UK, pp. 84–97. Originally published in L. Shulman and E. Keislar (eds) *Learning by Discovery*, Rand McNally, Chicago, IL, 1966.

Bunge, M.: 1991, 'A Critical Examination of the New Sociology of Science: Part 1', *Philosophy of the Social Sciences* 21(4), 524–560.

Bunge, M.: 1992, 'A Critical Examination of the New Sociology of Science: Part 2', *Philosophy of the Social Sciences* 22(1), 46–76.

Coady, C.A.J.: 1992, *Testimony: A Philosophical Study*, Oxford University Press, Oxford, UK.

Cobb, P.: 1994, 'Where is the Mind? Constructivist and Sociocultural Perspectives on Mathematical Development', *Educational Researcher* 23(7), 13–20.

Cochran-Smith, M., Feiman-Nemser, S. and McIntyre, D.J. (eds): 2008, *Handbook of Research on Teacher Education* (3rd edn), Routledge/Taylor & Francis, New York.

Cole, S.: 1992, *Making Science: Between Nature and Society*, Harvard University Press, Cambridge, MA.

Cole, S.: 1996, 'Voodoo Sociology: Recent Developments in the Sociology of Science'. In P.R. Gross, N. Levitt and M.W. Lewis (eds) *The Flight from Science and Reason*, Johns Hopkins University Press, Baltimore, MD, pp. 274–287.

Conant, J. and Haugeland, J.: 2000, *The Road Since Structure: Thomas S. Kuhn*, University of Chicago Press, Chicago, IL.

Cross, R.T. (ed.): 2003, *A Vision for Science Education: Responding to the Work of Peter Fensham*, RoutledgeFalmer, London.

Curren, R. (ed.): 2003, *A Companion to the Philosophy of Education*, Blackwell Publishing, Oxford, UK.

Darling-Hammond, L.: 1999, 'The Case for University-Based Teacher Education'. In R. Roth (ed.) *The Role of the University in the Preparation of Teachers*, Routledge/Falmer, New York, pp. 13–30.

Dearden, R.F., Hirst, P.H. and Peters, R.S. (eds): 1972, *Education and the Development of Reason*, 3 volumes, Routledge & Kegan Paul, London.

DeBoer, G.E.: 2014, 'Joseph Schwab: His Work and His Legacy'. In M.R. Matthews (ed.) *International Handbook of Research in History, Philosophy and Science Teaching*, Springer, Dordrecht, The Netherlands, pp. 2433–2458.

Dennett, D.C.: 2006, *Breaking the Spell: Religion as a Natural Phenomenon*, Penguin, New York.

Driver, R., Squires, A., Rushworth, P. and Woods-Robinson, V.: 1994, *Making Sense of Secondary Science*, Routledge, London.

Duschl, R.A.: 1985, 'Science Education and Philosophy of Science, Twenty-five Years of Mutually Exclusive Development', *School Science and Mathematics* 87(7), 541–555.

Fensham, P.J.: 2004, *Defining an Identity: The Evolution of Science Education as a Field of Research*, Kluwer Academic Publishers, Dordrecht, The Netherlands.

Fleer, M.: 1999, 'Children's Alternative Views: Alternative to What?', *International Journal of Science Education* 21(2), 119–135.

Fraser, J.W.: 1992, 'Preparing Teachers for Democratic Schools: The Holmes and Carnegie Reports Five Years Later – A Critical Reflection', *Teachers College Record* 94(1), 7–40.

Gil-Pérez, D. *et al.*: 2002, 'Defending Constructivism in Science Education', *Science & Education* 11(6), 557–571.

Haack, S.: 1996, 'Towards a Sober Sociology of Science'. In P.R. Gross, N. Levitt and M.W. Lewis (eds) *The Flight from Science and Reason*, Johns Hopkins University Press, Baltimore, MD, pp. 259–265.

Hirst, P.H.: 2008, 'Philosophy of Education in the UK. The Institutional Context'. In L.J. Waks (ed.) *Leaders in Philosophy of Education. Intellectual Self-Portraits*, Sense Publishers, Rotterdam, The Netherlands, pp. 305–310.

Jacob, M.C.: 1998, 'Reflections on Bruno Latour's Version of the Seventeenth Century'. In N. Koertge (ed.) *A House Built on Sand: Exposing Postmodernist Myths About Science*, Oxford University Press, New York, pp. 240–254.

Kuhn, T.S.: 1962/1970, *The Structure of Scientific Revolutions*, 2nd edn, Chicago University Press, Chicago, IL (1st edition, 1962).

Labaree, D.F.: 2008, 'An Uneasy Relationship: The History of Teacher Education in the University'. In M. Cochran-Smith, S. Feiman-Nemser and D.J. McIntyre (eds) *Handbook of Research on Teacher Education*, Routledge, New York, pp. 290–306.

Lakin, S. and Wellington, J.: 1994, 'Who will Teach the "Nature of Science"?: Teachers' Views of Science and Their Implications for Science Education', *International Journal of Science Education* 16(2), 175–190.

Latour, B. and Woolgar, S.: 1979/1986, *Laboratory Life: The Social Construction of Scientific Facts*, 2nd edn, SAGE, London.

Lemke, J.L.: 2001, 'Articulating Communities: Sociocultural Perspectives on Science Education', *Journal of Research in Science Teaching* 38(3), 296–316.

Loving, C.C. and Cobern, W.A.: 2000, 'Invoking Thomas Kuhn: What Citation Analysis Reveals for Science Education', *Science & Education* 9(1–2), 187–206.

Loving, C.C.: 1991, 'The Scientific Theory Profile: A Philosophy of Science Model for Science Teachers', *Journal of Research in Science Teaching* 28(9), 823–838.

Mackenzie, J., Good, R. and Brown, J.R.: 2014, 'Postmodernism and Science Education: An Appraisal'. In M.R. Matthews (ed.) *International Handbook of Research in History, Philosophy and Science Teaching*, Springer, Dordrecht, The Netherlands, pp. 1057–1086.

Maheus, J.-F., Roth, W.-M. and Thom, J.: 2010, 'Looking at the Observer Challenges to the Study of Conceptions and Conceptual Change'. In W.-M. Roth (ed.) *Re/Structuring Science Education: ReUniting Sociological and Psychological Perspectives*, Springer, Dordrecht, The Netherlands, pp. 201–219.

Manuel, D.E.: 1981 'Reflections on the Role of History and Philosophy of Science in School Science Education', *School Science Review* 62(221), 769–771.

Martin, M.: 1972, *Concepts of Science Education: A Philosophical Analysis*, Scott, Foresman, New York (Reprinted by University Press of America, 1985).

Matthews, M.R.: 1990, 'History, Philosophy and Science Teaching: What Can Be Done in an Undergraduate Course?' *Studies in Philosophy and Education* 10(1), 93–97.

Matthews, M.R.: 1997a, 'Scheffler Revisited on the Role of History and Philosophy of Science in Science Teacher Education', *Studies in Philosophy and Education* 17(1–2), 159–173.

Matthews, M.R.: 1997b, 'James T. Robinson's Account of Philosophy of Science and Science Teaching: Some Lessons for Today From the 1960s', *Science Education* 81(3), 295–315.

Matthews, M.R.: 2000, 'Appraising Constructivism in Science and Mathematics Education'. In D.C. Phillips (ed.) *National Society for the Study of Education 99th Yearbook*, National Society for the Study of Education, Chicago, IL, pp. 161–192.

Matthews, M.R.: 2004, 'Thomas Kuhn and Science Education: What Lessons Can be Learnt?' *Science Education* 88(1), 90–118.

Matthews, M.R.: 2014, 'Discipline-based Philosophy of Education and Classroom Teaching', *Theory and Research in Education* 12(1), 98–108.

McCarthy, C.L.: 2014 'Cultural Studies in Science Education: Philosophical Considerations'. In M.R. Matthews (ed.) *International Handbook of Research in History, Philosophy and Science Teaching*, Springer, Dordrecht, The Netherlands, pp. 1927–1964.

McComas, W.F.: 1998, 'A Thematic Introduction to the Nature of Science: The Rationale and Content of a Course for Science Educators'. In W.F. McComas (ed.) *The Nature of Science in Science Education: Rationales and Strategies*, Kluwer Academic Publishers, Dordrecht, The Netherlands, pp. 211–222.

Niaz, M., Abd-el-Khalick, F., Benarroch, A., Cardellini, L., Laburú, E., Marín, N., Montes, L.A., Nola, R., Orlik, Y., Scharmann, L.C., Tsai, C.-C. and Tsaparlis, G.: 2003,

'Constructivism: Defense or a Continual Critical Appraisal – A Response to Gil-Pérez *et al.*', *Science & Education* 12(8), 787–797.

Nidditch, P.H.: 1973, 'Philosophy of Education and the Place of Science in the Curriculum'. In G. Langford and D.J. O'Connor (eds) *New Essays in the Philosophy of Education*, Routledge & Kegan Paul, London, pp. 234–258.

NRC (National Research Council): 2007, *Taking Science to School. Learning and Teaching Science in Grades K-8*, National Academies Press, Washington, DC.

Peters, R.S.: 1959, *Authority, Responsibility and Education*, George, Allen & Unwin, London.

Peters, R.S.: 1966, *Ethics and Education*, George Allen & Unwin, London.

Peters, R.S. (ed.): 1967, *The Concept of Education*, Routledge & Kegan Paul, London.

Peters, R.S. (ed.): 1973, *The Philosophy of Education*, Oxford University Press, Oxford, UK.

Peters, R.S.: 1974, *Psychology and Ethical Development*, George Allen & Unwin, London.

Phillips, D.C.: 1981, 'Conceptual Change: Muddying the Conceptual Waters – Research on Conceptual Change', *Philosophy of Education*, 60-72.

Phillips, D.C.: 2000, 'An Opinionated Account of the Constructivist Landscape'. In D.C. Phillips (ed.) *Constructivism in Education*, National Society for the Study of Education, Chicago, IL, pp. 1–16.

Pickering, A.: 1995, *The Mangle of Practice: Time, Agency and Science*, University of Chicago Press, Chicago, IL.

Posner, G.J., Strike, K.A., Hewson, P.W. and Gertzog, W.A.: 1982, 'Accommodation of a Scientific Conception: Toward a Theory of Conceptual Change', *Science Education* 66(2), 211–227.

Robinson, J.T.: 1968, *The Nature of Science and Science Teaching*, Wadsworth, Belmont, CA.

Robinson, J.T.: 1969, 'Philosophical and Historical Bases of Science Teaching', *Review of Educational Research* 39, 459–471.

Rosa, K. and Martins, M.C.: 2009, 'Approaches and Methodologies for a Course on History and Epistemology of Physics: Analyzing the Experience of a Brazilian University', *Science & Education* 18(1), 149–155.

Roth, M.-W.: 1993, 'Construction Sites: Science Labs and Classrooms'. In K. Tobin (ed.) *The Practice of Constructivism in Science Education*, AAAS Press, Washington, DC, pp. 145–170.

Roth, M.-W.: 1995, *Authentic School Science: Knowing and Learning in Open-Inquiry Science Laboratories*, Kluwer Academic Publishers, Dordrecht, The Netherlands.

Roth, R. (ed.): 1999, *The Role of the University in the Preparation of Teachers*, Routledge/Falmer, New York.

Roth, W.-M.: 2007, 'Identity in Scientific Literacy: Emotional–Volitional and Ethico-Moral Dimensions'. In W.-M. Roth and K. Tobin (eds) *Science, Learning, Identity. Sociocultural and Cultural–Historical Perspectives*, Sense Publishers, Rotterdam, The Netherlands, pp. 153–184.

Roth, W.-M.: 2011, *Passibility: At the Limits of the Constructivist Metaphor*, Springer, Dordrecht, The Netherlands.

Roth, W.-M.and Tobin, K.: 2007, 'Introduction: Gendered Identities'. In W.-M. Roth and K. Tobin (eds) *Science, Learning, Identity. Sociocultural and Cultural–Historical Perspectives*, Sense Publishers, Rotterdam, The Netherlands, pp. 99–102.

Saha, L.J. and Dworkin, A.G. (eds): 2009, *International Handbook of Research on Teachers and Teaching*, Springer, Dordrecht, The Netherlands.

Scheffler, I.: 1963a, 'Is Education a Discipline?' In his *Reason and Teaching*, Routledge, London, 1973, pp. 45–57.

Scheffler, I.: 1963b, *The Anatomy of Inquiry*, Bobbs-Merrill, Indianapolis, IN.

Scheffler, I.: 1966/1982, *Science and Subjectivity*, 2nd edn, Hackett, Indianapolis, IN (1st edition, 1966).

Scheffler, I.: 1970, 'Philosophy and the Curriculum'. In his *Reason and Teaching*, London, Routledge, 1973, pp. 31–44. Reprinted in *Science & Education* 1(4), 385–394.

Scheffler, I.: 1973, *Reason and Teaching*, Bobbs-Merrill, Indianapolis, IN.
Schulz, R.M.: 2009, 'Reforming Science Education: Part I. The Search for a Philosophy of Science Education', *Science & Education* 18(3–4), 225–249.
Schulz, R.M.: 2014a, 'Philosophy of Education and Science Education: A Vital but Underdeveloped Relationship'. In M.R. Matthews (ed.) *International Handbook of Research in History, Philosophy and Science Teaching*, Springer, Dordrecht, The Netherlands, pp. 1259–1315.
Schulz, R.M.: 2014b, *Rethinking Science Education: Philosophical Perspectives*, Information Age Publishing, Charlotte, NC.
Science Council of Canada (SCC): 1984, *Science for Every Student: Educating Canadians for Tomorrow's World*, Report 36, SCC, Ottawa, Canada.
Sears, J.T.: 2003, 'From Margin to Centre: On the "Other" Side of the Curriculum Renaissance', *Curriculum Inquiry* 33(4), 427–439.
Shapin, S.: 2005, 'Hyper-professionalism and the Crisis of Readership in the History of Science', *Isis* 96(2), 238–243.
Shackel, N.: 2005, 'The Vacuity of Postmodernist Methodology', *Metaphilosophy* 36(3), 295–320.
Shimony, A.: 1976, 'Comments on Two Epistemological Theses of Thomas Kuhn'. In R.S. Cohen, P.K. Feyerabend and M.W. Wartofsky (eds) *Essays in Memory of Imre Lakatos*, Reidel, Dordrecht, The Netherlands, pp.569–588.
Shulman, L.S.: 1986, 'Those Who Understand: Knowledge Growth in Teaching', *Educational Researcher* 15(2), 4–14.
Shulman, L.S.: 1987, 'Knowledge and Teaching: Foundations of the New Reform', *Harvard Educational Review* 57(1), 1–22.
Siegel, H.: 1978, 'Kuhn and Schwab on Science Texts and the Goals of Science Education', *Educational Theory* 28, 302–309.
Siegel, H.: 1979, 'On the Distortion of the History of Science in Science Education', *Science Education* 63, 111–118.
Siegel, H.: 1989, 'The Rationality of Science, Critical Thinking, and Science Education', *Synthese* 80(1), 9–42. Reprinted in M.R. Matthews (ed.) *History, Philosophy and Science Teaching: Selected Readings*, OISE Press, Toronto and Teachers College Press, New York, 1991.
Siegel, H.: 1993, 'Naturalized Philosophy of Science and Natural Science Education', *Science & Education* 2(1), 57–68.
Siegel, H. (ed.): 1997, *Reason and Education: Essays in Honor of Israel Scheffler*, Kluwer Academic Publishers, Dordrecht, The Netherlands.
Siegel, H. (ed.): 2009, *The Oxford Handbook of Philosophy of Education*, Oxford University Press, Oxford, UK.
Silberman, C.E.: 1970, *Crisis in the Classroom: The Remaking of American Education*, Random House, New York.
Slezak, P.: 1994, 'Sociology of Science and Science Education. Part 11: Laboratory Life Under the Microscope', *Science & Education* 3(4), 329–356.
Slezak, P.: 2014, 'Constructivism in Science Education'. In M.R. Matthews (ed.) *International Handbook of Research in History, Philosophy and Science Teaching*, Springer, Dordrecht, The Netherlands, pp. 1023–1055.
Sprat, T.: 1667/1966, *The History of the Royal Society of London for the Improving of Natural Knowledge*, I.C. Jackson and H.W. Jones (eds), Routledge & Kegan Paul, London.
Staver, J.: 1998, 'Constructivism: Sound Theory for Explicating the Practice of Science and Science Teaching', *Journal of Research in Science Teaching* 35(5), 501–520.
Steinberg, S.R. and Kincheloe, J.: 2012, 'Employing the Bricolage as Critical Research in Science Education'. In B. Fraser, K. Tobin and C. McRobbie (eds) *International Handbook of Science Education*, 2nd edn, Springer, Dordrecht, The Netherlands, pp. 1485–1500.

Stenhouse, D.: 1985, *Active Philosophy in Education and Science*, Allen & Unwin, London.
Strike, K.A. and Posner, G.J.: 1992, 'A Revisionist Theory of Conceptual Change'. In R. Duschl and R. Hamilton (eds) *Philosophy of Science, Cognitive Psychology, and Educational Theory and Practice*, State University of New York Press, Albany, pp. 147–176.
Sullenger, K. and Turner, S.: 1998, 'Nature of Science: Implications for Education: An Undergraduate Course for Prospective Teachers'. In W.F. McComas (ed.) *The Nature of Science in Science Education: Rationales and Strategies*, Kluwer Academic Publishers, Dordrecht, The Netherlands, pp. 243–253.
Taylor, P.C.S.: 1998, 'Constructivism: Value Added'. In B.J. Fraser and K.G. Tobin (eds) *International Handbook of Science Education*, Kluwer Academic Publishers, Dordrecht, The Netherlands, pp. 1111–1123.
Thompson, J.J. (ed.): 1918, *Natural Science in Education* (known as the Thompson Report), HMSO, London.
Tibble, J.W. (ed.): 1966, *The Study of Education*, Routledge & Kegan Paul, London.
Tobin, K.: 1998, 'Sociocultural Perspectives on the Teaching and Learning of Science'. In M. Larochelle, N. Bednarz and J. Garrison (eds) *Constructivism and Education*, Cambridge University Press, Cambridge, UK, pp. 195–212.
Tozer, S., Anderson, T.H. and Armbruster, B.B. (eds): 1990, *Foundational Studies in Teacher Education: A Reexamination*, Teachers College Press, New York.
von Glasersfeld, E.: 1995, *Radical Constructivism. A Way of Knowing and Learning*, The Falmer Press, London.
Wagner, P.A. and Lucas, C.J.: 1977, 'Philosophic Inquiry and the Logic of Elementary School Science Education', *Science Education* 61(4), 549–558.
Waks, L.J. (ed.): 2008, *Leaders in Philosophy of Education: Intellectual Self-Portraits*, Sense Publishers, Rotterdam, The Netherlands.
Yager, R.E. and Penick, J.E.: 1990, 'Science Teacher Education'. In W.R. Houston (ed.) *Handbook of Research on Teacher Education*, Macmillan, New York, pp. 657–673.

Author Index

Subject Index